MW00760383

Lecture Notes in Electrical Engineering

Volume 524

Lecture Notes in Electrical Engineering (LNEE) is a book series which reports the latest research and developments in Electrical Engineering, namely:

- Communication, Networks, and Information Theory
- Computer Engineering
- Signal, Image, Speech and Information Processing
- Circuits and Systems
- Bioengineering
- Engineering

The audience for the books in LNEE consists of advanced level students, researchers, and industry professionals working at the forefront of their fields. Much like Springer's other Lecture Notes series, LNEE will be distributed through Springer's print and electronic publishing channels.

More information about this series at http://www.springer.com/series/7818

Ashish Khare · Uma Shankar Tiwary
Ishwar K. Sethi · Nar Singh
Editors

Recent Trends in Communication, Computing, and Electronics

Select Proceedings of IC3E 2018

 Springer

Editors
Ashish Khare
University of Allahabad
Allahabad, India

Ishwar K. Sethi
Oakland University
Rochester, MI, USA

Uma Shankar Tiwary
Indian Institute of Information Technology
Allahabad, Uttar Pradesh, India

Nar Singh
University of Allahabad
Allahabad, Uttar Pradesh, India

ISSN 1876-1100   ISSN 1876-1119 (electronic)
Lecture Notes in Electrical Engineering
ISBN 978-981-13-2684-4   ISBN 978-981-13-2685-1 (eBook)
https://doi.org/10.1007/978-981-13-2685-1

Library of Congress Control Number: 2018955717

This Springer imprint is published by the registered company Springer Nature Singapore Pte Ltd.
The registered company address is: 152 Beach Road, #21-01/04 Gateway East, Singapore 189721, Singapore

IC3E 2018—Committees

Hony General Chair

Prof. R. L. Hangloo, Vice-Chancellor, University of Allahabad, Allahabad, India

Chairman

Prof. Nar Singh, Head, J.K. Institute of Applied Physics and Technology, University of Allahabad, Allahabad, India

Co-chairs

Prof. R. R. Tewari, Dean, Faculty of Science, University of Allahabad, Allahabad, India
Prof. J. A. Ansari, J.K. Institute of Applied Physics and Technology, University of Allahabad, Allahabad, India
Prof. N. K. Shukla, J.K. Institute of Applied Physics and Technology, University of Allahabad, Allahabad, India

Convener

Dr. Ashish Khare, J.K. Institute of Applied Physics and Technology, University of Allahabad, Allahabad, India

Advisory Committee

Prof. P. Nagabhushan, Director, Indian Institute of Information Technology, Allahabad, India

Prof. Rajeev Tripathi, Director, Motilal Nehru National Institute of Technology Allahabad, India

Dr. Shamim Ahmed, Ex-VC, Jamia Hamdard University, New Delhi, India

Prof. K. K. Bhutani, Director, UPTEC, Allahabad, India

Prof. Madhu S. Gupta, San Diego State University, USA

Prof. Ishwar K. Sethi, Oakland University, USA

Prof. Vangalur Alagar, Concordia University, Canada

Prof. Moongu Jeon, Gwangju Institute of Science and Technology, Korea

Prof. Kaushik Deb, Chittagong University of Engineering and Technology, Bangladesh

Prof. B. K. Kanaujia, Jawaharlal Nehru University, India

Prof. Rajeev Srivastava, Indian Institute of Technology (BHU), Varanasi, India

Prof. Srikanta Patnaik, SOA University, India

Prof. U. S. Tiwary, Indian Institute of Information Technology, Allahabad, India

Prof. M. M. Gore, Motilal Nehru National Institute of Technology Allahabad, India

Prof. R. Z. Khan, Aligarh Muslim University, Aligarh, India

Prof. Suneeta Agarwal, Motilal Nehru National Institute of Technology Allahabad, India

Dr. N. P. Pathak, Indian Institute of Technology Roorkee, India

Dr. P. C. Vinh, Nguyen Tat Thanh University, Vietnam

Dr. V. Prameela, DRDL Hyderabad, India

Mr. Rajneesh Agarwal, Director, STPI, New Delhi, India

Mr. Praveen Roy, Scientist E, DST, New Delhi, India

Organizing Committee

Prof. C. K. Dwivedi, J.K. Institute of Applied Physics and Technology, University of Allahabad, Allahabad, India

Mr. P. N. Gupta, J.K. Institute of Applied Physics and Technology, University of Allahabad, Allahabad, India

Dr. Rajeev Singh, J.K. Institute of Applied Physics and Technology, University of Allahabad, Allahabad, India

Mr. Rajeev Srivastava, J.K. Institute of Applied Physics and Technology, University of Allahabad, Allahabad, India

Dr. R. S. Yadav, J.K. Institute of Applied Physics and Technology, University of Allahabad, Allahabad, India

Dr. T. J. Siddiqui, J.K. Institute of Applied Physics and Technology, University of Allahabad, Allahabad, India

Dr. R. K. Srivastava, J.K. Institute of Applied Physics and Technology, University of Allahabad, Allahabad, India

Dr. Om Prakash, CCE, IPS, University of Allahabad, Allahabad, India

Technical Program Committee

Mr. Abhay Chaturvedi, GLA University, Mathura, India.

Dr. Abhay Kumar Rai, University of Allahabad, Allahabad, India

Dr. Abhishek Gandhar, Bharati Vidyapeeth's College of Engineering, New Delhi, India

Mr. Akhilesh Kumar Gupta, Chang Gung University, Taiwan

Mr. Alkesh Kumar Patel, Software Engineer, Apple Inc., USA

Dr. Alok Kumar Singh Kushwaha, I. K. Gujral Punjab Technical University, Jalandhar, India

Prof. Alok Singh, University of Hyderabad, Hyderabad, India

Mr. Amaresh Kumar Pandey, AI and Machine Learning Researcher, HighQ, India

Dr. Anand Prakash Shukla, Krishna Institute of Engineering and Technology, Ghaziabad, India

Dr. Anand Sharma, Rajkiya Engineering College, Sonbhadra, India

Dr. Anchal Jain, Inderprastha Engineering College, Ghaziabad, India

Dr. Ankita Vaish, Banaras Hindu University, Varanasi, India

Dr. Anupam Das, Cotton University, Guwahati, India

Dr. Anurag Mishra, Ishwar Saran Degree College, Allahabad, India

Dr. Anurag Singh Baghel, Gautam Buddha University, Noida, India

Dr. Aprna Tripathi, GLA University, Mathura, India

Dr. Arun Prakash, Motilal Nehru National Institute of Technology, Allahabad, India

Dr. Ashish Khare, University of Allahabad, Allahabad, India

Mr. Ashish Mishra, Indian Institute of Technology Madras, Chennai, India

Dr. Ashish Singh, Nitte Mahalinga Adyanthaya Memorial Institute of Technology, Karkala, Karnataka, India

Dr. Ashish Srivastava, National Council of Educational Research and Training (NCERT), New Delhi, India

Mr. Ashutosh Kumar Singh, Babasaheb Bhimrao Ambedkar University, Lucknow, India

Dr. Awadhesh Kumar, Madan Mohan Malaviya University of Technology, Gorakhpur, India

Prof. B. H. Shekar, Mangalore University, Mangaluru, Karnataka, India

Prof. Benlian Xu, Changshu Institute of Technology, China

Mr. Brajesh Singh, Madan Mohan Malaviya University of Technology, Gorakhpur, India

Mr. Brij Bansh Nath Anchal, Indian Institute of Technology (Banaras Hindu University), Varanasi, India

Dr. Brijesh Bakariya, I. K. Gujral Punjab Technical University, Jalandhar, India

Dr. Brijesh Mishra, Shambhunath Institute of Engineering and Technology, Allahabad, India

Mr. Chetan Fadnis, Shri Vaishnav Institute of Technology and Science, Indore, India

Dr. Chintan Kumar Mandal, Jadavpur University, Kolkata, India

Dr. Deepa Raj, Babasaheb Bhimrao Ambedkar University, Lucknow, India

Dr. Devendra Kumar Tripathi, Rajkiya Engineering College, Sonbhadra, India

Mr. Dheeraj Kalra, GLA University, Mathura, India

Dr. Dhirendra Pal Singh, Lucknow University, Lucknow, India

Dr. Divakar Yadav, Madan Mohan Malaviya University of Technology, Gorakhpur, India

Ms. Divya Sharma, Motilal Nehru National Institute of Technology, Allahabad, India

Dr. Durgesh Singh, Birla Institute of Technology, Mesra, Ranchi, India

Dr. Emil Vassev, University of Limerick, Ireland

Ms. Farzeen Munir, Gwangju Institute of Science and Technology, Gwangju, Korea

Dr. Gagandeep Bharti, Madan Mohan Malaviya University of Technology, Gorakhpur, India

Mr. Gaurav Upadhyay, Motilal Nehru National Institute of Technology, Allahabad, India

Dr. Gyanendra Verma, National Institute of Technology Kurukshetra, India

Dr. H. M. Singh, Sam Higginbottom University of Agriculture, Technology and Sciences, Allahabad, India

Prof. Hongkook Kim, Gwangju Institute of Science and Technology, Korea

Prof. J. A. Ansari, University of Allahabad, Allahabad, India

Mr. Javed Khan, Integral University, Lucknow, India

Dr. Jeonghwan Gwak, Seoul National University Hospital, Seoul, Korea

Ms. Jyoti Srivastava, Marwadi Education Foundation Group of Institutions, Rajkot, India

Dr. Kamlesh Lakhwani, Vivekananda Institute of Technology, Jaipur, India

Prof. Kanak Saxena, Samrat Ashok Technological Institute, Vidisha, India

Dr. Kanojia Sindhuben Babulal, United College of Engineering and Research, Allahabad, India

Prof. Kaushik Deb, Chittagong University of Engineering and Technology, Chittagong, Bangladesh

Dr. Khyati Vachhani, Nirma University, Ahmedabad, India

Dr. Kulwinder Singh Parmar, I. K. Gujral Punjab Technical University, Jalandhar, India

Dr. Kurmendra, Rajiv Gandhi Central University, Itanagar, India

Dr. Maheshkumar H. Kolekar, Indian Institute of Technology Patna, India

Ms. Maneesha Dwivedi, Galgotias University, Greater Noida, India

Dr. Manjari Gupta, Banaras Hindu University, Varanasi, India

Mr. Manish Gupta, Moradabad Institute of Technology, Moradabad, India

Dr. Manish Khare, Dhirubhai Ambani Institute of Information and Communication Technology, Gandhinagar, India

Mr. Manish Kumar, GLA University, Mathura, India

Dr. Manoj Kumar Singh, Banaras Hindu University, Varanasi, India

Dr. Mexhid Ferati, Linnaeus University, Sweden

Mr. Mohd. Shahid Husain, Integral University, Lucknow, India

Dr. Mrigank Dwivedi, University of Allahabad, Allahabad, India

Dr. Mu Zhou, Stanford University, USA

Mr. Mudassar Majgaonkar, Senior Member, Technical Staff, Oracle, USA

Dr. Mukesh Saraswat, Jaypee Institute of Information Technology, Noida, India

Dr. Munawwar Alam, Integral University, Lucknow, India

Dr. Nagendra P. Singh, Madan Mohan Malaviya University of Technology, Gorakhpur, India

Mr. Nand Kishor, Motilal Nehru National Institute of Technology, Allahabad, India

Prof. Navita Srivastava, Awadhesh Pratap Singh University, Rewa, India

Mr. Navneet Kumar Singh, Motilal Nehru National Institute of Technology, Allahabad, India

Dr. Neelam Bharadwaj, Hindustan Institute of Technology and Management, Agra, India

Dr. Neeraj Gupta, Oakland University, USA

Mr. Neeraj Varshney, Indian Institute of Technology Kanpur, India

Dr. Nguyen Thanh Binh, Ho Chi Minh City University of Technology, Ho Chi Minh City, Vietnam

Dr. Nilay Khare, Maulana Azad National Institute of Technology, Bhopal, India

Prof. Nilesh Patel, Oakland University, USA

Dr. Nitin Pandey, Feroze Gandhi Institute of Engineering and Technology, Raebareli, India

Dr. Om Prakash, University of Allahabad, Allahabad, India

Mr. Palash Jain, GLA University, Mathura, India

Mr. Prabhir Kumar Sethy, Sambalpur University Institute of Information Technology, Sambalpur, Odisha, India

Dr. Pradeep Tomar, Gautam Buddha University, Noida, India

Dr. Prashant Srivastava, NIIT University, Neemrana, Rajasthan, India

Mr. Prateek Keserwani, Indian Institute of Technology Roorkee, Roorkee, India

Dr. Raghav Yadav, Sam Higginbottom University of Agriculture, Technology and Sciences, Allahabad, India

Mr. Raghavendra Pal, Motilal Nehru National Institute of Technology, Allahabad, India

Dr. Rahul Kaushik, Jaypee Institute of Information Technology, Noida, India

Mr. Rajesh Kumar, Government Polytechnic, Aurai, India

Mr. Rajesh Kumar Upadhyay, Mangalayatan University, Aligarh, India

Dr. Rajesh Prasad, American University of Nigeria, Nigeria

Prof. Rajeev Srivastava, Indian Institute of Technology (BHU), Varanasi, India

Mr. Rajeev Srivastava, University of Allahabad, Allahabad, India

Dr. Rajneesh Kumar Srivastava, University of Allahabad, Allahabad, India

Dr. Rajitha Bakthula, Motilal Nehru National Institute of Technology, Allahabad, India

Dr. Rajiv Singh, Banasthali Vidyapith, Banasthali, India

Dr. Rakhi Garg, Banaras Hindu University, Varanasi, India

Dr. Ravi Shankar Saxena, St. Martin's Engineering College, Hyderabad, India

Dr. Ravindra Kumar Purwar, Guru Gobind Singh Indraprastha University, New Delhi, India

Dr. Rejaul Karim Barbhuiya, National Council of Educational Research and Training, New Delhi, India

Mr. Roshan Singh, Indian Institute of Technology (BHU), Varanasi, India

Prof. S. G. Prakash, University of Allahabad, Allahabad, India

Prof. S. K. Udgata, University of Hyderabad, Hyderabad, India

Mr. Sachin Yele, Shri Vaishnav Vidyapeeth Vishwavidyalaya, Indore, India

Dr. Sanjeev Karmakar, Bhilai Institute of Technology, Durg, India

Ms. Santi Kumari Behera, Veer Surendra Sai University of Technology, Burla, Odisha, India

Dr. Sarika Yadav, University of Allahabad, Allahabad, India

Dr. Satish Kumar Singh, Indian Institute of Information Technology, Allahabad, India

Dr. Satyendr Singh, BML Munjal University, Gurgaon, India

Dr. Satyakesh Dubey, National Physical Laboratory, New Delhi, India

Mr. Saurabh Kumar, Madan Mohan Malaviya University of Technology, Gorakhpur, India

Mr. Selvaraj Karthikeyan, Galgotias University, Greater Noida, India

Dr. Shadab Alam, Jazan University, Saudi Arabia

Mr. Shaishav Agrawal, Indian Institute of Information Technology, Allahabad, India

Dr. Shalni Agarwal, Shri Ramswaroop Memorial University, Lucknow, India

Dr. Shams Tabrez Siddiqui, Jazan University, Saudi Arabia

Ms. Sharmistha Adhikari, National Institute of Technology Sikkim, India

Dr. Shashank Gupta, Bundelkhand Institute of Engineering & Technology, Jhansi, India

Mr. Shashi Bhushan Kumar, Bharati Vidyapeeth's College of Engineering, New Delhi, India

Dr. Sheo Kumar Mishra, Central Institute of Plastics Engineering and Technology, Lucknow, India

Dr. Shekhar Yadav, Madan Mohan Malaviya University of Technology, Gorakhpur, India

Ms. Shikha Dubey, Gwangju Institute of Science and Technology, Gwangju, Korea

Dr. Shikha Jaiswal, Feroze Gandhi Institute of Engineering and Technology, Raebareli, India

Prof. Shirshu Verma, Indian Institute of Information technology, Allahabad, India

Prof. Shishir Kumar, Jaypee University of Engineering and Technology, Guna, India

Ms. Shivani Mishra, Motilal Nehru National Institute of Technology, Allahabad, India

Dr. Shivendra Shivani, Trinity College Dublin, Ireland

Dr. Somak Bhattacharyya, Indian Institute of Technology (BHU), Varanasi, India

Ms. Sonam Agarwal, Motilal Nehru National Institute of Technology, Allahabad, India

Mr. Subodh Kumar, Mangalayatan University, Aligarh, India

Prof. Sudip Sanyl, BML Munjal University, Gurgaon, India

Mr. Sudipta Roy, Ganpat University, Mehsana, Gujarat, India

Prof. Suneeta Agarwal, Motilal Nehru National Institute of Technology, Allahabad, India

Dr. Sushila Maheshkar, Indian School of Mines, Dhanbad, India

Dr. Swati Nigam, SP Memorial Institute of Technology, Allahabad, India

Dr. T. J. Siddiqui, University of Allahabad, Allahabad, India

Mr. Utkarsh Sharma, GLA University, Mathura, India

Dr. Vikash Pandey, École polytechnique fédérale de Lausanne (EPFL), Switzerland

Dr. Vijay Shanker Tripathi, Motilal Nehru National Institute of Technology, Allahabad, India

Mr. Vinay Kumar, GLA University, Mathura, India

Prof. Vivek Kumar Singh, Banaras Hindu University, Varanasi, India

Dr. Yogendra Kumar Prajapati, Motilal Nehru National Institute of Technology, Allahabad, India

Mr. Yogesh Tripathi, Motilal Nehru National Institute of Technology, Allahabad, India

Organizer

J.K. Institute of Applied Physics and Technology
Department of Electronics and Communication,
University of Allahabad, Allahabad, India

Sponsors

Department of Science and Technology
Ministry of Science and Technology
Government of India   Council of Scientific & Industrial Research

Technical Co-sponsor

Preface

We are very pleased to introduce the proceedings of the International Conference on Emerging Trends in Communication, Computing, and Electronics (IC3E 2018) organized by the Department of Electronics and Communication, University of Allahabad, Allahabad, India. The conference was organized on the occasion of the 60th Anniversary of J.K. Institute of Applied Physics and Technology, Department of Electronics and Communication, University of Allahabad. This conference gave researchers an excellent opportunity and provided an internationally respected forum for the discussion of scientific research in the technologies and applications of different fields of communication, computing, and electronics.

IC3E 2018 received a total of 206 submissions. Technical program committee took tremendous efforts to review the submitted papers in a short time, and almost every paper was reviewed by at least three reviewers. On the recommendation of the technical program committee, 67 papers were accepted for the presentation in the main conference; out of them, 61 papers were presented in the conference, and we have shortlisted 56 papers and included them in the conference proceedings. These papers are divided into nine main categories of communication, computing, and electronics. The aim/objective of IC3E 2018 was to make it an excellent academic conference with quality papers and the involvement of experts.

We are thankful to Springer for their technical co-sponsorship and the sponsorship of the Department of Science and Technology (DST), India; Council of Scientific & Industrial Research (CSIR), India; and Defence Research and Development Organization (DRDO), India.

We sincerely thank our Vice-Chancellor Prof. R. L. Hangloo for his wholehearted support for this conference. This conference would not have been successful without the support of the Department of Electronics and Communication, University of Allahabad. Our sincere thanks to the entire family of J.K. Institute of Applied Physics and Technology, including faculty, staff, and students for their assistance, support, and hard work in organizing the conference and its proceedings.

And last but not least, we express our deep sense of gratitude to all the invited speakers, authors, and members of the advisory committee, technical program committee, and organizing committee; without their support, this conference could not have been successful.

Allahabad, India   Ashish Khare
Allahabad, India   Uma Shankar Tiwary
Rochester, USA   Ishwar K. Sethi
Allahabad, India   Nar Singh

Contents

Editors and Contributors

About the Editors

Dr. Ashish Khare completed his master's in computer science and his Ph.D. at the University of Allahabad in 1999 and 2007, respectively. He pursued postdoctoral research at Gwangju Institute of Science and Technology, Korea. Currently, he is working as Associate Professor in the Department of Electronics and Communication, University of Allahabad, India. His research interests include applications of wavelet transforms, computer vision, cyber security, and understanding human behavior. He has worked as Principal Investigator for research projects funded by UGC and DST. He has authored three books and published more than 100 articles in international peer-reviewed journals and conference proceedings.

Dr. Uma Shankar Tiwary obtained his Ph.D. in electronics engineering from the Indian Institute of Technology, Banaras Hindu University, Varanasi, in 1991. He is currently Professor at the Indian Institute of Information Technology (IIIT), Allahabad, where he has various academic and administrative responsibilities. His research interests include image processing, language processing, human–computer interaction, and cognitive computing. He has published seven books and more than 100 research articles in international journals and conference proceedings. He is also Senior Member of IEEE.

Dr. Ishwar K. Sethi obtained his M.Tech. and Ph.D. from the Indian Institute of Technology (IIT) Kharagpur in 1971 and 1978, respectively. Currently, he is Professor in the Department of Computer Science and Engineering, Oakland University. Prior to joining Oakland University, he was with the Wayne State University, IIT Delhi, and IIT Kharagpur. His major areas of research include data mining, image and video databases, document image processing, pattern recognition, and machine learning. He has published more than 250 articles in international journals and conference proceedings. He has served on the editorial boards of several reputed journals and also as a guest editor of numerous conference proceedings. He is Fellow of IEEE.

Dr. Nar Singh received his M.Tech. from the University of Allahabad in 1975 and his Ph.D. from Indian Institute of Technology (IIT) Delhi, in 1989. Currently, he is Professor in the Department of Electronics and Communication, University of Allahabad, where he has various academic and administrative responsibilities. His major research interests include optical fiber communication system, coherent systems, digital communication, and communication systems. He has worked as Principal Investigator for four research projects funded by AICTE. He has published more than 100 research articles in leading international journals and conference proceedings. He has also edited five conference proceedings. He is Fellow of IETE and several international and national societies.

Contributors

Sharmistha Adhikari Department of Computer Science and Engineering, National Institute of Technology Sikkim, Sikkim, India

Neelesh Agrawal Department of Electronics & Communication, University of Allahabad, Allahabad, UP, India

Brij Bansh Nath Anchal Department of Ceramic Engineering, Indian Institute of Technology (BHU), Varanasi, India

Jamshed A. Ansari Department of Electronics and Communication, University of Allahabad, Allahabad, UP, India

Vanya Arun Department of Electronics and Communication, University of Allahabad, Allahabad, Uttar Pradesh, India

B. Balamurugan School of Computing Science and Engineering, Galgotias University, Greater Noida, Uttar Pradesh, India

Sasmita Behera VSSUT, Burla, India

Gagandeep Bharti Department of Electronics and Communication Engineering, Madan Mohan Malaviya University of Technology, Gorakhpur, India

Kusum Kumari Bharti Design and Manufacturing, PDPM-Indian Institute of Information Technology, Jabalpur, MP, India

Bandan Kumar Bhoi Department of Electronics & Telecommunication, Veer Surendra Sai University of Technology, Burla, India

Mantosh Biswas Department of Computer Engineering Kurukshetra, National Institute of Technology Kurukshetra, Kurukshetra, India

Inderveer Chana CSED, Thapar University, Patiala, Punjab, India

Chandni Department of CSE, IKGPTU, Kapurthala, India

Satish Chandra Department of CSE & IT, JIIT University, Noida, UP, India

Abhay Chaturvedi GLA University, Mathura, India

Tusarjyoti Das Department of Electronics & Telecommunication, Veer Surendra Sai University of Technology, Burla, India

Devesh Department of Electronics and Communication, University of Allahabad, Allahabad, UP, India

Saurabh Dixit Central Institute of Plastics and Engineering Technology, Lucknow, India

C. K. Dwivedi Department of Electronics and Communication, (J K Institute of Applied Physics), University of Allahabad, Allahabad, India

Savita Gautam University Women's Polytechnic, F/O Engineering & Technology, Aligarh Muslim University, Aligarh, India

Abhinav Gupta Electronics Engineering Department, Rajkiya Engineering College, Sonbhadra, India

Neeraj Gupta School of Computer and Engineering Science, Oakland University, Rochester, MI, USA

Utkarsh Gupta University of Petroleum and Energy Studies, Dehradun, India

Jeonghwan Gwak Biomedical Research Institute, Seoul National University Hospital (SNUH), Seoul, Korea; Department of Radiology, Seoul National University Hospital (SNUH), Seoul, Korea

Misbahul Haque Department of Computer Engineering, Aligarh Muslim University, Aligarh, India

Thi Kieu Khanh Ho Biomedical Research Institute, Seoul National University Hospital, Seoul, Korea; School of Electrical Engineering and Computer Science, Gwangju Institute of Science and Technology, Gwangju, Korea

Mohd. Imran Department of Computer Engineering, Aligarh Muslim University, Aligarh, India

Ankit Kumar Jaiswal Computing and Vision Lab, Department of Computer Science and Engineering, Indian Institute of Technology (BHU) Varanasi, Varanasi, India

Komal Jaiswal Department of Electronics and Communication, University of Allahabad, Allahabad, Uttar Pradesh, India

Manish Jaiswal Department of Electronics and Communication, University of Allahabad, Allahabad, India

Vinit Jaiswal Department of Electronics and Communication Engineering, Motilal Nehru National Institute of Technology Allahabad, Allahabad, UP, India

Tasleem Jamal Department of Computer Science & Engineering, Harcourt Butler Technical University, Kanpur, India

Moongu Jeon School of Electrical Engineering and Computer Science, Gwangju Institute of Science and Technology, Gwangju, Korea

Hyunsu Jeong Department of Electrical Engineering and Computer Science, Gwangju Institute of Science and Technology, Gwangju, Korea

Dheeraj Kalra GLA University, Mathura, India

Kamakshi IMS Engineering College, Ghaziabad, India

Ankita Kar Department of Electronics and Communication, (J K Institute of Applied Physics), University of Allahabad, Allahabad, India

Nivedita Kar Department of Electronics and Communication, (J K Institute of Applied Physics), University of Allahabad, Allahabad, India

S. Karthikeyan School of Computing Science and Engineering, Galgotias University, Greater Noida, Uttar Pradesh, India

Manpreet Kaur Department of EIE, SLIET, Longowal, India

Vishal Kesari Microwave Tube Research and Development Centre, Bangalore, India

Ashish Khare Department of Electronics and Communication, University of Allahabad, Allahabad, Uttar Pradesh, India

Manish Khare Dhirubhai Ambani Institute of Information and Communication Technology, Gandhinagar, Gujarat, India

Mahdi Khosravy Electrical Engineering Department, Federal University of Juiz de Fora, Juiz de Fora, Brazil

Rajat Khurana Department of CSE, IKGPTU, Kapurthala, India

Pranjali M. Kokare Indian Institute of Technology Patna, Bihta, Patna, Bihar, India

Maheshkumar H. Kolekar Indian Institute of Technology Patna, Bihta, Patna, Bihar, India

Amrish Kumar Department of Electronics and Communication Engineering, Motilal Nehru National Institute of Technology Allahabad, Allahabad, India

Arvind Kumar Motilal Nehru National Institute of Technology Allahabad, Allahabad, India

Awadhesh Kumar Department of Electrical Engineering, Madan Mohan Malaviya University of Technology, Gorakhpur, UP, India

Divesh Kumar GLA University, Mathura, India

Lalit Kumar Design and Manufacturing, PDPM-Indian Institute of Information Technology, Jabalpur, MP, India

Manish Kumar GLA University, Mathura, India

Rajeev Kumar School of Computer and Systems Sciences, Jawaharlal Nehru University, New Delhi, India

Rajesh Kumar Department of Electronics & Communication Engineering, North Eastern Regional Institute of Science & Technology, Itanagar, India

Ravinder Kumar CSED, Thapar University, Patiala, Punjab, India

Rohit Kumar Department of Electronics and Communication Engineering, Motilal Nehru National Institute of Technology Allahabad, Allahabad, India

Sunil Kumar Motilal Nehru National Institute of Technology Allahabad, Allahabad, India

Kurmendra Department of Electronics & Communication Engineering, Rajiv Gandhi University (A Central University), Itanagar, India; Department of Electronics & Communication Engineering, North Eastern Regional Institute of Science & Technology, Itanagar, India

Alok Kumar Singh Kushwaha Department of CSE, IKGPTU, Kapurthala, India

Arati Kushwaha Department of Electronics and Communication, University of Allahabad, Allahabad, Uttar Pradesh, India

Rahul Chandra Kushwaha DST-Centre for Interdisciplinary Mathematical Sciences, Institute of Science, Banaras Hindu University, Varanasi, India

Sejeong Lee School of Electrical Engineering and Computer Science, Gwangju Institute of Science and Technology, Gwangju, Korea

Nikhlesh K. Mishra Department of Electronics and Communication, University of Allahabad, Allahabad, Uttar Pradesh, India

Rishabh Bhooshan Mishra Smart Sensors Area, CSIR - Central Electronics Engineering Research Institute, Pilani, Rajasthan, India

Sheo Kumar Mishra Central Institute of Plastics and Engineering Technology, Lucknow, India

Shivani Mishra Motilal Nehru National Institute of Technology Allahabad, Allahabad, India

Neeraj Kumar Misra Bharat Institute of Engineering and Technology, Hyderabad, India

Ravindra Mukhiya Smart Sensors Area, CSIR - Central Electronics Engineering Research Institute, Pilani, Rajasthan, India

Navendu Nitin Department of Electronics and Communication, University of Allahabad, Allahabad, UP, India

Raghavendra Pal Department of Electronics and Communication Engineering, Motilal Nehru National Institute of Technology Allahabad, Allahabad, India

Akhilesh Kumar Pandey Department of Electronics and Communication, University of Allahabad, Allahabad, Uttar Pradesh, India

Chang Min Park Department of Radiology, Seoul National University Hospital, Seoul, Korea; Department of Radiology, Seoul National University College of Medicine, Seoul, Korea; Institute of Radiation Medicine, Seoul National University Medical Research Center, Seoul, Korea; Seoul National University Cancer Research Institute, Seoul, Korea

Cheolbin Park Department of Electrical Engineering and Computer Science, Gwangju Institute of Science and Technology, Gwangju, Korea

Rizwan Patan School of Computing Science and Engineering, Galgotias University, Greater Noida, Uttar Pradesh, India

Ankit Kumar Patel Department of Electronics and Communication, University of Allahabad, Allahabad, Uttar Pradesh, India

Nilesh Patel School of Computer and Engineering Science, Oakland University, Rochester, MI, USA

Osor Pertin Department of Electronics & Communication Engineering, Rajiv Gandhi University (A Central University), Itanagar, India

Shashi Prabha Department Computer Science and IT, Sam Higginbottom University of Agriculture, Technology and Sciences, Allahabad, UP, India

Y. K. Prajapati Department of Electronics and Communication Engineering, Motilal Nehru National Institute of Technology Allahabad, Allahabad, UP, India

Arun Prakash Department of Electronics and Communication Engineering, Motilal Nehru National Institute of Technology Allahabad, Allahabad, India

Om Prakash Department of Computer Science and Engineering, Nirma University, Ahmedabad, India; Centre of Computer Education, Institute of Professional Studies, University of Allahabad, Allahabad, India

Ram Pyare Department of Ceramic Engineering, Indian Institute of Technology (BHU), Varanasi, India

M. A. Qadeer Department of Computer Engineering, Aligarh Muslim University, Aligarh, India

Sanjeev Rai Department of Electronics and Communication Engineering, Motilal Nehru National Institute of Technology Allahabad, Allahabad, India

Shalini Rai Department of Electronics and Communication, University of Allahabad, Allahabad, India

Prateek Raj Gautam Motilal Nehru National Institute of Technology Allahabad, Allahabad, India

Tarique Rashid Motilal Nehru National Institute of Technology Allahabad, Allahabad, India

Sangram Ray Department of Computer Science and Engineering, National Institute of Technology Sikkim, Sikkim, India

B. Reddy Mounika Department of Electronics and Communication, University of Allahabad, Allahabad, Uttar Pradesh, India

Rashmishree Rout Department of Electronics & Telecommunication, Veer Surendra Sai University of Technology, Burla, India

Roopam Sadh School of Computer and Systems Sciences, Jawaharlal Nehru University, New Delhi, India

Gagandeep Sahib Electronics and Communication Engineering Department, UIET, Panjab University, Chandigarh, India

Himanshu Sahu University of Petroleum and Energy Studies, Dehradun, India

Abdus Samad University Women's Polytechnic, F/O Engineering & Technology, Aligarh Muslim University, Aligarh, India

S. Santosh Kumar Smart Sensors Area, CSIR - Central Electronics Engineering Research Institute, Pilani, Rajasthan, India

Abhishek Kumar Saroj Department of Electronics and Communication, University of Allahabad, Allahabad, UP, India

Saiyed Salim Sayeed Department of Electronics and Communication, University of Allahabad, Allahabad, UP, India

R. Seshadri Microwave Tube Research and Development Centre, Bangalore, India

Ishwar K. Sethi School of Computer and Engineering Science, Oakland University, Rochester, MI, USA

Divya Sharma Department of Electronics and Communication Engineering, Motilal Nehru National Institute of Technology Allahabad, Allahabad, UP, India

Manoj Kumar Sharma Electrical and Electronics Engineering Department, UIET, Panjab University, Chandigarh, India

Neha Sharma University of Petroleum and Energy Studies, Dehradun, India

Prateek Sharma Indian Institute of Technology Patna, Bihta, Patna, Bihar, India

Narendra K. Shukla Department of Electronics and Communication, University of Allahabad, Allahabad, Uttar Pradesh, India

Mohd. Gulman Siddiqui Department of Electronics and Communication, University of Allahabad, Allahabad, UP, India

Tanveer J. Siddiqui Department of Electronics and Communication, University of Allahabad, Allahabad, India

Sikandar Electronics Engineering Department, Rajkiya Engineering College, Sonbhadra, India

Arun Kumar Singh Rajkiya Engineering College, Kannauj, India

Awadhesh Kumar Singh ECE Department, University of Allahabad, Allahabad, India

Birmohan Singh Department of CSE, SLIET, Longowal, India

Brajesh Kumar Singh Department of Electrical Engineering, Madan Mohan Malaviya University of Technology, Gorakhpur, UP, India

Gurwinder Singh Department of CSE, SLIET, Longowal, India

Muskaan Singh Research Labs, CSED, Thapar University, Patiala, Punjab, India

Nagendra Pratap Singh Department of Computer Science and Engineering, MMM University of Technology, Gorakhpur, UP, India

Nar Singh ECE Department, University of Allahabad, Allahabad, India

Preetam Singh Department of Ceramic Engineering, Indian Institute of Technology (BHU), Varanasi, India

Rajeev Singh Department of Electronics and Communication, University of Allahabad, Allahabad, Uttar Pradesh, India

Shiksha Singh Department of Computer Science and Engineering, MMM University of Technology, Gorakhpur, UP, India

Shivani Singh Department of Electronics and Communication Engineering, Madan Mohan Malaviya University of Technology, Gorakhpur, India

Sweta Singh Department of Electronics and Communication, University of Allahabad, Allahabad, Uttar Pradesh, India

Achintya Singhal Department of Computer Science, Institute of Science, Banaras Hindu University, Varanasi, India

Jong-In Song School of Electrical Engineering and Computer Science, Gwangju Institute of Science and Technology, Gwangju, Korea; Department of Electrical Engineering and Computer Science, Gwangju Institute of Science and Technology, Gwangju, Korea

Anima Srivastava Department of Electronics and Communication, University of Allahabad, Allahabad, India

Gargi Srivastava Indian Institute of Technology (BHU) Varanasi, Varanasi, UP, India

Karunesh Srivastava Department of Electronics and Communication, University of Allahabad, Allahabad, Uttar Pradesh, India

Rajeev Srivastava Department of Electronics and Communication, University of Allahabad, Allahabad, India

Rajeev Srivastava Computing and Vision Lab, Department of Computer Science and Engineering, Indian Institute of Technology (BHU) Varanasi, Varanasi, India; Indian Institute of Technology (BHU) Varanasi, Varanasi, UP, India

Rajneesh Kumar Srivastava Department of Electronics and Communication, University of Allahabad, Allahabad, Uttar Pradesh, India

Saumya Srivastava Department of Electronics and Communication, University of Allahabad, Allahabad, Uttar Pradesh, India

S. K. Swain Faculty of Education, Banaras Hindu University, Varanasi, India

Rajeev Tripathi Department of Electronics and Communication Engineering, Motilal Nehru National Institute of Technology Allahabad, Allahabad, UP, India

Yogesh Tripathi Department of Electronics and Communication Engineering, Motilal Nehru National Institute of Technology Allahabad, Allahabad, India

Kamal K. Upadhyay Department of Electronics and Communication, University of Allahabad, Allahabad, Uttar Pradesh, India

Pawan Kumar Updhyay Department of CSE & IT, JIIT University, Noida, UP, India

Atul Kumar Uttam GLA University, Mathura, UP, India

Akshay Verma Motilal Nehru National Institute of Technology Allahabad, Allahabad, India

Renu Vig Electronics and Communication Engineering Department, UIET, Panjab University, Chandigarh, India

Pratima Yadav Department of CSE, MMM University of Technology, Gorakhpur, UP, India

Raghav Yadav Department Computer Science and IT, Sam Higginbottom University of Agriculture, Technology and Sciences, Allahabad, UP, India

Rama Shankar Yadav Department of Computer Science and Engineering, Motilal Nehru National Institute of Technology, Allahabad, India

Sarika Yadav Department of Computer Science and Engineering, Motilal Nehru National Institute of Technology, Allahabad, India

Shekhar Yadav Department of Electronics and Communication, University of Allahabad, Allahabad, Uttar Pradesh, India

Sugandha Yadav School of VLSI Design & Embedded System, National Institute of Technology Kurukshetra, Kurukshetra, India

Ujjawala Yati Department of Computer Engineering Kurukshetra, National Institute of Technology Kurukshetra, Kurukshetra, India

Part I
Communication Systems

Design of An Improved Micro-Electro-Mechanical-Systems Switch for RF Communication System

Kurmendra ⓘ, Rajesh Kumar and Osor Pertin

Abstract This paper presents the design of improved MEMS shunt switch for RF communication applications. The switch was designed to provide a better performance in 10–100 GHz range. The switch was optimized in terms of width of the beam and air gap between the fixed type beam and dielectric layer to improve the isolation, insertion, and return loss. This study concludes that materials with high k-dielectrics and high Young's modulus are desirable for better performance in high-frequency range. The isolation, insertion, and return loss for the designed switch are obtained as –12 dB, –0.05 dB, and –45 dB, respectively.

Keywords MEMS · RF · Switch · Insertion · Isolation

1 Introduction

FET-based switches have failed to provide satisfactory performance when they are used for high-frequency applications such as microwave applications due to high power consumption and RF power-handling capability. Also, the demand of miniaturized devices is the reason for failing these FET-based switches. Micro-electromechanical systems (MEMSs)-based switches have been overcome all the problems associated with FET-based switches. MEMS switches have replaced FET switches due to their low power consumption with better RF performance in terms of iso-

Kurmendra (✉) · O. Pertin
Department of Electronics & Communication Engineering,
Rajiv Gandhi University (A Central University), Itanagar, India
e-mail: kurmendra.nits@gmail.com

O. Pertin
e-mail: pertinosor@gmail.com

Kurmendra · R. Kumar
Department of Electronics & Communication Engineering,
North Eastern Regional Institute of Science & Technology, Itanagar, India
e-mail: itsrk2006@gmail.com

© Springer Nature Singapore Pte Ltd. 2019
A. Khare et al. (eds.), *Recent Trends in Communication, Computing,
and Electronics*, Lecture Notes in Electrical Engineering 524,
https://doi.org/10.1007/978-981-13-2685-1_1

Table 1 Design parameters of the proposed switch

Serial number	Design parameter	Value (in μm)
1	Beam length	100
2	Beam width	5
	Beam thickness	1
3	Coplanar waveguide	50-10-1
4	Dielectric thickness	0.4
5	Substrate thickness	5
6	Air gap	0.6

lation, insertion, and return losses at GHz frequencies. These switches are also in demand since the fabrication of these MEMS-based switches does not require any new fabrication technology; they are compatible enough to the existing technology for the fabrication. MEMS switches are most successful designs among other MEMS-based designed devices, and they are capable of working efficiently in the range of 20–100 GHz [1–7]. MEMS switches with different designs are presented in the articles [3–7]. There are two types of MEMS switches which exist: (i) series MEMS switch and (ii) shunt MEMS switch. MEMS switches with cantilever beams are considered as series MEMS switch and those who employ fixed–fixed type of beams is considered as shunt type of MEMS switches. MEMS shunt-type switches are much popular due to their high performance in RF range, and many researchers have investigated that MEMS shunt-type switches are capable of providing a low pull-in voltage which is very important for RF MEMS switch designed for low-power applications [7–12]. Capacitance analysis of these shunt switches shows that they can also be used for tunable capacitor application. MEMS shunt type of switches is also designed with perforated holes in the beam for better Cd/Cu ratio. Cd/Cu ratio depends on width of the beam and the air gap between dielectric layer and beam. Many different ways of increasing capacitance ratio are given in the articles [12–20].

2 Design of the MEMS Switch

Figure 1 shows the designed MEMS switch. It consists of substrate (polysilicon), CPW (gold), dielectric (Si_3N_4), anchor (gold) both the side to support the beam (gold). The switch is designed along with CPW line to study the high-frequency characteristics. The CPW transmission line consists of two parts: (i) center conductor for signal transmission and (ii) two conductors which are connected to ground. The dimensions of each part of the switch are mentioned in Table 1.

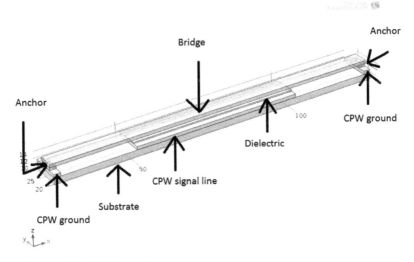

Fig. 1 Design of the MEMS switch

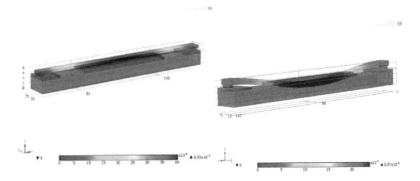

Fig. 2 Switch in upstate position and downstate position (FEM simulation using COMSOL)

2.1 Working Principle of the Designed Switch

When the switch is in upstate position, no electrostatic actuation is applied and the RF signal can transmit through the CPW transmission line. When an electrostatic voltage (actuation voltage) is applied, it creates electrostatic force to bend the beam in downstate. When the switch is biased with a voltage, the beam will bend down toward the substrate and keep in touch with dielectric, and then the RF signal will be connected to the beam and it will be transmitted to the ground lines in downstate position. In Fig. 2, the simulated graph for both the position can be seen. The equation for the actuation voltage can be given as [8]

$$V_p = \sqrt{\frac{8\,kg^3}{27\varepsilon_0\varepsilon_r Ww}} \qquad (1)$$

where g is the air gap, k is the spring constant of the fixed–fixed beam, W is the beam width, w is the width of CPW signal line, and ε_0, ε_r are the permittivity of free space and dielectric, respectively.

2.2 RF Parameters' Analysis

The S parameters of CPW transmission line characteristics can be employed to do the performance analysis of the switch. In the upstate position, S_{11} parameter gives the return loss of the switch while S_{21} parameter decides the insertion loss [7, 12, 13]. In the downstate position, S_{21} gives the isolation of the switch. The S_{11} and S_{21} are calculated by using the following expressions [10]

$$S_{11} = \frac{j\omega CZ_0}{2 + \omega CZ_0} \qquad (2)$$

$$S_{21} = \frac{2}{2 + \omega CZ_0} \qquad (3)$$

The CPW characteristic impedance Z_0 is taken as 65.64 Ω, calculated by coplanar waveguide calculator [21]. The resonant frequency is 61.5 GHz calculated using the formula $\omega = \sqrt{\frac{K}{m}}$, where K is the spring constant and m is mass of the beam. The return loss, isolation, and insertion loss can be calculated in dB value by taking log 10 of S_{11} and S_{21} and multiplying by 20.

3   Result and Discussions

The MEMS shunt type of switch was designed considering the specification provided in Table 1. The switch was actuated by applying pull-in voltage and the study was done specifically for radio-frequency analysis such as insertion loss, isolation loss, and return loss. Figure 3 depicts the effect of up capacitance on insertion loss, and the insertion loss decreases as the formed-up capacitance increases. So, it is desirable to have high up capacitance for low value of insertion loss, and at the same time, Cu value should be less (for High Cd/Cu value), so trade-off should be there for a better performance.

Figure 4 shows the dependency of switch insertion loss on frequency, and it can be concluded that the insertion loss is low at higher frequency for the designed switch which is very much desirable for high-frequency application. Figure 5 shows the effect of beam width on insertion loss, and the switch was designed by considering

Fig. 3 Insertion loss as
function of up capacitance

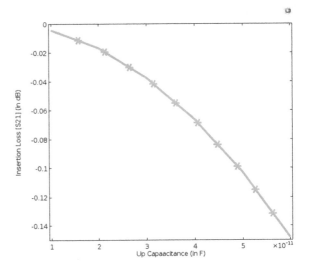

beam width as 5 and 7 μm. For high-frequency operation, switch designed with 5 μm is desirable compared to that of 7 μm. It also concludes that the trade-off between frequency of operation and insertion loss required. Figure 6 shows that the isolation loss is having a higher value for very low value of down capacitance (Cd). It is highly desirable to have a lower value of isolation loss (Cd should be high) since it assures that switch does not allow any signal in OFF state. Figure 7 depicts the isolation curve versus frequency, and the switch provides good isolation for the frequency range of 70–100 GHz with minimum loss and at frequencies lower than this range downgrades the isolation property of the switch. In Fig. 8, the effect of beam width on isolation can be seen. For better isolation, the switch should be optimized with shorter beam length.

Figure 9 shows the effect of up capacitance on return loss of the designed switch. The return loss should ideally be as minimum as possible. It can be concluded that the return loss increases as the up capacitance increases and since up capacitance is required to have a lower value for high Cd/Cu ratio. The trade-off is required to be done between return loss and up capacitance. The effect of frequency and beam width has also been shown in Fig. 10 and Fig. 11, respectively. They show a better performance for high frequency and smaller beam width.

Fig. 4 Insertion loss as
function of frequency

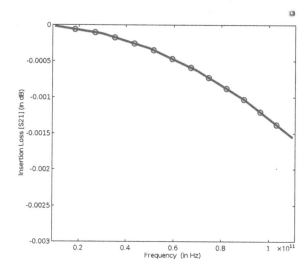

Fig. 5 Insertion loss as
function of frequency at
different beam widths

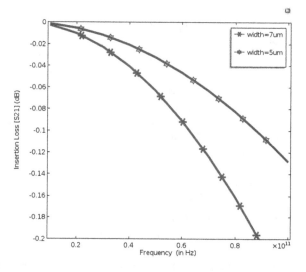

Fig. 6 Isolation as function of down capacitance

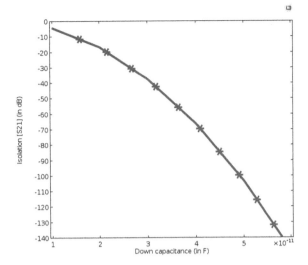

Fig. 7 Isolation as function of frequency

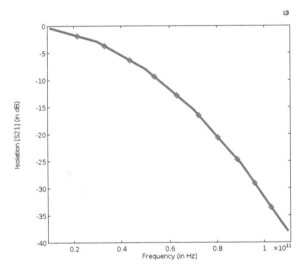

Fig. 8 Isolation as function
of frequency at different
beam widths

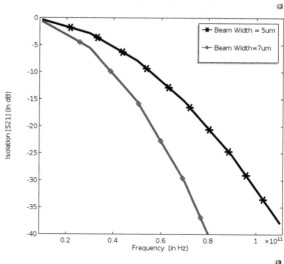

Fig. 9 Return loss as
function of up capacitance

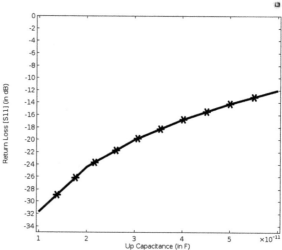

Fig. 10 Return loss as
function of frequency

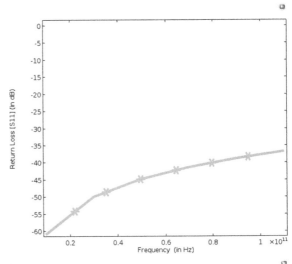

Fig. 11 Return loss as
function of frequency at
different beam widths

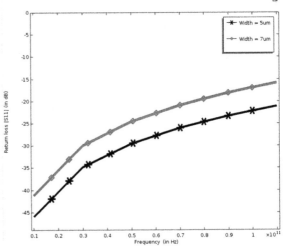

Table 2 Comparison of RF performance achieved by different MEMS switches

Authors	Used dielectric materials	Insertion loss	Isolation loss	Return loss	Pull-in voltage
Muhua Li et al. (2017) [22]	Si_3N_4	0.29 dB at 35 GHz	20.5 dB at 35 GHz	–	18.3 V
Guha et al. (2016) [23]	Si_3N_4 Hfo_2	−0.4 dB at 20 GHz	80 dB at 20 GHz	–	2.45 V 2.7 V
Our work	Si_3N_4	−0.05 dB at 62 GHz	−12 dB at 62 GHz	−45 dB at 62 GHz	10.5 V

4 Conclusion

In this paper, a MEMS switch was designed and study for high frequency was carried out. When the switch is supplied with actuation voltage of 10.5 V, then the S parameters can be plotted for different parameters such as frequency, beam width, and capacitance. It can be observed that isolation of switch in the downstate position of bridge decreases for increasing frequency sweep. It is also observed that as the frequency increases, insertion loss of the switch decreases. Insertion loss varies from −0.02 to −0.13 dB for a frequency sweep from 10 to 100 GHz. Comparison of RF performances for different designed switches is shown in Table 2. Proposed switch shows acceptable value of insertion, isolation, and return loss and operating frequency greater compared to other designed switches.

Acknowledgements This research work has been carried out in MEMS laboratory, Department of ECE, Rajiv Gandhi Central University, Itanagar, India.

References

1. Persano, A., Quaranta, F., Martucci, M. C., Siciliano, P., & Cola, A. (2015). On the electrostatic actuation of capacitive RF MEMS switches on GaAs substrate. *Sensors and Actuators A-Physical, 232*, 202.
2. Kenny, T. W. (2013). Experimental validation of topology optimization for RF MEMS capacitive switch design. *Journal of Microelectromechanical Systems, 22*, 1296.
3. Rebeiz, G. M. (2003). *RF MEMS, Theory Design and Technology*. Hoboken, New Jersey: Wiley.
4. Philippine, M. A., Sigmund, O., Rebeiz, G. M., & Kenny, T. W. (2013). Topology optimization of stressed capacitive RF MEMS switches. *Journal of Microelectromechanical Systems, 22*, 206.
5. Persano, A., Tazzoli, A., Farinelli, P., Meneghesso, G., Siciliano, P., & Quaranta, F. (2012). K-band capacitive MEMS switches on GaAs substrate: design, fabrication, and reliability. *Microelectronics Reliability, 52*, 2245.
6. Lin, C., Hsu, C., & Dai, C. (2015). Fabrication of a micromachined capacitive switch using the CMOS-MEMS technology. *Micromachines, 6*, 1645.

7. Angira, M., & Rangra, K. (2015). A low insertion loss, multi-band, fixed central capacitor based RF-MEMS switch. *Microsystem Technologies, 21,* 2259.
8. Mishra, B., Panigrahi, R., & Alex, Z. C. (2009). Design of RF MEMS switch with high stability effect at the low actuation voltage. *Sensors & Transducers, 111,* 58.
9. Badia, M. F., Buitrago, E., & Ionescu, A. M. (2012). RF MEMS shunt capacitive switches using AlN compared to Si_3N_4 dielectric. *Journal of Microelectromechanical Systems, 21,* 1229.
10. Zhu, Y., Han, L., Qin, M., & Huang, Q. (2014). Novel DC-40 GHz MEMS series-shunt switch for high isolation and high power applications. *Sensors and Actuators A, 101.*
11. Fernandez-Bolanos, M., Perruisseau-Carrier, J., Dainesi, P., & Ionescu, A. M. (2008). RF MEMS capacitive switch on semi-suspended CPW using low-loss high-resistivity silicon substrate. *Microelectronic Engineering, 85,* 1039.
12. Angira, M., & Rangra, K. (2015). Design and investigation of a low insertion loss, broadband, enhanced self and hold down power RF-MEMS switch. *Microsystem Technologies, 21,* 1173.
13. Demirel, K., Yazgan, E., Demir, S., & Akinodotn, T. (2015). Cantilever type radiofrequency micromechanical systems shunt capacitive switch design and fabrication. *Journal of Micro/Nanolithography, MEMS and MOEMS, 14,* 35005.
14. Koutsoureli, M., et al. (2016). An in depth analysis of pull-up capacitance-voltage characteristic for dielectric charging assessment of MEMS capacitive switches. *Microelectronics Reliability.* http://dx.doi.org/10.1016/j.microrel.2016.07.027.
15. Larson, L. E., Hackett, R. H., Melendes, M. A., & Lohr, R. F. (1991). Micromachined microwave actuator (MIMAC) technology a new tuning approach for microwave integrated circuits. In *Procedings IEEE Microwave and Millimeter-Wave Monolithic Circuits Symposium Digest* (pp. 27–30).
16. Yao, J. J., & Chang, M. F. (1995). A surface micromachined miniature switch for telecommunications applications with signal frequencies from DC up to 4 GHz. In *Proceedings. International Conference on Solid-State Sensors and Actuators Digest* (pp. 384–387).
17. Agarwal, S., Kashyap, R., Guha, K., & Baishya, S. (2017). Modeling and analysis of capacitance in consideration of the deformation in RF MEMS shunt switch. *Superlattices and Microstructures, 101.*
18. Kurmendra, & Kumar, R. (2017). Design analysis, modeling and simulation of novel rectangular cantilever beam for MEMS sensors and energy harvesting applications. *International Journal of Information Technology.* https://doi.org/10.1007/s41870-017-0035-6.
19. Guo, Z., Fu1, P., Liu, D., & Huang, M. (2017). Design and FEM simulation for a novel resonant silicon MEMS gyroscope with temperature compensation function. *Microsystem Technologies.* https://doi.org/10.1007/s00542-017-3524-4.
20. Sravani, K. G., & Rao, K. S. (2017). Analysis of RF MEMS shunt capacitive switch with uniform and non-uniform meanders. *Microsystem Technologies.* https://doi.org/10.1007/s00542-017-3507-5.
21. Coplanar Waveguide Calculator. http://www.microwave101.com/encyclopedia.
22. Li, M., Zhao, J., You, Z., & Zhao, G. (2017). Design and fabrication of a low insertion loss capacitive RF MEMS switch with novel micro-structures for actuation. *Solid-State Electronics,127,* 32–37.
23. Guha, K., Kumar, Mi, Parmar, A., & Baishya, S. (2016). Performance analysis of RF MEMS capacitive switch with non uniform meandering technique. *Microsystem Technologies, 22,* 2633.

Comparative Response Evaluation of Multilevel PSK and QAM Schemes

Awadhesh Kumar Singh and Nar Singh

Abstract The comparative response evaluation of M-ary phase-shift keying and M-ary quadrature amplitude modulation schemes is presented in this paper. The constellation diagrams for 16-PSK and 16-QAM are plotted to show the exact location of symbols, and the expressions for error probability are derived for both signaling schemes. The performance is compared by way of bit error probability and bandwidth efficiency. The simulated bit error rate graph plotted for different value of M are presented and compared with theoretical bit error rate plots separately. The combined BER graphs are also plotted for M = 16 and M = 64 for both signaling schemes and combined bandwidth efficiency graph plotted for M-ary PSK and M-ary QAM.

Keywords Additive white Gaussian noise · BER · M-ary phase-shift keying
M-ary quadrature amplitude modulation

1 Introduction

The digital communication systems have been used in lieu of previously used analog communication systems. These digital communication systems have several attainments such as band width efficiency, more data transfer capability, delivering better performance at low power, and design flexibility. The basic digital communication systems such as binary phase-shift keying, amplitude-shift keying, and frequency-shift keying were used [1]. For better data transfer rate and spectrum efficiency, the M-ary signaling systems are being used, such as M-ary amplitude-shift keying/digital pulse amplitude modulation, M-ary frequency-shift keying, M-ary phase-shift key-

A. K. Singh (✉) · N. Singh
ECE Department, University of Allahabad, Allahabad, India
e-mail: aksjkau5356@gmail.com

N. Singh
e-mail: nsjk53@rediffmail.com

© Springer Nature Singapore Pte Ltd. 2019
A. Khare et al. (eds.), *Recent Trends in Communication, Computing,
and Electronics*, Lecture Notes in Electrical Engineering 524,
https://doi.org/10.1007/978-981-13-2685-1_2

ing, and M-ary quadrature amplitude modulation [2]. It is assumed that the ideal channel is used to derive expression for symbol error probability. In an ideal channel, attenuation coefficient is unity and there is negligible time delay between transmitted and received signals.

2   M-ary PSK Signaling Scheme

M-ary PSK scheme is better than BPSK scheme. The modulator and demodulator structures are discussed here, and performance is measured in terms of BER and spectrum efficiency.

The M-ary PSK modulated signal is written as

$$S_m(t) = \left(\sqrt{\frac{2E_s}{T_s}}\right) \cos (2\pi f_c t + \phi_m), 0 < t < Ts \tag{1}$$

where E_s is signal energy, f_c is carrier frequency, T_s is symbol duration, and phase ϕ_m is different for different symbol. It is given by

$$\phi_m = \frac{2\pi (m-1)}{M} + \lambda \tag{2}$$

where $m = 1, 2, \ldots, M$ and fixed phase offset is $\lambda = \frac{\pi}{M}$.

For $M = 16$, phases are $11.25°, 33.75°, 56.25°, 78.75°, 101.25°, 123.75°, 146.25°, 168.75°, 191.25°, 213.75°, 236.25°, 258.75°, 281.25°, 303.75°, 326.25°$ and $348.75°$.

These phases are shown in constellation diagram for 16-PSK in Fig. 1.

Expanding the cosine term, $S_m(t)$ may be written as

$$S_m(t) = \sqrt{\frac{2E_s}{T_s}} \cos\phi_m \cos 2\pi f_c t - \sqrt{\frac{2E_s}{T_s}} \sin\phi_m \sin 2\pi f_c t, 0 < t < Ts \tag{3}$$

This is further written as

$$S_m(t) = A_{cm}\cos 2\pi f_c t - A_{sm}\sin 2\pi f_c t \tag{4}$$

where

$$A_{cm} = \sqrt{\frac{2E_s}{T_s}}\cos\phi_m \tag{5a}$$

and

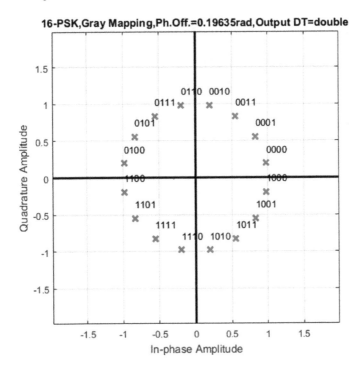

Fig. 1 Constellation diagram of 16-PSK

$$A_{sm} = \sqrt{\frac{2E_s}{T_s}} \sin\phi_m \tag{5b}$$

M-ary PSK is equivalent to the modulation of two quadrature carriers with A_{cm} and A_{sm}. The value of A_{cm} and A_{sm} depends on the symbol.

2.1   M-ary Phase-Shift Keying Transmitter

The transmitter structure is given in Fig. 2. A_{cm} and A_{sm} are obtained from input data. A_{cm} and A_{sm} are applied to balance modulators 1 and 2, respectively. Outputs from modulator 1 and 2 are combined.

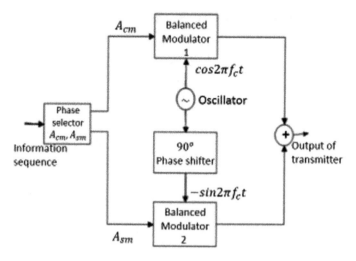

Fig. 2 Block diagram of M-PSK transmitter

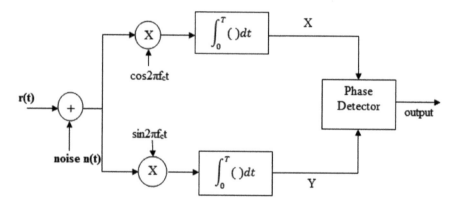

Fig. 3 Block diagram of M-ary phase-shift keying receiver

2.2 *M-ary Phase-Shift Keying Receiver (Fig. 3)*

The received signal plus noise is applied together to two cross-correlator receiver. The outputs of both are X and Y. The vector V is given by $X + jY$. The phase of V is compared with known phase of symbols. The decision is made in favor of closest [3].

The probability density function of $P(\theta)$ can be taken as

$$p(\theta) \approx \sqrt{\frac{\gamma}{\pi}} \cos\theta \, e^{-\gamma \sin^2\theta} \tag{6}$$

where $\gamma = \alpha^2 E_s/N_o$ is SNR per symbol, α is channel attenuation, and $N_o/2$ is PSD of white Gaussian zero mean noise.

For correct decision, the phase will be within the range of $-\pi/M \leq \theta \leq \pi/M$; otherwise, the error is made. The probability of symbol error is given

Thus

$$P_M = 1 - \int_{-\pi/M}^{\pi/M} p(\theta) d\theta \tag{7}$$

Substituting $p(\theta)$ in Eq. (7)

$$P_M \approx 1 - \int_{-\pi/M}^{\pi/M} \sqrt{\frac{\gamma}{\pi}} \cos\theta \, e^{-\gamma \sin^2\theta} d\theta \tag{8a}$$

$$P_M \approx 1 - \left(1 - \frac{2}{\sqrt{\pi}} \int_{\sqrt{\gamma} \sin \pi/M}^{\infty} e^{-u^2} du \right) \tag{8b}$$

$$P_M \approx \mathrm{erfc}\left(\sqrt{\gamma} \sin \frac{\pi}{M} \right) = \mathrm{erfc}\left(\sqrt{k\gamma_b} \sin \frac{\pi}{M} \right) \tag{8c}$$

where $k = \log_2 M$. It is noted that the error probability approximation is good for all values of $M > 4$.

The simulation and theoretical plots of signal-to-noise ratio versus bit error rate plots are depicted in Fig. 4, and it is found that for least BER, the simulated and theoretical plots are almost same. It is also found that the consumption of signal-to-noise ratio increases with increase in M; e.g., at BER $= 10^{-9}$, the additional power is required (over BPSK); for $M = 4$, SNR is 13 dB; for $M = 8$, SNR is 16 dB; for $M = 16$, SNR is 21 dB; and for $M = 32$, SNR is 26 dB.

3   M-ary Quadrature Amplitude Modulation Scheme

The M-ary quadrature amplitude modulation scheme is broadly used in practical applications because it has some special qualities. These special properties are consuming less signal-to-noise ratio and better bandwidth efficiency [4]. In this signaling scheme, both phase and amplitude vary with respect to the input symbol [5]. The M-ary quadrature amplitude modulated signal is specified as

$$S_m(t) = A_m \cos(2\pi f_c t + \Phi_m), 0 \leq t \leq Ts \tag{9}$$

On expanding Eq. (9), it is obtained that

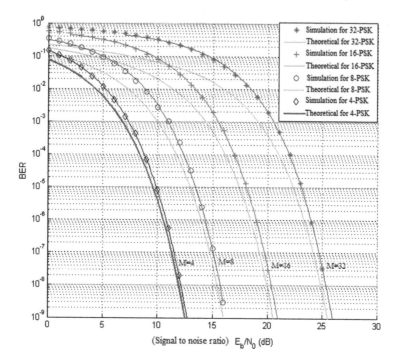

Fig. 4 Eb/No versus BER plots of M-ary phase-shift keying

$$s_m(t) = A_m \cos\phi_m \cos 2\pi f_c t - A_m \sin\phi_m \sin 2\pi f_c t$$
$$s_m(t) = A_{cm} \cos 2\pi f_c t + A_{sm} \sin 2\pi f_c t \qquad (10)$$

where $A_{cm} = A_m \cos\phi_m$ and $A_{sm} = -A_m \sin\phi_m$.

It is clearly seen from Eq. (10) that quadrature amplitude modulation can be viewed as digital pulse amplitude modulation of two phase quadrature sinusoidal carriers A_{cm} and A_{sm}.

3.1 M-ary Quadrature Amplitude Modulation (Fig. 5)

The binary input data is applied to the amplitude and phase detector to evaluate different values of A_{cm} and A_{sm}. After that, the quadrature carriers $\cos(2\pi f_c t)$ and $\sin(2\pi f_c t)$ are then modulated by A_{cm} and A_{sm}, respectively. These modulated signals are added together to obtain QAM modulated signal.

The constellation diagram for 16-QAM is depicted in Fig. 6. Gray coding is used for the representation of code word.

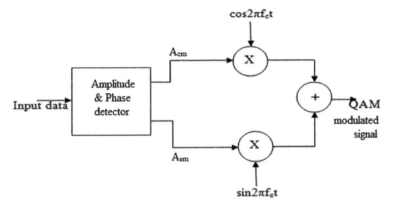

Fig. 5 Block diagram of M-ary QAM modulator

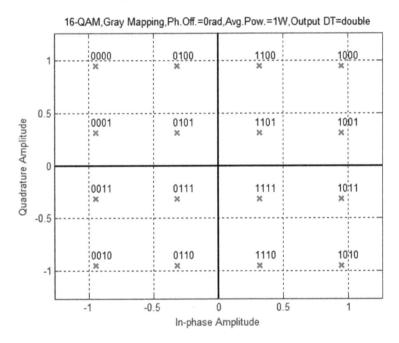

Fig. 6 Constellation diagram of 16-QAM

3.2 M-ary QAM Receiver

The M-ary QAM receiver block diagram is depicted in Fig. 7. The signal corrupted with noise n(t) is received at receiver end and processed at cross-correlator receiver. This noise is white Gaussian zero mean with power spectral density of $N_o/2$. The receiver computes the distance between observed values of A_{cm} and A_{sm} with differ-

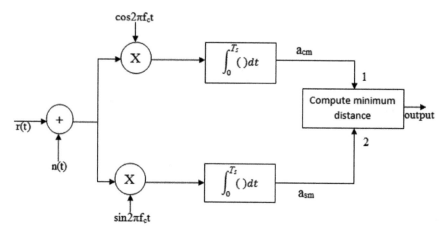

Fig. 7 Block diagram of M-ary QAM receiver

ent symbol points and decides in favor of minimum distance [6]. The received signal will be given by

$$r(t) = A_{cm}cos2\pi f_c t + A_{sm}sin2\pi f_c t \tag{11}$$

In Eq. (11), two terms are behaving like digital pulse amplitude modulation; therefore, A_{cm} and A_{sm} are recovered in the same manner as in the case of PAM.

The probability of symbol error expression for digital pulse amplitude modulation is given below.

$$P_M = \frac{M-1}{M} erfc\left(\sqrt{\frac{3}{M^2-1}\gamma_{avg}}\right) \tag{12}$$

Here, γ_{avg} is average SNR/symbol. The expression (12) can be used for QAM by replacing M by $\sqrt{M}$ and γ_{avg} by $\gamma_{avg}/2$. After replacing these values, the probability of symbol error at point 1 and 2 is given below

$$P_{\sqrt{M}} = \frac{\sqrt{M}-1}{\sqrt{M}} erfc\left(\sqrt{\frac{3}{M-1}\gamma_{avg}/2}\right) \tag{13}$$

The probability of correct decision at point 1 and 2 is given by

$$P_c = 1 - P_{\sqrt{M}}$$

And at the output of minimum distance detector, it will be

$$P_c = \left(1 - P_{\sqrt{M}}\right)^2$$

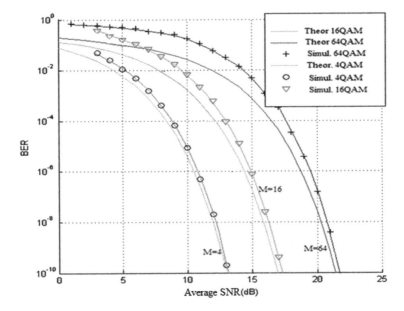

Fig. 8 (γ_{avg}) Average SNR versus BER plots of M-ary QAM

Finally, the probability of symbol error is written as

$$P_M = 1 - P_c = 1 - \left(1 - P_{\sqrt{M}}\right)^2$$

This expression can be simplified as

$$P_M \leq 2\left(1 - \frac{1}{\sqrt{M}}\right) erfc\left(\sqrt{\frac{3}{2(M-1)}\gamma_{avg}}\right) \qquad (14)$$

This expression holds the equal sign when k (k = log$_2$ M) is even. For odd value of k, the P_M with equal sign will represent upper bound on probability.

The simulated and theoretical results are depicted in Fig. 8 for M-ary QAM. It is found that the simulated BER plots are almost same with theoretical BER plots at very low bit error rate. It is also found that the demand of signal-to-noise ratio increases with increment in M for a given BER; for example BER at 10^{-10}, the power requirement for M = 4 is 13 dB, for M = 16 is 17 dB, and for M = 64 is 22 dB.

4 Bandwidth Efficiency of M-ary PSK and M-ary QAM

The transmitted data rate is denoted by R, and required value of bandwidth is denoted by W. The bandwidth efficiency is the ratio of data rate and required bandwidth (R/W)

(a) For M-ary phase-shift keying; $R = 1/T_b$ and $W = 1/T_s = 1/kT_b$, where bit duration is denoted by T_b and symbol duration is denoted by T_s.

$$\text{For M-ary PSK,} \qquad R/W = \log_2 M \qquad (15)$$

(b) For M-ary quadrature amplitude modulation; $R = 1/T_b$ and $W = 1/2T_s = 1/2kT_b$

$$\text{For M-ary QAM,} \qquad R/W = 2\log_2 M \qquad (16)$$

M-ary QAM is two times bandwidth efficient than M-ary PSK (Fig. 9).

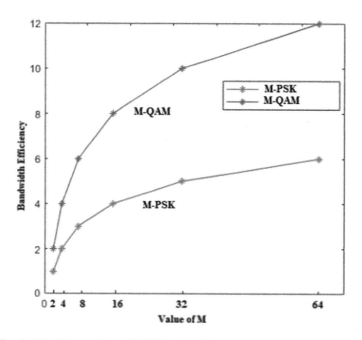

Fig. 9 Bandwidth efficiency plots of M-PSK and M-QAM

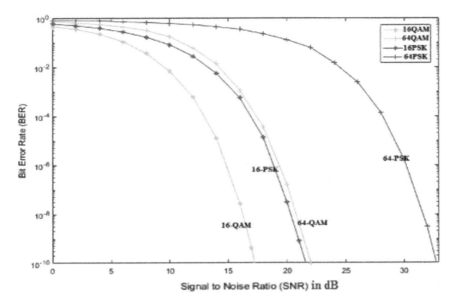

Fig. 10 Comparative SNR versus BER plots of 16-PSK, 64-PSK, 16-QAM, and 64-QAM

5 Comparative Result of M-PSK and M-QAM

In Fig. 10, the combined BER plot is shown for 16-PSK, 16-QAM, 64-PSK, and 64-QAM. It is observed the M-ary QAM is the most efficient at very low SNR. For BER at 10^{-10}, the required SNR for 16-QAM is 17 dB and for 64-QAM is 22 dB.

6 Conclusion

The performance of both M-PSK and M-QAM is analyzed here. Simulated results are almost identical with theoretical results. The bandwidth efficiency increases with increase in M which is very significant advantage over other signaling schemes. However, SNR requirement increases with M but this can be tolerated. It is observed that the bandwidth requirement for M-ary QAM signaling scheme is half than M-ary PSK signaling scheme. It is also observed that the M-ary QAM signaling scheme is very efficient scheme in comparison with M-ary PSK signaling scheme. Presently, 64-QAM is broadly used in communication. The higher-order QAM signaling schemes are only solution of ever-increasing communication demand. The communication demand will increase in near future; therefore, 256-QAM and higher-order QAM signaling schemes will be certainly employed.

References

1. Haykin, S. (2001). *Communication Systems* (4th ed.). Wiley.
2. Taub, H., Schilling, D., & Saha, G. (2008). *Principles of Communication Systems* (3rd ed.). New Delhi: Tata McGraw-HILL Publishing Company Limited.
3. Proakis, J. G. (2001). *Digital Communications* (4th ed.). McGraw Hill.
4. Bernard, S., & Ray, P. K. (2013). *Digital Communications: Fundamentals and Applications*. (2nd ed.). Pearson, India.
5. Singh, A. K., & Singh, N. (2017). Performance evaluation of digital M-ary quadrature amplitude signalling scheme. IJARECE, *6*(10), 1070–1073.
6. Li, J., Dazhang, X., Gao, Q., Luo, Y., & Gu, D. (2008). Exact BEP analysis for coherent M-ary PAM and QAM over AWGN and Rayleigh fading channels. In *IEEE Conference Publications*, pp. 390–394.

Performance Optimization of Carving Signal RZ-DQPSK Modulation Scheme

Divya Sharma, Vinit Jaiswal, Y. K. Prajapati and Rajeev Tripathi

Abstract This paper depicts a modified return-to-zero differential quadrature phase shift keying (RZ-DQPSK) modulation scheme. A simulative analysis has been performed for an optimization of the proposed scheme. The analysis has also been extended for two different types of commercial fibers; standard single-mode fiber (SSMF) and true wave-reduced slope (TW-RS) fiber to report the similar qualitative performance characteristics. For the sake of better understanding of the proposed modified carving signal RZ-DQPSK scheme, optical spectrum and eye pattern plots are also investigated.

Keywords Return-to-zero differential quadrature phase shift keying (RZ-DQPSK)
True wave-reduced slope fiber (TW-RS) · Bit error rate (BER) · Quality factor
Eye pattern

1 Introduction

Deployment of the optical fiber technology entirely revolutionized the telecommunication world. To meet the ever-increasing demand of high-capacity optical networks for data transmission at 40 Gb/s and beyond with acceptable signal quality, a vast research is being carried with an aim to explore new multiplexing, modulation, and optical signal processing techniques [1–3]. As the data rate increases, polarization

D. Sharma (✉) · V. Jaiswal · Y. K. Prajapati · R. Tripathi
Department of Electronics and Communication Engineering, Motilal Nehru National
Institute of Technology Allahabad, Allahabad 211004, UP, India
e-mail: divya.fgiet@gmail.com

V. Jaiswal
e-mail: jaiswalvinit27@gmail.com

Y. K. Prajapati
e-mail: yogendrapra@mnnit.ac.in

R. Tripathi
e-mail: rt@mnnit.ac.in

© Springer Nature Singapore Pte Ltd. 2019
A. Khare et al. (eds.), *Recent Trends in Communication, Computing,
and Electronics*, Lecture Notes in Electrical Engineering 524,
https://doi.org/10.1007/978-981-13-2685-1_3

mode dispersion (PMD) and nonlinear impairments (Kerr nonlinearities) such as self-phase modulation (SPM), cross-phase modulation (CPM), and four-wave mixing (FWM) become significant across the optical fiber, which severely affect the successful transmission of information signal through the fiber channel [4, 5]. Hence in order to successfully transmit the optical signal at high data rate, we have to choose some alternative schemes such as dispersion management system to compensate the deterioration of the affected signal or different modulation techniques to enhance the response of the system [6, 7]. Performance optimization of return-to-zero differential quadrature phase shift keying (RZ-DQPSK) modulation scheme for dispersion compensated optical link has already been examined with three different fiber types such as standard single-mode fiber (SSMF), true wave-reduced slope (TW-RS) fiber, and the large effective area fiber (LEAF) [8]. In 2016, a 40 Gb/s DWDM system is designed, in which DQPSK modulation scheme-based results have been compared with some other existing advance modulation schemes, e.g., 33 and 66% differential phase shift keying (DPSK), carrier suppressed return to zero (CSRZ), duobinary return to zero (DRZ), and modified duobinary return to zero (MDRZ) [9]. Generally, DQPSK modulation technique doubles the spectral efficiency rather than amplitude shift keying (ASK) or DPSK, due to being two signal quadrature of a single optical carrier. Henceforth, two message carrying bits can be transmitted per symbol, which leads to optical phase changes at four positions between successive symbol periods [10]. Therefore, we found DQPSK as a suitable scheme, and in this communication, we modified the architecture of RZ-DQPSK modulation scheme in order to get the improved quality factor (Q factor) performance of the system if compared with previous existing work [8].

The organization of this paper is as follows: Sect. 2 gives an idea of simulation setup of the proposed modified RZ-DQPSK modulation scheme. Section 3 deals with obtained result and discussions while Sect. 4 ends up in fruitful conclusion and future prospects.

2 Simulation Setup

Figure 1 exhibits the simulation setup for 10 Gb/s RZ-DQPSK modulated optical signal transmission for amplified spontaneous emission (ASE) noise-limited optical channel. Simulation setup remains identical for whole analysis, and only fiber types are varied.

DQPSK precoder encodes the data signal from pseudo-random sequence generator (PRBS) sources in order to match the received signal from transmitted one. Return-to-zero (RZ) driver converts the input logical data signal into electrical signal and these electrical signals pass through the low-pass filter (LPF) of center frequency 15 GHz for further transmission. DQPSK modulator modulates these electrical data signal with optical signal by means of carving modulator. Here, optical laser source is used with output power −10 dBm, linewidth 10 MHz, and center frequency 194 THz. In the proposed work, instead of directly introducing optical laser source out-

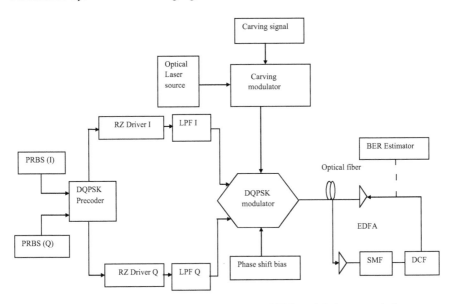

Fig. 1 Proposed simulation setup for carved signal RZ-DQPSK modulation transmission

put in DQPSK modulator, firstly it is modulated with the aid of a carving signal in order to avoid the chirp effect so that Q factor can be improved in comparison with previous work [8]. Chirp effect is a very significant phenomenon, which occurs in a pulse transmitted through a transparent medium due to effect of dispersion and Kerr nonlinearities, especially due to SPM [11]. This chirp effect can be reduced by controlling the extinction ratio and optimizing the optical modulation waveform [11]. Now, RZ-DQPSK modulated optical signal is transmitted through the optical fiber loop. This optical fiber loop consists of an erbium-doped fiber amplifier (EDFA) and single-mode fiber (SMF) followed by dispersion compensation fiber (DCF). EDFA provides the maximum gain of 35 dB with a 4.5 dB ASE noise. Bit error rate (BER) estimator is used to measure the performance of the proposed system. In this paper, OPTSIM commercial software is deployed for analysis of the proposed work over single span of each type of SMF fiber followed by a single segment of respective DCF. Details of parameter values of different fiber types considered in RZ-DQPSK modulation setup are given in Table 1.

3 Result and Discussion

Figure 2 exhibits the performance of RZ-DQPSK system in the presence of SSMF fiber followed by DCF. Here, Q factor is plotted against optical signal-to-noise ratio (OSNR) along with varying duty cycles from 40 to 90%. From Fig. 2, it is clearly observed that the Q factor performance of the system increases as duty cycle increases

Table 1 Simulation parameters of RZ-DQPSK modulation

Fiber type	Fiber attenuation α	D (ps/nm/km) λ @ 1550 nm	Dispersion slope S (ps/km)	Nonlinear refractive index	Effective core area Aeff (μm)	Single span length (km)
SSMF	0.22	17	0.016	3	80	25
DCF (SSMF)	0.55	−80	−0.076	2.5	20	5.3125
TW-RS	0.2	4.5	0.045	2.5	55	25
DCF (TW-RS)	0.55	−80	−0.8	2.5	20	1.40625
LEAF	0.2	4	0.1	2.5	20	1.25
DCF (LEAF)	0.55	−80	−2	2.5	20	1.25

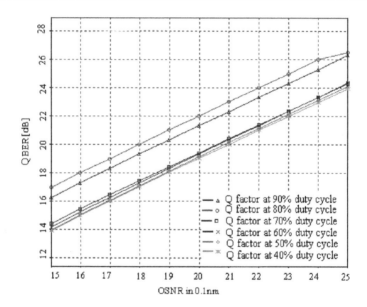

Fig. 2 Proposed Q factor of SSMF fiber for different duty cycles of RZ-DQPSK

from 40 to 80%, in a linear fashion. While on further increasing duty cycle from 80 to 90%, Q factor value decreases. Obtained results deliver an acceptable performance (greater than min. Q factor value of 6 dB) even at 40% duty cycle.

Figure 3 exhibits the performance of RZ-DQPSK system corresponding to a special type of TW-RS fiber. During analyzing the system behavior on introducing TW-RS fiber with SSMF, it is observed that the system follows the similar linear type of pattern with different duty cycles (40–90%) as in the case of SSMF fiber. In Figs. 2

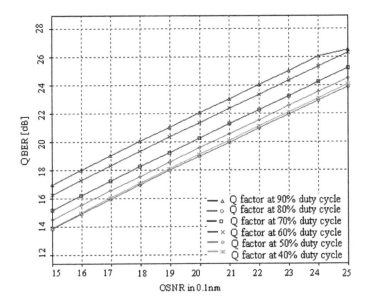

Fig. 3 Proposed Q factor of TW-RS fiber for different duty cycles of RZ-DQPSK

and 3, the highest Q factor performance is achieved at optimized duty cycle 80% for SSMF and TW-RS fiber-based proposed systems, respectively. But the key finding is that an improvement in Q factor value is examined in the presence of SSMF and TW-RS fibers on comparing with existing literature [8]. From the simulation results obtained so far, it can be observed that all different fibers deliver their best performance when RZ-DQPSK pulse has 80% duty cycle in an ASE noise-limited system. Also, reason behind such improvement in the Q factor value is introducing a carving modulator, through which a generated laser optical signal is passed. It decreases the chirp effect by providing a perfect carving over the laser source optical signal. Hence with controlling this chirp effect, transmitted pulse quality is getting improved with a good strength.

Figure 4 exhibits obtained optical spectrum of the proposed RZ-DQPSK modulation transmission without and with carving signal. A significant change can be observed on the elimination of side lobes due to carving signal. Similarly, eye pattern plot of proposed RZ-DQPSK modulation transmission is illustrated in Fig. 5 in the absence and presence of carving signal. A slight increment in eye amplitude, reduction in jitter, and improvement in sampling period can be better observed in the presence of carving signal.

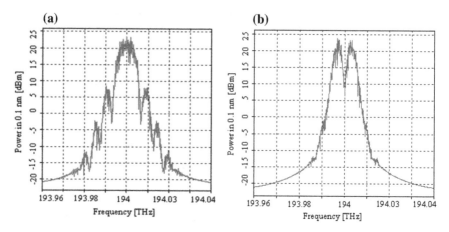

Fig. 4 Optical spectrum of proposed RZ-DQPSK modulation transmission. **a** Without carving signal; **b** with carving signal

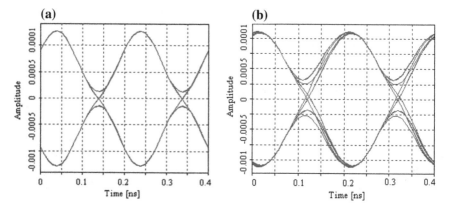

Fig. 5 Eye pattern plot of proposed RZ-DQPSK modulation transmission. **a** Without carving signal; **b** with carving signal

4 Conclusion

This paper proposes modified RZ-DQPSK system with enhanced Q factor. Hence even 40% duty cycle of RZ-DQPSK is acceptable at high value of OSNR. We performed analysis for two different fiber types, i.e., SSMF and TW-RS fiber, and observed the same characteristics. Finally, it is concluded that RZ-DQPSK delivers optimum performance at 80% duty cycle. Also, significance of introducing carving signal in the proposed RZ-DQPSK system is being analyzed by presenting optical spectrum and eye pattern plot. Future research direction may include suitability of type of fiber while designing a long-haul communication link for RZ-DQPSK modulated signal with 80% duty cycle. Future scope also includes analysis of multi-span system by varying different modulation techniques and terabit transmission [12].

Acknowledgements Authors are highly thankful to the institute Motilal Nehru National Institute of Technology, Allahabad, for providing licensed simulation software "RSOFT OPTSIM."

References

1. Mohamed, M. E., & Halawany, E. L. (2011). Efficient Raman amplifiers within propagation and multiplexing techniques for high capacity and ultra long haul transmission systems. *International Journal of Computer Science and Telecommunications, 2,* 16–23.
2. Sharma, D., & Prajapati, Y. K. (2016). Performance analysis of DWDM system for different modulation schemes using variations in channel spacing. *Journal of Optical Communications, 37*(4), 401–413.
3. Singh, A., Sharma, D., & Prajapati, Y. K. (2016). Comparison of DPSK and QAM modulation schemes in Passive optical network. In *International Conference on Fibre Optics and Photonics*, Optical Society of America, IIT Kanpur, India (pp. Tu4A-56).
4. Agrawal, G. P. (2001). *Nonlinear Fiber Optics*. New York: Academic.
5. Agrawal, G. P. (2002). *Fiber-Optic Communication Systems*. New York: Wiley.
6. Winzer, P. J., Raybon, G., Song, H., Adamiecki, A., Corteselli, S., Gnauck, A. H., et al. (2008). 100 Gb/s DQPSK transmission: From laboratory experiments to field trials. *IEEE Journal of Lightwave Technology, 26,* 3388–3402.
7. Winzer, P. J., & Essiambre, R. J. (2006). Advanced optical modulation formats. *Proceedings of the IEEE, 94*(5), 952–985.
8. Tiwari, V., Sikdar, D., & Chaubey, V. K. (2013). Performance optimization of RZ-DQPSK modulation scheme for dispersion compensated optical link. *Optik-International Journal for Light and Electron Optics, 124*(17), 2593–2596.
9. Sharma, D., & Prajapati, Y. K. (2016). Comparative aspect of different multi channel DWDM optical network. In *Advances in Computing, Control and Communication Technology,* Allahabad University, India (vol. 1, pp. 155–160).
10. Chien, C. C., Wang, Y. H., Lyubomirsky, I., & Lize, Y. K. (2009). Experimental demonstration of optical DQPSK receiver based on frequency discriminator demodulator. *IEEE Journal of Lightwave Technology, 27*(19), 4228–4232.
11. Pathak, V. B., Vieira, J., Fonseca, R. A., & Silva, L. O. (2012). Effect of the frequency chirp on laser wakefield acceleration. *New Journal of Physics, 14*(2), 023057.
12. Sharma, D., Prajapati, Y. K., & Tripathi, R. (2018). Spectrally efficient 1.55 Tb/s Nyquist-WDM superchannel with mixed line rate approach using 27.75 Gbaud PM-QPSK and PM-16QAM. *Optical Engineering, 57*(7), 076102. https://doi.org/10.1117/1.OE.57.7.076102.

Design and Performance Analysis of Reversible XOR Logic Gate

Kamal K. Upadhyay, Vanya Arun, Saumya Srivastava, Nikhlesh K. Mishra and Narendra K. Shukla

Abstract The main goal of this article is to design an optimized optical logic gates which are a key point to improve and enhance the speed of next era photonic integrated circuit. These kinds of integrated circuits are much more capable for high-speed transmission of information and very low energy loss. The energy dissipation by any integrated circuit in form of heat reduces the life span of device. This kind of design protects it and improves the life expectancy of the integrated circuit. In this article, opto-electronic conversion and vice versa are taken out.

Keywords Reversible logic · Semiconductor optical amplifier (SOA)
Extinction ratio · Quality factor (Q)

1 Introduction

As we all know that in the present century, huge amount of data is being transferred from one place to another and the medium used for the transportation of this bulk data is optical fiber. For low data transfer in less dense circuit, energy loss was negligible but with growing need of technology and circuits becoming more dense and when they are being used for high rate of data transfer the amount of energy lost will be of significant amount, and there comes the concept of reversible logic for circuit rescue which was discussed right back in 1960 by the R. Landauer who says that the loss of 1 bit of information will produce some amount of energy in form of heat which is KT * ln(2) joules, where K is Boltzman's constant and T is the absolute temperature [1]; after few years in 1973, Bennett [2] proposed that the energy lost against 1 bit of information lost—Ktln2 joules. It can be reduced by making reversible circuits.

K. K. Upadhyay (✉) · V. Arun · S. Srivastava · N. K. Mishra · N. K. Shukla
Department of Electronics and Communication, University of Allahabad, Allahabad, Uttar Pradesh, India
e-mail: kamal.kishoresiet@gmail.com

© Springer Nature Singapore Pte Ltd. 2019
A. Khare et al. (eds.), *Recent Trends in Communication, Computing, and Electronics*, Lecture Notes in Electrical Engineering 524,
https://doi.org/10.1007/978-981-13-2685-1_4

Fig. 1 Block diagram of reversible XOR gate

Moreover as per the concept of MOORE law, circuits designed are getting denser day by day with increased load of transfer of data processing, Moore law suggested that the transistor used in any integrated circuit may be double at every 18 months.

It is known fact that the base for the photonic processing circuits is the optical logic gates. It is classified in two categories reversible and irreversible optical logic gates. In the reversible logic gate, no information has been lost at the time of operation so the problem of heat dissipation has been reduced. Pustie and Blow [3] use nonlinear loop mirror with SOA as nonlinear element to propose optical Fredkin Gate. He used terahertz optical asymmetric demultiplexer in his design. Forsati et al. [4] develop Toffoli gate. Mandal et al. [5] and Garai [6] used frequency conversion property of SOA to implement reversible gates. Taraphdar et al. [7] implement Toffoli and Feynman gate using SOA-based MZI. With the use of SOA-based MZI, we get maximum switching speed and it increases speed of the system. Arun et al. [8] proposed a model of Toffoli and AND gate on single photonic circuit.

In this paper, we have proposed a novel model of reversible XOR logic gate on a single photonic circuit. At data rate of 10 Gbps with SOA-MZI setup. It is one of the first reversible 2-input $A \oplus B$ logic gate which was designed to achieve minimum value of bit error rate and high value of quality factor. It would be found that these gates XOR hold a novelty in itself, since no work bearing similarity in design to these have not been brought out yet. It can also be learnt that no other design bearing resemblance in a similar accord can be found in literary surveys and findings. It is the first step to develop an optical processor because the gates are the buildings blocks of any integrated circuit design (Fig. 1).

2 Principle of Operation

A reversible logic gate is a n × n device, where n is the number of input and number of output. It fulfills the criteria of one-to-one mapping. It helps to determine the output from the given input and also help in retrieval of input from the output. First time the reversible logic gate was introduced by Tommaso Toffoli in 1980 who invented the Toffoli gate. Garai (2011) said that any Boolean function can be designed by using reversible logic. Figure 2 represents the block diagram of reversible XOR logic gate. OUTPUT P = INPUT $A \oplus B$, and OUTPUT Q = INPUT $\bar{A}$. The proposed design implements reversible XOR logic gate by integrating two different units, namely unit 1, unit 2, as shown in Fig. 2. SOA-based MZI is the main building block of the proposed design. The internal architecture and all the parameters in both the MZI are identical. Unit 1 consists of identical MUX and SOA in both their arms with 2 × 2 couplers positioned at input and output. In the whole design, different type

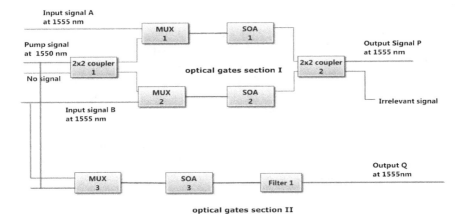

Fig. 2 Proposed model reversible XOR logic gate

Table 1 Functional table

S. no	Input signal	Output signal
	A B	Y
1	0 0	0
2	0 1	1
3	1 0	1
4	1 1	0

Table 2 Functional table of proposed model

S. no	Input signal		Output signal	
	A	B	P = A⊕B	Q = Ā
1	0	0	0	1
2	0	1	1	1
3	1	0	1	0
4	1	1	0	0

of signals are divided in two categories, i.e., pump (all one) and input signal A and Input signal B. All the input data values are given in Table 3. The assigned pre-bit values are 00000 and the value assigned for post-bit 0000000. All input signal have same pre-bit and post-bit values. Here the representation of the single bit is in form of four bit, i.e., 1 is written as 1111 and 0 as 0000. Table 1 represent the functional table of conventional XOR logic gate.

In above design, 4 bit is taken as a single bit, for example, 0000 is considered as 0 and 1111 considered as 1 for the ease of reading output graphs.

Unit 1

To design XOR reversible gate, we used SOA-based MZI architecture. The Boolean function that it implements is $P = A \oplus B$. The unit one consists of two SOA and two MUX connected with 2×2 coupler in input and output end. Signal named as pump at 1550 nm, i.e., nothing but all one (1111 1111 1111 1111) given to coupler 1, at the other input of the 2×2 coupler we put no signal. Two input signal named A at 1555 nm and B at 1555 nm are given to MUX. The output of multiplexer is input for the SOA 1. Similarly, MUX 2 has two input port, port 1 is input signal B at 1555 nm and port 2 is output-port 2 of 2×2 coupler 1. Coupler 1 has two input, first one is pump signal at 1550nm and second input is no signal. The output of MUX 2 is treated as input for SOA 2. At different output port of coupler 2 induced a phase difference of π at individual wavelength. In this unit, our motive is to remove the pump signal at the first output port of 2×2 coupler 2. If the data value is 1, the pump signal saturates the gain of SOA so the available gain is decreased for the signal 2. Therefore, the binary value comes to zero. On the other side, if pump signal value is zero so the high carrier density amplifies the signal, which shows the binary value 1.

Unit 2

It is a MUX-based architecture which is designed to implement output Q. Unit 2 contains MUX 3. It has one input A at 1555 nm and second input as pump signal. The output of MUX 3 is directly given as input to SOA 3. A filter is attached with SOA which passes the output of SOA. It gives the relevant output $Q = \bar{A}$ at 1555 nm.

3 Results and Discussion

This model is designed for data rates 10 Gbps. All SOAs used in above model have similar values and specifications. Figure 3 consists of all the input and output signals of the proposed model to design any reversible logic we have to use few more signal different to input signal those signal are known as pump signal and probe signal which was described as above. Figure 3a, b shows the input spectrum of proposed model. Figure 3a represents signal A while signal B in Fig. 3b. We have varied some parameters of SOA based on those parameters we observed some graph which was shown in Fig. 4a–c. Figure 3a represents the data value A which presents in Table 3 and Fig. 3b shows the spectrum of data B while Fig. 3c shows the spectrum of all ones, which represents pump signal Fig. 3d represents the output P which is verified by truth Table 2, and Fig. 3e shows the result Q which is just reverse of spectrum of data value A. This kind of design must play a good role in the field of optical fiber communication to develop optical processors. This kind of optical processors works with a very good speed and efficient data communication and also reduces the problem of heat dissipation at the receiving end (Table 4).

All above spectrum of input and output signal have been drawn. According to the parameter variation of SOA quality factor and extinction ratio plots are given. Figure 4a shows the variation of extinction ratio and quality factor due to change in pump current. Figure 4b shows variation due to changes in length of SOA while in Fig. 4c based on variation in current injection efficiency the value of quality factor and extinction ratio are calculated. Figure 4d shows the variation in extinction ratio due to change in confinement factor.

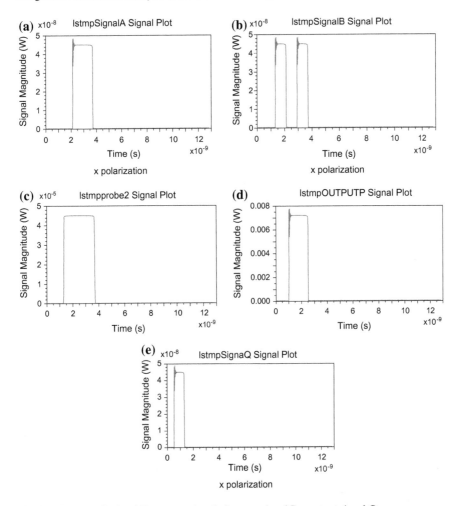

Fig. 3 **a** Signal A, **b** signal B, **c** pump signal, **d** output signal P, **e** output signal Q

Table 3 General inputs with their data values

S. no.	Signal	Pre-bits	Data values	Post-bits
1	A	00000	0000 1111 0000 1111	0000000
2	B	00000	0000 0000 1111 1111	0000000
3	Pump signal	00000	1111 1111 1111 1111	0000000

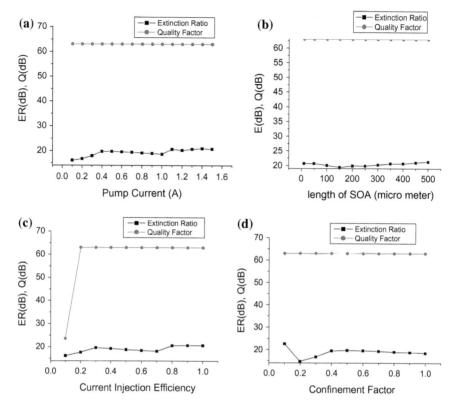

Fig. 4 **a** Quality factor, ER versus pump current, **b** quality factor, ER versus length of SOA, **c** quality factor, ER versus current injection efficiency, **d** quality factor, ER versus confinement factor

Table 4 Parameters of SOA

S. no.	Parameter	Value	Unit
1	Current injection efficiency	0.54	–
2	Width of active region	3.0	μm
3	Length of the active region	100.0	μm
4	Thickness of the active region	0.010	μm
5	Pump current	0.3	A
6	Confinement factor	0.34	–

4   Conclusion

The design of all-optical reversible XOR logic gate is achieved and based on above analysis the results gives good enough. Here, we get bit error rate negligible while quality factor is much better. The average quality factor for both the output P and Q is 53.03 dB. The value of average extinction ratio is 19.43 which is quite good. The work presented in manuscript is achieved targeted objectives.

Acknowledgements This simulation is done by using software Rsoft OptSim^{TM}.

References

1. Landauer, R. (1961). Irreversibility and heat generation in computing process. *IBM Journal of Research and Development, 5,* 183–191.
2. Bennett, C. H. (1973). Logical reversibility of computation. *IBM Journal of Research and Development, 17,* 525–532.
3. Paustie, A. J., & Blow, K. J. (2000). Demonstration of an all optical fredkin gate. *Optics Communications, 174,* 317–320.
4. Forasti, R., Ebrahimi, S. V., Navi, K., Mojoherani, E., & Jashnsaz, H. (2013). Implementation of all optical reversible logic gate based on holographic laser induced grating using azo-dye doped polymers. *Optics & Laser Technology, 45,* 565–570.
5. Mandal, D., Mandal, S., & Garai, S. K. (2015). Alternative approach of developing all optical fredking and Toffoli gate. *Optics & Laser Technology, 72,* 33–41.
6. Garai, S. K. (2014). A novel method of developing all optical frequency encoded fredking gates. *Optics Communications, 313,* 441–447.
7. Taraphdar, C., Chattopadhyay, T., & Roy, J. N. (2010). Mach-zdhnder interferometer based all optical reversible logic gate. *Optics and Laser Technology, 42,* 249–259.
8. Arun, V., Singh, A. K., Shukla, N. K., & Tripathi, D. K. (2016). Design and performance analysis of SOA MZI based reversible toffoli and irreversible AND logic gates in a single photonic circuit. *Optical and Quantum Electronics.*

Analysis of Photonic Crystal Fiber-Based Micro-Strain Sensor

Divesh Kumar, Dheeraj Kalra and Manish Kumar

Abstract Photonics crystal fiber-based sensors have small size, high sensitivity, flexibility, robustness, and ability of remote sensing. They can be used in unfavorable environmental conditions such as strong electromagnetic field, nuclear radiation, noise, and high voltages for explosive or corrosive media at high temperature. Fabrication of these sensors is so simple, and altogether makes them as a very efficient sensing solution for medical and industrial applications. A simple configuration of hollow-core photonic crystal fiber (HC-PCF) is presented for application as a micro-strain sensor to exhibit better sensitivity then the typical fiber Bragg grating (FBG)-based fiber optic strain sensors. Also, cross-sensitivity to changes in surrounding refractive index is avoided. The performance of the designed sensor is investigated for different strain levels ranging from 0 to 2000 $\mu\varepsilon$ at the wavelength, 1550 nm. Additionally, due to the air-hole structure of HC-PCF, the sensitivity of strain measurement remains unaffected of the changes in surrounding refractive index (SRI). Sensitivity of HC-PCF also depends on the fiber parameters like pitch and air-filling fraction.

Keywords Photonic crystal · Fiber Bragg grating · Sensitivity
Strain measurement

1 Introduction

Sensing is a key technology for application areas like industrial, health, transport, and entertainment uses. Many such applications require the concept of remote sensing, where data is acquired from places which are not easily accessible. Such data is transmitted from the remote location to the monitoring and control station. While transmitting the electrical signal received from conventional sensors, these electrical signals suffer from the problem of external interference from electromagnetic fields.

D. Kumar (✉) · D. Kalra · M. Kumar
GLA University, Mathura, India
e-mail: divesh.kumar@gla.ac.in

© Springer Nature Singapore Pte Ltd. 2019
A. Khare et al. (eds.), *Recent Trends in Communication, Computing, and Electronics*, Lecture Notes in Electrical Engineering 524,
https://doi.org/10.1007/978-981-13-2685-1_5

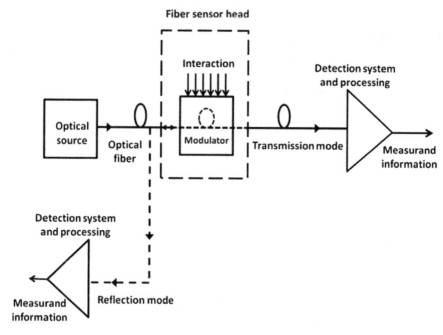

Fig. 1 Basic block diagram of fiber optic sensor

Optical fibers, in contrast, are immune to such kind of interference. Optical fibers are small and light weighted which make it more suitable where miniaturization is crucial, e.g., in spacecraft of aircraft industry. Optical fibers are also able to withstand extreme conditions of temperatures (Fig 1).

Optical fiber sensors have emerged as a unique solution in particular cases (e.g., electrically hazards or potentially explosive environments). PCF is a recently developed class of optical fiber [4] which is characterized by periodic arrangement of air-filled capillaries running along entire length of the fiber, centered on a solid or a hollow core. The mode shape, dispersion, transmission spectrum, birefringence, etc., can be optimized by changing the position and varying size of the holes on the cladding and size of the core. PCFs turn out to be a promising and enhanced solution for practical applications where tracking of strain-induced changes is crucial, such as in the fields of experimental mechanics, metallurgy, aeronautics, and significantly, in health monitoring of complex structures [5] (Fig 2).

Shi et al. [6] obtained a strain sensitivity of 1.55 pm/$\mu\varepsilon$ and insensitivity to temperature and bend, with the help of Fabry–Perot-type strain sensor based on HC-PCF.

Gong et al. [8] presented a modal interferometer based on short length PCF for strain measurement to achieve a strain sensitivity of 1.83 pm/$\mu\varepsilon$.

In spite of the vast research in the field of strain sensing using PCF, it has not lead to commercialization of PCF sensors. So, techniques have to be developed so as to make PCF sensing more cost effective. Also, strain sensitivity needs to be increased in

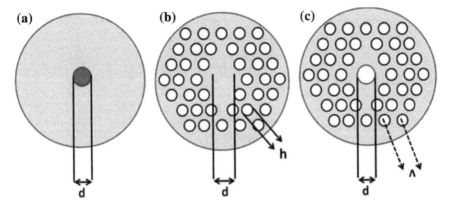

Fig. 2 Cross-sectional geometry of **a** SMF, **b** SC-PCF, **c** HC-PCF

order to measure comparatively lower magnitudes of applied strain. This can be done by manipulating the geometry of the PCF, e.g., by varying the core diameter or by changing the pitch of the fiber in order to obtain a higher sensitivity [1]. Insensitivity to temperature is an issue which has mostly been researched upon in the past. PCF sensors which can also avoid the cross-sensitivity to other parameters, e.g., changes in SRI, vibration, bend have to be designed [3].

Two different structures of HC-PCF are proposed which differ in their core diameters (large core and small core) and demonstrated their application as a strain sensor. The FDTD method is used to simulate operation of sensor at different strain levels. Comparison of the strain sensitivities of the two fibers has been done. Insensitivity of this sensor to the changes in SRI has also been verified.

2 Theory and Simulation

A PCF is made up of two materials which have a high contrast of refractive index. Thus, it can be referred to as a 3D structure having a 2D refractive index distribution such that the light remains confined to the core. When light is incident into a HC-PCF, the fundamental core mode is widely spread, and higher order modes are excited in the PCF including the cladding modes. The total intensity can be calculated as [9]:

$$I(\lambda) = I_{co}(\lambda) + I_{cl}(\lambda) + 2[I_{co}(\lambda) I_{cl}(\lambda)]^{1/2} \cos\left(\frac{2\Pi \Delta n L}{\lambda}\right) \tag{1}$$

where $I_{co}(\lambda)$ and $I_{cl}(\lambda)$ are the powers of the interfering core and cladding modes at wavelength λ, and L is the length of the HC-PCF. The transmission spectrum is directly related to the optical path difference φ which depends on the differential

Table 1 Parameters of
HC-PCF

Parameter	Large-core PCF	Small-core PCF
Pitch, Λ(µm)	2	7
Air-filling fraction, d/Λ	0.75	0.45

effective index Δn of the two modes. Dips in the transmission spectrum can be observed when the phase matching condition is satisfied [10]:

$$\varphi = \frac{2\Pi \Delta n L}{\lambda} = \frac{2\Pi (n_{co} - n_{cl})L}{\lambda} = (2n + 1)\Pi \qquad (2)$$

where n is some positive integer and φ is the phase difference between the two interfering modes. n_{co} and n_{cl} denote the effective modal refractive indices of the fundamental core mode and the higher order modes in cladding, respectively.

Application of micro-strain results in the extension in length of the PCF or a transverse contraction of the PCF. Since, the phase difference between the two interfering modes depends on the effective index Δn between the modes as well as the length, L of the PCF, there will be a shift in the resonance wavelength in the spectrum due to change in length of the HC-PCF. The resonance wavelength is given by [10]

$$\lambda = \frac{(n_{co} - n_{cl})L}{2n + 1} \qquad (3)$$

The shift in the resonance wavelength due to the change in the length of the fiber can be calculated by differentiating (3) with respect to L. This is given by [9, 11]

$$\Delta \lambda = 1 + \frac{L}{\Delta n} \frac{d(\Delta n)}{dL} \lambda \qquad (4)$$

As can be seen from the above equation, there is a linear relation between the shift in resonance wavelength and the strain applied. Thus, this shift can be employed for the measurement of applied strain.

Two different geometries of hollow-core PCF as strain sensors are investigated as large-core PCF and small-core PCF. Both of these are hexagonal lattice structures designed with silica as the base material comprising of circular air holes centered on a hollow circular core. The parameters of the two fibers used in our investigation are listed in Table 1.

The shift in resonance wavelength is higher at longer wavelengths as per the relation given in (4), thus the operating wavelength is chosen as 1550 nm in order to achieve higher strain sensitivity. Different levels of strain ranging from 0 to 2000 µε are applied on both the structures (assuming uniform deformation of the holes throughout the structure), and the corresponding shift in resonance wavelength is

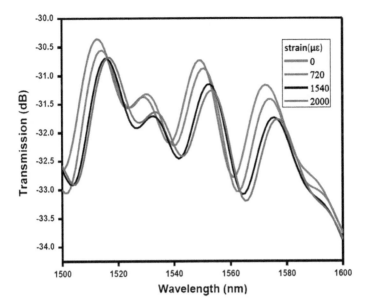

Fig. 3 Transmission spectra of small-core HC-PCF with different levels of applied strain

observed. Strain sensitivities of both the sensor structures are calculated and compared. Effect of changes in SRI on the strain sensitivity is also analyzed.

3 Results

The mentioned sensor geometries are analyzed individually for different strain levels ranging up to 2000 $\mu\varepsilon$. Generally, micro-level of strain induces minute changes in the length of the sensor as well as in its cross-sectional geometry. This affects the position of the dips obtained in the spectrum, as compared to the condition when no strain was applied. By correlating the position of these dips in the spectrum, a calibration curve can be computed. The transmission spectrum of small-core HC-PCF when it is subjected to four different levels of axial strain can be seen in Fig. 3. Transmission dips wavelength shift is observed near wavelengths 1540 and 1560 nm.

A similar shifting of the transmission spectrum around wavelength 1560 nm is observed for the large-core HC-PCF when subjected to varying strain levels as shown in Fig. 4. By measuring the shift in the resonance wavelength, the applied strain can be detected. From Figs. 3 and 4, it is observed that the transmission dips shift to a longer wavelength without a change in shape. Only a variation in intensity is observed which indicates participation of more than two modes in the process of interference.

The wavelength of the dips and the applied strain relationship for small-core PCF and large-core PCF geometries of the sensor are shown in Fig. 5a, b, respectively. A

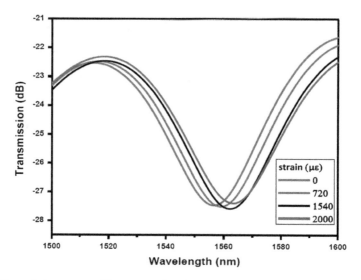

Fig. 4 Transmission spectra of large-core HC-PCF with different levels of applied strain

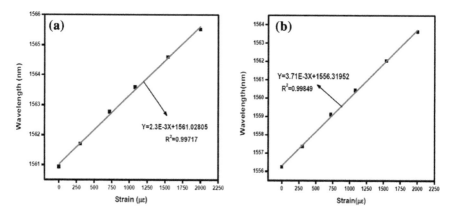

Fig. 5 Wavelength shift versus strain graph for **a** small-core HC-PCF and **b** large-core HC-PCF

linear behavior can be seen in both cases. The value of correlation factor $R_2 = 0.99717$ and $R_2 = 0.99849$ are obtained from Fig. 5a, b, respectively. High strain sensitivities of 2.3 and 3.71 pm/$\mu\varepsilon$ are obtained for small-core HC-PCF and large-core HC-PCF, respectively.

Strain sensitivity achieved with large-core HC-PCF is comparatively higher than that achieved with small-core HC-PCF. This observation shows that lower pitch or higher air-filling fraction in a HC-PCF results in a greater sensitivity for strain. The sensitivity of large-core HC-PCF, i.e., 3.71 pm/$\mu\varepsilon$ is also higher as compared to the earlier reported PCF-based strain sensors. The dependency of resonance wavelength shift on changes in the surroundings refractive index is shown in Fig. 6.

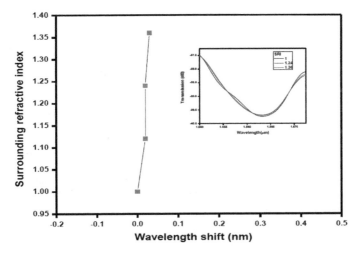

Fig. 6 Resonance wavelength dependence of large-core HC-PCF on SRI

The wavelength shift is negligible for variations in SRI, hence, the strain sensitivity for HC-PCF remains unaffected by refractive index changes in the environment. This property of HC-PCF makes it an attractive solution as a sensor for environments with variable refractive index.

4 Conclusion

Usually the sensitivity of FBG-based sensors is around 1 μm/$\mu\varepsilon$. A high strain sensitivity of 3.71 pm/$\mu\varepsilon$ is observed with large-core HC-PCF as compared to 2.3 pm/$\mu\varepsilon$ with small-core PCF signifying the role of higher air-filling fraction in sensing. The sensor shows linear response to strain. It is also insensitive to change in the surrounding refractive index. Due to higher sensitivity, the presented sensor is an attractive alternative for strain measurement in hazardous environments and medical applications.

References

1. Villatoro, J., & Joseba, Z. (2016). New perspectives in photonic crystal fibre sensors. *Optics & Laser Technology, 78*, 67–75.
2. Serker, N. H. M. K. (2010). Structural health monitoring using static and dynamic strain data from long-gage distributed FBG sensor. In *IABSE-JSCE Joint Confrence on Advances in Bridge Engineering-II*, Dhaka, Bangladesh.
3. Nidhi, R. S. K., & Kapur, P. (2014). Enhancement of sensitivity of the refractive index using ITO coating on LPG. *Journal of Optoelectronics and Advanced Materials, 8*(1–2), 45–48.

4. Russell, P. S. J. (2006). Photonic-crystal fibers. *Journal of Lightwave Technology, 24*(12), 4729–4749.
5. Pinto, A. R., & Lopez-Amo, M. (2012). Photonic crystal fibers for sensing applications. *Journal of Sensors, 2012,* 1–21.
6. Shi, Q., Lv, F., Wang, Z., Jin, L., Hu, Z. J. J., Kai, G., et al. (2008). Environmentally stable Fabry-Perot-type strain sensor based on hollow-core photonic bandgap fiber. *IEEE Photonics Technology Letters, 20*(4), 237–239.
7. Dong, B., & Hao, E. J. (2011). Temperature-insensitive and intensity-modulated embedded photonic crystal fiber modal interferometer based microdisplacement sensor. *Journal of the Optical Society of America B, 28*(10), 2332–2336.
8. Gong, H., Li, X., Jin, Y., & Dong, X. (2012). Hollow-core photonic crystal fiber based modal interferometer for strain. *Sensors and Actuators, A: Physical, 187*(1), 95–97.
9. Zheng, J., Yan, P., Yu, Y., Ou, Z., Wang, J., Chen, X., et al. (2013). Temperature and index insensitive strain sensor based on a photonic crystal fiber inline Mach–Zehnder interferometer. *Optics Communications, 297*(1), 7–11.
10. Xu, F., Li, C., Ren, D., Lu, L., Lv, W., Feng, F., et al. (2012). Temperature-insensitive Mach-Zehnder interferometric strain sensor based on concatenating two-waist-enlarged fiber tapers. *Chinese Optics Letters, 10*(7), 070603–070606.
11. Frazao, O., Santos, J. L., Araujo, F. M., et al. (2008). Optical sensing with photonic crystal fibers. *Laser & Photonics Reviews, 2*(6), 449–459. https://doi.org/10.1002/lpor.200810034.
12. Rajan, G. (2015). Introduction to optical fiber sensors. In *Optical fiber sensors: Advanced techniques and applications* (pp. 1–12). CRC press.
13. Cerqueira, S. A. (2010). Recent progress and novel applications of photonic crystal fibers. *Reports on Progress in Physics, 73*(2), 1–21.
14. Singh, S., Kaur, H., & Singh, K. (2017). Far field detection of different elements using photonic crystals. *Journal of Nanoelectronics and Optoelectronics, 12*(4), 400–403.

Part II
Microwave and Antenna Technology

Magnetic Field Sensitivity in Depressed Collector for a Millimeter-Wave Gyrotron

Vishal Kesari and R. Seshadri

Abstract The electron beam trajectories were simulated in a single-stage depressed collector for a millimeter-wave gyrotron. This collector was designed to handle the spent beam obtained after beam-wave interaction in the 100 kW gyrotron (accelerating voltage 55 kV and beam current 5 A). Similar to other high-power gyrotrons, this collector has larger volume considered at ground potential. The cathode and the beam-wave interaction cavity were considered at -40 kV and $+15$ kV, respectively. The collector sees the depression of 15 kV. The geometry of the collector is considered as three sections: (i) the open entrance conical section, (ii) the smooth cylindrical section, and (iii) the closed conical section. In order to simulate the beam trajectory from nonlinear taper to collector, the electron trajectories and spent beam power distribution data obtained from large-signal analysis have been fed at the entrance of mode converter of the gyrotron with required potentials applied. The collector geometry and the magnetic field are profiled to ensure the landing of the gyrating electrons to the wider smooth cylindrical section for better thermal management. The sensitivity of the magnetic field profile is studied and observed that for $\pm 5\%$ variation in the magnetic field profile would not shift the electron beam landing to conical sections, and the spent electron beam has no interception. The power dissipation on the collector is found to be 80.55 kW. The collector efficiency is calculated as ~48% for 120 kW RF output. The maximum thermal loading on collector inner surface is estimated as 0.38 kW/cm^2.

Keywords Depressed collector · Electron trajectory simulation
Millimeter-wave gyrotron · W-band gyrotron

V. Kesari (✉) · R. Seshadri
Microwave Tube Research and Development Centre, Bangalore, India
e-mail: vishal_kesari@rediffmail.com

© Springer Nature Singapore Pte Ltd. 2019
A. Khare et al. (eds.), *Recent Trends in Communication, Computing, and Electronics*, Lecture Notes in Electrical Engineering 524,
https://doi.org/10.1007/978-981-13-2685-1_6

53

1 Introduction

Gyrotron is the most popular oscillator in gyro-devices for the generation of mil-limeter waves for the power range from few kilowatts to megawatt level. They find applications in the electron cyclotron resonance heating (ECRH), electron cyclotron current drive (ECCD), stability control, and diagnostics of magnetically confined fusion plasmas in controlled thermonuclear reactor, tokamak, stellarator, active denial system, material processing, spectroscopy, plasma chemistry, etc. [1–5]. A gyrotron has a number of subassemblies such as magnetron injection gun, beam compres-sion region, interaction cavity, nonlinear taper, internal mode converter, window and collector, as part of vacuum envelop assembly, and a superconducting magnet as focusing structure. Out of all these seven subassemblies of vacuum envelop assembly of gyrotron, the collector is the most responsible for the better device efficiency under the consideration of depressed collector. In the configuration of depressed collector, the spent electron beam is collected at a negative potential than that of interaction cavity and thus recovers the beam energy. This allows soft landing of the electrons at the collector that felicitate the better thermal management [1–5]. Thirty-five percent-age of device efficiency has been achieved in gyrotrons without any energy recovery from the spent electron beam, i.e., no collector depression [1–6]. The improvement in device efficiency can be achieved by recovering the kinetic energy available with electron beam after the beam-wave interaction. This reduces the requirement of DC input power and cooling unit. Implementation of single-stage collector depression leads to ~50% device efficiency of the gyrotron [6–11]. In case of 165 GHz, 1.3 MW coaxial gyrotron, the single-stage collector depression increased the device efficiency to 41% from 27.3% (electronic) [9].

In the present paper, a single-stage depressed collector (Fig. 1) is designed for a millimeter-wave gyrotron, in which the electron beam trajectories are simulated. A tread off has been made for achieving an accelerating voltage of 55 kV with an annular cathode at −40 kV. The beam-wave interaction cavity is considered at +15 kV and the bulkier mass collector at ground potential. This implements a single-stage depressed collector with electron beam experiencing a 15 kV depression (negative with respect to beam-wave interaction region). The paper has been organized into four sections. Section 2 includes the modeling for electron beam trajectory simulation in CST Studio Suite [12]. Section 3 brings out the outcome of the simulation and related discussion, and the results are concluded in Sect. 4.

2 Modeling for Simulation

The shape of the collector electrode is considered in three sections: first, a conical section with an open entrance for electron beam; second, the straight smooth cylin-drical section for soft landing of electrons; and third, a closed conical section for vacuum enclosure (Fig. 1). For the purpose of electron beam trajectory simulation

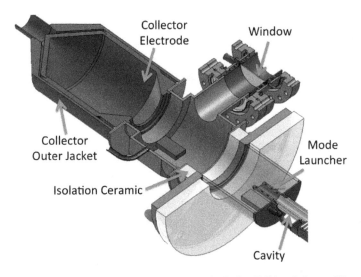

Fig. 1 3D model of the collector with mode converter including DC break for a millimeter-wave gyrotron

in CST Studio Suite, we consider the geometry from the end of the nonlinear taper to collector and the profile of magnetic field for the same axial length.

The cathode is considered at −40 kV with respect to ground, cavity region at +15 kV with respect to ground, and the collector electrode at ground potential. Effectively, the electron beam sees an accelerating potential of 55 kV and the collector depression of 15 kV (negative with respect to cavity). More precisely, the model includes an internal mode converter with a ceramic DC break of 15 kV. In this model, the spent beam power distribution data obtained after the beam-wave interaction in the large-signal analysis are fed at the end of the nonlinear taper. Due to symmetrical azimuthal distribution of the beamlets at the location of the hollow beam radius, only one beamlet is assumed carrying full beam current. This electron beam carrying the current of 5 A is further allowed to proceed toward collector under the influence of the applied magnetic field profile (Fig. 2).

3 Results and Discussion

The electrostatic simulation has been carried out after assigning the required potentials in the model (Sect. 2). The collector geometry and the magnetic field are profiled to ensure the landing of the gyrating electrons to the wider smooth cylindrical section for better thermal management (Fig. 3). The collector geometry is decided so as not to exceed maximum allowable power density of 1 kW/cm^2. The collector geometry and magnetic field profile have been optimized for maximum axial beam spread

Fig. 2 Optimized, 5% higher, and 5% lower magnetic field profiles for a millimeter-wave gyrotron

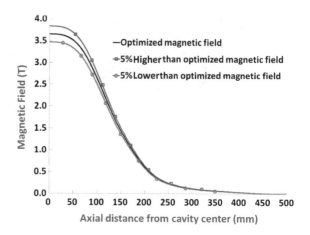

on the collector inner wall. The length of the smooth cylindrical section is taken as approximately three times of the axial beam spread over the smooth cylindrical section. The sensitivity of the magnetic field profile is studied and observed that for ±5% variation in the magnetic field profile (Fig. 2) would not shift the electron beam landing to conical sections, and the spent electron beam does not intercept anywhere while moving from cavity to collector (Fig. 3). The power dissipation on the collector is found to be 80.55 kW. The collector efficiency is calculated as ~48% for 120 kW RF output. The maximum thermal loading on collector inner surface is estimated as 0.38 kW/cm^2 (<1 kW/cm^2).

4 Conclusion

A single-stage depressed collector (-15 kV with respect to cavity) has been designed for a W-band gyrotron assuming a gyrating electron beam (accelerating potential 55 kV, current 5 A) using electron beam through trajectory simulation in CST Studio Suite. The trajectories have been simulated after feeding spent beam power distribution data obtained from large-signal analysis. Collector geometry and magnetic field have been profiled to ensure the landing of the gyrating electrons to the wider smooth cylindrical section for better thermal management. About ±5% variation in magnetic field profile does not shift the electron beam landing to conical sections. The power dissipation on the collector and the collector efficiency are found to be 80.55 kW and ~48%, respectively. The maximum thermal loading on collector inner surface has been estimated as 0.38 kW/cm^2 well within the limit (<1 kW/cm^2) of thermal loading for high-power collector.

(a)   **(b)**

(c)   **(d)**

(e)   **(f)**

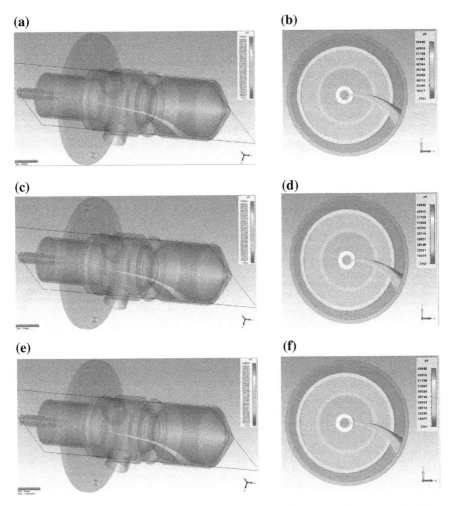

Fig. 3 Electron beam trajectories along the length (**a**, **c**, and **e**); and over the cross-section (**b**, **d**, and **f**); from cavity to collector under the influence of optimized (**a** and **b**), 5% higher (**c** and **d**), and 5% lower (**e** and **f**) magnetic field profiles. Optimized, 5% higher, and 5% lower magnetic field profiles for a millimeter-wave gyrotron

References

1. Gilmour, A. S. (2011). *Klystrons, traveling wave tubes, magnetrons, crossed-field amplifiers, and gyrotrons*. Boston, London: Artech House.
2. Thumm, M. (2017). *State-of-the-art of high power gyro- devices and free electron masers, Update 2016*. Karlsruhe, Germany: KIT.
3. Nusinovich, G. S. (2004). *Introduction to the Physics of Gyrotron*. Maryland, USA: JHU.
4. Edgecombe, C. J. (1993). *Gyrotron Oscillators: Their principles and practice*. London: Taylor & Francis Ltd.

5. Kesari, V., & Basu, B. N. (2018). *High power microwave tubes: Basics and trends*. California for IOP Concise Physics, London: Morgan and Claypool Publishers.
6. Sakamoto, K., Tsuneoka, M., Kasugai, A., Takahashi, K., Maebava, S., Imai, T., et al. (1994). Major improvement of gyrotron efficiency with beam energy recovery. *Physical Review Letters, 73*(26), 3532–3535.
7. Ling, G., Piosczyk, B., & Thumm, M. K. (2000). A new approach for a multistage depressed collector for gyrotrons. *IEEE Transactions on Plasma Science, 28*(3), 606–613.
8. Piosczyk, B., Iatrou, C. T., Dammertz, G., & Thumm, M. (1996). Single-stage depressed collectors for gyrotrons. *IEEE Transactions on Plasma Science, 24,* 579–585.
9. Piosczyk, B., Braz, O., Dammertz, G., Iatrou, C. T., Illy, S., Kuntze, M., et al. (1999). Thumm, M.:165 GHz, 1.5 MW—Coaxial gyrotron with depressed collector. *IEEE Transactions on Plasma Science, 27*(2), 484–489.
10. Saraph, G. P., Felch, K. L., Feinstein, J., Borchard, P., Cauffman, S. R., & Chu, S. (2000). A comparative study of three single-stage, depressed-collector designs for a 1-MW, CW gyrotron. *IEEE Transactions on Plasma Science, 28*(3), 830–840.
11. Pagonakis, I. G., Illy, S., Ioannidis, Z. C., Rzesnicki, T., Avramidis, K. A., Gantenbein, G., et al. (2018). Numerical investigation on spent beam deceleration schemes for depressed collector of a high-power gyrotron. *IEEE Transactions on Electron Devices*.
12. User Manual Computer Simulation Technology (CST) Studio Suite 2016, Darmstadt, GmbH, Germany (2016).

A Compact Inverted V-Shaped Slotted Triple and Wideband Patch Antenna for Ku, K, and Ka Band Applications

Ankit Kumar Patel, Komal Jaiswal, Akhilesh Kumar Pandey, Shekhar Yadav, Karunesh Srivastava and Rajeev Singh

Abstract A novel and compact inverted V-shaped slot loaded patch antenna is presented. The proposed antenna comprises of an inverted V-shaped slot, two rectangular slots, and one vertical notch which are connected to each other on the patch, two via hole for shorting ground and patch. Intermediate and final designs are simulated for comparison. The performance analysis of these antennas is compared and analyzed in terms of return loss and gain. The proposed antenna resonates at three frequency bands at resonating frequency of 12.5 GHz, 25.8 GHz, and 28.8 GHz having impedance bandwidth of 6%, 19.45%, and 20.61% with peak gain of 4 dBi, 8 dBi, and 9 dBi, respectively. The first resonating band is useful for Ku band, second is useful for K band, and third is useful for Ka band and 5G applications.

Keywords Wideband · Impedance bandwidth · Shorting pins · 5G applications
Ku and ka band

1 Introduction

The necessity to meet large network demand for 5G wireless communication band, requirement of multiband antenna is indispensible. Over the past decades, significant development has taken place in wideband patch antenna designs. However, conventional patch antennas suffer from very narrow bandwidth, typically only about several percentages. There are various enhancement techniques investigated and reported to achieve wideband [1–6]. These techniques are able to increase the bandwidth dramatically up to 59%. Meanwhile, the compact size of antenna is necessary for the mobile communications. The size of an antenna can be miniaturized by many different techniques such as shorting pins to provide short circuit path between patch and ground, increasing dielectric constant of substrate material, using meander line

A. K. Patel · K. Jaiswal · A. K. Pandey · S. Yadav · K. Srivastava · R. Singh (✉)
Department of Electronics and Communication,
University of Allahabad, Allahabad, Uttar Pradesh, India
e-mail: rsingh68@allduniv.ac.in

© Springer Nature Singapore Pte Ltd. 2019
A. Khare et al. (eds.), *Recent Trends in Communication, Computing,
and Electronics*, Lecture Notes in Electrical Engineering 524,
https://doi.org/10.1007/978-981-13-2685-1_7

shaped patch reported in literatures [7, 8]. Shorting pins technique has the merit of gain enhancement and also it shifts the resonating frequencies at lower frequency range.

The antenna designed for high frequencies applications (Ku, K, and Ka bands) suffers by high path loss and atmospheric absorption which degrade the performance of antenna. To compensate the higher path loss, the antenna is reported [9] to have been designed having higher gain, high radiation efficiency, and low cross-polarization.

This paper presents a novel inverted V-shaped slot loaded patch antenna with a notch and two shorting post fed by coaxial probe. The antenna is designed in several steps by changing the patch structure and position of feed and shorting post for optimization. The proposed antenna is designed by loading slots, notch, and shorting pins to achieve multiband and wideband operations with high gain for Radar, Satellite, and 5G communications. For this purpose, antenna is miniaturized to the dimension of $12 \times 12 \times 1.6$ mm^3 with the help of two shorting pins that also enhances the gain and bandwidth.

2   Antenna Design

Figure 1 shows the geometry of proposed antenna. The designing of the proposed antenna is initiated by the simulation of inverted V-shaped slot loaded square patch fed by the coaxial probe feed. The size of the antenna is kept very small which is compatible with the handheld devices and 5G applications. The FR4 material of dimension $(12 \times 12$ mm$^2)$ with relative permittivity $\varepsilon_r = 4.4$, loss tangent $\delta = 0.025$, and height h $= 1.6$ mm is used as a substrate of the antenna. The inverted V-shaped slot is designed by using of two equilateral triangles of sides 7 mm and 4.6 mm, respectively which are placed on the patch at a center position of $(-1, 0$ mm) shown in Fig. 1. The dimension of patch, ground, and substrate are same. In this antenna, the probe feed position is optimized to obtained multiband and wide bandwidth with high gain. The probe feed is partially connected to patch as shown in Fig. 1. The inverted V-shaped slot provides the electromagnetic gap coupled effect which produces multiple resonances.

The proposed antenna is designed in five steps by modifying the reference antenna (antenna A_0) structure shown in Fig. 2a. In the beginning, a rectangle-1 of dimension $(9.5 \times 0.3$ mm$^2)$ connected to inverted V shape at position $(-1.3, -0.9$ mm) with angle 140° from Y-axis is loaded as slot on the patch named as antenna A_1 and shown in Fig. 2b. In the next step, a horizontal rectangle-2 of dimension $(9.5 \times 0.3$ mm$^2)$ is placed as a slot to form antenna A_2 as shown in Fig. 2c. Again a vertical rectangle-3 of dimension $(10.7 \times 0.3$ mm$^2)$ is cut from the patch as a notch to form antenna A_3 as shown in Fig. 2d.

Furthermore, two shorting post S_1 and S_2 are loaded at positions $(3.4, -1$ mm) and $(3.4, 4.8$ mm), respectively to form antenna A_4 and A_5 as shown in Fig. 2e, f. The antenna A_5 is chosen as proposed antenna. The proposed antenna A_5 exhibits multiband and wideband operation.

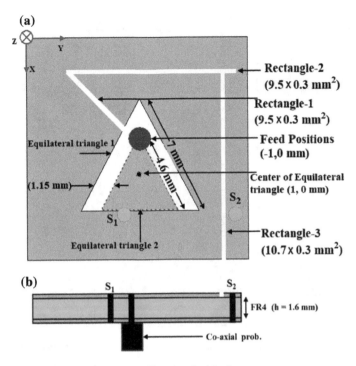

Fig. 1 Geometry of proposed antenna. **a** Top view, **b** side view

3 Result and Discussion

The proposed antenna and other four antennas which are designed at different stages of designing of the proposed antenna have been simulated and analyzed by commercial finite element method (FEM) software, namely HFSS simulator tool. The parametric shape analysis of these five antennas have been performed in terms of return loss and gain as shown in Fig. 3 and Fig. 4, respectively. From Fig. 3, it is seen that the simple structured inverted V-shaped slot loaded antenna A_0 exhibits five resonating bands at resonating frequencies 11.57 GHz, 13.82 GHz, 17.70 GHz, 27.12 GHz, and 29.91 GHz with narrow bandwidth. The gains of upper three resonating bands are very high up to 12 dBi. Due to slot loaded rectangle-1, Antenna A_1 shows seven resonating bands but the gains of these bands are decreased. When rectangle-2 is cut, it forms antenna A_2 which resonates at three resonating bands with resonating frequencies of 12.5, 25.8, and 28.8 GHz with higher gain. Antenna A_2 is more useful for both K band and 5G applications. When a rectangle-3 cut as a notch it forms antenna A_3. This additional notch creates multiple resonances at 15.0 GHz, 17.22 GHz, 19.87 GHz, and ringing resonances at 27.27 GHz and 28.36 GHz. In A_3 a wide band (from 26.59 GHz to 29.78 GHz) of 3.19 GHz is observed with an average gain of 7dBi.

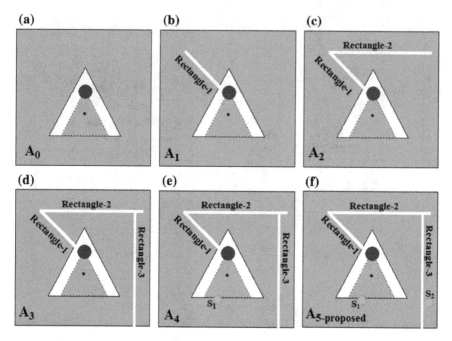

Fig. 2 Different patch shape. **a** Antenna A_0, **b** antenna A_1, **c** antenna A_2, **d** antenna A_2, **e** antenna A_2, **f** antenna A_5-(proposed)

The antenna A_4 is obtained by shorting the patch and ground of antenna A_3 through shorting post S_1. The shorting post is used for reducing the size and enhancing the gain of antenna and also it may shift the resonating bands toward the lower frequency range [1, 10]. Due to shorting post S_1, the bandwidth of the antenna A4 is enhanced at resonating frequency 27 GHz from 3.23 to 3.66 GHz as shown in Fig. 3, and the peak gain for this frequency band is also increased from 7 to 11.8 dBi as shown in Fig. 4.

The proposed antenna (A_5) is obtained by loading another shorting post S_2 on antenna A_4. By loading S_2, antenna achieves three resonating bands at resonating frequencies 12.5 GHz, 18.8 GHz, and 27.28 GHz with impedance bandwidth of 6%, 19.45%, and 20.61% and peak gain of 4 dBi, 8 dBi, and 9 dBi, respectively. The first band is useful for satellite communication, second band is useful for K band applications, and third band is useful for Ka band and 5G applications.

The input impedance at resonating frequencies 12.5 GHz, 18.8 GHz, and 25.8 GHz is 38.41 ohm, 45.45 ohm, and 50.15 ohm, respectively as shown in Fig. 5. It is observed from Fig. 5 that the antenna is approximately matched to 50 ohm at resonating frequencies. Figure 6 illustrates the group delay of the proposed antenna lies between −0.2 and +0.93 ns which is good and accounts for the fact that the transmission from such antennas can be achieved with least distortion.

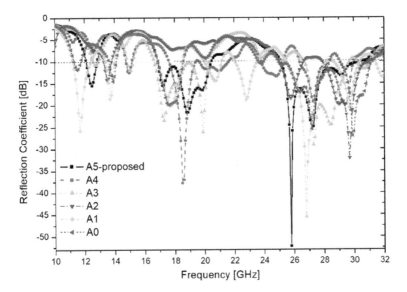

Fig. 3 Reflection coefficient versus frequency plot

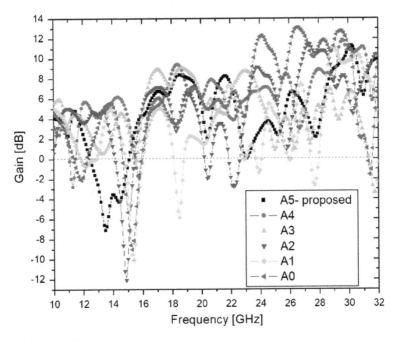

Fig. 4 Gain versus frequency plot

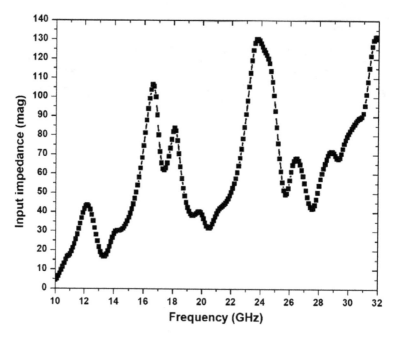

Fig. 5 Input impedance versus frequency plot

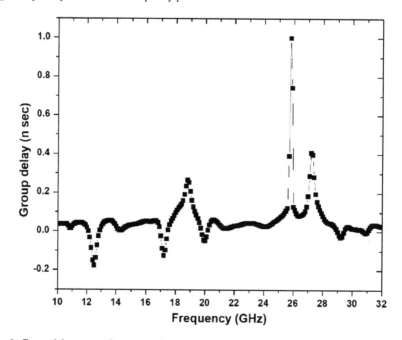

Fig. 6 Group delay versus frequency plot

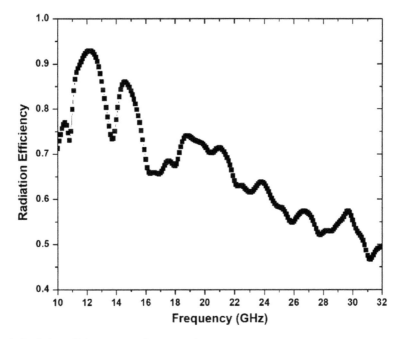

Fig. 7 Radiation efficiency versus frequency plot

From Fig. 7 it is observed that the radiation efficiency at lower frequency range is very high and at higher frequency range it degraded due to high path loss and atmospheric absorption. At resonating frequencies 12.5 GHz, 18.8 GHz, and 25.8 GHz, the radiation efficiencies are 92.1%, 74.5%, and 54.8%, respectively which are reasonably good.

The simulated radiation pattern for E-plane and H-plane of proposed antenna at 12.5 GHz, 18.8 GHz, and 28.8 GHz are shown in Fig. 8. The cross-polarization level is minimum as compared to co-polarization level at 12.5 GHz frequency as shown in Fig. 8a, b. From Fig. 8c, d, it is observed that co-polarization is maximum at angle theta $\theta = 90°$ and $\theta = 270°$ as compared to cross polarization at 18.8 GHz. At 25.8 GHz co-polarization and cross polarization are approximately at the same level for both E-plane and H-plane as shown in Fig. 8e, f. The overall radiation pattern behavior at 12.5 GHz, 18.8 GHz, and 25.8 GHz for both E-plane and H-plane are omni-directional as shown in Fig. 8.

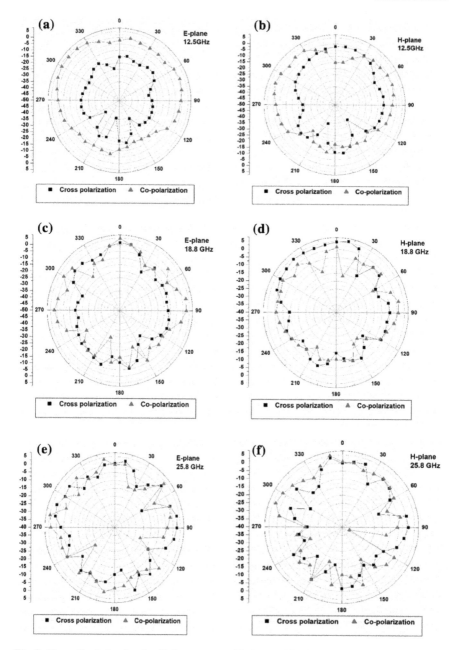

Fig. 8 Normalized simulated radiation pattern. **a** E-plane at 12.5 GHz, **b** H-plane at 12.5 GHz, **c** E-plane at 18.8 GHz, **d** H-plane at 18.8 GHz, **e** E-plane at 25.8 GHz, **f** E-plane at 25.8 GHz

4 Conclusion

Design of a compact inverted V shape slotted patch antenna with a volume of 230.4 mm^3 is presented. On analysis, we observe that the antenna parameters are affected by the introduction of slots and notches. We achieve reasonably enhanced bandwidth (6, 19.45, and 20.6%) and gain (4, 8, and 9 dBi) and the proposed antenna is useful for Ku, K, and Ka band and 5G applications.

References

1. Wu, J., Ren, X., Wang, Z., et al. (2014). Broadband polarized antenna with L-shaped strip feeding and shorting pin loading. *IEEE Antennas and Wireless Propagation Letters, 13,* 1733–1736.
2. Guo, Y. X., Mak, C. L., Luk, K. M., & Lee, K. F. (2001). Analysis and design ofL probe proximity fed patch antennas. *IEEE Transactions on Antennas and Propagation., 49*(2), 145–149.
3. Lai, H. W., & Luk, K. M. (2006). Design and study of wideband patch antennafed by meandering probe. *IEEE Transactions on Antennas and Propagation., 54*(2), 564–571.
4. Lai, H. W., & Luk, K. M. (2005). Wideband stacked patch antenna fed bymeandering probe. *Electronics Letters, 41*(6).
5. Mak, C. L., Lee, K. F., & Luk, K. M. (2000). Broadband patch antenna with aT-shaped probe. *IEE Proc. Microw Trans. Antennas Propagat., 147*(2), 73–76.
6. Ooi, B. L., Lee, C. L., Kooi, P. S., & Chew, S. T. (2001). A novel F-probe fedbroadband patch antenna. *IEEE Antennas and Propagation Society International Symposium.*
7. Nasimuddin, N., Qing, X. M., & Chen, Z. N. (2014). A compact circularly polarizedslotted patch antenna for GNSS applications. *IEEE Transactions on Antennas and Propagation, 62*(12), 6506–6509.
8. Chen, K., Yuan, J., & Luo, X. (2017). Compact dual-band dual circularly polarized annular-ring patch antenna for Bei Dou navigation satellite system application. *IET Microwaves, Antennas & Propagation, 11*(8), 1079–1085.
9. Mak, K. M., Lai, H. W., Luk, K. M., & Chan, C. H. (2014). Circularly polarized patch antenna for future 5G mobile phones. *IEEE Access, 2,* 1521–1529, 108–121.
10. Zhang, Xiao, & Zhu, Lei. (2016). Gain-enhanced patch antennas with loading of shorting pins. *IEEE Transactions on Antennas and Propagation, 64,* 3310–3318.

Analysis of Plus Shape Slot Loaded Circular Microstrip Antenna

Sikandar and Kamakshi

Abstract This paper presents the analysis of plus shape slot loaded patch antenna. It is observed that wideband property of antenna is found by suitably selecting the positions and dimensions of the slot on the patch. The operating frequency band of the proposed antenna is achieved from 6.60 to 8.15 GHz with bandwidth of 21.01% (simulated) and 6.50–8.33 GHz with 24.67% bandwidth (theoretical). The characteristics of the antenna are also observed by varying parameters like slot width, substrate height, and dielectric constant. The gain of the antenna is about 6.5 dBi in the operating frequency range. The simulated and theoretical results are compared which are in good agreement. The proposed antenna can be used in C-band applications.

Keywords Circular patch · Slot loading · Coaxial feed · RT-Duroid

1 Introduction

Wireless technology provides less expensive alternative and a flexible way for communication. Antenna is one of the important elements of the wireless communications systems. Microstrip antennas are widely used in the microwave frequency regions because of their attractive features such as low profile, lightweight, easy fabrication, and conformability to mounting hosts. The advantages of microstrip antennas make them suitable for applications in many versatile fields, viz. medical, military, aircraft, remote sensing, biomedical telemetry, satellite, cell phone, and other very high frequency to microwave applications. The main limitations, however, of the microstrip antenna are narrow bandwidth and low gain that restricts its applications [1, 2]. There are different techniques available for enhancing the bandwidth such as loading of

Sikandar (✉)
Electronics Engineering Department, Rajkiya Engineering College, Sonbhadra, India
e-mail: sikandar.jkiapt@gmail.com

Kamakshi
IMS Engineering College, Ghaziabad, India
e-mail: kamakshi.kumar21@gmail.com

© Springer Nature Singapore Pte Ltd. 2019
A. Khare et al. (eds.), *Recent Trends in Communication, Computing, and Electronics*, Lecture Notes in Electrical Engineering 524,
https://doi.org/10.1007/978-981-13-2685-1_8

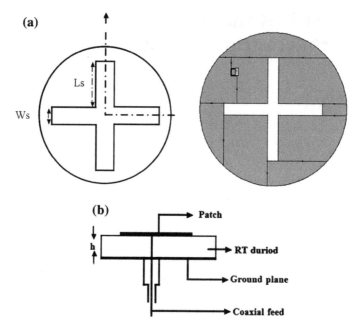

Fig. 1 Plus shape slot loaded circular patch antenna. **a** Top view, **b** side view

notch, by stacked structure, by coplanar structure, by loading of active devices and loading of slots. But loading of slot helps in simple design and easy loading and improves bandwidth without increase in volume of the antenna [3–7].

In the present paper, bandwidth improvement is achieved by loading slots on the patch. Here, the effect of antenna parameters is also analyzed. The details of the paper are discussed in the next section.

2 Antenna Configuration and Analysis

The proposed antenna of dimension (effective radius, a_e) is designed on RT-Duroid substrate having relative dielectric constant of 2.32. The proposed structure consists of a circular patch loaded with a plus shape slot of dimension $L_s \times W_s$. Here, coaxial feeding is used to excite the patch. The geometry of the antenna is shown in Fig. 1, and design parameters values are listed in Table 1.

Figure 2 shows the current distribution of the proposed patch structure. Maximum current strength is obtained along the circumference of the radiating patch which covers longer path in comparison to inner portion. This distribution of large amount of current along circumference is responsible for the generation of lower- and higher-order modes. A number of nearly excited modes are combined and give wider bandwidth.

Table 1 Design Parameters

Substrate material used	RT-Duroid
Relative permittivity of the substrate (ε_r)	2.32
Thickness of the dielectric substrate (h)	15 mm
Radius of the patch (a)	10 mm
Length of the first slot (Ls1)	13.5 mm
Width of the first slot (Ws1)	1.5 mm
Length of the second slot (Ls2)	13.5 mm
Width of the second slot (Ws2)	1.5 mm
Feed location (x, y)	(−5.35 mm, 4.825 mm)

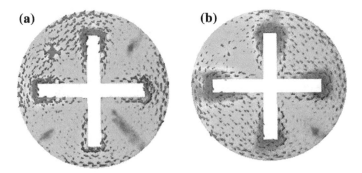

Fig. 2 Current distribution at **a** lower resonance frequency, **b** upper resonance frequency

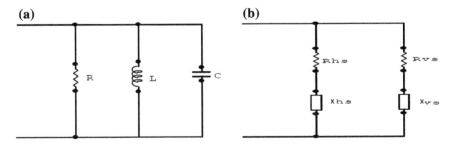

Fig. 3 Equivalent circuit of **a** circular patch antenna, **b** slot loaded circular patch antenna

3 Equivalent Circuits of the Proposed Antenna

The equivalent circuits of the slot loaded patch antenna are shown in Figs. 3 and 4.

The total input impedance of the proposed antenna can be calculated with the help of [3].

Reflection coefficient can be calculated as

Fig. 4 Equivalent circuit of
the plus shape slot loaded
antenna

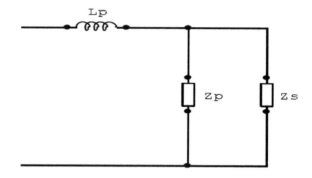

$$\Gamma = \frac{Z_{in} - Z_0}{Z_{in} + Z_0}$$

where Z_0 = characteristic impedance of probe feed
$\quad Z_{in}$ = total input impedance
And the return loss is calculated by

$$R_L = 20 \log|\Gamma|$$

4 Results and Discussion

The simulation of the proposed antenna has done by method of moment-based simulator IE3D [8], and simulated results are in good agreement with the theoretical results. The detailed observations of the proposed antenna characteristics are discussed in this section. Frequency versus reflection coefficient characteristic of the proposed antenna is shown in Fig. 5. The theoretical characteristic shows two resonating modes at 6.85 GHz and 7.70 GHz that combine to give broader bandwidth of 24.67%, whereas the simulated result presents the two closed resonating modes appeared at 6.95 GHz and 7.65 GHz giving bandwidth of 21.01%. This shows that the simulated result is in acceptable agreement with the theoretical result.

The effects of different substrate height "h" on the antenna performance, having all other parameters constant, are shown in Fig. 6. The different values of substrate height "h" are taken from 15 to 27 mm, and its effects on resonance and bandwidth are studied on the antenna characteristics. It is observed that lower resonance shifts more toward lower frequency side in comparison to upper resonance with increase in height of the substrate, and hence, corresponding bandwidth increases and reflection coefficient decreases.

The changes in the reflection coefficient are obtained by changing the value of slot width "W_S" as shown in Fig. 7. From the figure, it is found that on increasing the slot width from 1.5 to 3.0 mm the corresponding bandwidth of the proposed antenna increases and the value of reflection coefficient also increases. It is also observed that

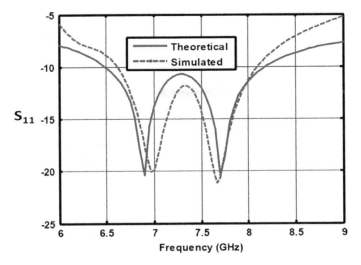

Fig. 5 Theoretical and simulated result of the plus shape slot loaded patch antenna

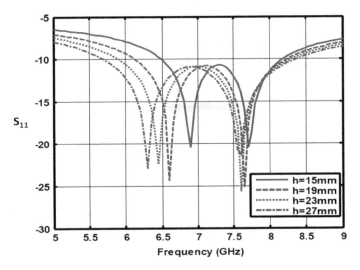

Fig. 6 Reflection coefficient versus frequency for different substrate height

after a certain value of slot width (Ws = 2.0 mm) antenna does not show wideband behavior.

The effects of varying relative dielectric constant of material "εr" on the antenna performance, keeping all other parameters are constant, are shown in Fig. 8. The values of dielectric constant "εr" of the substrate are varied from 2.32 to 4.8 and its effects are obtained. Figure shows that there are significant shift in the center frequency and change in BW with different dielectric constant. Therefore, we can optimize BW and cost of dielectric material in the antenna design.

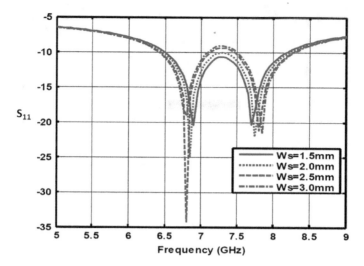

Fig. 7 Reflection coefficient versus frequency for different slot width

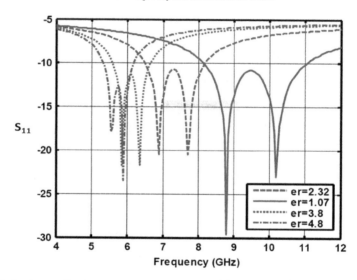

Fig. 8 Reflection coefficient versus frequency for variable dielectric constant

Figure 9a shows the characteristic of simulated gain versus frequencies, and the antenna gain varies from 6.5 to 5.4 dBi over an operating frequency range from 6.50 to 8.33 GHz. The maximum gain is observed to be 6.7 dBi at frequency 7.25 GHz. Figure 9b shows the total efficiency of the antenna. The maximum efficiency of the antenna is observed 70% at frequency 7.25 GHz. These observations of the proposed antenna meet the requirements of some wireless communication devices.

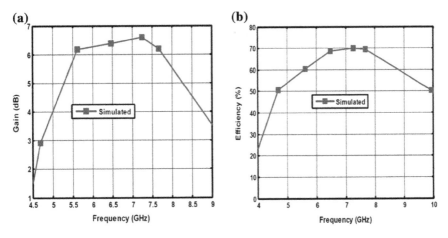

Fig. 9 Plot for proposed antenna. **a** Gain versus frequency, **b** efficiency versus frequency

5 Conclusion

A slot loaded circular wideband patch antenna with increased bandwidth has been discussed and presented successfully. The proposed antenna gives a wide impedance bandwidth of 24.67%. The performance of the antenna depends on antenna parameters like slot width, substrate thickness, and substrate materials. Antenna performance is improved for increased substrate thickness and lower dielectric constant. The proposed antenna has a maximum gain of 6.7 dBi at frequency 7.25 GHz and shows 70% efficiency. The proposed antenna can be used in C-band applications.

References

1. Bahl, I. J., & Bhartia, P. (1980). *Broadband microstrip antennas*. Norwood: Artech House.
2. Garg, R., Bhartia, P., Bahl, I., & Ittipiboon, A. (2001). *Microstrip antenna design handbook*. Norwood: Artech House.
3. Ansari, J. A., & Ram, R. B. (2008). Analysis of compact and broadband microstrip patch antenna. *Microwave and Optical Technology Letters, 50*(8), 2059–2063.
4. Deshmukh, A. A., & Ray, K. P. (2009). Compact broadband slotted rectangular microstrip antenna. *IEEE Antennas and Wireless Propagation Letters, 8,* 1410–1413.
5. Deshmukh, A. A., & Kumar, G. (2005). Compact broadband Uslot-loaded rectangular microstrip antennas. *Microwave and Optical Technology Letters, 46*(6), 556–559.
6. Albooyeh, M., Komjani, N., & Shobeyri, M. (2008). A novel crossslot geometry to improve impedance bandwidth of microstrip antennas. *Progress in Electromagnetics Research Letters, 4,* 63–72.
7. Shivnarayan, S. S, Vishkarma, B. R. (2005). Analysis of slot loaded rectangular microstrip patch antenna. Indian Journal of Radio & Space Physics, *34,* 424–43.
8. Zeland Software. (2008). IE3D (full wave) Simulation Software, Version 14.05, Zeland Software, Fremont, Calif, USA.

Study the Effect of Ground on Circular Loop Patch Antenna (CLPA)

Abhishek Kumar Saroj, Mohd. Gulman Siddiqui, Devesh and Jamshed A. Ansari

Abstract This proposed article presents multiband circular ring loop patch antenna (CLPA). The design is simulated using HFSS tool and studied by varying the ground patch size (area). The substrate material FR-4 is used and studied for the frequency range from 1 to 30 GHz in the proposed article. After analysing the effect of change in ground (GND), it is found that multiband characteristics slightly change in terms of S_{11} (dB) parameter below 20 GHz and above 20 GHz bandwidth increases. The CLPA shows eight to ten resonating frequency bands. Simulated data show -40.98 (dB) reflection coefficient at the frequency 15.88 GHz in design 1; similarly, design 2, design 3, design 4, design 5, and design 6 obtain maximum reflection coefficient -33.76 (dB) at 10.70 GHz, -37.27 (dB) at 27.65 GHz, -36.43 (dB) at 10.70 GHz, -37.08 (dB) at 21.33 GHz, and -29.11 (dB) at 15.35 GHz. The proposed antenna can be applied in numerous wireless applications by selecting different ground size.

Keywords Multiband · HFSS · Microstrip patch antenna · Reflection coefficient

1 Introduction

Microstrip patch antenna (MPA) is a sub-branch of antenna and introduced in the early 1970s. In current scenario, it is an appropriate candidate to meet the desired requirement because it can be applied in broad range of microwave and antenna applications. It can be used in Bluetooth, Wi-Fi, WLAN, WiMAX, RADAR, navigation, GPS, satellite applications, mobile phones, cordless phones, and some portable smart applications. MPA is also used in smart biomedical applications. It is broadly used due to small size, compact, ease of fabrication, low manufacturing cost, and simultaneously manufacturing of microwave devices. Due to these facts, MPA has gathered centre of attraction both from the research students and scientists; their

A. K. Saroj (✉) · Mohd. G. Siddiqui · Devesh · J. A. Ansari
Department of Electronics and Communication,
University of Allahabad, Allahabad 211002, UP, India
e-mail: abhisheksaroj@gmail.com

© Springer Nature Singapore Pte Ltd. 2019
A. Khare et al. (eds.), *Recent Trends in Communication, Computing, and Electronics*, Lecture Notes in Electrical Engineering 524,
https://doi.org/10.1007/978-981-13-2685-1_9

primary focus is to reduce the size of antenna. In [1], microstrip antenna has been designed that shows triple band for WLAN/WiMAX application by using microstrip line feed. For portable and small devices, i.e. laptop, mobile pager, and energy harvesting devices use multiple frequencies. A compact antenna with narrow slits for multiband operation has been proposed in [2, 3]. Multiple operating frequencies are also achieved by creating slots on radiating patch [4–6]. A multiband dual-polarized omnidirectional antenna for 2G/3G/LTE mobile communication is proposed in [7] using multiple feed points. In [8], a broadband circularly polarized antenna with compact size is presented using a loop feeding structure. A quad-rectangular-shaped microstrip antenna for multiband operation is also reported in [9] with the help of slots and notches. In the proposed antenna, multiband characteristics have been studied by changing the size and area of ground (GND).

In the proposed antenna, design 1 has been selected as reference antenna while other designs are studied by varying the size of ground area. Design 1 has circular loop patch as a radiator that radiates electromagnetic wave into air, having 60×60 mm^2 as ground area. The other designs have same radiating patch, whereas change in ground area has been studied.

2   Proposed Antenna Design

2.1   Antenna Structure

The proposed antenna is implemented on a FR-4 ($\varepsilon_r = 4.4$, loss tangent $= 0.02$) with a dimension of $60 \times 60 \times 1.6$ mm^3. The design consists of a circular loop patch antenna (CLPA) with thickness of 1 mm on FR-4 substrate. In CLPA, circular loop was designed using outer circular patch of radius 20 mm and inner circular patch of radius 19 mm on the surface of substrate. Circular loop was formed after subtracting or removing inner circular patch of size 19 mm radius from outer circular patch of size 20 mm radius. Microstrip feed line is connected to female SMA connector. This circular loop was fed by 2×10 mm^2 microstrip line feed. Circular loop patch antenna (CLPA) behaves as a radiating patch that radiates the electromagnetic waves outside the radiating patch (air). In this antenna, multiband behaviour has been studied by changing the dimension of ground (GND) over frequency range from 1 to 30 GHz. The geometrical configuration of circular loop patch antenna is shown in Fig. 1, and various parameters of the proposed antenna are given in Table 1.

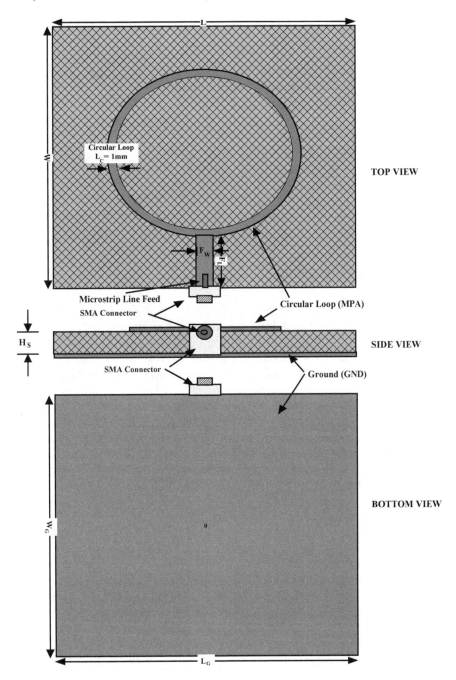

Fig. 1 Geometry of the proposed antenna—top view, front view, and bottom view

Table 1 Geometrical dimensions of the proposed antenna

Parameter	Value (mm)
Length of substrate (L)	60
Width of substrate (W)	60
Length of ground (L_G)	60
Width of ground (W_G)	60
Diameter of inner circle (D_i)	38
Radius of inner circle (R_i)	19
Thickness of circular loop radiator (L_c)	1
Length of microstrip line feed (F_L)	10
Width of microstrip line feed (F_W)	2
Height of substrate (H_S)	1.6

3   Result and Discussion

The proposed design was simulated using HFSS tool from frequency range 1 GHz to 30 GHz range. It shows multiband behaviour. The results are analysed step by step. The CLPA was simulated on 60×60 mm^2 FR-4 substrate with 1.6 mm height. The circular loop radiator was fed by 2×10 mm^2 microstrip 1 feed line connected through to edge mount SMA connector. In this proposed design, circular loop was unchanged and only analysing the effect of ground (GND) in antenna performance. The size of ground is taken as 60×60 mm^2 initially and then after reducing 5 mm from each side up to the size 35×35 mm^2. In the proposed CLPA, design 1 has been taken as a reference antenna. Design 1 (ground area 60×60 mm^2) shows ten resonating frequencies below 10 dB line from 1 to 30 GHz. The comparative study of reflection coefficients of the proposed antenna is given in Table 2a, b and shown in Fig. 2 a, b, c, d, e, f.

4   Conclusion

In this proposed design, a circular loop patch antenna (CLPA) has been simulated and studied the effect of ground (GND) by altering the area of ground patch. Here, only reflection coefficients are analysed by HFSS tool. On the basis of results, reflection

Table 2 (a, b) Comparative analysis of S_{11} parameter in dB for different designs

a

No.	Comparative analysis of reflection coefficient					
	Design 1 Area = 60 × 60 (mm²)		Design 2 Area = 55 × 55 (mm²)		Design 3 Area = 50 × 50 (mm²)	
	Freq (GHz)	RL	Freq (GHz)	RL	Freq (GHz)	RL
1	9.51	−11.53	9.51	−11.60	9.46	−12.16
2	10.66	−25.16	10.70	−33.76	10.60	−31.31
3	12.00	−21.68	12.09	−27.08	12.00	−30.42
4	13.19	−23.35	13.32	−18.83	13.21	−20.74
5	14.59	−18.27	14.67	−19.43	14.47	−18.16
6	15.88	−40.98	15.91	−25.50	15.84	−29.73
7	17.10	−15.99	17.21	−19.00	17.12	−12.28
8	19.32	−13.04	21.16	−23.49	18.18	−11.31
9	21.16	−20.80	27.02	−21.86	21.34	−19.84
10	27.93	−17.00	–	–	27.65	−37.27

b

No.	Comparative analysis of reflection coefficient					
	Design 4 Area = 45 × 45 (mm²)		Design 5 Area = 40 × 40 (mm²)		Design 6 Area = 35 × 35 (mm²)	
	Freq (GHz)	RL	Freq (GHz)	RL	Freq (GHz)	RL
1	9.49	−10.93	9.44	−11.32	3.02	−20.80
2	10.70	−36.43	10.71	−34.56	7.99	−10.69
3	12.07	−27.22	12.02	−30.38	9.15	−14.84
4	13.33	−16.80	13.26	−20.81	10.56	−16.97
5	14.63	−14.22	14.63	−17.88	13.34	−22.85
6	15.97	−26.38	15.82	−35.03	15.35	−29.11
7	17.09	−13.74	17.09	−12.72	20.70	−26.90
8	18.35	−11.60	21.33	−37.08	27.58	−18.92
9	20.50	−18.09	27.39	−26.19	–	–
10	27.31	−30.59	–	–	–	–

coefficients are accepted below 10 dB line and useful for commercial applications. Results show that minimum eight and maximum ten resonating frequency bands are obtained in an acceptable range. After analysing the result, it is found that FR-4 substrate material is good for below 20 GHz frequency range-based applications.

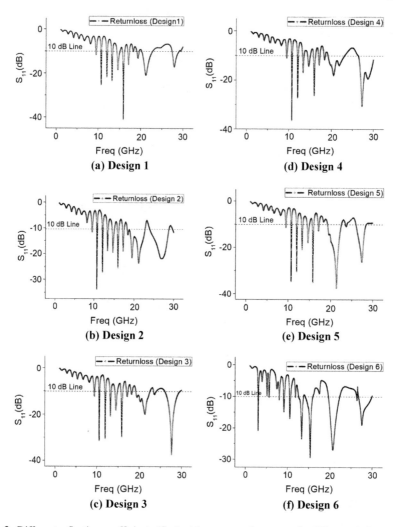

Fig. 2 Different reflection coefficients (S_{11}) with respect to frequency for different designs

References

1. Thomas, K. G., & Sreenivasan, M. (2009). Compact triple band antenna for WLAN/WiMAX applications. *IEEE Electronics Letters, 45*(16).
2. Verma, S., Ansari, J. A., & Verma, M. K. (2013). A novel compact multi-band microstrip antenna with multiband narrow slits. *Microwave and Optical Technology Letters, 55*(6), 1196–1198.
3. Liu, C. S., Chiu, C. N., & Deng, S. M. (2008). A compact disc-slit monopole antenna for mobile devices. *IEEE Antennas and Wireless Propagation Letters, 7*, 251–254.
4. Khajepour, S., Ghaffarian, M. S., & Moradi, G. (2017). Design of novel multiband folded printed quadrifilar helical antenna for GPS/WLAN applications. *IEEE Electronics Letters, 53*(2), 58–60.

5. Liu, Y. F., Lau, K. L., Xue, Q., & Chan, C. H. (2004). Experimental studies of printed wide-slot antenna for wide-band applications. *IEEE Antennas and Wireless Propagation Letters, 3,* 273–275.
6. Chen, W. S., & Ku, K. Y. (2008). Band-rejected design of printed open slot antenna for WLAN/WiMAX operation. *IEEE Transactions on Antennas Propagation, 56*(4), 1163–1169.
7. Dai, X. W., Wang, Z. Y., Liang, C. H., Chen, X., & Wang, L. T. (2013). Multiband dual-polarized omnidirectional antenna for 2G/3G/LTE applications. *IEEE Antenna and Wireless Propagation Letters, 12,* 1492–1495.
8. Ding, K., Gao, C., Qu, D., & Yin, Q. (2017). Compact broadband circularly polarized antenna with parasitic patches. *IEEE Transactions on Antenna and Propagation, 65*(9), 4854–4857.
9. Saroj, A. K., Siddiqui, M. G., Kumar, M., & Ansari, J. A. (2017). Design of multiband quad-rectangular shaped microstrip antenna for wireless applications. *Progress in Electromagnetics Research M, 59,* 213–217.

Design and Analysis of W-Slot Microstrip Antenna

Neelesh Agrawal, Jamshed A. Ansari, Navendu Nitin,
Mohd. Gulman Siddiqui and Saiyed Salim Sayeed

Abstract In this paper, W-slot microstrip antenna is explored to obtain multiband resonance. The proposed design shows three resonance frequencies which pivot on the structure of W-slot in the radiating patch. The frequency range of the proposed antenna is between 5 and 10 GHz. The operating bands are at 5.1, 8.3, 9.5 GHz and is suitable for C and X bands. The results are analysed on HFSS simulator.

Keywords W-slot · Truncated corners · Gap-coupled feeding

1 Introduction

The call for compact multiband microstrip antennas (MSAs) is increasing progressively. In order to fulfil the demands for wireless communication application having frequency bands such as S band (2–4 GHz), C band (4–8 GHz) and X band (8–12 GHz), many designs were proposed by various scientists/designers.

Some of the designs are a semi-circular dish with half U-slot patch antenna producing two resonance frequencies [1], shorted rectangular patch antenna having half U-slot producing broadband operation [2], broadband stacked having U-slot MSA producing two resonating frequencies [3], truncated corner rectangular microstrip patch antenna producing broadband operation [4], microstrip antenna with U-shaped parasitic elements producing wide bandwidth [5, 6], W-slot microstrip antenna having stacked configuration increases bandwidth of microstrip antenna [7], notched small monopole antenna with novel W-shaped conductor-backed plane and novel T-shaped slot offering wide bandwidth [8], monopole antenna with three band-notched characteristics producing wideband [9]. Compact and small planar monopole antenna with symmetrical L- and U-shaped slots proved

N. Agrawal (✉) · J. A. Ansari · N. Nitin · Mohd. G. Siddiqui · S. S. Sayeed
Department of Electronics and Communication, University
of Allahabad, Allahabad 211002, UP, India
e-mail: agrawalneelesh@gmail.com

© Springer Nature Singapore Pte Ltd. 2019
A. Khare et al. (eds.), *Recent Trends in Communication, Computing,*
and Electronics, Lecture Notes in Electrical Engineering 524,
https://doi.org/10.1007/978-981-13-2685-1_10

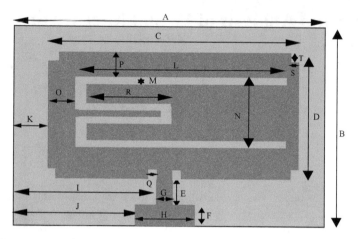

Fig. 1 Geometry of the W-slot MSA

useful for WLAN/WiMAX applications [10]. Above-discussed antennas have some limitations such as number of resonating frequencies and limited bandwidth.

In this paper, W-slot loaded rectangular patch MSA with inset feeding along with four truncated corners on the four corner of radiating patch is proposed. The simulation is carried out in HFSS. The results show that the antenna is capable of operating at multiband frequencies due to inculcation of four truncated corners.

2 Antenna Design

Simulated design using HFSS have been presented in Fig. 1, which determines the proposed W-slot MSA with inset feeding having two truncated corners at the upper side and two truncated corners at the lower side of the radiating patch having FR4 as dielectric substrate. The relative permittivity (ε_r) of the substrate is 4.4 having thickness as 1.6 mm. The microstrip antenna is excited by 50 Ω microstrip feed. The dimensions of microstrip feed are represented as length (F) and width (H). The microstrip feed uses quarter wavelength transformer having dimensions as length (E) and width (G). The different parameters are shown in Table 1.

3 Results and Discussions

The simulation of the designed antenna is carried out in high-frequency structure simulation (HFSS) tool. The performance can be seen in Fig. 2 which shows the variation of return loss versus frequency. It is observed that three resonating fre-

Table 1 Antenna parameters

Antenna parameters	Size in (mm)
Length of ground plane (B)	40
Width of ground plane (A)	40
Length of patch (D)	11.33
Width of patch (C)	15.24
Length of quarter wave transformer (E)	4.92
Width of quarter wave transformer (G)	0.5
Length of microstrip feed (F)	6.18
Width of microstrip feed (H)	3.5
(I)	19.75
(J)	18.25
(K)	12.38
Length of the horizontal arms of W slot (L)	8
Width of the horizontal and vertical arms of W slot (M)	1
Width of W slot (N)	9
(O)	3.97
(P)	1.09
(R)	5
Width of gap-coupled feeding (Q)	0.2
Length of truncated corners (T)	0.5
Width of truncated corners (S)	0.2

quencies are obtained. These resonating frequencies are at 5.1, 8.3, 9.5 GHz having peak return loss up to −20.1163, −21.0485 and −16.8299 dB, respectively. Figure 3 shows the parametric analysis with respect to different positions of W slot. Figure 3 shows the variation of the return loss with different frequencies for different positions of W slot. The W slot is moved from left (wslot = 14.35 mm) to the right (wslot = 18.35 mm). wslot = 16.35 mm is taken as the best position, and other parameters are also calculated for wslot = 16.35 mm.

VSWR of the proposed antenna for frequencies 5.1, 8.3 and 9.5 GHz as shown in Fig. 4 is 1.7197, 1.5437, 2.5199.

Radiation pattern of the proposed antenna for frequencies 5.1, 8.3 and 9.5 GHz is shown in Figs. 5, 6, 7. Antenna has broadside radiation at frequency 5.1 GHz for $\Phi = 0$, 90 further radiation pattern is not broadside at resonating frequencies, i.e. 8.3 and 9.5 GHz.

2D gain for frequencies at 5.1, 8.3 and 9.5 GHz is shown in Figs. 8, 9, 10. 3D gain for frequencies at 5.1, 8.3 and 9.5 GHz is shown in Figs. 11, 12, 13. The antenna has a maximum gain of 2.0768, 8.4222 and 26.548 dB at frequencies 5.1, 8.3 and 9.5 GHz, respectively.

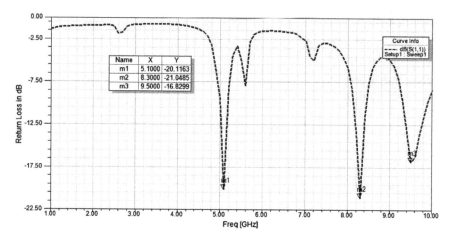

Fig. 2 Variation of return loss with frequency

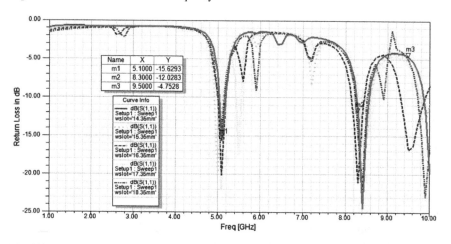

Fig. 3 Variation of return loss with frequency for different positions of W slot

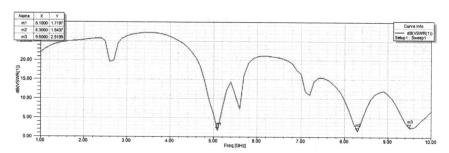

Fig. 4 Variation of with frequency for proposed antenna VSWR

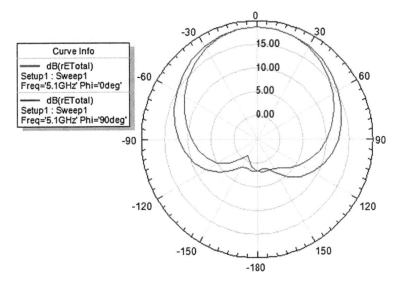

Fig. 5 Radiation pattern of the proposed antenna for frequency at 5.1 GHz

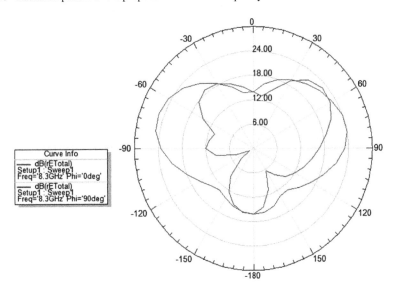

Fig. 6 Radiation pattern of the proposed antenna for frequency at 8.3 GHz

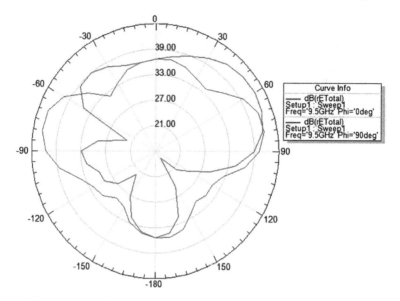

Fig. 7 Radiation pattern of the proposed antenna for frequency at 9.5 GHz

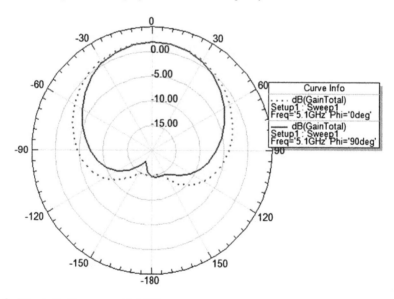

Fig. 8 2D gain for frequency at 5.1 GHz

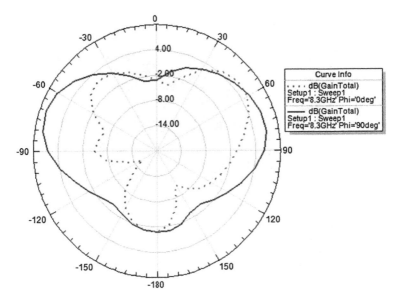

Fig. 9 2D gain for frequency at 8.3 GHz

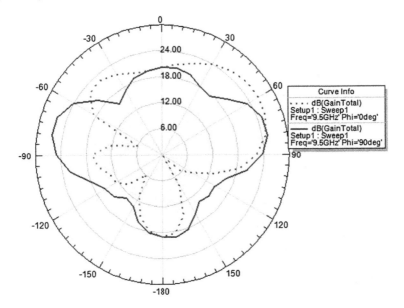

Fig. 10 2D gain for frequency at 9.5 GHz

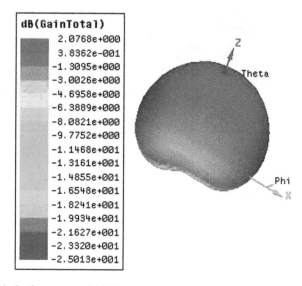

Fig. 11 3D gain for frequency at 5.1 GHz

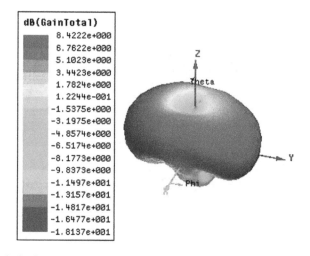

Fig. 12 3D gain for frequency at 8.3 GHz

Fig. 13 3D gain for
frequency at 9.5 GHz

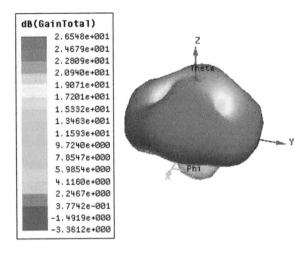

4 Conclusion

The proposed antenna has two truncated corners at the upper and lower side of the
radiating patch and W slot which plays vital role in generating three resonating
frequencies. The resonating frequencies are obtained 5.1, 8.3, and 9.5 GHz. It has
good radiation pattern at 5.1 GHz frequency and maximum gain at 9.5 GHz frequency.
The proposed antenna can be utilized for different wireless communication bands.

References

1. Ansari, J. A., & Mishra, A. (2011). Half U-slot loaded semicircular disk patch antenna for
   GSM mobile phone and optical communications. *Progress in Electromagnetics Research C,
   18,* 31–45.
2. Ansari, J. A., Yadav, N. P., Singh, P., & Mishra, A. (2009). Compact half U-slot loaded shorted
   rectangular patch antenna for broadband operation. *Progress in Electromagnetics Research C,
   9,* 215–226.
3. Ansari, J. A., & Ram, R. B. (2008). Broadband stacked U-slot microstrip patch antenna.
   Progress in Electromagnetics Research Letters, 4, 17–24.
4. Verma, S., & Ansari, J. A. (2015). Analysis of U-slot loaded truncated corner rectangular
   microstrip patch antenna for broadband operation. *International Journal of Electronics and
   Communications,* 1434–8411.
5. Aneesh, M., Ansari, J. A., & Singh, A. (2015). Kamakshi: Effect of shorting wall on compact
   2*4 MSA array using artificial neural network. *TELKOMNIKA Indonesian Journal of Electrical
   Engineering, 13*(3), 512–520.
6. Wi, S. H., Lee, Y. S., & Yook, J. G. (2007). Wideband Microstrip patch antenna with U-shaped
   parasitic elements. *IEEE Transactions on Antennas and Propagation, 55*(4).
7. Ali, Z., Singh, V. K., Singh, A. K., Ayub, S. (2012). Bandwidth enhancement of W slot
   microstrip antenna using stacked configuration. In *International Conference on Communication Systems and Network Technologies.*

8. Ojaroudi, N., Ojaroudi, M., Ghadimi, N. (2013). Dual band-notched small monopole antenna withnovel W-shaped conductor backed-plane and novel T-shaped slot for UWB applications. *IET Microwaves, Antennas & Propagation, 7,* 8–14.

9. Cai, Y. Z., Yang, H. C., & Cai, L. Y. (2014). Wideband monopole antenna with three band-notched characteristics. *IEEE Antennas and Wireless Propagation Letters, 13.*

10. Moosazadeh, M., & Kharkovsky, S. (2014). Compact and small planar monopole antenna with symmetrical L- and U-shaped slots for WLAN/WiMAX applications. *IEEE Antennas and Wireless Propagation Letters, 13.*

A Multiband Antenna with Enhanced Bandwidth for Wireless Applications Using Defected Ground Structure

Shivani Singh and Gagandeep Bharti

Abstract This paper describes a patch antenna with slotted ground to operate at public domain frequency bands. Proposed design has a ground and a patch on the lower and upper planes of the PCB, respectively. FR4 material is used as substrate for PCB having thickness 0.8 mm. Proposed antenna offers a wide frequency ranges from (2.17–2.53) GHz, (3.0–3.72) GHz, and (5.05–6.5) GHz for Bluetooth, WiFi, WiMAX, and WLAN applications. Ansoft HFSS software is used for simulation process of the proposed design.

Keywords Multiband · Defected ground · WiMAX · WIFI · WLAN · GPS Bluetooth

1 Introduction

With the increasing demand in various wireless standards in public domain, it is required to integrate different standards in a single device. Since antennas can be considered as eyes of communication system, antenna needs to be multifunctional with wideband characteristics. Different wireless standards include Global Positioning System (GPS), Bluetooth, WiFi, WiMAX, WLAN, etc. Already many antennas are proposed in the literature to provide multifunctional characteristics. Detail study is done on existing antennas for wireless communication, e.g., monopole antenna for dual band [1, 2], slot antennas for multiband operation in [3, 4], planar inverted-F antenna in [5], dual-band loop antenna in [6]. Along with multiband characteristic, wideband characteristic is also important to cover the entire frequency band. There are various techniques to obtain wideband characteristic like use of different structure of patches [7], defected ground [8], use of slots in patches [9], and parasitic strips [10]. Antenna should be designed in such a way that it should have simple

S. Singh (✉) · G. Bharti
Department of Electronics and Communication Engineering, Madan Mohan
Malaviya University of Technology, Gorakhpur, India
e-mail: shivanisjic@gmail.com

© Springer Nature Singapore Pte Ltd. 2019
A. Khare et al. (eds.), *Recent Trends in Communication, Computing,
and Electronics*, Lecture Notes in Electrical Engineering 524,
https://doi.org/10.1007/978-981-13-2685-1_11

95

structure and planar geometry for ease in fabrication. From the past years, use of printed patch antenna is in demand because of its compatibility and low cost. Printed antennas with easy structure are intended directly on PCB and gives suitable radiation characteristics.

This paper presents a semicircular patch antenna which is prepared to operate for multiband wireless applications. This paper is a modified version of [11] which offers three frequency bands of few MHz. Antenna is mounted on FR4 substrate of dimension 62×67 mm^2 with thickness of 1.58 mm. The radiating portion and the ground are located on the upper and lower face of the substrate, respectively. A defected ground structure along with one rectangular slot and two open slots in the patch is used to provide wideband characteristics. Proposed antenna provides three bands for Bluetooth, WiFi, WiMAX, and WLAN applications. This paper is modified version of antenna in [11] which covers narrow bandwidths.

2 Antenna Design and Analysis

Design of the antenna proposed is shown in Fig. 1, and its detail geometry with various design parameters can be seen in Fig. 2. Figure 2a shows the semicircular patch of radius 22 mm which is printed on the upper plane of the substrate; it consists of a rectangular slot of length (w3) of 12.2 mm and width 0.5 mm. Two open slots of 5 mm are introduced at the diameter of the patch to obtain better return loss at 3.3–3.7 GHz. Figure 2b shows the back side of the antenna consisting of the defected ground plane with various optimized parameters. The proposed antenna without the defected ground plane gives the bandwidth of few mega hertz, but after the introduction of defected ground, bandwidth enhancement is obtained covering the entire range for desired applications. Slot in the ground improves the input impedance resulting in better S11.

For designing the antenna, FR4 material of $\varepsilon_r = 4.44$ and $\tan\delta = 0.02$ is used as the substrate with dimensions of $62 \times 67 \times 1.58$ mm^3. The main ground plane is of dimension 62×67 mm^2, denoted by length L_{sub} and width W_{sub}. Coaxial probe feed is used to excite the patch and is given at a rectangular structure which is connected to the patch via tapered structure. Proposed antenna consists of three parts semicircular patch of radius W_4, tapered feed of length L_2, and a rectangular structure of width W_1 and length L_1. Feed through tapered structure gives wideband resonance at 2.5 GHz. All the design parameters are shown in Figs. 1 and 2 with their values illustrated in Table 1, respectively.

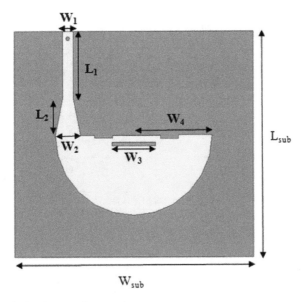

Fig. 1 Antenna design (proposed antenna)

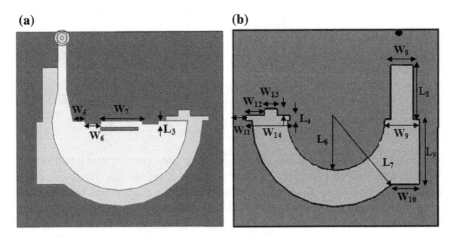

Fig. 2 Design of the proposed antenna: **a** front view and **b** back view

3 Results and Discussion

Using the ANSYS HFSS tool, simulation and optimization of the design is done to obtain desired results. In order to obtain wideband, feed is given through tapered structure. Concept of defected ground is studied to enhance the bandwidth of the antenna. Defected ground structure is implemented by etching simple geometry in the ground plane resulting into disturbance in shield current in such a way that

Table 1 All values are in mm

Lsub	62	W_{12}	6
Wsub	67	W_{13}	4
W_1	3	W_{14}	14
W_2	7	L_1	18.5
W_3	12.2	L_2	10.13
W_4	22	L_3	0.75
W_5	4	L_4	1.5
W_6	5	L_5	2
W_7	13.5	L_6	16
W_8	7.1	L_7	26.3
W_9	11	L_8	17
W_{10}	8.86	L_9	20
W_{11}	4.5	h	1.56

Fig. 3 Reflection coefficient of antenna without defected ground (reference antenna)

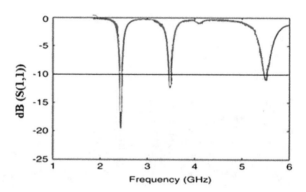

propagation of electromagnetic waves can be controlled via substrate layer. Firstly, the study and analysis of return loss without defected ground was done, and result is plotted in Fig. 3. From the analysis it can be observed that antenna operates for a narrow band frequency range. In order to improve the bandwidth to cover the entire frequency band for particular applications, defected ground structure is employed, and simulated results are studied. Figure 4 shows simulated reflection coefficient of the antenna, which indicates that the proposed antenna with DGS covers complete frequency range for Bluetooth, WiFi, WiMAX, and WLAN. Obtained bands lie from 2.17–2.53 GHz, 3.0–3.72, and 5.05–6.5 GHz.

Antenna provides good impedance matching as reflection coefficient up to −33 dB, −17 dB, and −25 dB is achieved in the first, second, and third band, respectively. The radiation efficiency (in %) and the simulated peak gain (in dBi) can be analyzed in Fig. 5. The value of peak gain lies between 2.5 dBi to 3.8 dBi in the first operating band, 3.2 dBi to 4.4 dBi in the second band, and 2.7 dBi to 4.9 dBi in the third operating band. Highest values of radiation efficiency are 86%, 84%,

Fig. 4 Reflection coefficient of antenna with defected ground (proposed antenna)

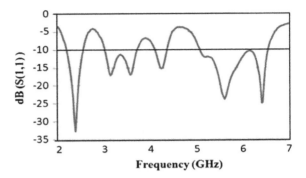

Fig. 5 Simulated peak gain and radiation efficiency

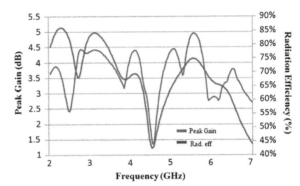

and 74.9% in all the three bands, respectively. Radiation patterns in all the operating bands can be seen in Fig. 6 at YZ and XZ planes.

4 Conclusion

An antenna with easy semicircular structure and plain geometry is designed on low-cost FR4 substrate. With the use of Ansoft HFSS software, optimization of various parameters and design are done. −10 dB bandwidth of simulated antenna is obtained for Bluetooth and WiFi (2.17–2.53) GHz, second band is for WiMAX (3.0–3.72) and WLAN (5.05–6.5) GHz. Defected ground structure is used to enhance the bandwidth and other radiation performance of the antenna. Simulated results of the proposed antenna like peak gain, radiation efficiency, and radiation pattern imply that antenna can be used for wireless applications. Same antenna can be further modified using other performance-enhancing techniques and can be made to operate for other applications like GPS and satellite communication.

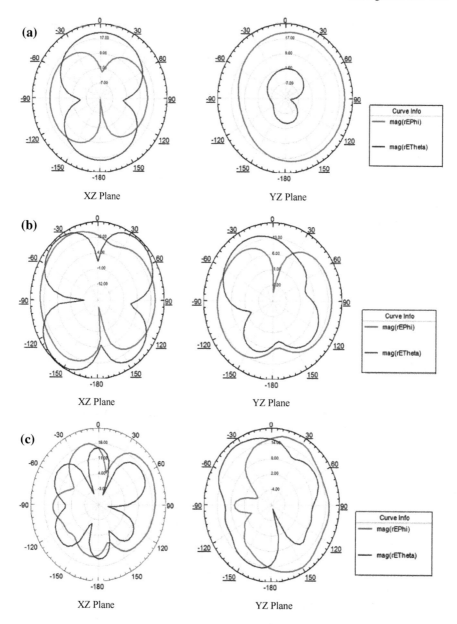

Fig. 6 Simulated radiation patterns at **a** f = 2.35 GHz, **b** f = 3.65 GHz, **c** f = 5.6 GHz

References

1. Sun, X. L., Cheung, S. W., & Yuk, T. I. (2013). Dual-band monopole antenna with frequency-tunable feature for WiMAX applications. *IEEE Antennas and Wireless Propagation Letters, 12,* 100–103.
2. Pan, S. C., & Wong, K. L. (1997). Dual frequency triangular microstrip antenna with a shorting pin. *IEEE Transactions Antennas Propagation Letters, 45,* 1889–1891.
3. Wong, K. L., & Lee, L. C. (2009). Multiband printed monopole slot antenna for WWAN operation in the laptop computer. *IEEE Transactions on Antennas and Propagation, 57*(2), 324–330.
4. Cao, Y., Yuan, B., & Wang, G. (2011). A compact multiband open-ended slot antenna for mobile handsets. *IEEE Antennas and Wireless Propagation Letters, 10,* 911–914.
5. Chang, C. H., & Wong, K. L. (2009). Printed $lambda/8$-PIFA for penta-band WWAN operation in the mobile phone. *IEEE Transactions on Antennas and Propagation, 57*(5), 1373–1381.
6. Su, S. W. (2010). High-gain dual-loop antennas for MIMO access points in the 2.4/5.2/5.8 GHz bands. *IEEE Transactions on Antennas and Propagation, 58*(7), 2412–2419.
7. Panusa, S., & Kumar, M. (2014). Triple-band H-slot microstrip patch antenna for WiMAX application. In *2014 International Conference on Advances in Engineering & Technology Research (ICAETR—2014),* Unnao (pp. 1–3).
8. Aris, M. A., Ali, M. T., Rahman, N. H. A., & Ramli, N. (2015). Frequency reconfigurable aperture-coupled microstrip patch antenna using defected ground structure. In *2015 IEEE International RF and Microwave Conference (RFM),* Kuching (pp. 200–204).
9. Chakraborty, U., Kundu, A., Chowdhury, S. K., & Bhattacharjee, A. K. (2014). Compact dual-band microstrip antenna for IEEE 802.11a WLAN application. *IEEE Antennas and Wireless Propagation Letters, 13,* 407–410.
10. Bharti, P. K., Singh, H. S., Pandey, G. K., & Meshram, M. K. (2015). Thin profile wideband printed monopole antenna for slim mobile handsets applications. *Progress in Electromagnetics Research C, 57,* 149–158.
11. Jhamb, K., Li, L., & Rambabu, K. (2011). Novel-integrated patch antennas with multi-band characteristics. *IET Microwaves, Antennas and Propagation, 5*(12), 1393–1398.

Notch-Loaded Patch Antenna with Multiple Shorting for X and Ku Band Applications

Shekhar Yadav, Komal Jaiswal, Ankit Kumar Patel, Sweta Singh, Akhilesh Kumar Pandey and Rajeev Singh

Abstract A compact two notch-loaded patch antenna with multiple shorting pins for X and Ku band applications is presented. Two different substrates FR-4 and RT Duroid 5880 are used to compare the antenna characteristics. Multiple shorting pins are used to enhance the bandwidth and gain of antenna. Volume of the proposed antenna is 785 mm^3, and it resonates at 10.5 GHz and 14.65 GHz with impedance bandwidth of 10.2% and 6.49% and gain of 7.42 dBi and 12 dBi, respectively. This antenna is useful for X band and Ku band applications.

Keywords Microstrip patch antenna (MSA) · Multiple shorting pins Notch loaded · X and Ku band

1 Introduction

Rapid development in wireless communication has increased the requirement of compact antennas resonating at multiple frequencies. X band and Ku band microstrip antennas are widely used for point-to-point wireless communication such as on-road traffic control, air traffic control, imaging radar satellite communication, industrial scientific medical, small electronic devices for radio communication to aircraft, spacecraft, and military [1–4]. To improve bandwidth and to reduce size of the antenna, techniques such as loading of slots, notches, stacking of patches, and by using a thicker substrate, or by reducing the dielectric constant, or by using gap-coupled multiresonator and loading of shorting pins are reported [5–7]. By using shorting pins, a short circuit path between patch and ground is created which results into the change in overall current distribution which enhances the gain and bandwidth.

S. Yadav · K. Jaiswal · A. K. Patel · S. Singh · A. K. Pandey · R. Singh (✉)
Department of Electronics and Communication, University of Allahabad, Allahabad,
Uttar Pradesh, India
e-mail: rsingh68@allduniv.ac.in

© Springer Nature Singapore Pte Ltd. 2019
A. Khare et al. (eds.), *Recent Trends in Communication, Computing, and Electronics*, Lecture Notes in Electrical Engineering 524,
https://doi.org/10.1007/978-981-13-2685-1_12

Table 1 Comparison of parameters of different patch antennas for X and Ku band applications

Refs	Patch area (mm^2)	Resonating frequencies (GHz)	Antenna gain (dBi)	Band
[8]	960	14.1	7.80	Ku
[9]	900	15.0	6.30	Ku
[10]	225	12.2	7.60	Ku
[11]	900	9.5	4.60	X
Proposed antenna	500	10.50	7.42	X and Ku
		14.65	12.00	

In this work, we propose a dual-band compact micro-strip patch antenna structure by introducing the triangular and semi-circle notches on the radiating patch. The multiband characteristics of the designed patch make it suitable for X and Ku band applications.

A comparative analysis of different antennas is presented in terms of antenna size, resonating frequency, antenna gain, and its applications as shown in Table 1. From the perusal of the table, it is observed that the proposed antenna achieves higher gain than other antennas operating in X and Ku bands [8–11].

2   Antenna Design

Two different substrates, namely FR-4 (flame retardant-4) and RT Duroid® 5880 with dielectric constants of 4.4 and 2.2, respectively, have been used to design the proposed antenna. The antenna size is $(20 \times 25 \times 1.57)$ mm^3. The basic rectangular radiating patch is (23×16) mm^2. An arc shape notch is cut out from one side of the rectangular patch, and a triangular notch is cut out from the other side of the patch. Seven shorting pins on x-y plane named as S_1 $(-4, -9)$, S_2 $(1, -9)$, and S_3 $(4, -9)$ placed on the left side of the patch and S_4, S_5, S_6 and S_7 are placed near coaxial feed at $(-0.82, -0.70)$, $(1, 1)$, $(-0.82, 8.29)$ and $(5, 9)$, respectively. The rectangular patch antenna is excited by 50 Ω coaxial probe feed with inner radius of 0.7 mm at $(4, 7)$ position. Geometrical top and side views of the proposed antenna are shown in Figs. 1 and 2 respectively.

3   Results and Discussion

The proposed antenna is designed, simulated, and optimized using commercial finite element method solver HFSSv.13 (high-frequency structure simulator) software. Antenna parameters like return loss, VSWR, gain, group delay, and radiation effi-

Fig. 1 Top view of the proposed antenna

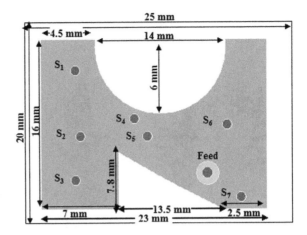

Fig. 2 Side view of the proposed antenna

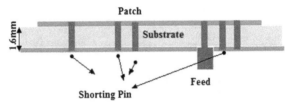

ciency are simulated, analyzed, and compared for two different substrates FR-4 and RT Duroid® 5880. Figures 3 and 4 demonstrates simulated return loss versus frequency behavior of the FR-4 and RT Duroid® 5880 substrates with rectangular patch (without notches), rectangular patch (with shorting pins) and patch with arc and triangular shaped notches.

The behavior of the antenna is analyzed in a sequential manner as the design proceeds. At the first stage, we analyze the simple rectangular structure (without shorting pins) and observe two resonating frequencies at 10.15–12.45 GHz (cf. Fig. 4) for RT Duroid® 5880 substrate and two resonating frequencies at 8.9–10.4 GHz (cf. Fig. 3) for FR-4 substrate.

At the second stage, we have analyzed the complete rectangular structure (without introducing notches) shorted with seven pins at the locations already described in the preceding section. When the above design structure is implemented with FR-4 substrate, it resonates at 5.1, 8.6, 10.9, and 14.4 GHz and exhibits multiband characteristics (cf. Fig. 3), whereas when implemented using RT Duroid® 5880 substrate, it resonates only at 6.9 and 14.4 GHz (cf. Fig. 4).

However, at the third stage, when the arc-shaped and triangular-shaped notches are introduced (with shorting pins), four resonating frequencies are observed for FR-4 substrate at 8.6 GHz, 9.9 GHz, 12.2 GHz, and 15 GHz (cf. Fig. 3). In case of RT Duroid® 5880 substrate, we observe only single resonating frequency at 12.81 GHz (cf. Fig. 4).

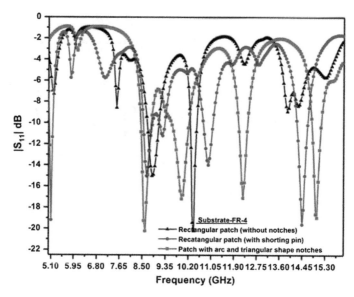

Fig. 3 |S₁₁| versus frequency of proposed antennas on FR-4 substrate with rectangular patch (without notches), rectangular patch (with shorting pin), patch with arc- and triangular-shaped notches

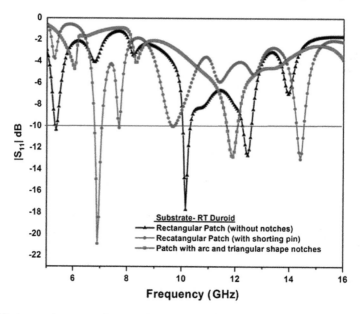

Fig. 4 |S₁₁| versus frequency of proposed antennas on RT Duroid substrate with rectangular patch (without notches), rectangular patch (with shorting pin), patch with arc- and triangular-shaped notches

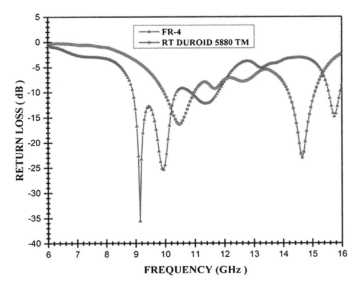

Fig. 5 |S$_{11}$| versus frequency of proposed antenna on FR-4 and RT Duroid substrates

The proposed antenna is designed using both techniques, i.e., using shorting pins and by introducing notches. It is observed from Fig. 5 that the antenna designed using FR-4 substrate shows a "ringing resonating effect" between 8.89 and 10.37 GHz with impedance bandwidth of 16.67% and peak gain of 8.75 dBi (cf. Figs. 5 and 6), and at 11.45 GHz (10.90–11.70 GHz), it exhibits an impedance bandwidth of 7.07% and gain of 4.65 dBi (cf. Fig. 6). The "ringing resonating effect" degrades the gain of the antenna. The proposed antenna designed using RT Duroid® 5880 resonates at 10.46 GHz and 14.6 GHz. The first band lies between 9.97 GHz to 10.95 GHz and the second band lies between 14.18 GHz to 15.08 GHz with impedance bandwidths of 10.05% and 9.1% and gain of 7.42 dBi and 12 dBi respectively as calculated and observed from Figs. 5 and 6. Figure 7 shows variation of simulated gain with frequency for FR-4 and RT Duroid® 5880 substrates with rectangular patch (without notches), rectangular patch (with shorting pin), patch with arc and triangular-shaped notches.

The group delay of antennas refers to the degree of distortion between transmitted and received pulses. The observed group delay (cf. Fig. 8) between −1.0 and 1.0 ns shows that the antenna exhibits minimum distortion, whereas antennas designed using both substrates (FR-4 and RT Duroid® 5880) exhibit a radiation efficiency of around 70% which is reasonably good and is suggestive of the fact that the antenna will radiate efficiently (Fig. 9).

The radiation pattern of the antennas designed on FR-4 and RT Duroid® 5880 substrates is shown in Figs. 10, 11, 12 and 13. Vertical axis is representing magnitude of gain in dB. Proposed antenna is useful for X band and Ku band applications.

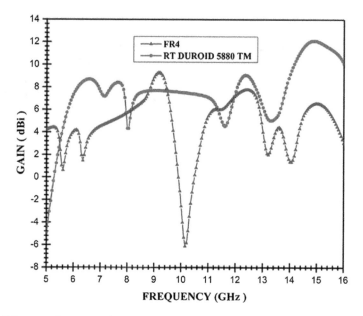

Fig. 6 Gain versus frequency of proposed antennas on FR-4 and RT Duroid substrates

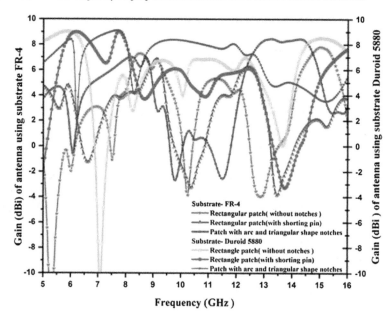

Fig. 7 Gain versus frequency of proposed antennas for FR4 and RT Duroid substrates with rectangular patch (without notches), rectangular patch (with shorting pin), patch with arc- and triangular-shaped notches

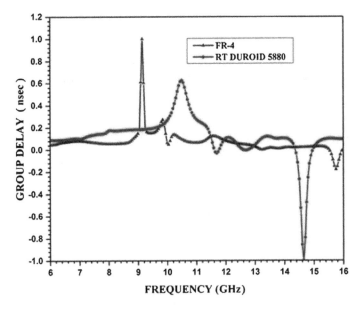

Fig. 8 Group delay versus frequency of proposed antennas on FR4 and RT Duroid substrates

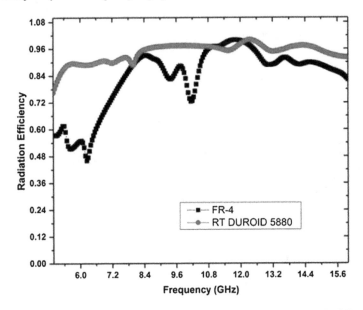

Fig. 9 Radiation efficiency versus frequency of proposed antennas on FR4 and RT Duroid substrates

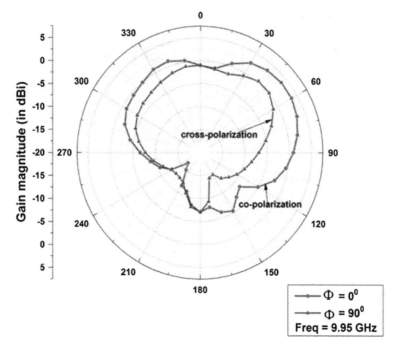

Fig. 10 Radiation pattern for FR-4 at 9.95 GHz

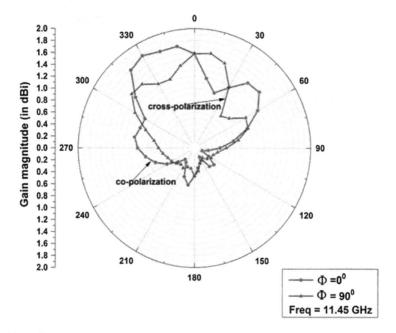

Fig. 11 Radiation pattern for FR-4 at 11.45 GHz

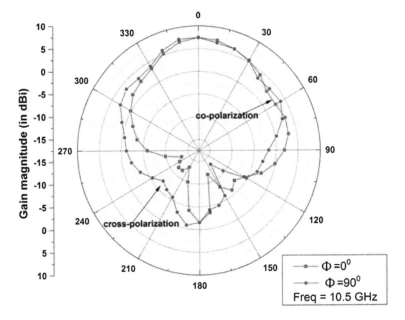

Fig. 12 Radiation pattern for RT Duroid at 10.5 GHz

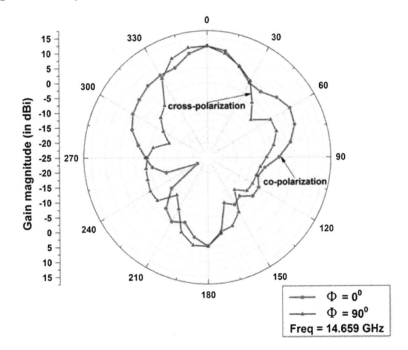

Fig. 13 Radiation pattern for RT Duroid at 14.65 GHz

4 Conclusions

A compact two notch-loaded patch antenna with multiple shorting pins on two different substrates have been investigated. The antenna resonates at dual frequency band for both substrates FR-4 and RT Duroid® 5880. Antenna with RT Duroid 5880 substrate resonates at frequency bands of 10.46 GHz and 14.6 GHz with peak gain of 7.42 dBi and 12 dBi, respectively. Hence, antenna with RT Duroid® 5880 is chosen as proposed antenna due to dual-band operation with high gain which are applicable for X and Ku band applications.

References

1. Balanis, C. A. (2005). Antenna theory analysis and design (3rd ed) (pp. 816). Hoboken, NJ, USA: Wiley
2. Bahl, I. J., & Bhartia, P. (1980). *Microstrip antennas*. Dedham, MA: Artech House.
3. Garg, R. (2001). *Microstrip Design Handbook*. Norwood: Artech House. Inc.
4. Kumar, G., & Ray, K. P. (2003). *Broadband microstrip antennas*. Artech: Artech House.
5. Deshmukh, A. A., Ray, K. P., & Kadam, A. (2014). Analysis of slot cut broadband and dual band rectangular microstrip antennas. *IETE Journal of Research, 59,* 193–200.
6. Kula, J., Psychoudakis, D., Liao, W.-J., Chen, C.-C., Volakis, J., & Halloran, J. (2006). Patch antenna miniaturization using recently available ceramic substrates. *IEEE Antennas and Propagation Magazine, 48*(6), 13–20.
7. Wong, K. L., & Chen, W. S. (1997). Compact microstrip antenna with dual frequency operation. *Electronics Letters, 33*(8), 646–647.
8. Chen, Yu., Wei, H., Zhenqi, K., & Haiming, W. (2012). Ku band linearly polarized omnidirectional planar filtenna. *IEEE Antennas Wireless Propag. Lett., 11,* 310–313.
9. Prasad, P. C., Chattoraj, N. (2003). Design of compact Ku band micro-strip antenna for satellite communication. In *International Conference on Communication and Signal Processing*. IEEE, pp. 196–200.
10. Azim, N. M. R., Islam, M. T. (2011). Dual polarized micro-strip patch antenna for Ku band application. *Informacije MIDEM*. 114–117.
11. Singh, V., Mishra, B., Singh, R. (2015). A compact and wide band microstrip patch antenna for X-band applications. In *Second International Conference on Advances in Computing and Communication Engineering*, pp. 296–300.

Part III
Wireless Sensor Networks and IoT

A Delay-Oriented Energy-Efficient Routing Protocol for Wireless Sensor Network

Yogesh Tripathi, Arun Prakash and Rajeev Tripathi

Abstract Wireless sensor network (WSN) has become a prominent technology in order to access data from the remote or non-remote areas, e.g., forests, battlefields, hospitals, homes. As WSN has energy constraints, energy-efficient routing protocols are required to prolong the network lifetime. Delay is one of the important parameters for WSN because of it data losses its importance of data and creates congestion in the network also. In this paper, a delay-aware energy-efficient routing protocol is proposed. Delay is minimized with the help of mobile base station, and optimized numbers of hops improve the energy efficiency of proposed routing protocol. Simulation results show the improvement in performance over the existing routing protocol. Extensive simulation study is carried out to evaluate the performance of the proposed protocol with respect to delay, throughput, average residual energy, and network lifetime.

Keywords WSN · Routing · Energy efficiency · Hop count

1 Introduction

WSN is known as a network of tiny and low power motes which have limited memory and processing capability. WSN is an infrastructure-based ad hoc network comprised of sensing, computing, and communicating elements which enable it to give the information about the environment where it is deployed. WSNs can be deployed in different applications such as environmental monitoring, healthcare, military surveil-

Y. Tripathi (✉) · A. Prakash · R. Tripathi
Department of Electronics and Communication Engineering, Motilal Nehru National
Institute of Technology Allahabad, Allahabad 211004, India
e-mail: rel1706@mnnit.ac.in

A. Prakash
e-mail: arun@mnnit.ac.in

R. Tripathi
e-mail: rt@mnnit.ac.in

© Springer Nature Singapore Pte Ltd. 2019
A. Khare et al. (eds.), *Recent Trends in Communication, Computing,
and Electronics*, Lecture Notes in Electrical Engineering 524,
https://doi.org/10.1007/978-981-13-2685-1_13

lance, habitat monitoring, industrial plant monitoring [1–3]. The information sensed by the nodes is transmitted to sink node for further processing via direct or multi-hop packet forwarding mechanism. Unlike the sensor nodes, sink node is powerful device in terms of processing capability, power, and memory.

WSN has energy constraints so that energy-efficient routing protocol is one of essential requirements that prolong the network lifetime. So, it is required to develop energy-efficient routing protocols for the data transmission from the source node to sink node. Data can be transmitted in two ways (i) direct-hop transmission and (ii) multi-hop transmission [4]. In direct-hop data transmission, farther nodes from the sink consume more energy as compared with the hop by hop transmission. This results in the disconnection of the network due to the early death of farther nodes from the sink. In multi-hop, data is routed with the help of multiple nodes, and selections of hop to hop forwarding nodes create delay. To solve this problem, the combination of mixed-hop transmission and movable sink node may be one of the promising solutions. This will reduce the end-to-end delay, energy consumption, and improve the network lifetime also.

The recent research related to mobile WSN proves that it has advantages over the conventional WSN [5–7], which are as follows:

(i)   Mobile sensor network improves the network lifetime [8].
(ii) Minimization of energy consumption.
(iii) It can improve the data conformity with the help of reducing number of hops.

In this paper, energy-efficient, delay-aware multi-hop routing protocol having mobile sink node. In this mixed-hop routing is proposed; the number of hops for data transmission from the source node to sink node is calculated on the basis of optimized distance.

The rest of the paper is organized as follows: Related work is summarized in Sect. 2. In Sect. 3, proposed routing protocol is described. Performance evaluation of the proposed routing protocol is presented in Sect. 4. Finally, Sect. 5 concludes the paper along with some future work.

2   Related Work

This section presents some data collection and forwarding mechanism in mobile sink, and energy-efficient routing protocols. The recent advances, requirements, and challenging issues have been described and explained in [9].

In [10], the authors have investigated different types of energy-efficient routing protocols for heterogeneous and homogeneous as well as static and mobile sink or source node. They have also investigated WSN impairments in routing design and its applications and present a comparative analysis of each class. In this paper, authors suggested some open issues in designing of routing protocols like new routing metrics, QoS routings, secure routing.

In [11], authors have proposed delay-aware routing protocol for mobile sink nodes. In this paper, reduction in delay is achieved with the help of the mobile sink nodes. If mobile sink node is in the range of sensor node, it will transmit its data otherwise it follows the multi-hop data transmission based on the movement of sink node. The major drawback of the protocol is that it consists of multiple sink nodes due to which same data is transmitted through multiple routes which will increase the energy consumption.

In [12], authors have proposed a routing protocol for WSNs with mobile sink which improves network lifetime. In this algorithm, route is set up based on termite-hill approach as per the requirement. Packet loss increases as the number of sensor nodes increases in transmission range of the sink node in pause time.

In [13], authors have proposed a routing protocol for prolonging the network lifetime with the help of predicted path of mobile sink. It improves the delivery ratio of data, where the sink stays temporarily at the sojourn points and the stay time is higher than the movement time between two sojourn points. Sojourn point is defined as where sink node stays after moving some distance. Due to sink mobility, all nodes must have to aware of topological changes.

3 Proposed Protocol

In this section, energy-efficient delay-aware routing protocol with single sink node is described. The main QoS parameters of the proposed routing protocol are network lifetime and delay. There are some assumptions made for the proposed routing protocol, which is described below:

(i)   Homogeneous sensor nodes are deployed in grid.
(ii) All sensor nodes are static except the sink node.
(iii) The sensor nodes can estimate the sink distance with the help of RSSI.
(iv) Initial energy and transmission range of the sensor nodes are same except for the sink nodes.

3.1 Network Model

There is single mobile sink node moving in the predefined path. It has sojourn point in its moving direction and pause time at the sojourn point such that all nearby nodes can transmit their data to sink. The network topology of proposed routing protocol is shown in Fig. 1.

When the sink node stays at sojourn point after moving certain distance, it broadcasts its presence and sensor node estimates their distance to sink for the data forwarding. Data can be forwarded via single hop or multi-hop based on distance to the sink.

Fig. 1 Network topology

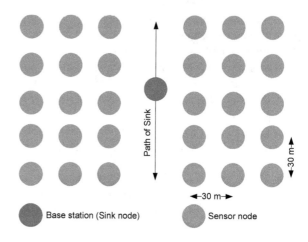

Base station (Sink node)   Sensor node

3.2 Energy Model

In this proposed routing protocol, a simplified energy model [14, 15] is used for transmission of data. The energy consumption, e_t, in transmission of B bit message over the distance d is expressed below:

$$e_t = (e_{ele} + \varepsilon_{amp} \cdot d^\tau) \cdot B \tag{1}$$

where e_{ele} is energy consumption due to sensor node's circuitry like transmitter and receiver and ε_{amp} is the energy dissipation in the transmitter amplifier. τ is the channel path-loss exponent of the environment and satisfies the condition $2 \leq \tau \leq 4$.

Energy consumption, e_r, in receiving B bit message at the receiver side is given by

$$e_r = e_{ele} \cdot B \tag{2}$$

3.3 Proposed Routing Algorithm

In this proposed routing protocol, data is transmitted to the sink via direct transmission or multi-hop transmission based on the distance to the mobile sink. As mobile sink reaches to its sojourn point and broadcasts control packet for its availability. As soon as sensor nodes receive this control packet, nodes estimate that whether sink is in the transmission range of themselves or not. Sensor nodes can work as source node or forwarding node based on the distance to the sink. If sink is in transmission range of the nodes, it can broadcast the data directly to the sink after exchanging control packets. The sensor nodes wait for the random time, after exchanging control packets, based on the residual energy of the nodes if there are multiple nodes contending for the data transmission to the sink.

If sink is not in the transmission range of the nodes, then based on optimum transmission distance (d_{op}) forwarder list is created [16]. It will reduce the number of hops from source to sink so that hops number can be optimized. According to priority of the nodes in the forwarder list, the source node selects the forwarder node for the next hop transmission. The optimum transmission distance is given by (3)

$$d_{op} = \{(2e_{elec})/[(\tau - 1)\varepsilon_{amp}]\}^{1/\tau} \tag{3}$$

4 Performance Evaluation

The performance evaluation of the proposed routing protocol is done on network simulator (ns) version 2.35 [17]. The performance is evaluated for end-to-end delay, throughput, packet delivery ratio, total energy consumption, average residual energy, and network lifetime under the mobility consideration of the sink node. The mobility of the sink is such that the simulation time is equal to the round trip time of the sink. The round trip time of the sink is taken as time in which it completes one round (starting-end-starting). The simulation parameters used while evaluating the performance evaluation are given in Table 1 (Figs. 2 and 3).

The results of the proposed algorithm are compared with ad hoc on-demand distance vector (AODV) [18] routing protocol. Figure 4 shows variation of delay with number of nodes. As number of node increases, delay also increases due to contention among nodes.

The proposed algorithm shows improved results over AODV because of mobile node and optimized number of hops. The mobile node moves in straight line, and nearby nodes send their data directly to sink. The optimized hop distance reduces number of hops for the nodes which are not in transmission range of sink node. It reduces delay as travelled distances by the data packets get optimized.

Table 1 Simulation parameters

Parameter	Value
Number of nodes	20–100
Simulation time	200 s
Simulation area	5000×5000 m^2
MAC protocol	802.15.4
Traffic types	CBR
Packet transmission rate	Random
Base station's speed	2 m/s
Base station's pause time	05 s

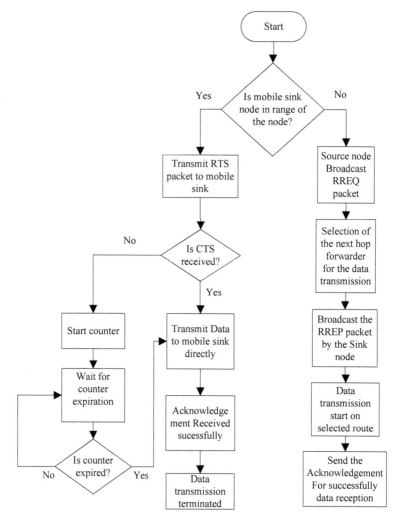

Fig. 2 Flow chart of the proposed routing algorithm

Packet delivery ratio (PDR) represents success rate of receiving of packet of any routing algorithm. In Fig. 5, as number of nodes increase, PDR decreases due to congestion. The improved result is due to mobility of sink, and it stays at sojourn point.

Throughput is a measure of how fast data can be processed over the given bandwidth. In Fig. 6, as the numbers of node increase, throughput decreases, but it shows the improvement over the existing one. The improvement has been achieved due to the fact that the remaining nodes send their data to nodes near the sojourn point which send data to mobile sink directly.

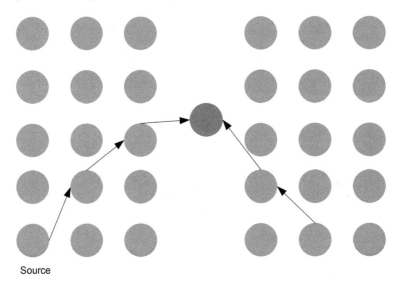

Source

Fig. 3 Data forwarding mechanism to mobile sink node

Fig. 4 Delay versus number of nodes

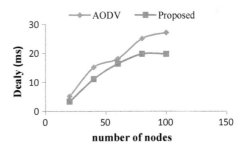

Fig. 5 PDR versus number of nodes

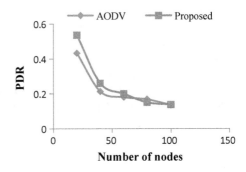

Fig. 6 Throughput versus number of nodes

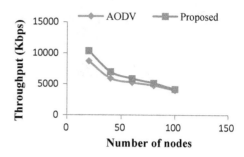

Fig. 7 Total energy consumption versus number of nodes

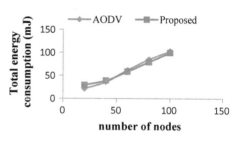

Fig. 8 Average residual energy versus number of nodes

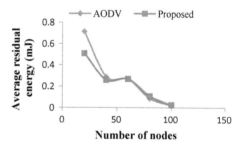

Figure 7 shows the energy consumption of proposed routing protocol. As the number of node increases, energy consumption increases because nodes are transmitting its own data as well as previous hop data. Improvement in energy consumption is achieved due to less number of hops from source to sink.

Average residual energy represents how much energy is remaining in nodes. Due to optimized number of hops, there is reduction in proposed routing protocol over the AODV which results in the improvement of residual energy (Fig. 8).

Network lifetime represents at which time the first node has not sufficient energy for further communication. As the residual energy increases, results in improved network lifetime (Fig. 9).

Fig. 9 Network lifetime versus number of nodes

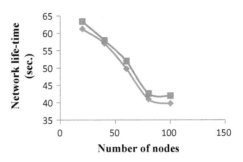

5 Conclusion and Future Work

This paper presented a delay-oriented and energy-efficient routing protocol. Mobile sink node and optimized number of hops are responsible for the improvement of the performance parameters over the existing routing protocol like AODV. Simulation studies show the improved result which is carried out on ns-2.35. It is envisaged that this work can be extended for the random topology of the node deployment and duty cycling approach for further improvement in network lifetime.

Acknowledgements This work is supported by the council of science and technology under the project entitled "wireless sensor network (WSN) routing protocol for industrial applications: algorithm design and hardware". Project grant number is CST/2872.

References

1. Sohraby, K., Minoli, D., & Znati, T. (2007). *Wireless sensor networks: Technology, protocols, and applications*. Willey.
2. Bhuyan, B., & Sarma, N. (2014). A delay aware routing protocol for wireless sensor networks. *International Journal of Computer Science, 11*(6), 60–65.
3. Basagni, S., Carosi, A., Melachrinoudis, E., Petrioli, C., & Maria Wang, Z. (2008). Controlled sink mobility for prolonging wireless sensor network life time. *Wireless Networks, 14*(6), 831–858.
4. Kulshrestha, J., & Mishra, M. K. (2017). An adaptive energy balanced and energy efficient approach for data gathering in wireless sensor networks. *Ad Hoc Networks, 54,* 130–146.
5. Sara, G. S., & Sridharan, D. (2014). Routing in mobile wireless sensor network: a survey. *Telecommunication Systems, 57*(1), 51–79.
6. Anastasi, G., Conti, M., Di Francesco, M., & Passarella, A. (2009). Energy conservation in wireless sensor networks: A survey. *Ad Hoc Networks, 7,* 537–568.
7. Yang, Y., Fonoage, M. I., & Cardei, M. (2010). Improving network lifetime with mobile wireless sensor networks. *Computer Communications, 33*(4), 409–419.
8. Olariu, S., & Stojmenovic, I. (2006). Design guidelines for maximizing lifetime and avoiding energy holes in sensor networks with uniform distribution and uniform reporting. In *Proceedings of IEEE INFOCOM*.
9. Tunca, C., Isik, S., Donmez, M. Y., & Ersoy, C. (2014). Distributed mobile sink routing for wireless sensor networks: A survey. *IEEE Communications Surveys & Tutorials, 16*(2), 877–897.

10. Yan, J., Zhou, M., & Ding, Z. (2016). Recent advances in energy-efficient routing protocols for wireless sensor networks: A review. *IEEE Access, 4,* 5673–5686.
11. Bhuyan, B., & Sarma, N. (2016). A QoS aware routing protocol in wireless sensor networks with mobile base stations. In *Proceedings of the International Conference on Internet of things and Cloud Computing*. Cambridge, United Kingdom, Article No. 15.
12. Zungeru, A. M., Ang, L. M., & Seng, K. P. (2012). Termite-hill: Routing towards a mobile sink for improving network lifetime in wireless sensor networks. In *Third International Conference on Intelligent Systems Modelling and Simulation*.
13. Luo, J., Panchard, J., Piorkowski, M., Grossglauser, M., & Hubaux, J. (2006). MobiRoute: Routing towards a mobile sink for improving lifetime in sensor networks. In *Proceedings of Second IEEE/ACM International Conference Distributed Computing in Sensor Systems* (pp. 480–497).
14. Bhardwaj, M., Garnett, T., & Chandrakasan, A. P. (2001). Upper bounds on the lifetime of sensor networks. In *Proceeding of the IEEE International Conference on Communication (ICC'01)* (Vol. 3, pp. 785–790).
15. Min, R., Bhardwaj, M., Ickes, N., Wang, A., & Chandrakasan, A. (2002). The hardware and the network: Total-system strategies for power aware wireless microsensors. In *Proceeding of the IEEE CAS Workshop Wireless Communication Networks*, Pasadena, CA, USA, (36–12).
16. Luo, J., Hu J., Wu, D., Renfa, L. (2015). Opportunistic routing algorithm for relay node selection in wireless sensor networks. *IEEE Transaction on Industrial Informatics, 11*(1), 112–121.
17. The Network Simulator NS-2. http://www.isi.edu/nsnam/ns/index.html.
18. Perkins, C., Belding-Royer, E., & Das, S. (2003). *Ad hoc on-demand distance vector (AODV) routing*. RFC Editor.

LTE Network: Performance Analysis Based on Operating Frequency

Tasleem Jamal, Misbahul Haque, Mohd. Imran and M. A. Qadeer

Abstract LTE is a widespread technology for high-speed wireless communication that is used in our mobile networks and various data terminals. With the increase in number of users for broadband communication, the requirement for data rate has increased significantly due to the evolution of LTE took place. LTE is the evolvement of high-speed packet access (HSPA) that was assimilated by third-generation partnership project (3GPP) release 8, in order to accomplish widening demand for a very high-speed and proficient access of data. LTE promoted such a high-speed data by enabling larger bandwidth. This paper focuses on the performance of LTE networks at changing spectral frequency bands. In our work, we have tried to simulate LTE networks using NS-2 simulator. Our primary focus is on specifications that greatly affect the behavior of LTE networks. The key specifications that we have used in our work are throughput, average throughput, and jitter. After simulation of our LTE networks, we have concluded that with changing frequency bands, throughput is not affected until and unless bandwidth and modulation type are not varied. We have observed same behavior in case of average throughput but it is smaller in comparison to throughput as bandwidth is shared among multiple broadband users; hence, it can be noted that average throughput falls off with the rise in number of nodes. Jitter does not show any distinct behavior with changing spectral frequency band. It may rise for some instance of time and diminish for other instants.

T. Jamal
Department of Computer Science & Engineering, Harcourt
Butler Technical University, Kanpur, India
e-mail: tasleem.jamal@zhcet.ac.in

M. Haque · Mohd. Imran (✉) · M. A. Qadeer
Department of Computer Engineering, Aligarh Muslim University, Aligarh, India
e-mail: mimran.ce@amu.ac.in

M. Haque
e-mail: misbahul.haque@zhcet.ac.in

M. A. Qadeer
e-mail: maqadeer@gmail.com

© Springer Nature Singapore Pte Ltd. 2019
A. Khare et al. (eds.), *Recent Trends in Communication, Computing, and Electronics*, Lecture Notes in Electrical Engineering 524,
https://doi.org/10.1007/978-981-13-2685-1_14

Keywords LTE · E-UTRAN · OFDMA · SC-FDMA · NS-2 · UMTS · HSPA MIMO · eNB · Thpt · Freq · Jtr

1 Introduction

1.1 Long-Term Evolution (LTE)

Mobile communication has brought a drastic change in consumption of data for Internet facility. Due to this, a high data rate is required by the service providers that caused the development of LTE [1]. The previous 3G technology was insufficient to provide high data rate for users; hence, LTE was prompted as an emerging technology that has become popular among the users as compared to its preceding technology. LTE, which is an important project of 3GPP, a third-generation partnership project, was put forward in 2004 but actual works of LTE was started in 2006. As we know, it is a release 8 project, but some enhancements have been done. It is enhanced to release 9 by incorporating some more features; that is, it can support additional operating frequency bands and voice calls through LTE.

Release 9 enhancement occurred on HSPA+ side by incorporating HSPA+ with MIMO+DC-HSDPA [2].

Release 10 incorporates the standardization of LTE-Advanced, i.e., 4G. It tries to facilitate 4G services by modifying the existing LTE technology. The target for 4G was to develop a system with a speed of 1 Gbps in downward transmission and 500 Mbps in the upward transmission.

The ultimate goal of LTE is to increase the system capacity, to enlarge coverage area, to increase extreme data rate, to maintain multiple antennas, and to minimize latency.

LTE supports both frequency-division duplex and time-division duplex individual spectrum and provides connections for mobile networks [3, 4]. LTE supports orthogonal frequency-division multiplexing scheme during downlink transmission and single-carrier frequency-division multiple access (SC-FDMA) scheme during uplink transmission of packets [5]. In OFDMA, different number of sub-carriers can be assigned to different users in order to support quality of services, i.e., to control the data rate of each user.

2 Correlated Work

Use of cell phones has become very popular nowadays. Mobile users require more data rate for an effective communication around the world. Haider et al. [1] discussed the performance of LTE networks at different spectrum bands. He took unique carrier in two distinct operating frequencies, i.e., 800 MHz and 2.6 GHz. After analyzing the performance, the author found that higher system throughput is not necessarily a result of increase bandwidth as sometimes, lower operating frequency systems are superior to higher frequency system. Adinoyi and Alshaalan [6] tried to find out defiance that he faced during the implementation of LTE networks in his paper titled "present performance evaluation of the LTE technology through experimental setup of the radio access and the evolved packet core." Abed et al. [4] tried to find out the throughput, packet, and queue size that are lost during LTE transmission.

Throughout the remaining paper, we have discussed the following contents as follows: Sect. 3 illustrates the interface and architecture of LTE networks. Section 4 illustrates about simulation study. Performance analysis at operating frequency is discussed in Sect. 5. Section 6 deals with the conclusion of the paper.

3   Interface and Architecture of LTE Networks

LTE network is based on evolved packet system (EPS). The EPS consists of radio access network known as E-UTRAN and IP core networks.

Previous cellular system mainly focused on circuit-switched network, then long-term evolution (LTE) was designed whose main focus is to support only packet-switched network. The main target of LTE network is to provide seamless Internet protocol (IP) connections between packet data network (PDN) and user equipment (UE) without undergoing any disturbance to users during mobility.

Long-term evolution (LTE) is the main source of evolution of universal mobile telecommunications system (UMTS) radio access through the evolved UTRAN (E-UTRAN). It also encompasses evolution of non-radio access defined under "System Architecture Evolution" which contains evolved packet core (EPC). Evolved packet system (EPS) consists of LTE and SAE together. EPS is used to route the traffic to the user equipment. The concept of bearer is Internet protocol packet flow that has a defined quality of service between the gateway and the user equipment. LTE has a flat architecture because it consists of only evolved node base station (eNB), whereas central core is absent in its core; hence, it is different from UMTS. To reduce the latency up to 10 ms, the number of eNB is minimized and MIMO is introduced to make higher uplink data rate around 50 Mbps and downlink rate around 100 Mbps [7]. LTE has been designed to support packet-switched network; hence, we can say that core technology has been migrated toward evolved packet core (EPC) form LTE RAN architecture is depicted in Fig. 1.

In order to meet the requirements of LTE networks, the evolved UTRAN (E-UTRAN) architecture has been improved dramatically from the 3G/3.5G radio access network. The function of eNB in E-UTRAN systems not only includes base station (Node B) functions that terminate the radio interface, it also works as radio network controller (RNC) to handle resources.

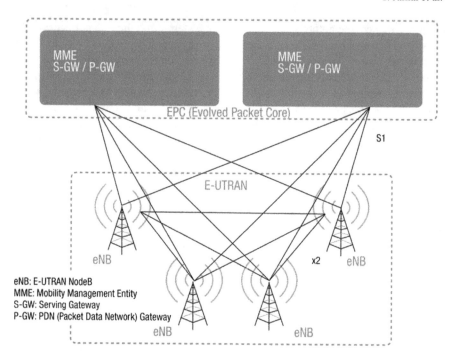

Fig. 1 LTE RAN architecture

E-UTRAN release 10 is a further modification in release 8, and our main focus is on higher data rates and lower latency. Evolved node B (eNB) is a component of E-UTRAN which supervises the interaction between mobile and EPC [8]. eNB has replaced Node B that was used in 3G technology. Now, eNB is a combined form of Node B and radio network controller. eNB acts as an interface with user equipment (UE) that can help several cells at a time [5]. eNB is connected to S-GW and MME by means of S1–U interface and S1–MME interface, respectively [9]. It connects with other eNB using X2 interface which forwards packets during handover [10]. E-UTRAN architecture is explained in Fig. 2 and overall LTE architecture is shown in Fig. 3.

SAE architecture consists of one important component, i.e., evolved packet core (EPC) whose functionalities are as follows: (a) Serving Gateway (SGW) acts as a interface between E-UTRAN and PDN Gateway and it routes and forwards packets between user equipment and PDN. (b) PDN Gateway (PGW) provides connectivity for the UE to external packet data networks, fulfilling the function of entry and exit point for UE data. (c) Mobility management entity (MME) is the main control node

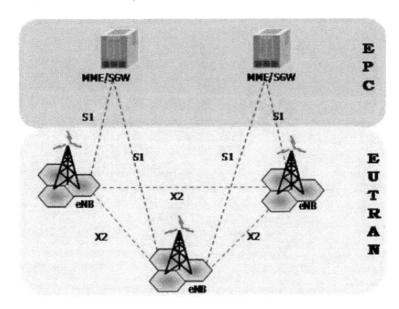

Fig. 2 E-UTRAN architecture

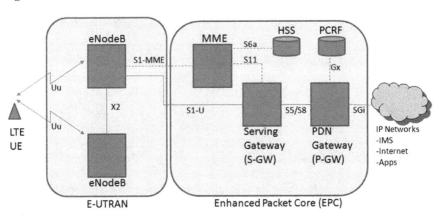

Fig. 3 Overall LTE architecture

for the LTE SAE access network handling a number of features, i.e., idle mode UE tracking, choice of SGW for a UE, interacting with HSS to authenticate user on attachment and implement roaming restrictions. SAE system architecture offers a number of key advantages, i.e., improved data capacity, all IP architectures, reduced latency, reduced OPEX, and CAPEX.

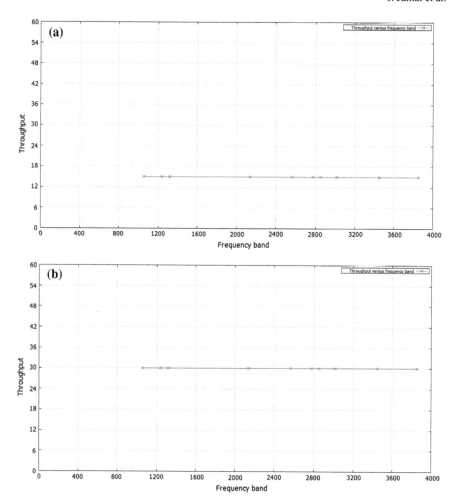

Fig. 4 **a** Thpt. versus operating freq. band at 15 MHz channel bandwidth. **b** Thpt. versus operating freq. band at 30 MHz channel bandwidth. **c** Thpt. versus operating freq. band at 45 MHz channel bandwidth

4 Simulation Study

We have used NS-2 simulator to analyze the performance of LTE at various operating frequencies. NS-2 is an open-source simulation tool that runs on Linux. It is a discrete event simulator targeted at networking research and provides substantial support for simulation of routing, multicast protocols, and IPs such as UDP, TCP over wired and wireless networks [11, 12]. NS-2 comprises two tools, one of the tools support for simulation of routing, multicast protocols, and IPs such as UDP, TCP, whereas another tool helps to imagine the simulations. NS-2 consists of two key languages:

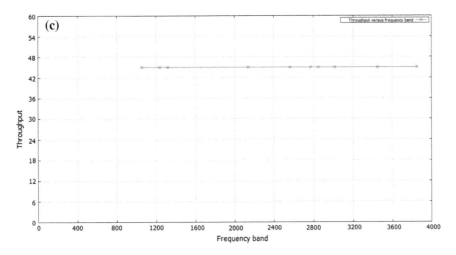

Fig. 4 (continued)

C++ and object-oriented tool command language (OTcl), while C++ defined backend of simulation objects and OTcl sets up simulation. It simulates wired and wireless network. It uses TCL as its scripting language. OTcl is used for object-oriented support of tool command language [12].

5 Performance Analysis at Operating Frequency Bands

We have taken throughput, avg. throughput, and jitter as our main parameter to analyze the performance of LTE at operating frequency [12]. We have taken different conditions depending upon parameters, i.e., bandwidth, and we have plotted the graph at different operating frequency bands.

In data transmission, throughput is the amount of data moved successfully from one place to another in a given time period and typically measured in bits per second (bps). In every network, higher throughput is considered as an absolute choice. We have taken three different bandwidths, i.e., 15, 30, and 45 MHz, at which we have plotted the graph to evaluate the performance.

Figure 4 represents the graph plotted between the throughput and operating frequency bands in which we have taken bandwidths as 15, 30, and 45 MHz. We have considered 9 operating spectral frequency bands, i.e., 400 MHz, 800 MHz, 1200 MHz, 1600 MHz, 2000 MHz, 2400 MHz, 2800 MHz, 3200 MHz, and 3600 MHz, respectively.

With increasing operating frequency bands, we have observed that throughput remains same as long as type of modulation and bandwidth remains the same.

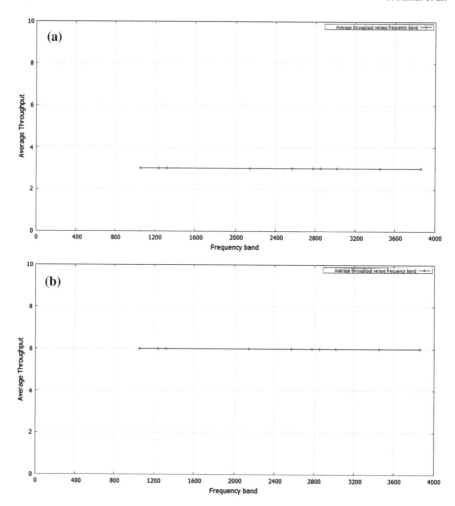

Fig. 5 **a** Avg. Thpt. versus operating freq. at 15 MHz channel bandwidth. **b** Avg. Thpt. versus operating freq. at 30 MHz channel bandwidth. **c** Avg. Thpt. versus operating freq. at 45 MHz channel bandwidth

It is shown in Fig. 4a, b, and c. But as bandwidth is increased, the throughput also increases because as long as bandwidth is smaller, less.

In unit time, traffic will drift, and with increasing channel width, more traffic will be allowed to flow per unit time.

The graph plotted between avg. throughput and operating frequency bands is shown in Fig. 5a, b, and c We have taken 9 operating frequency bands, i.e., 400 MHz, 800 MHz, 1200 MHz, 1600 MHz, 2000 MHz, 2400 MHz, 2800 MHz, 3200 MHz, and 3600 MHz, respectively. After keen observation, we have found that increase

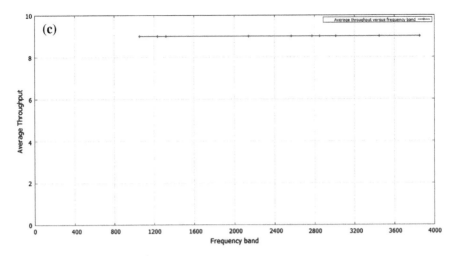

Fig. 5 (continued)

in operating frequency band does not cause any effect on average throughput. It is shown in Fig. 5a, b, and c, respectively.

We have found that average throughput is lower than throughput because in average throughput bandwidth for jitter is shared among multiple nodes of the network.

The plotted graph is shown in Fig. 6. We have taken 10 different operating frequency bands, i.e., 400 MHz, 800 MHz, 1200 MHz, 1600 MHz, 2000 MHz, 2400 MHz, 2800 MHz, 3200 MHz, and 3600 MHz, respectively.

We have taken only two channel bandwidth, i.e., 15 and 30 MHz. After keen observation, we have found that jitter shows anomalous behavior. At the beginning, it has increased, then it became constant; further, it increased with the continuous increase in operating frequency bands. Finally, we have deduced that since jitter is the variation in the latency on a packet flow between two systems; hence, it does not show any specific behavior when operating frequency band is changed.

6 Conclusion

The ultimate goal of our work is to evaluate and analyze the performance of LTE networks at various operating frequency bands. We have taken basically three parameters, i.e., throughput, average throughput, and jitter. We have plotted graphs, and on the basis of these graphs, it is evaluated that as long as bandwidth and modulation types remain same, throughput does not change with increase in operating frequency bands. It remains constant, but if bandwidth is increased, throughput also increases because as bandwidth increases, more traffic would flow per unit time as depicted in TCP. Average throughput follows the same trend as throughput. It

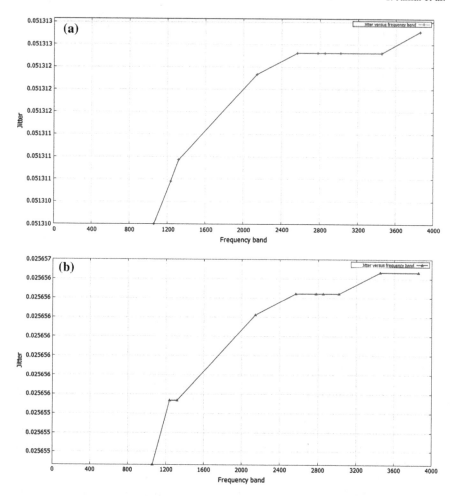

Fig. 6 **a** Jtr. versus operating freq. at 15 MHz channel bandwidth. **b** Jtr. versus operating freq. at 30 MHz channel bandwidth

remains same with increase in operating frequency bands as long as the modulation type and bandwidth are same. Average throughput is lower than throughput because in average throughput, bandwidth is shared among the various nodes of network. Jitter shows anomalous behavior with increase in operating frequency bands. It first increases with increase in operating frequency band. Afterward, it remains constant with increase in operating frequency band, then it further increases with increase in operating frequency band because jitter is termed as packet delay variation, so it does not have any specific trend with operating frequency band.

References

1. Haider, F., Hepsaydir, E., & Binucci, N. (2011). Performance analysis of LTE-advanced networks in different spectrum bands. In *Wireless Advanced (WiAd)* (pp. 230–234).
2. Vieira, R. D., Paiva, R. C. D., Hulkkonen, J., Jarvela, R., Iida, R. F., Saily, M., et al. (2010). GSM evolution importance in re-farming 900 MHz band.
3. Mukharjee, T., & Biswas, S. (2014). Simulation and performance analysis of physical downlink shared channel in long term evolution (LTE) cellular networks. In *India Conference (INDICON), 2014 Annual IEEE*.
4. Abed, G. A., Ismail, M., & Jumari, K. Modeling and performance evaluation of LTE networks with different TCP variants. *World Academy of Science, Engineering and Technology, 75,* 1401–1406.
5. Damnjanovic, A., Montojo, J., Wei, Y., Ji, T., Luo, T., Vajapeyam, M., et al. (2011). A survey on 3GPP heterogeneous networks. *IEEE Wireless Communication, 18*(3), 10–21.
6. Adinoyi, A., & Alshaalan, F. On the performance and experience of the long term evolution (LTE) technology. In *1st International Conference on Computing and Information Technology* (pp. 866–870).
7. Fu, Z., Zerfos, P., Luo, H., Lu, S., Zhang, L., & Gerla, M. (2003). The impact of multi hop wireless channel on TCP throughput and loss. In *IEEE INFOCOM'03,* San Francisco.
8. Alcatel-Lucent. (2009). *The LTE network architecture: A comprehensive tutorial. A technical overview.* Strategic white paper. http://www.cse.unt.edu/Brdantu/FALL_2013_WIRELESS_NETWORKS/LTE_Alcatel_White_Paper.pdf.
9. Sassan, A. (2014). *LTE-advanced: A practical systems approach to understanding the 3GPP LTE releases 10 and 11 radio access technologies.* San Diego, CA: Academic Press, An imprint of Elsevier.
10. Cox, C. (2012). *An introduction to LTE: LTE, LTE-advanced, SAE and 4G mobile communications.* Hoboken, NJ: Wiley.
11. Ezreik, A., & Gheryani, A. (2012). Design and simulation of wireless networks using NS-2. In *Second International Conference on Computer Science and Information Technology (ICCSIT 2012),* Singapore (pp. 157–61).
12. Obaidat, M. S., & Boudriga, N. (2010). *Fundamentals of performance evaluation of computer and telecommunications systems.* Hoboken, NJ: Wiley.

Time of Arrival Positioning with Two and Three BTSs in GSM System

Atul Kumar Uttam and Sasmita Behera

Abstract In recent years, location-based services like emergency, rescue, and response, and location-based marketing are gaining popularity. In all such services, location of involved entity should be located accurately and timely. GSM system is the most popular communication network around the world. Hence, such location detection techniques which require very less or no modification in the existing infrastructure are required. This paper talks about two enhanced ToA-based localization methods using two and three base transreceiver stations. The comparative study between standard ToA techniques and proposed techniques shows encouraging results in various regions like urban area, suburban area, hilly area, and open area.

Keywords Location-based services (LBSs) · GSM network
Time of arrival (ToA) · Base transceiver station (BTS)

1 Introduction

Localization is the process of locating an unknown point location with respect to some reference points, whose locations are known. The major problem in location-based services LBSs is to find the position of MS. Some major examples of LBSs [1] are (1) emergency/rescue services, (2) navigation services, (3) monitoring logistic services, (4) location/context-based event services, etc. To provide these LBSs and several other services, it is very necessary to locate the MS accurately in a short

A. K. Uttam (✉)
GLA University, Mathura, UP, India
e-mail: atul.uttam@gla.ac.in

S. Behera
VSSUT, Burla, India

© Springer Nature Singapore Pte Ltd. 2019
A. Khare et al. (eds.), *Recent Trends in Communication, Computing, and Electronics*, Lecture Notes in Electrical Engineering 524,
https://doi.org/10.1007/978-981-13-2685-1_15

137

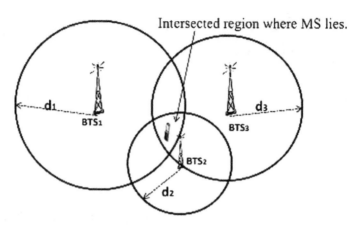

Fig. 1 MS positioning in NLOS environment

span of time. The GSM is the most popular digital cellular system used around the world. It was first developed in the European context. Initially, the GSM system was not designed for the positioning, hence such positioning techniques which require less modification in existing network infrastructure and less deployment cost as well as provide high accuracy required. For the wireless network, several positioning techniques [2, 3] have been proposed by several researchers. Cell identification (Cell ID) [4] based on BTS range, receive signal strength (RSS) [5] based on signal strength of wireless signal, angle of arrival (AoA) based on direction of arrival of signal, time based: time of arrival (ToA) [6–9] and time difference of arrival (TDOA) [10], and fingerprinting-based [11] positioning are major positioning techniques for locating the MS. In time-based, ToA and TDOA positioning techniques, TDOA requires synchronization of BTSs, while ToA requires synchronization of MS and BTS but can be avoided if round-trip time from BTS to MS and MS to BTS or vice versa is used for time of arrival calculation. The standard ToA [7–9] positioning technique is based on trilateration principle in which at least three BTSs require for position estimation. Trilateration principle uses the distance between MS to three BTSs, and form three circles having radii equal to the distance between MS to three BTSs and center at these BTSs. The intersection of these three circles provides the location of MS. Due to the multipath propagation/non-line of sight (NLOS) environment effect these three circles will not collide at a single point rather they form an intersected region (as shown in the Fig. 1), where MS lies.

Rest of the paper is organized as follows: Sect. 2 gives detail about system model; Sect. 3 presents our proposed methods; Sect. 4 presents the simulation results; and finally Sect. 5 concludes with future possibilities of the ToA positioning technique.

2 System Model

In GSM network, an MS is connected to a BTS, called home BTS for communication purpose. When an MS/user asks for a location-based service, for position calculation, home BTS sends a packet to MS, and MS responds to this packet to the BTS. The BTS calculates the total round-trip time (RTT) of this packet. Half of the RTT is called the ToA for corresponding MS–BTS. On multiplying this ToA value to the speed of the electromagnetic wave gives the distance between MS and BTS. For calculating the ToA from other BTSs to MS, force handover/handoff as MS can connect to a single BTS at a time is required.

3 Proposed Method

3.1 Assumptions

MS remains stationary during the localization process. Coordinates of BTSs are fixed and known. An MS scans the GSM spectrum for available base stations. It stores the information (signal strength) of the six neighboring BTSs. These six BTSs are stored in their signal strength order, i.e., highest signal strength BTS is the home BTS. All this information is collected by the MS and sent to the BTS periodically, for handover process as stated in [12]; generally, a user does not have access to this information as stated in [13], and we assume that we have access to this information.

3.2 Two-BTS Method

Let (x, y), (x_1, y_1), and (x_2, y_2) are the coordinates of the unknown MS and the two BTSs, BTS1 and BTS2, respectively. Let d_1 and d_2 are the true distances between MS to BTS_1 and MS to BTS_2, which can be calculated by measuring the ToA value from the respective MS–BTS pair. So the distance between MS and BTS_i can be calculated as (where $i = 1, 2$.):

$$d_i = t_i * c. \tag{1}$$

where t_i is the time of arrival value between MS and BTS_i in line of sight (LOS) condition, and c is the speed of the electromagnetic wave. But due to multipath propagation effect [14] and other errors present in the environment, the electromagnetic wave takes a longer path to reach MS from BTS or vice versa. Hence, while doing ToA value calculation, the multipath propagation/delay spread error e will be added. So the (1) can be expressed as:

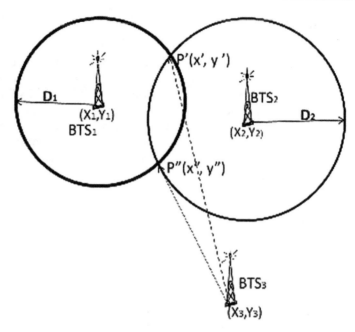

Fig. 2 MS positioning via Two-BTS method

$$T_i = t_i + e \tag{2}$$

$$D_i = T_i * c. \tag{3}$$

where T_i and D_i are the time of arrival value and the distance between MS and BTS_i, in non-line of sight (NLOS) condition.

For $i = 1, 2$:

$$(x - x_1)^2 + (y - y_1)^2 = (D_1)^2 \tag{4}$$

$$(x - x_2)^2 + (y - y_2)^2 = (D_2)^2. \tag{5}$$

Equations (4) and (5) can be solved by using any mathematical tools (MATH-MATICA, etc.) easily. The (4) and (5) will give an ambiguous location of MS $P'(x', y')$, and $P''(x'', y'')$ as shown in the Fig. 2.

To remove the ambiguous location of MS, we use BTS_3 and measure the distance between the P' (X', Y') and BTS_3 (X_3, Y_3), and between P'' (X'', Y'') and BTS_3 (X_3, Y_3). Among P' and P'', whichever will give the minimum distance from BTS_3 be the estimated location of the MS.

Thus, the Two-BTS method requires the time of arrival value calculation from only two BTSs, each from, home BTS and the BTS whose signal strength is just less than the home BTS. A third BTS is required only for removing the ambiguity in MS

Table 1 ToA values for BTSs

BTS$_1$	BTS$_2$	BTS$_3$	–	BTS$_{N-1}$	BTS$_N$
$1T_1$	$2T_1$	$3T_1$	–	$(N-1)T_1$	NT_1
$1T_2$	$2T_2$	$3T_2$	–	$(N-1)T_2$	NT_2
–	–	–	–	–	–
$1T_{M-1}$	$2T_{M-1}$	$3T_{M-1}$	–	$(N-1)T_{M-1}$	NT_{M-1}
$1T_M$	$2T_M$	$3T_M$	–	$(N-1)T_M$	NT_M

location. The proposed method requires only three force handovers to measure the ToA values from the two BTSs.

3.3 *Three-BTS Dual Circle Method*

In this method, time of arrival values are calculated between MS and the home BTS and the neighboring BTSs (BTSs listed in signal strength order for handover purpose). Let M number of time of arrival values are calculated for each N number of BTSs. Then jT_k is the kth time of arrival for jth BTS, where $j = 1$ to N, and $k = 1$ to M. For N BTSs, we get a table like as shown in Table 1.

On performing the first, vertical sorting and selecting the two first minimum ToA values for each N BTS and then performing the horizontal sorting and selecting the first three BTSs whose time of arrival values are less than rest of BTSs. Thus, for each three BTSs, we will get two ToA values, let say these values are (t_{11}, t_{12}), (t_{21}, t_{22}), and (t_{31}, t_{32}). Similarly as in (3), $D_{11}, D_{12}, D_{21}, D_{22},$ and D_{31}, D_{32} can be calculated.

Let (X, Y), (X_1, Y_1), (X_2, Y_2), and (X_3, Y_3) are the coordinates of the MS, and the tree selected BTSs, above. The problem can be formulated as:

$$(X - X_1)^2 + (Y - Y_1)^2 = (D_{11})^2 \tag{6}$$

$$(X - X_2)^2 + (Y - Y_2)^2 = (D_{21})^2 \tag{7}$$

$$(X - X_3)^2 + (Y - Y_3)^2 = (D_{31})^2 \tag{8}$$

$$(X - X_1)^2 + (Y - Y_1)^2 = (D_{12})^2 \tag{9}$$

$$(X - X_3)^2 + (Y - Y_3)^2 = (D_{22})^2 \tag{10}$$

$$(X - X_3)^2 + (Y - Y_3)^2 = (D_{32})^2 \tag{11}$$

Equations (6), (7), (8), (9), (10), and (11) form the three inner circles and three outer circles. A least square solution has been performed for each equation pair {(6), (7), (8)} and {(9), (10), (11)} similar as in [15]; from these two pairs, we get a pair of x coordinate and y coordinate, and the mean of x coordinate and y coordinate gives the (X, Y) MS location. The proposed method requires several time of arrival (ToA)

Table 2 Multipath delay

Environment	Multipath delay (μs)
Open area	<0.2
Suburban area	<1
Urban area	1–3
Hilly area	3–10

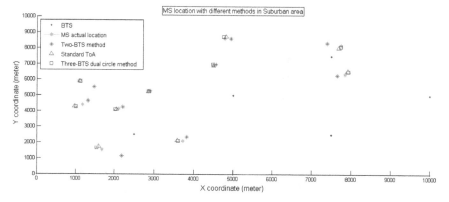

Fig. 3 MS localization in suburban area (ten iterations has been shown.)

values calculation from each BTS, so it takes more time as well as more numbers of force handovers than Two-BTS method.

4   Simulation and Discussion

We have done a computer-based simulation of the above two proposed techniques and standard ToA positioning technique and considered the macrocell for each environment (open area, suburban area, urban area, and hilly area). For simulation purpose, we had considered the respective rms delay spread as stated in [16] (Table 2).

For simulation purpose, we have considered an area of (10000 m x 10000 m), and 13 BTSs are placed at fixed location (0, 0), (5000, 0), (10000, 2500), (2500, 2500), (7500, 5000), (0, 5000), (5000, 5000), (10000, 7500), (2500, 7500), (7500, 7500), (0, 10000), (5000, 10000), and (10000, 10000). For each environment, 1000 iteration has been performed. Let MS accurate location is (Xk, Yk), and the calculated location of MS is (X, Y); then error is given by:

$$\text{error (R)} = \sqrt{\left((X - Xk)^2 + (Y - Yk)^2\right)}. \tag{12}$$

In Fig. 3, ten iterations have been shown for suburban area. On analyzing the 1000 iteration for the each environment areas, we get following results:

Table 3 Suburban macrocell

Error range (m)	ToA localization methods		
	Two-BTS (% MSs)	Standard ToA (% MSs)	Three-BTS dual circle (% MSs)
0–100	12.2	24.2	41.4
100–200	12.1	46.3	47.4
200–300	20.0	25.4	9.9
>300	55.7	4.1	4.3

Table 4 Urban macrocell

Error range (m)	ToA localization methods		
	Two-BTS (% MSs)	Standard ToA (% MSs)	Three-BTS dual circle (% MSs)
0–100	0	3.1	7.0
100–200	0.3	8.7	16.6
200–300	1.1	11.5	19.4
>300	98.6	76.7	57.0

Table 3 shows that our proposed Three-BTS dual circle method can locate 41.4% MSs in 0–100 m, 47.4% MSs in 100–200 m, 9.9% MSs in 200–300 m, and 1.3% MSs located in >300 m. While standard ToA method locates 24.2% MSs in 0–100 m, 46.3% MSs in 100–200 m, 25.4% MSs in 200–300 m, and 4.1% MSs in >300 m. The Two-BTS method proposed by us locates 12.2% MSs in 0–100 m, 12.1% MSs in 100–200 m, 20.0% MSs in 200–300 m, and 55.7% MSs in >300 m. The results show that Three-BTS dual circle method provides higher accuracy than Two-BTS method and standard ToA method in every environment. Though Two-BTS method provides less accuracy than standard ToA method and Three-BTS dual circle method, it requires only two BTSs for time of arrival calculation, and hence, it takes less time as well as less number of force handovers required. The Two-BTS method in open area gives significant amount of accuracy as it locates 96% MSs in 0–300 m area. In hilly and urban areas where the multipath delay is very high, the method proposed by us and standard ToA method provide less accuracy. But Three-BTS dual circle method gives better accuracy than Two-BTS method and standard ToA method in every environment area (Tables 4 and 5).

Table 5 Hilly macrocell

Error range (m)	ToA localization methods		
	Two-BTS (% MSs)	Standard ToA (% MSs)	Three-BTS dual circle (% MSs)
0–100	0	0.2	4.5
100–200	0	0.7	14.0
200–300	0	1.7	14.4
>300	100	98.4	67.1

Table 6 Open macrocell

Error range (m)	ToA localization methods		
	Two-BTS (% MSs)	Standard ToA (% MSs)	Three-BTS dual circle (% MSs)
0–100	42.7	96.4	98.8
100–200	35.0	3.6	1.2
200–300	18.3	0	0
>300	4.0	0	0

5 Conclusion

This paper presents two modified time of arrival-based methods, Two-BTS method and Three-BTS with dual circle method. The numerical result shows that Three-BTS method can locate a greater number of MSs with less error, than Two-BTS method as well as standard ToA method. Though Two-BTS method provides less accuracy, it requires only two BTSs for the time of arrival value measurement, so definitely it takes less time for localization than other two methods. Table 6 shows that Two-BTS method gives a reasonable amount of accuracy in the open area, where the multipath delay is very less.

References

1. Kuhn, P. J. (2004). Location-based services in mobile communication infrastructures. AEU Int. J. Electron. Commun. *58*(3), 159–164. ISSN 1434-8411.
2. Roxin, A., Gaber, J., Wack, M., & Nait-Sidi-Moh, A. (2007). Survey of wireless geolocation techniques. In *Globecom Workshops* (pp. 1–9), 26–30 Nov 2007. IEEE
3. Drane, C., Macnaughtan, M., & Scott, C. (1998). Positioning GSM telephones. *IEEE Communication Magazine 36*(4).
4. Borenovic, M. N., Simic, M. I., Neskovic, A. M., Petrovic, M. M. (2005). Enhanced cell-ID+TA GSM positioning technique. In *EUROCON*, Serbia & Montenegro, Belgrade, 22–24 Nov 2005.
5. Markoulidakis, J. G. (2010). Receive signal strength based mobile terminal positioning error analysis and optimization. *Computer Communications, 33*(10), 1227–1234.

6. Pent, M., Spirito, M. A., & Turco, E. (1997). Method for positioning GSM mobile stations using absolute time delay measurements. *IEEE Electronics Letters, 33*(24), 20.

7. Zhang Y.-H., Cui Q.-M., Ll. Y.-X., & Zhang P. (2008). A novel TOA estimation method with effective NLOS error reduction. *The Journal of China Universities of Posts and Telecommunications, 15*(1).

8. Lin, L., & Pingzhi, F. (2006). An improved NLOS error mitigation TOA reconstruction method. In *2006 IET International Conference on Wireless, Mobile and Multimedia Networks* (pp. 1–3), 6–9 Nov 2006.

9. Chan, Y.-T., Tsui, W.-Y., So, W.-Y., & Ching, P.-C. (2006). Time-of-arrival based localization under NLOS conditions. *IEEE Transactions on Vehicular Technology, 55*(1).

10. Gustafsson, F., Gunnarsson, F. (2003). Positioning using time-difference of arrival measurements. In *2003 IEEE International Conference on Acoustics, Speech, and Signal Processing, 2003, Proceedings. (ICASSP '03)* (Vol. 6, pp. VI 553–556) 6–10 April 2003.

11. Bshara, M., Orguner, U., Gustafsson, F., & Van Biesen, L. (2010). Fingerprinting localization in wireless networks based on received-signal-strength measurements: A case study on WiMAX networks. *IEEE Transactions on Vehicular Technology, 59*(1).

12. Farley, T., & van der Hoek, M. (2006). GSM History. http://www.privateline.com/mt_gsmhistory/2006/01/handover.html.

13. Jain, S., Ghosh, R. K., & Shyamsundar, R. K. (2010). Engineering location based pathfinding on Indian road networks over low end mobile phones. In *2010 Second International Conference on Communication Systems and Networks (COMSNETS)* (pp. 1–9), 5–9 Jan 2010.

14. Yost, G. P., & Panchapakesan, S. (1998). Improvement in estimation of time of arrival (TOA) from timing advance (TA). In *IEEE 1998 International Conference on Universal Personal Communications, 1998, ICUPC '98* (Vol. 2, pp. 1367–1372), 5–9 Oct 1998.

15. Zheng, Y., Wang, H., Wan, L., & Zhong, X. (2009, May 11–13). A placement strategy for accurate TOA localization algorithm. In *Seventh Annual Communication Networks and Services Research Conference, CNSR'09* (pp. 166–170).

16. Willis, M. (2007). Propagation. http//www.mike-willis.com/Tutorial/propagation/.html.

Energy Efficiency in Wireless Sensor Networks: Cooperative MIMO-OFDM

Arun Kumar Singh, Sheo Kumar Mishra and Saurabh Dixit

Abstract This paper investigates the use of cooperative multiple-input-multiple-output orthogonal frequency-division multiplexing (MIMO-OFDM) technique to limit energy consumption used to set up communications among distant hubs in a remote wireless sensor network (WSN). As energy exhaustion is a pertinent issue in WSN field, various methods aim to safeguard such asset, particularly by means for harvesting energy amidst communication among sensor hubs. One such widely utilized strategy is multi-hop communication to reduce the energy required by a single hub to transmit a given message, giving a homogeneous utilization of the energy assets among the hubs in the system. The case of multi-hop communication is not continuously more effective than single-hop. In multi-hop communication, energy efficiency will depend upon the distance between transmit and receive clusters. If the distance is large, then the energy expended will be less than single-hop communication. In this paper, an agreeable MIMO-OFDM transmission strategy for WSN is exhibited, which is contrasted with single-hop. The cooperation among adjacent nodes is analyzed, highlighting its points of interest in connection with both. The inference drawn from performance improvement manifests itself in the utility of applying the proposed strategy for energy-sparing purposes.

Keywords Cooperative · MIMO · Multi-hop · OFDM · WSN

A. K. Singh (✉)
Rajkiya Engineering College, Kannauj, India
e-mail: aksingh_uptu@rediffmail.com

S. K. Mishra · S. Dixit
Central Institute of Plastics and Engineering Technology, Lucknow, India
e-mail: sheokumarmishra@gmail.com

S. Dixit
e-mail: saurabh2911@ieee.org

© Springer Nature Singapore Pte Ltd. 2019
A. Khare et al. (eds.), *Recent Trends in Communication, Computing,
and Electronics*, Lecture Notes in Electrical Engineering 524,
https://doi.org/10.1007/978-981-13-2685-1_16

1   Introduction

Wireless sensor networks (WSNs) are utilized as a part of various rising applications adhering to an essential innovation for the future [1]. With the ubiquitous proliferation of smartphones, heavily reliant on sensors, it is imperative to design energy efficient sensor nodes which are self-organized. Notwithstanding, a significant worry, in connection with the utilization of WSN is the energy utilization. Remote sensor hubs are typically asset compelled devices, driven by batteries, which constrains their energy spending plan. Moreover, these sensor hubs are generally conveyed in territories that are hard to be got to, hence making impracticable the substitution of such energy assets. Therefore, to defeat such issue, a keen energy resource management is an imperative requirement. Considering that the most energy devouring errand in the sensor hubs is communication, proficient communication components are very essential to diminish the energy consumption in remote sensor system [2]. Various strategies address this issue, including elective steering conventions, energy sensitive communications, among others [3, 4]. A typical part of these methodologies is the investigation of multi-hop communication to spread the energy consumption among the hubs in the system so that no single hub suffers an extraordinary reduction in its energy spending plan because of costly long separation single-jump transmissions. Be that as it may, multi-hop does not speak of a "silver bullet" to take care of the issue, as there are cases in which even a solitary hop transmission can perform superior to a multi-hop one. In [5], the authors have demonstrated that over specific distance ranges, the total energy consumption can be reduced. However, in short ranges, the performance of single-input-single-output (SISO) may outperform multiple-input-multiple-output (MIMO) or cooperative MIMO as far as energy efficiency is concerned. In [6], the author(s) have investigated the performance of nodes in a soft handover region, receiving the signal from distributed base stations cooperating among themselves. In [7], the authors have analyzed the performance of orthogonal frequency-division multiple access (OFDMA), which is a combination of orthogonal frequency-division multiplexing (OFDM) and frequency-division multiple access (FDMA) in a multiple antennas environment. However, implementing multiple antennas is constrained due to size, cost, and hardware limitations. Cooperative diversity provides an effective solution aimed at improving spectral and power efficiency of the wireless networks doing away with the additional complexity of multiple antennas. The basic premise behind cooperative diversity is the multipath propagation nature in a wireless environment. The signal transmitted by the source nodes is often heard by other nodes, which can be deemed as cooperative partners. The source and their partners coherently process and transmit their information, thereby creating a virtual antenna array. In [8] the authors have provided a vivid overview of cooperative communication technique. As long as the fading path is uncorrelated, the benefits of spatial diversity are reaped. The trade-off between total power consumed and sum rate is analyzed. A comparison of performance in cooperation signaling methods is provided. The results demonstrate that coded cooperation and decode and forward scheme perform better than without cooperation.

1.1 Wireless Networks

Energy utilization is the fundamental issue that hinders the application of WSN [2]. Because of the compelled energy spending that the sensor hubs face, a watchful utilization of this asset in every individual hub must be considered. At the same time, the end goal to extend the life expectancy of the whole system should be adhered to. As every single disseminated framework, WSN has their essential functionalities profoundly reliant on the communication among their hubs. Notwithstanding, as remote interchanges are expensive regarding energy utilization, it prompts an impasse about the use of the communications. The arrangement of this impasse needs to consider a productive utilization of the communication to limit the misuse of energy. Remote sensor arrangements typically display a planar or various-leveled design [9]. In the first step, sink hubs disperse data in the system, which are transmitted from hub to hub as indicated by the sort of the data being scattered. Various leveled-based WSNs limit the more costly interchanges to exceptional hubs that trade messages among them and are in charge of multiple hubs, as delegates. Cases of such WSNs are cluster-based WSN, in which those unique sensor hubs are called cluster heads, and they can be all the more capable hubs that are in control for long-range communications with other bunch heads and sink hubs. Different arrangements of grouped WSN are conceivable, in which the cluster heads and the group individuals are similarly intense, however, contrasts, for example, staying accessible assets, the geological situation among other criteria can be utilized to choose a given sensor hub as the group head. In spite of the fact that multi-trusting is viewed as a proficient communication answers for WSN, there are cases in which single-hop caters to a superior option as far as energy utilization is concerned [10]. Then again, in a few situations, the utilization of single-hop for long separation communication may trade off the whole system lifetime. This is the situation, for example, when the single-hop communication brings about unpaired exhaustion of the energy assets of individual hubs, which is exceptionally undesired. Watching the essential attributes of a WSN, in which a few hubs in an area give comparative information, elective arrangements can be made. One of them investigates the idea of progressive WSN, in which diverse sorts of sensor total or sensor combination methods are utilized [11]. Despite the advantages of such strategies, they may require a few communications among the group individuals, contingent upon the assertion convention that is utilized, consequently expanding the energy utilization because of communication. Other than single and multi-hop communication in WSN, agreeable different information yielding MIMO-OFDM plans are likewise critical option arrangements that can be considered. As neighboring sensors need to send identical information in an instant of time, cooperative MIMO-OFDM can be utilized, as quickly portrayed in the following subsection, and further investigated in next segments of this paper.

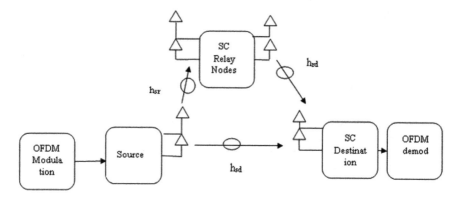

Fig. 1 Cooperative relay nodes

2 Cooperative MIMO-OFDM

Orthogonal frequency-division multiplexing (OFDM) is resilient to inter-symbol interference (ISI) while multiple-input-multiple-output (MIMO) gives capacity and performance advantage [12]. However, implementation of multiple antennas is restrained by factors such as size, cost, and hardware limitations. Figure 1 displays a MIMO-OFDM cooperative relay system where single-hop and multi-hops are portrayed from source to destination. If the channels are independent and identically distributed (iid), then the benefits of MIMO system are reaped. Hence, a virtual array of antennas can be realized by the use of cooperative relay nodes. The cooperative MIMO-OFDM communication considered in this work depends on two stages. The initial step evaluates the cooperative MIMO-OFDM channel by utilizing pilot signals. Once the channel is assessed, the new data can be transmitted. In Sect. 3, this cooperative MIMO-OFDM approach for communication in WSN is displayed point by point, while the outcomes acquired with this option arrangement are contrasted and those obtained with single and multi-hop are discussed in Sect. 4. The results obtained from MIMO-OFDM cooperative relays point out the energy efficiency of the cooperative scheme.

3 Results

3.1 Simulation Setup

An irregular patch of sensors that are around 100 m separated from each other is chosen exhibiting the likelihood for multi-hop communication using different hubs that are 30 m separated from each other, i.e., 30 m separate for each jump. The single-hop approach includes just the sensor that has the information that should be transmitted and the sensor that will get this information. For the multi-hop approach, the procedure is fundamentally the same. However, the separation of the information needs to travel is divided among them, with sensors coordinating en route to transmit the information over littler separations, along these lines requiring less power. Thirty pilot images are transmitted between each match of sensors, so the channel pick up can be assessed, after that 1000 information images are transmitted comparable as in the single-hop approach. The BER is evaluated contrasting the transmitted images, and the images got at the last sensor hop. The sensors are conveyed agreeing on an irregular example following Poisson appropriation in two measurements, for all reenactment runs. For the cooperative MIMO-OFDM case, the images to be transmitted are first transmitted to the collaborating sensors of the transmit group.

3.2 Simulation Results

The efficiency of energy depends on different factors, and hence, the expression often changes from one model to the other. The energy efficiency (η_{ee}) means that the system needs to consume minimum energy in transferring the bits when the value is small, energy often increases and is then given by [13]:

$$\eta_{ee} = \frac{P_t T_t}{N_{good}} \qquad (1)$$

where P_t represents the amount of power transferred, T_t represents the amount of time, N_{good} represents the number of bits received. For this paper, the formula is modified in the following form:

$$\eta_{ee} = \frac{P_t T_t}{(1 - BER) * Notb} \qquad (2)$$

where P_t represents the amount of power transferred, T_t represents the amount of time, BER represents Bit Error Rate, $Notb$ represents the number of transmitted bits. Table 1 depicts the system parameters for simulation. The results in Fig. 2 are based on the erasure of packet scheme that is applied to diverse configurations such as SISO, 2×2 MIMO, 4×4 MIMO. These results can be compared with those of Fig. 3 where the output of the model specifies that this new MIMO-OFDM approach with cooperative relays ensures that the system consumes little energy, particularly at high signal-to-noise-ratio (SNR) regime. Alternatively, this approach is more energy competent at higher SNR. The EE of MIMO-OFDM is given by [14] as the ratio of capacity and expectation of total transmit power.

$$\eta_{ee} = \frac{C_{total}}{E(P_{total})} \qquad (3)$$

Table 1 System parameters
for simulation

Parameters	Value
Symbol rate R_s	1 million/sec
White noise, N_o	−174 dBm/Hz
Modulation	16-QAM
Useful symbol duration	22.472 μs
Bandwidth BW	10 MHz
FFT point	1024
Carrier frequency, f_c	400 MHz
Transmission power P_t	500 Mw

Fig. 2 Packet erasure
method to identify energy
efficiency

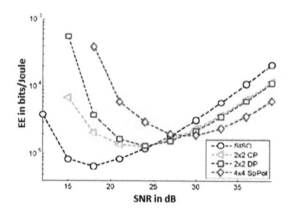

4 Discussion

Multi-hop communication requires substantially less-transmitting power compared
to the single-hop communication. However, the distance should be large. The benefits
of multi-hop are more pronounced for large distances. The energy consumed becomes
progressively high as the number of hops increases, because receiving information
can consume up to twice as much power as transmitted from a low-power node.
In the MIMO-OFDM case, information can be transmitted by at least two sensors
and decoded by at least one sensor; signals can be transmitted at a lower control,
along these lines requiring less energy. This further enhances energy productivity
over the system when long separation transmissions should be made, and coordinate
transmission is impractical. This difficulty can be either because of a high number of
multi-way parts or because of the way that there are no delegate hubs accessible to
transfer the bundles over the system. For this situation, the energy efficiency of the
MIMO-OFDM case outperforms the productivity of the multi-hop approach, which
turns out to be progressively exorbitant mainly because of the quantity of aggregation
that should be performed. Data aggregation can consume twice the energy in the
transmission of a low-power node than in transmission of high-power node.

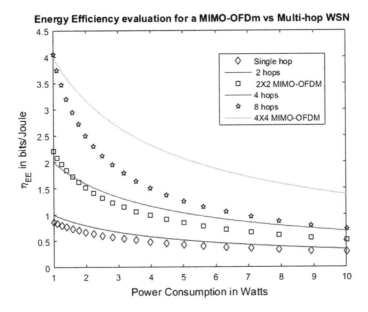

Fig. 3 Energy efficiency of energy using MIMO-OFDM with Alamouti transmit diversity

5 Conclusion

This paper shows an investigation contrasting single-hop, multi-jump, and cooperative MIMO-OFDM-based transmissions over WSN. Simulation values were displayed, examined and compared with hypothetical ones. As seen by the examination of the results, the cooperative MIMO-OFDM approach turns into the best choice as the quantity of multi-hops increases. For four hops, MIMO-OFDM, the procedure costs 8.7% more power. When the number of hops is higher, the cooperative MIMO-OFDM approach begins to be more power productive than the multi-hop plan or if the information is sensitive to delay. Our outcomes validate that cooperative MIMO-OFDM is a capable alternative for information transmission over long separations over the WSN. Bearings of future work are the examination of conceivable upgrades in the proposed cooperative MIMO-OFDM method, specifically by considering sensor hubs versatility, which was not considered in the present paper.

References

1. Business Week: 21 ideas for the 21st century, Aug. 30, pp. 78–167 (1999).
2. Mini, R. A. F., & Loureiro, A. A. F. (2009). Energy in wireless sensor networks. In B. Garbinato, H. Miranda, & L. Rodrigues (Eds.), *Middleware for network eccentric and mobile applications* (pp. 3–24). Springer.

3. Goyal, D., & Tripathy, M. R. (2012). Routing protocols in wireless sensor networks: A survey. In *Second International Conference on Advanced Computing & Communication Technologies* (pp. 474–480).
4. Durresi, A., Paruchuri, V., Barolli, L., & Raj, J. (2005). QoS-energy aware broadcast for sensor networks. In *Proceedings of 8th ISPAN* (p. 6).
5. Cui, S., Goldsmith, A. J., & Bahai, A. (2004). Energy-efficiency of MIMO and cooperative MIMO techniques in sensor networks. *IEEE Journal on Selected Areas in Communications, 22*(6), 1089–1098.
6. Tolli, A., Codreanu, M., & Juntti, M. (2008). Cooperative MIMO-OFDM cellular system with soft handover between distributed base station antennas. *IEEE Transactions on Wireless Communications, 7*(4), 1428–1440.
7. Baek, M. S., & Song, H. K. (2008). Cooperative diversity technique for MIMO-OFDM uplink in wireless interactive broadcasting. *IEEE Transactions on Consumer Electronics, 54*(4), 1627–1634.
8. Nosratinia, A., Hunter, T. E., & Hedayat, A. R. (2004). Cooperative communication in wireless networks. *IEEE Communications Magazine, 42*(10), 74–80.
9. Akyildiz, I. F., Su, W., Sankarasubramaniam, Y., & Cayirci, E. (2002). A Survey on Sensor Networks. *IEEE Communications Magazine, 40*(8), 102–114.
10. Chen, C., Ma, J., & Yu, K. (2006). Designing energy-efficient wireless sensor networks with mobile sinks. In *Sensys'06, ACM*.
11. Nakamura, E. F., Loureiro, A. A. F., &, A. C. (2007). Information fusion for wireless sensor networks: Methods, models, and classifications. *ACM Computing Surveys (CSUR), 39*(3), 9–12.
12. Bolcskei, H. (2006). MIMO-OFDM wireless systems: Basics, perspectives, and challenges. *IEEE Transactions on Wireless Communications, 13*(4), 31–37.
13. Han, C., et al. (2011). Green radio: Radio techniques to enable energy efficient wireless networks. *IEEE Communications Magazine, 49*(6), 46–54.
14. Li, G. F., Xu, Z., Xiong, C., Yang, C., Zhang, S., Chen, Y., et al. (2011). Energy-efficient wireless communications: Tutorial, survey and open issues. *IEEE Wireless Communications, 18*(6), 28–35.

An Energy-Balanced Cluster-Based Routing Protocol for Wireless Sensor and Actuator Network

Yogesh Tripathi, Arun Prakash and Rajeev Tripathi

Abstract Wireless sensor and actuator network is the extension of wireless sensor network. Wireless sensor and actuator network is made of sensor and actuator nodes. To overcome energy constraints of sensor networks, some energy-rich nodes (actuator nodes) are deployed in surveillance field. Due to energy heterogeneity of deployed nodes, sensor and actuator nodes have different responsibility. Sensor nodes gather data from surveillance field, and actuator nodes are responsible for transmission of collected data from sink. In this paper, an energy-balanced cluster-based routing protocol for wireless sensor and actuator networks is proposed. The network is divided into clusters, and the most energy-rich node among sensor nodes is selected as cluster head. Cluster head selection and actuator node selection metrics balance the energy consumption of network. Simulation results show the improvements in the performance parameters over the existing routing protocol. Simulation is carried out on network simulator (ns-2.35).

Keywords Wireless sensor and actuator networks · Clustering · Routing
Energy balancing

1 Introduction

Due to the advancements in wireless technologies and microelectronics mechanical systems (MEMS), incorporating with the very-large-scale integration (VLSI) design technique leads to materialization of the wireless sensor networks (WSNs). WSNs

Y. Tripathi (✉) · A. Prakash · R. Tripathi
Department of Electronics and Communication Engineering, Motilal Nehru National
Institute of Technology Allahabad, Allahabad 211004, India
e-mail: rel1706@mnnit.ac.in

A. Prakash
e-mail: arun@mnnit.ac.in

R. Tripathi
e-mail: rt@mnnit.ac.in

© Springer Nature Singapore Pte Ltd. 2019
A. Khare et al. (eds.), *Recent Trends in Communication, Computing,
and Electronics*, Lecture Notes in Electrical Engineering 524,
https://doi.org/10.1007/978-981-13-2685-1_17

are proficient in keeping an eye on physical world, processing over data accordingly, taking decisions based on inspected data, and performing suitable actions, but sensor networks have energy constraints and limited hardware capability [1]. So, it is required to the inclusion of actuators in WSNs which are very affluent in energy and hardware capability to improve network lifetime of sensor network. Wireless sensor and actuator network (WSAN) is the extension of the WSNs. In WSANs, sensor nodes are responsible for event detection and actuator nodes are responsible for taking the prompt decision over data sensed by sensor nodes. In the Internet of things (IoT) era, WSANs are playing a very important role. So, these networks are going to be an essential part for industrial and home applications like battlefield surveillance, building automation [2], environmental monitoring, and nuclear and chemical reaction detection [3]. To take advantage of actuator nodes in the above-mentioned applications, it requires proper coordination in communication among nodes. In the present literature, there are three types of cooperation, namely sensor–sensor, sensor–actuator, and actuator–actuator. The sensed information of environment is collected through sensor–sensor coordination, and sensor–actuator coordination is responsible for transmission of sensor's collected data to actuator for further processing. Finally, gathered data at actuator is transmitted to sink through actuator–actuator coordination [4].

In WSNs, energy saving is critical issue and there are several protocols proposed in literature to save the energy consumption for prolonging network lifetime [5, 6]. To overcome the energy constraints of WSNs, actor nodes are deployed in the surveillance field. Actuator nodes are deployed where the requirement is timely delivery of sensed data because it has large transmission distance. Due to large transmission distance, a less number of hopes are required for transmitting data from source to sink which results in improved network lifetime.

In this paper, an energy-balanced cluster-based routing protocol is proposed. The network is divided into clusters, and each cluster is supervised by the cluster head. The cluster head (CH) is selected by a novel metric which includes the energy of the neighbor nodes, actuator nodes, and its distance. The energy consideration in metric is due to the most energy-efficient node which is selected as the cluster head. In this proposed algorithm, actor nodes are responsible for the data routing. Actor nodes are energy-rich, and it has more transmission range. So, it reduces the end-to-end delay which reduces the energy consumption of the sensor nodes and prolongs network lifetime of the network as compared to the conventional routing protocol for WSNs. Simulation shows improved results of the proposed routing protocol over existing routing protocols.

The rest of the paper is organized as follows: In Sect. 2, related work is summarized. In Sect. 3, description of the proposed routing protocol is detailed. In Sect. 4, discussion of performance evaluation of the proposed routing protocol is presented. Finally, Sect. 5 includes conclusion and future work.

2   Related Work

Although WSAN is an extension of WSN and various energy-efficient routing protocols that exist in literature but these routing protocols cannot be directly applied to WSAN's architecture due to heterogeneity. WSANs are more reliable and energy-efficient over WSNs. Still, some open research issues and challenges exist [4].

This section presents different cluster- and non-cluster-based routing protocols related to WSANs.

In [7], the authors proposed multi-tier clustering approach for data transmission from source node to sink node. The protocol uses the genetic algorithm to select the cluster head of sensor nodes as well as actor nodes. CH has more caching capability so that energy efficiency can take place.

In [8], authors proposed an adaptive clustering mechanism. Clusters are self-regulating, and configuration depends on node movement. In this protocol, frequency reuse concept is applied in different cluster and frequency of each cluster is distributed in centralized manner. This protocol is robust to topological changes.

In [9], authors proposed an energy-efficient cluster-based routing protocol. Energy consumption is reduced by multi-level hierarchical structure. Each cluster selects some nodes as intermediate nodes (INs), and neighbor nodes are managed by (INs). INs make a sub-cluster and transmit aggregated data to CH. A minimum energy path is selected for the data transmission. Energy optimization can be achieved by mixed-integer linear programming.

In [10], authors proposed an asymmetric routing protocol for WSANs. This protocol supports single-path source routing and multipath graph routing. Source routing is used for sensing and graph routing for actuation. An extended Kalman filter with model predictive controller and actuator buffer are used for improving data loss in WSANs.

3   The Proposed Routing Structure

This section gives detail description of network model and cluster head selection to make routing algorithm energy-efficient.

3.1 Network Model

In this paper, sensor and actuator nodes make a network which can communicate wirelessly. Sensor nodes have hardware and energy constraints, deployed for monitoring of field and reporting sensed data to actuator nodes. The actuator nodes are deployed regularly in field so that it can communicate with other actuator nodes as well as sensor nodes (CH). The actuator nodes also have constraints but are more

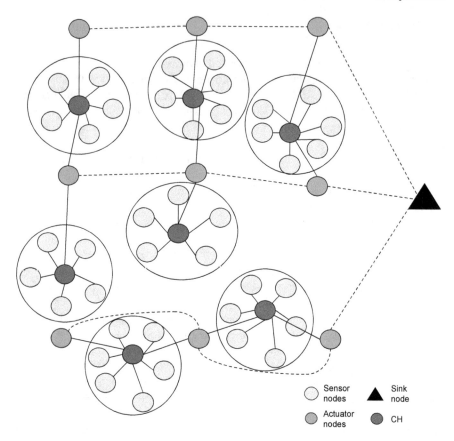

Fig. 1 Network model

rich compared to sensor nodes. The sensor nodes are arranged in cluster, and each cluster has a CH. All nodes (sensor and actuator) deployed in field are static which reduces complexities in cluster membership. The network model used in this paper is shown in Fig. 1. The sensed data of sensor nodes in deployed region is transmitted to the corresponding CH. The CH gathers all data and forwards its data to the best actuator nodes which are in its transmission range.

3.2 Energy-Balanced Clustering Approach

The proposed approach takes care of energy of node to become as CH while cluster formation. Each node is aware of its cluster and works as cluster member (CM) in respective cluster. To achieve energy balancing in the network, selection of CH is based on energy of node. Each node broadcasts a beacon message which consists of the residual energy of nodes and its neighborhood distance. As node receives the beacon message, it calculates its weight function for the CH selection. The node pair which has highest weight function and higher residual energy will be selected as CH, and another node is selected as backup CH. After completion of CH selection process, it will broadcast a message to all nodes which are in its transmission range to become CM. As weight function of CH drops below the threshold, it sends a beacon message to backup CH and it will act as CH. The selection process of cluster head based on energy and distance balances the energy consumption in the network. The selection of CH is governed by the equation given below:

$$CH = \frac{Ei \times Ej}{dij} \tag{1}$$

where $i = 1, 2, 3 \ldots, n$ and $i \neq j$ E_i and E_j are the energy of sensor node i, j, respectively, d_{ij} is distance between node i and j.

3.3 Routing Algorithm

The proposed routing algorithm is based on the selection of efficient actuator node for data transmission. CH broadcasts an alert message to all actuator nodes. If any actuator nodes are present in the range of CH, they will reply an alert message. CH selects the best actuator node based on weight function. The pseudocode of the proposed routing algorithm is given below:

Algorithm 1:

---
1.   Broadcast RREQ packet with Source Address and Destination Address
2.   **While** $(t < t_{max})$
3.   **if** RREP received then
4.   Collect RREP packet and select route having best actuator node
5.   Send data packet
6.   **if** ACK is received **then**
7.   Transmit next data packet
8.   **else** resend data on back up route
9.   **end if**
10.   **end if**

Selection of best actuator node will be governed by given formula:

$$\text{Max}\{E_{a1}/d_{ia1}, E_{a2}/d_{ia2}, E_{a3}/d_{ia3}, \ldots, E_{aj}/d_{iaj}\}$$

where E_{aj} is residual energy of actuator nodes
d_{iaj} is distance between CH i and actuator nodes j.

The number of CHs and actuator nodes depends on the topology. As size of the network increases, requirement of CHs and actuator nodes in surveillance field increases, and as the network size decreases, requirement of CHs and actuator nodes in surveillance field decreases.

4 Simulation Results

Simulation of the proposed routing protocol is carried on network simulator (ns-2.35) [11] and the performance is evaluated. The performance evaluation has been done for different routing metrics like energy consumption, end-to-end delay. In this paper, results of the proposed algorithm are compared with the existing routing protocols like AODV [12]. The proposed algorithm has improved results over the existing routing algorithms. It is due to the actor nodes which are more energy-rich as compared to the sensor nodes. The parameters used in simulation are given in Table 1.

Throughput is the measure of how the message is successfully transmitted over given communication channel. Figure 2 shows the graph between throughput versus number of nodes. In the proposed algorithm, data transmission takes place in two steps. In first step, sensor data is transmitted to CH and collected data at CH is routed to sink with the help of actuator nodes in the second step. The actuator nodes have more energy and processing capability, so success rate of data transmission increases which improves the network throughput.

Table 1 Simulation parameters

Parameter	Value
Number of nodes (n)	20–100
Simulation time	200 s
Simulation area	$5000 \times 5000 \text{ m}^2$
Mac protocol	802.15.4
Traffic types	CBR
Packet transmission rate	Random
Actuator nodes	9

Fig. 2 Throughput versus number of nodes

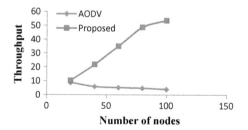

Fig. 3 PDR versus number
of nodes

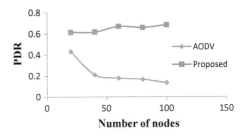

Fig. 4 Delay versus number
of nodes

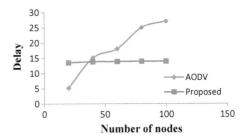

Packet delivery ratio (PDR) gives information about successful reception of packet at receiver side. In Fig. 3, graph is shown between PDR and number of nodes. Sensor nodes have energy constraints due to which it cannot efficiently handle packet transmission and packet loss happens in the network. In the proposed algorithm, sensor nodes are responsible only for sensing the environmental phenomenon and sensed data is routed through actuator nodes. Actuator nodes are energy-rich, and it handles packet efficiently which improves packet reception at receiver.

Delay is defined as how much time is consumed by a packet to travel a distance from source to destination. Figure 4 shows variation of delay with number of nodes. In the proposed algorithm, all data traffics are routed with the help of CH and actuator nodes. Delay of the proposed algorithm does not depend on the number of nodes because sensor nodes are not responsible for data transmission to sink node. Sensor nodes only sense data and send to CH for further transmission.

Average residual energy is the remaining energy of nodes. As shown in Fig. 5, residual energy decreases as the number of nodes increases. More number of nodes generates more data traffic in the network which leads to congestion. So, increment in number of nodes results in more energy consumption.

Network lifetime represents time at which first node drains out its energy. It depends on the residual energy of the sensor nodes. Figure 6 shows the graph between network lifetime and number of nodes. There is improvement in residual energy of the proposed algorithm due to which there is improvement in network lifetime over the existing routing protocols.

Fig. 5 Average residual energy versus number of nodes

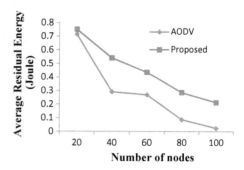

Fig. 6 Network lifetime versus number of nodes

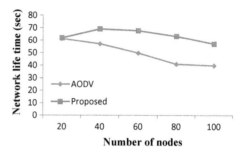

5   Conclusion and Future Work

In this paper, energy balancing is achieved with the help of clustering among sensor nodes. The clustering approach takes into consideration the residual energy of nodes and distances of sensor nodes. The proposed algorithm minimizes energy consumption and delay which improves network lifetime. It is possible due to the selection of best node for CH, and selected CH is near to best actuator nodes. The simulated results on ns-2.35 of the proposed algorithm verified that it has improved results over the existing routing algorithms. This work can be extended toward mobile WSANs, mathematical modeling of a general WSAN scenario, load balancing, and cognitive radio technology.

Acknowledgements This work is supported by the council of science and technology under the project entitled "wireless sensor network (WSN) routing protocol for industrial applications: Algorithm design and Hardware." Project grant number is CST/2872.

References

1. Sohraby, K., Minoli, D., & Znati, T. (2007). *Wireless sensor networks: Technology, protocols, and applications*. Willey Publication.
2. Petriu, E. M., Georganas, N. D., Petriu, D. C., Makrakis, D., & Groza, V. Z. (2000). Sensor-based information appliances. *IEEE Instrumentation and Measurement Magazine, 3*(4), 31–35.
3. Akyildiz, I. F., Su, W., Sankarasubramaniam, Y., & Cayirci, E. (2002). Wireless sensor networks: A survey. *Computer Networks, 38*(4), 393–422.
4. Akyildiz, I. F., & Kasimoglu, I. H. (2004). Wireless sensor and actor networks: Research challenges. *Ad Hoc Networks, 2*(4), 351–367.
5. Yan, J., Zhou, M., & Ding, Z. (2016). Recent advances in energy-efficient routing protocols for wireless sensor networks: A review. *IEEE Access, 4,* 5673–5686.
6. Ibrahim, A. H., Han, Z., & Ray Liu, K. J. (2008). Distributed energy-efficient cooperative routing in wireless networks. *IEEE Transaction Wireless Communication, 7*(10), 3930–3941.
7. Kumar, A., & Ranga, V. (2014). A cluster-based coordination and communication framework using GA for WSANs. In *Proceedings of Ninth International Conference on Wireless Communication and Sensor Networks* (pp. 111–124).
8. Lin, C. R., & Gerla, M. (1997). Adaptive clustering for mobile wireless networks. *IEEE Journal in Selected Areas Communication, 15*(7), 1265–1272.
9. Quang, P. T. A., & Kim, D. S. (2015). Clustering algorithm of hierarchical structures in large-scale wireless sensor and actuator networks. *Journal of Communication and Networks, 17*(5), 473–481.
10. Li, B., Ma, Y., Westenbroek, T., Wu, C., Gonzalez, H., & Lu, C. (2016). Wireless routing and control: A cyber-physical case study. In *ACM/IEEE 7th International Conference on Cyber-Physical Systems*.
11. The network simulator NS-2. http://www.isi.edu/nsnam/ns/index.html.
12. Perkins, C., Belding-Royer, E., & Das, S. (2003). *Ad hoc on-demand distance vector (AODV) routing*. RFC Editor.

A Collective Scheduling Algorithm for Vehicular Ad Hoc Network

Rohit Kumar, Raghavendra Pal⊙, Arun Prakash and Rajeev Tripathi

Abstract Vehicular ad hoc network deals with traffic congestions. It also ensures the safety and convenience of drivers and passengers. Safety is one of the most important aspects in VANETs, and to make it safe and reliable, its performance needs to be enhanced. To enhance the performance, there is a need to control the congestion. In the high-density region, congestion control becomes a challenging task and special characteristics of VANETs (e.g. high rate of topology changes, high mobility) make it more challenging. In this paper, collective scheduling strategy is proposed. In this strategy, priority of the message is defined which mainly depends on three factors, i.e. size of messages, type of messages and state of networks. Based on these factors, messages priorities are calculated and then these messages are scheduled. Collective scheduling uses the concept of clustering while assigning priority of information. Simulation is carried out to demonstrate improvement in comparison with collective scheduling with tabu search scheduling.

Keywords Vehicular ad hoc network · Intelligent transportation systems
Vehicle to vehicle · Applications · Congestion control · Message priority
Message scheduling

R. Kumar · R. Pal (✉) · A. Prakash · R. Tripathi
Department of Electronics and Communication Engineering, Motilal Nehru
National Institute of Technology Allahabad, Allahabad 211004, India
e-mail: raghavendra.pal3@gmail.com

R. Kumar
e-mail: rohitpayoj@gmail.com

A. Prakash
e-mail: arun@mnnit.ac.in

R. Tripathi
e-mail: rt@mnnit.ac.in

© Springer Nature Singapore Pte Ltd. 2019
A. Khare et al. (eds.), *Recent Trends in Communication, Computing,
and Electronics*, Lecture Notes in Electrical Engineering 524,
https://doi.org/10.1007/978-981-13-2685-1_18

1 Introduction

Vehicular ad hoc network (VANET) [1] has unique characteristics such as high mobility of nodes, high rate of topology changes, high density, sharing of wireless channels and frequently broken routes [2]. During rush hour, due to very high density of nodes, congestion arises, and since a few channels are shared among several vehicles, channels get saturated. Therefore, there is an increase in delay and packet loss, and as a result, the performance of the system degrades, and hence, the system becomes less reliable. To improve its reliability, there must be some means to provide a fair channel access to each node available. To control the congestion, several works have been done on MAC layer. Strategies based on MAC layer deal with prioritizing of message signals and schedule them to access wireless channels. Prioritizing and scheduling of messages reduce delay and packet loss by prioritized sharing of the channel.

Dedicated short-range communication (DSRC) [3] defines standards and protocols for VANET. The main motivation for deploying DSRC is to prevent a collision by two methods of frequent data exchange, i.e. between vehicle to vehicle (V2V) and vehicle to infrastructure (V2I). DSRC employs many standards like IEEE 802.11p and IEEE 1609 standards for enhancing the performance by wireless access in vehicular environment (WAVE) [2]. DSRC uses 5.9 GHz frequency band having a bandwidth of 75 MHz (5.85–5.925 GHz) [4], and communication takes place over 100–300 m. DSRC bandwidth contains service channels (SCHs) and a control channel (CCH). Six SCHs, each of 10 MHz, are used for non-safety applications, while one 10 MHz CCH is used for safety application.

IEEE 1609.4 develops two main concepts: control channel (CCH) and time division. CCH transmits safety or high priority messages and beacon signals, while SCH is used to transmit non-safety or low priority messages. The time division concept assumes that devices are synchronized to Universal Coordinate Time (UTC). In time division concept, segmented time is called as "synchronization period" which has a default value of 100 ms each. IEEE 1609.4 follows periodically switching between the channels. Hence, CCH interval (CCHI) is of 50 ms followed by SCH interval (SCHI) of 50 ms. CCHI and SCHI start with 4 ms guard intervals.

Major contributions of this work are as follows:

A collective scheduling algorithm (CoSch) is proposed which contains two subdivisions:

- The priority assignment unit assigns priorities to the messages by taking care of static factor, size of message and dynamic factor. Dynamic factor is calculated with the help of clustering. Static factor classifies messages as the safety and non-safety.
- Message scheduling unit reschedules these messages into control channel queue and service channel queues based on their priority. High priority message or safety messages are rescheduled into control channel queue, while low priority messages or service messages are rescheduled into service channel queues.

The performance is evaluated through simulation using various parameters like number of packet loss, throughput, delay with respect to the number of vehicles and simulation time [5].

This paper further includes related work in Sect. 2, CoSch strategy in Sect. 3, and obtained results are discussed and compared in Sect. 4. Section 5 concludes the paper.

2 Related Works

Congestion control is necessary to make VANET an efficient, safe and reliable system. We must also take care of effective utilization of available bandwidth. In a network, there are two types of congestion control mechanism: (1) open-loop mechanism—it avoids the congestion before it happens, and (2) closed-loop mechanism—it controls the congestion after it happened. In VANET, three classifications of congestion control strategies are as follows: (1) controlling the rate of transmission, (2) controlling the power of transmission and (3) prioritizing and scheduling the messages over a communication channel. This paper discusses third type which is an open-loop mechanism.

Prioritizing is introduced firstly by enhanced distributed channel access (EDCA) [6]. In this, message is divided into four traffic categories from high to low priority order: voice, video, best effort and background, respectively. The high priority messages will have higher chances to be sent because of shorter contention window (CW) size and lesser Arbitration Inter-Frame Space (AIFS) value. EDCA also introduces Transmit Opportunity (TXOP), which is defined as an interval of time when a station has right to start the transmission, defined by a starting time and maximum duration.

A survey of MAC protocols has been done in [7]. Several scheduling algorithms are introduced, and some of them will be discussed here. Maximum request first (MRF) [8] allows scheduling of such messages which are repeatedly appearing in queues. In MRF, there is no valid concept of prioritization of safety over non-safety information. First deadline first (FDF) transmits the most urgent message first, but in this time taken by message while transmitting is not considered. Smallest data size first (SDF) schedules the message which has the smallest size, neglecting the urgency of message.

In context awareness beacon scheduling (CABS) [9], beacon overloading in high vehicle density is taken care by scheduling beacon signal dynamically. CABS appends information like velocity, direction and position in beacon messages for dynamic scheduling. In [10], Uni-Objective Tabu Search (UOTabu) and Multi-Objective Tabu Search (MOTabu) congestion control strategies are introduced in order to make VANET reliable. In this first channel, utilization is observed and Tabu Search algorithm is used for adjusting the transmission rate and transmission range. In UOTabu, minimum delay is considered, while in MOTabu, both minimum delay and minimum jitter are considered as main parameters for tabu search algorithm. MOTabu includes short-term, mid-term and long-term memories in order to determine near-optimal transmission rate and transmission range. These strategies reduce delay and packet loss more than the other strategies.

A congestion control strategy was introduced in which priorities were assigned based on validity and utility of messages and speed of sender and receiver nodes. Then, prioritized message was scheduled. Simulation result shows that the delay of the safety message decreased, but in a worst-case scenario, it is still larger than 50 ms which is a one-time interval for a control channel or a service channel [11].

Above summary of some of the congestion control strategies is discussed with their deficiencies. In this paper, some of the problems are trying to minimize to make a reliable and safe VANET system. Higher throughput, less delay and less packet drop ensure enhancement in existing strategies of congestion control.

3   Proposed Protocol: Collective Scheduling Algorithm (CoSch)

The proposed strategy consists of priority assignment unit and message scheduling unit. Priority assignment unit prioritizes messages by taking effect of static factor, dynamic factor and size of the message. For the calculation of dynamic factor, concept of clustering is used. Message scheduling unit reschedules these messages into service channel and control channel queues. The proposed strategy is open-loop strategy; i.e. it tries to avoid congestion before it happens. Schematics of CoSch are illustrated in Fig. 1.

3.1   Problem Statements and System Model

In VANET system, congestion is the main problem. Congestion control results in minimum delay, no loss in safety packets, minimum loss in non-safety packets, minimum jitter and high throughput. Maintaining quality of service (QoS) is another important task in VANET. An algorithm that takes local information into account is required to correctly assign priorities according to network situations. Hence, a mechanism is required that can prioritize messages on static and dynamic parameters which depends on neighbourhood.

In the proposed system model, each vehicle has an on-board unit (OBU) and contains a transceiver for transmission and reception of message. It works on IEEE802.11p/WAVE 1609 standard. Each OBU has a GPS that provides position, velocity and other GPS data. Cluster members (CM) can broadcast their messages only to CH. CH has two ranges, 300 m to communicate within cluster and 1000 m to communicate to other CHs. Only CHs have the capability of multihop transmission. VANET system model is shown in Fig. 2.

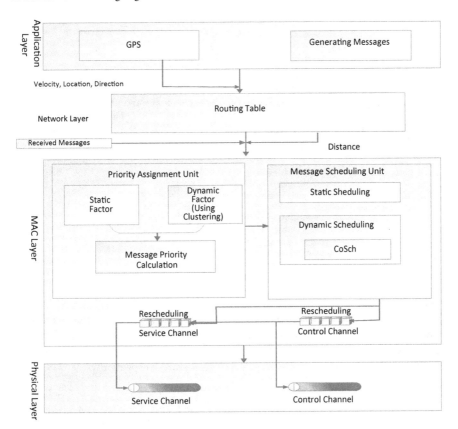

Fig. 1 Schematic of the CoSch

3.2 Priority Assignment Unit

In VANET, the message is generated by the vehicle itself and also received from vehicles which are within range. Then, relative transmission times for each message are defined based on its assigned priorities. Priorities are assigned based on static factor, dynamic factor and size of the message as in Eq. (1).

$$Priority\ of\ \text{message} = \frac{static\ factor \times dynamic\ factor}{size\ of\ the\ message} \tag{1}$$

Hence, the priority of the message is directly proportional to static factor and dynamic factor and inversely proportional to the size of the message. Emergency messages and high priority safety messages are smaller in size compared to service messages, hence high priority.

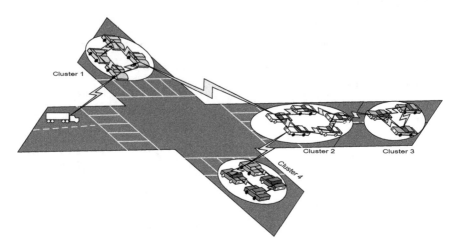

Fig. 2 System model for VANET

Static Factor:
Messages are divided into five categories based on the content of message and type of applications, and values from 1 to 5 are assigned to them [5, 12, 13]. Five categories are: $\text{Priority}_{\text{Service-low}}$, $\text{Priority}_{\text{Service-high}}$, $\text{Priority}_{\text{Safety-low}}$, $\text{Priority}_{\text{Beacon}}$, $\text{Priority}_{\text{Emergency}}$, and values assigned are 1, 2, 3, 4 and 5, respectively.

1 $\text{Priority}_{\text{Service-low}}$ are messages which have very low priority and generated by low priority service applications and rescheduled into service channel. Such messages are instant messaging, parking spot locator, electronic toll payment, etc.
2 $\text{Priority}_{\text{Service-high}}$ are messages generated by high priority service applications and have higher priority but still lesser than safety messages. It is also rescheduled into service channel. Such messages are GPS update, map download, intelligent traffic flow control, etc.
3 $\text{Priority}_{\text{Safety-low}}$ are safety messages, which are rescheduled into control channel. It comes under safety messages but has the lowest priority among all other safety messages. Such messages are alane change warning system, left turn to assist, stop sign assist, forward collision avoidance assist, etc.
4 $\text{Priority}_{\text{Beacon}}$ are messages which are periodically transmitted, and it comes under safety messages because it contains important information like speed, distance, direction, position of the vehicles. It is also rescheduled into control channel. This information is very important to safety and non-safety applications.
5 $\text{Priority}_{\text{Emergency}}$ are messages which have the highest priority, and these messages should be delivered instantly, and no drop in these messages are acceptable. Such messages are emergency brake lights, an emergency vehicle approaching warning, intersection collision warning, pedestrian crossing information, etc.

Dynamic Factor: It is defined according to the state of the network [5, 11]. State of network is based on the velocity of vehicles, usefulness, validity of messages,

Fig. 3 Velocity metric

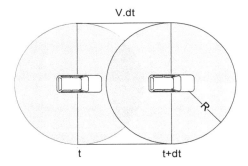

direction of the sender and receiver vehicles and the distance between vehicles. Dynamic factor uses clustering algorithm.

Clustering [14] is used to limit the channel contention and to provide channel access within the clusters. Clustering allows spatial reuse of resources; hence, it enhances network capacity. In clustering, cluster head (CH) is selected based on mobility. Since topology changes frequently hence to maintain a stable cluster, CH is selected. Each node in a cluster sends a status message in the CCH interval periodically. Status message contains some important parameters of corresponding nodes, such as vehicle ID, weighted stabilization factor, velocity, position, range, cluster head ID (CHID), backup cluster head ID (CHBK). The vehicle having higher weighted stabilization factor (WSF) is selected as CH of that cluster. Here, each node will relay its information to the nodes of the own cluster and CH relays that information to all neighbouring clusters. Neighbour cluster's CH will receive this information and broadcast within cluster and all neighbouring clusters and so on. Components of dynamic factor are as follows:

1. Average velocity metric (Vel): It is defined as the relative speed of a cluster moving with average velocity v in small time interval dt and which is relaying its information to nearby clusters which come in its range R. Figure 3 shows velocity metric for a CH. Velocity metric is given as

$$Vel = \frac{\left(\pi R^2 + 2.r.v.dt\right)}{\pi R^2} \qquad (2)$$

Here, R = communication range of cluster (i.e. 1000 m), v = average speed in time dt. If average velocity is high, cluster covers a large distance in unit time. Hence, the probability of disconnection with neighbouring clusters is more, resulting in high value of Vel. If average velocity is low, then there is enough time that this cluster is connected with its neighbour, resulting in low value of Vel.

2. Usefulness metric (Use): It can be seen as the probability of retransmission of message by neighbouring CHs.

$$Use = \frac{Communication\ area}{Overlapped\ area} \qquad (3)$$

Fig. 4 Usefulness metric

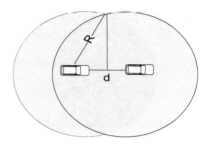

Use can be calculated by calculating overlapped area using Fig. 4.

$$Overlapped\ area = 4 \times \left(across\left(\frac{d}{2 \times R}\right) \times \frac{R^2}{2} - \frac{d}{4} \times \sqrt{R^2 - \left(\frac{d}{2}\right)^2}\right) \quad (4)$$

$$Communication\ area = \pi R^2 \quad (5)$$

Hence,

$$Use = \frac{\pi R^2}{4 \times \left(across\left(\frac{d}{2 \times R}\right) \times \frac{R^2}{2} - \frac{d}{4} \times \sqrt{R^2 - \left(\frac{d}{2}\right)^2}\right)} \quad (6)$$

If overlapped area is high, then there is a high probability that cluster will be within range for some time and message can be again received from neighbouring clusters. Hence, Use is low and for lower overlapped area value of Use is high.

3. Validity metric (Val): It is the ratio of remaining time to the deadline to the transferring time. Transferring time is the estimated time taken while transferring message between two CHs. If deadline time is low means, soon packet is going to be dropped. Hence, priority will be high to avoid packet loss.

$$Val = \frac{Remaining\ time\ to\ deadline}{Transferring\ time} \quad (7)$$

4. Distance metric (Dis): It is the distance between CHs. If distance among them is high, then the probability of disconnection is higher; hence, higher priority is assigned. If distance among them is low, then the probability of being in connection is higher. Hence, lower priority is assigned.

5. Direction metric (Dir): It has two values: (1) Dir = 0 means two clusters coming closer to each other; then, probability of being connected increases hence lower priority assigned, and (2) Dir = 1 means two clusters are driving away from each other; then, probability of being connected is very low so higher priority assigned.

Based on above metrics and their effects, dynamic factor is calculated as:

$$Dynamic\,factor = \frac{Vel \times Use}{(Val + 1) \times Dis}; Dir = 0$$

$$Dynamic\,factor = \frac{Vel \times Use \times Dis}{(Val + 1)}; Dir = 1 \qquad (8)$$

Equation 8 is calculated using Eqs. 2–7. Equation (8) states that dynamic factor or message priority (message priority is directly proportional to dynamic factor) is directly proportional to Vel, Use and inversely proportional to the Val. Val can have zero value, then dynamic factor will be infinite, and to avoid such ambiguities, 1 is added to Val. Dis cannot be zero, and Dis = 0 means accident has occurred. Dir = 0 means two clusters coming closer to each other; hence, distance decreases and thus increases the probability of being in connection, resulting in low priority. Therefore, Dis is taken as inversely proportional to the dynamic factor. Dir = 1 means two clusters are driving away so distance increases; hence, the probability of disconnection is more, resulting in high priority. Therefore, Dis is taken as directly proportional to the dynamic factor.

All the required information to calculate dynamic factor is provided by GPS and the routing table. Static factor, dynamic factor and message size are known to us; hence, priority is calculated and added to the header of packets.

3.3 Message Scheduling Unit

Message scheduling is a challenging task due to special characteristics of the network (i.e. high mobility, high rate of topology change, sharing of channel). An effective message scheduler makes the VANET system reliable and safe. In message scheduling unit, messages are rescheduled into control and service channel queues and sent to the channels. Message scheduling is performed mainly in two steps, i.e. static and dynamic scheduling.

In static scheduling, messages are rescheduled into control channel queues or service channel queues based on apriority calculated by static factor. $Priority_{Service-low}$ and $Priority_{Service-high}$ are rescheduled into service channel queues, while $Priority_{Safety-low}$, $Priority_{Beacon}$ and $Priority_{Emergency}$ are rescheduled into the control channel queues. If the control channel is full, then messages having $Priority_{Safety-low}$, $Priority_{Beacon}$ and $Priority_{Emergency}$ priorities will be rescheduled to service channel queue in order to improve reliability. Figure 5 illustrates the static scheduling process.

Dynamic scheduling can be performed in two ways. (1) using priority of the message: in this, packet is transferred to the control and service channel queues based on their priorities in descending order, and on arrival of the new packet, the queues will be rescheduled based on new and old packet priorities together. Then, the packet is dequeued from service channel queues or control channel queues and then relayed over the channel. This type of method of scheduling is called as dynamic

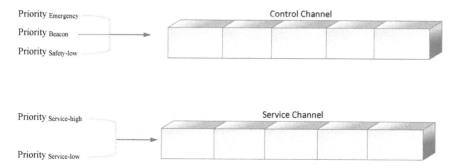

Fig. 5 Static scheduling process

scheduling (DySch). So, DySch is only based on the priority of messages. (2) Using meta-heuristic technique (one of the best meta-heuristic techniques is tabu search algorithm) [15]: in this work, tabu search algorithm is used for dynamic scheduling, hence called as TaSch. TaSch is based on the priority of messages, delay and jitter. Based on these parameters, TaSch finds near-optimal solution of the network quickly so delay and jitter of message delivery are reduced. The equation to calculate delay and jitter is:

$$D(i-1, i) = (R_{i-1} - R_i) - (S_{i-1} - S_i) = (R_{i-1} - S_{i-1}) - (R_i - S_i) \tag{9}$$

$$J(i) = J(i-1) + \frac{(|D(i-1, i)| - J(i-1))}{16} \tag{10}$$

Here, R_i = arrival time of ith packet, S_i = time stamp (when data was transferred) of ith packet, R_{i-1} = arrival time of (i − 1)th packet, S_{i-1} =time stamp of (i-1)th packet, $J(i)$ = jitter of ith packet, $J(i-1)$ = jitter of (i − 1)th packet, $D(i-1, i)$ = difference between transmission time of ith and (i − 1)th packet [2].

Tabu search can have short-term, mid-term and long-term memories. Short-term memories are used to keep best generated optimal solutions. Short-term memories are used in this paper that contains 50 optimal solutions. In case if tabu list is full, the old solution will be deleted from the list [16, 17].

In this paper, the dynamic factor is calculated using clustering technique. Metrics have collective effect of vehicles of a cluster to decide dynamic factor, and then, tabu search algorithm is used for message scheduling. Hence, proposed strategy is named as collective scheduling (CoSch). In CoSch, each vehicle of a cluster sends its information to the CH of that cluster, and then, CH broadcast messages within the cluster and neighbouring CHs are in range. Neighbouring CHs will receive these messages and forward them to CMs and CHs which are in range and so on. The proposed strategy (CoSch) operation is explained through flowchart (Fig. 6).

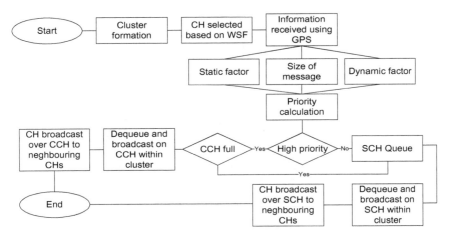

Fig. 6 Flowchart of proposed strategy

Algorithm for CoSch

Step 1: Vehicles form clusters using clustering algorithm.
Step 2: Vehicles receive information using GPS.
Step 3: Message priorities are calculated using equation (1).
Step 4: If high priority messages:
- If CCH queue is full
  - Message goes to SCH queue.
  - Broadcasted over SCH within cluster.
  - CH broadcast it over SCH to neighbouring CHs.
- If CCH queue is not full
  - Message goes to CCH queue.
  - Broadcasted over CCH within cluster.
  - CH broadcast it over CCH to neighbouring CHs.
Step 5: If low priority messages:
- Message goes to SCH queue.
- Broadcasted over SCH within cluster.
- CH broadcast it over SCH to neighbouring CHs.

---

4 Performance Evaluation

Performance is evaluated using network simulator (NS) version 2.34. The simulation parameters of the proposed protocol are presented in Table 1. Each parameter is evaluated five times, and the average is taken as final simulation result. For performance evaluation, these metrics are needed to be defined:

Table 1 Simulation parameters

Parameters	Value
Simulation area	3000 × 4000 m
Number of vehicles	10–100
Vehicle speed	30 km/h
Transmission rate	3 Mbps
Message size	500 bytes
MAC interface	Mac/802_11Ext
Routing protocol	AODV
Simulation time	50–200 s
Transmission range	300, 1000 m
Simulation runs	5

Fig. 7 Delay versus number of vehicles

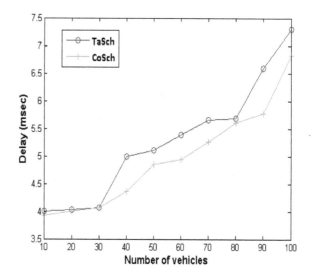

Average Delay: The average time taken by packets to reach from sender to receiver. It should be as minimum as possible.

Average Throughput: Information successfully transmitted per unit time. Its maximum value is desired.

Number of Packet Loss: Number of packets lost or dropped. It should be as minimum as possible.

Packet Loss Ratio: Ratio of number of packet loss to the transmitted number of packets. Its value is in between 0 and 1. Values closer to zero will be better.

Figure 7 shows comparison of simulation results of delay of TaSch and CoSch by varying the number of vehicles. Delay increases with increase in number of vehicles due to high congestion. CoSch is showing less delay than TaSch.

When control channel queue is full, emergency or high priority packets transferred to the service channel queues, results in lower delay and packet loss. Figure 8 shows

Fig. 8 Number of packet loss versus number of vehicles

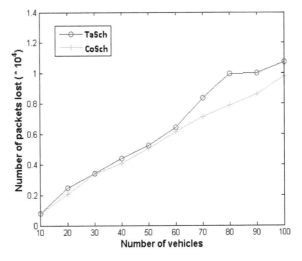

Fig. 9 Throughput versus number of vehicles

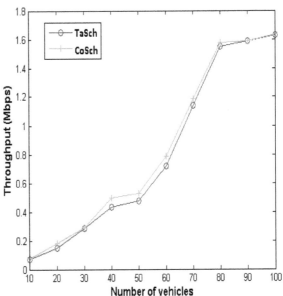

comparison of simulation results of number of packet loss of TaSch and CoSch by varying the number of vehicles. Due to high congestion, drop in message packets increases; hence, number of packet loss increases with increase in number of vehicles. CoSch is showing less packet drop than TaSch.

Figure 9 shows comparison of simulation results of throughput of TaSch and CoSch by varying the number of vehicles. As the number of vehicles increases, there is an increase in successful data transmission among vehicles, and hence, there is an increase in throughput. Minimization in delay and number of packet loss in

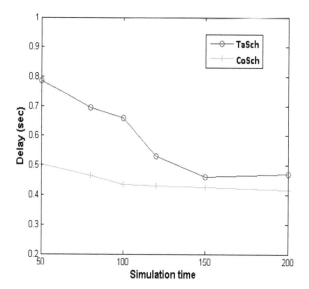

Fig. 10 Delay versus simulation time

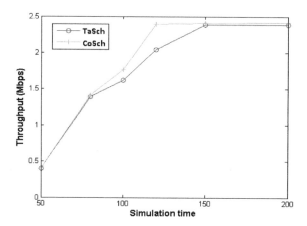

Fig. 11 Throughput versus simulation time

CoSch compared to TaSch results in high throughput in CoSch compared to TaSch as shown below. Figure 10 shows comparison of simulation results of delay of TaSch and CoSch by varying simulation time. CoSch is showing less delay than TaSch. Figure 11 shows comparison of simulation results of throughput of TaSch and CoSch by varying simulation time. CoSch is showing better result than TaSch.

5 Conclusion

In this paper, a scheduling algorithm is proposed to control the congestion. Better congestion control is achieved by collectively prioritizing the messages and scheduling. Message priority was calculated based on static factor and dynamic factor with the concept of clustering and size of the message. These prioritized messages are sent to the control channel queues or service channel queues. Then, these messages are dequeued as a packet using meta-heuristic technique, i.e. tabu search algorithm. Simulation is carried out to illustrate that CoSch is better than TaSch due to collective use of information. Results are compared in terms of number of packet loss, throughput and delay. Hence, more safe and reliable VANET system is implemented.

References

1. Yi, Q., Lu, K., & Moayeri, N. (2008). A secure VANET MAC protocol for DSRC applications. In *Global Telecommunications Conference*, Nov 2008 (pp. 1–5). https://doi.org/10.1109/glocom.2008.ecp.376.
2. IEEE draft amendments for wireless access in vehicular environment (WAVE). IEEE 802.11p/D5.0, April 2009.
3. Kenney, J. B. (2011). Dedicated short-range communications (DSRC) standards in the United States. *Proceedings of the IEEE, 99*(7), 1162–1182. https://doi.org/10.1109/JPROC.2011.2132790.
4. Ghosh, T., & Mitra, S. (2012). Congestion control by dynamic sharing of bandwidth among vehicles in VANET. In *12th International conference on Intelligent Systems Design and Applications (ISDA)* (pp. 291–296). https://doi.org/10.1109/isda.2012.6416553.
5. Taherkhani, N., & Pierre, S. (2016). Prioritizing and scheduling messages for congestion control in vehicular ad hoc networks. Comput. Netw. *108*, 15–28. https://doi.org/10.1016/j.comnet.2016.06.027.
6. Torrent-Moreno, M., Jiang, D., & Hartenstein, H. (2004). Broadcast reception rates and effects of priority access in 802.11-based vehicular ad-hoc networks. In *Proceedings of the 1st ACM International Workshop on Vehicular Ad Hoc Networks* (pp. 10–18). https://doi.org/10.1145/1023875.1023878.
7. Gupta, N., Prakash, A., & Tripathi, R. (2015). Medium access control protocols for safety applications in vehicular ad-hoc network: A classification and comprehensive survey. *Vehicular Communications, 2*(4), 223–237. https://doi.org/10.1016/j.vehcom.2015.10.001.
8. Kumar, V., & Chand, N. (2010). Data scheduling in VANETs: A review. *International Journal of Computer Science and Communication, 1,* 399–403.
9. Bai, S., Oh, J., & Jung, J.-I. (2013). Context awareness beacon scheduling scheme for congestion control in vehicle to vehicle safety communication. *Ad Hoc Network, 11,* 2049–2058. https://doi.org/10.1016/j.adhoc.2012.02.014.
10. Taherkhani, N., & Pierre, S. (2012). Congestion control in vehicular ad hoc networks using meta-heuristic techniques. In *Proceedings of the Second ACM International Symposium on Design and Analysis of Intelligent Vehicular Networks and Applications* (pp. 47–54). https://doi.org/10.1145/2386958.2386966.
11. Bouassida, M. S., & Shawky, M. (2008). On the congestion control within VANET. In *1st IFIP Wireless Days, WD'08* (pp. 1–5). https://doi.org/10.1109/wd.2008.4812915.
12. Meghdadi, V. (2011/2012). Vehicular ad-hoc networks (VANETs) applied to intelligent transportation system (ITS).

13. Kargl, F. (2006). Vehicular communications and VANETs. In *23rd Chaos Communication Congress*.
14. Hafeez, K. A., Zhao, L., Mark, J. W., Shen, X., & Niu, Z. (2013). Distributed multichannel and mobility-aware cluster-based MAC protocol for vehicular ad hoc networks. *IEEE Transactions on Vehicular Technology, 62*(8), 3886–3902. https://doi.org/10.1109/TVT.2013.2258361.
15. Zapfel, G., Braune, R., & Bogl, M. (2010). Meta-heuristics based on solution construction. *Meta-heuristics search concepts* (pp. 75–93) Springer. https://doi.org/10.1007/978-3-642-11343-7_5.
16. Glover, F., & Laguna, M. (1997). *Tabu search* (pp. 2–151). Boston, MA: Kluwer.
17. Gendreau, M., Glover, F., Kochenberger, G.(2003). An introduction to Tabu search. *Handbook of meta-heuristics* (pp. 37–54). Boston: Kluwer. https://doi.org/10.1007/0-306-48056-5_2.

Trusted-Differential Evolution Algorithm for Mobile Ad Hoc Networks

Shashi Prabha and Raghav Yadav

Abstract Mobile ad hoc networks are established and deployed spontaneously without any infrastructure in geographical area. The performance of network is satisfied only when all the member nodes have intensity to work in cooperative manner. But due to lack of any centralized unit, it is vulnerable to various attacks of malicious nodes. To overcome these types of attacks, the network has to be enhanced to provide secure delivery services. Our proposed Trusted-Differential Evolution algorithm deals with malicious node and inhibits them to become a member of data transmission path. It has two components: one to find the fittest path and other to deal with fluctuating credibility of nodes through trust. The dynamic of trust is handled by new trust-updation scheme along with punishment factor for malicious node. The proposed algorithm is compared with DSR and genetic algorithm.

Keywords Differential evolution · Trust · Punishment factor · Fitness function
Genetic algorithm · Objective function

1 Introduction

The aim of mobile ad hoc networks is to provide wireless network facility without any centralized unit, in distribution form. The functionality of mobile ad hoc networks is dependent on the collaboration of all participants in an honest form. Therefore, trust is an important driving factor for this collaboration in order to secure the MANET. Due to lack of centralized management unit and dynamic topology, MANET frequently undergoes various types of vulnerability by malicious node. The main targets of these attacks are data, bandwidth, routing protocols, and battery power. In actual, various

S. Prabha (✉) · R. Yadav
Department Computer Science and IT, Sam Higginbottom University of Agriculture,
Technology and Sciences, Allahabad 211007, UP, India
e-mail: shashi17feb@gmail.com

R. Yadav
e-mail: raghav.yadav@shiats.edu.in

© Springer Nature Singapore Pte Ltd. 2019   181
A. Khare et al. (eds.), *Recent Trends in Communication, Computing,
and Electronics*, Lecture Notes in Electrical Engineering 524,
https://doi.org/10.1007/978-981-13-2685-1_19

attacks are classified on the basis of attacker types such as external and internal attacker. The external attackers attempt to interrupt the network by injecting fallacious routing information. They can attempt to create partition in the network (wormhole attack), routing loops, or fake-functional routes. Another class of attacker is internal attacker wherein legitimate nodes intend to be jeopardized by malicious node. They promote imprecise routing statistic in order to distort the flow of information in wireless networks, formation of gray hole, black hole, Sybil attack and modification. One of the security challenges emerges due to mobility. Actually, mobile nodes very frequently fall in and out of the frequency range which results in having highly dynamic topology. In order to intensify the security of the network and minimize the threat from malicious node, it is significant to estimate the trustworthiness of other member nodes to have trusted environment without depending on any central unit. The trust mechanism empowers the node to rate the trust level of other nodes, which not only assist to enhance the security but also detect the presence of malicious node. In simple words, with the help trust this mechanism, nodes can decide whether and to which extent they can rely on other nodes for communication. Trust guides node to keep away any type of communication with low-trusted nodes. The proposed algorithm Trusted-Differential Evolution (TDE) provides secure data transmission path, and its components are fittest path model and trust-update model. The fittest path model is differential evolution-based methodology to search all possible feasible paths and opt for the best-suited solution with the assistance of effective fitness function based on not-to-trust factor. Another model updates trust on account of previous trust and current trust with punishment factor to associate the effect of malicious activity in terms of packet dropping.

Differential evolution (DE) was proposed by Storn and Price [1] in 1996. The basic idea is to retrieve the best-suited solution of an optimization problem. Actually, to solve any optimization problem, one has to identify factor variable on which whole system deviates, the selection of appropriate variable result to have best-suited optimization solution. These variables are framed into objective function, and the calculated objective function value decides whether the solution is best or not. DE algorithm proceeds in four steps: initialization, mutation, crossover, and selection. Here initialization takes place only once, whereas rest of three executes in iterative manner till the termination condition is met. DE owns two control parameters (F and Cr) which can affect the results. Both parameters have to be tuned according to the application; their values can be generated by some self-adaptive strategy or kept fixed. When it comes to MANET, to search secure data dissemination path becomes hectic procedure since it is vulnerable to various attacks. Differential evolution is a simple mathematical model of complex evolutionary process which can facilitate to search secure path for data dissemination. In this paper, a differential evolution-based approach effectively searches the fittest path; this optimization model tends to minimize the not-to-trust factor and quantifies path reliability.

The remaining paper is organized as in the following sections; Sect. 2 focuses background studies on trust and DE application in MANET, Sect. 3 explains the proposed Trusted-Differential Evolution algorithm, Sect. 4 discusses analysis of simulation results, and Sect. 5 concludes the proposed work. Note that objective function and fitness function have similar meaning.

2 Background Studies

In the last few years in recent scientific journals, it has been noticed that very small amount of research had been conducted on DE-based work to tackle mobile ad hoc networks issues. On the basis of conducted survey, the applications of DE-based algorithm are categorized as follows: optimum path, routing protocol, and topology control mechanism. In Table 1, DE-based applications in the field of mobile ad hoc networks are listed. DE has various mutation strategies, but according to the application, best-fitted strategy is applicable only. Such as Chakraborty et al. [2] selected DE/best/1 as a suitable mutation strategy for clustering to elect cluster head. Mutation strategy DE/rand/1 is another variant; it selects random vector candidates to undergo mutation. Table 1 has listed both the mutation variants in different perspective, that is, according to the application.

For mobile ad hoc networks, trust turns out to be a decision policymaker for secure communication. Eventually in context of MANET, trust can be understood as subjective probability, belief, and risk. Trust reveals the opinion or confidence on integrity, honesty, and truthfulness about target node's future behavior. Additionally, trust is a security aid against misbehavior of targeted node; it is classified into two types of trust: direct and indirect trust. Wherein trust computation varies in two ways: centralized or distributed. In centralized trust computation technique, a central agent computes the node trust, and in distributed trust computation, node calculates own trust on neighbor node. A survey article by Govindan and Mohapatra [3] on trust computation and trust dynamics significantly covered exhaustive amount of research. Xia et al. [4] proposed trust prediction model to identify malicious activity; it accumulates trust computation for the previous activity of node to predict the most secure route. Tremendous work has been done to reveal the significant influence of trust while selecting secure route for data delivery in mobile ad hoc networks. Table 2 shows noteworthy outlook of handling trust computation. This table focuses on the different trust solutions to handle malicious node activity; few authors have addressed direct trust calculation, few on indirect, and few have given solution using direct and indirect trust both. In our proposed work only direct trust is considered, with previous trust consideration, and punishment factor is used to tackle malicious activity.

3 The Proposed Trusted-Differential Evolution Algorithm

To imbibe security in MANET is a thorniest job, but this proposed algorithm concentrates on best possible trusted path for effective network performance in secure

Table 1 Differential evolution with MANET

Author, Description	Methodology		
	Fitness function	Chromosome encoding	Mutation strategy
Gundry et al. [7], TCM-DE (future placement of nodes), topology control mechanism	Inspired from diffusive properties of liquid and gases. Total Sum of virtual forces	Single node. Dimensions: speed and direction	DE/rand/1/bin
Gundry et al. [8], TCM-Y (fault-tolerant situation to maintain minimum number of connective nodes), topology control mechanism	Yao graph inspired fitness function	Single node. Dimensions: speed and direction	DE/rand/1/bin
Chakraborty et al. [2], (clustering) Novel cluster encoding technique	Distance, degree, and remaining battery power	Single cluster	DE/best/1
Ahn and Ramakrishna [5], optimal path (GA)	Link matrix	path	Flipping
Ren et al. [9], optimal path (GA-based)	Link matrix	path	Flipping
Anjum and Mohammed [10], optimal path (wireless sensor networks)	Transmission cost	path	DE/rand/1
Yetgin et al. [11], multi-objective-based DE algorithm	End-to-end delay and energy cost metrics	path	DE/rand/1

manner. The proposed model explores the secure path with the assistance of trust, to inhibit the participation of malicious node. The components of Trusted-Differential Evolution algorithm are fittest path model and trust-updation model.

3.1 Fittest Path Model

A cumbersome situation is to encode mobile nodes as the suitable input for differential evolution-based algorithm. The given MANET is considered as graph $G = (V, e)$, where V represents vertex and e represents links between vertexes. The all possible paths are encoded in the form of chromosome where first and last genes are allotted for source and destination; one node can participate once and further chromosome

Table 2 Trust model strategies

Author and year	Mechanism	Trust schemes	Attack	Result analysis and metrics involved
Subramaniam et al. [12]	Trust: improve network lifetime by new metrics that are residual battery power. Trust aging factor: exponential time decay	Direct and indirect	Selfish and malicious node	Metrics: expected energy cost metrics, EXA metrics
Wei et al. [13]	Trust to handle uncertain reasoning to enhance the security. Punishment factor: reputation fading	Direct trust: based on Bayesian inference rule. Indirect trust: derived by Dempster–Shafer theory (DST)	Malicious node	Metrics: delay, PDR, throughput, and routing overhead. More accuracy of trust can be identified
Sun et al. [14]	Associated mobility to minimize uncertainty in two different methods: proactive and reactive. Both methods collect all trust evidence and information	Trust to tackle uncertainty with three basic components in it {belief, disbelief, uncertainty}. Trust in the form of first hand (as opinion) and second hand (as a recommendation). Bayesian inference system for evaluation	Misbehaving node	Metrics: authentication probability, convergence time, average delay or cost, average uncertainty. Certainty-model identity misbehaving node by good detection rate mechanism
Xia et al. [4]	Trust-based source routing in MANET. Trust prediction model: based on Fuzzy logic	Historical trust, current trust, and route trust of node	Black hole attack and malicious node	Metrics: end-to-end delay, packet delivery ratio, route optimality, routing overhead, throughput, detection ratio. Reliable packet delivery by inhibiting malicious node from path
Xia et al. [15]	Predicts node's trust level by enhanced Gray–Markov chain model	Trust as subjective reputation and indirect reputation	Malicious node	Metrics: detection ration, trust prediction, intrinsic trust value. Effective against anti-jamming

length can vary as all paths cannot have similar number of intermediate node. A symmetric matrix is maintained to acquire all link information during network lifetime t. The link between node x and y is expressed by e_{xy} and described as:

$$e_{x,y} = \begin{cases} 1 * T_{x,y} & link\ exist \\ 0, & otherwise \end{cases} \tag{1}$$

where $T_{x,y}$ is trust of node x on node y; each node maintains the trust information of its own for neighbor nodes. Nodes x and y are neighbor nodes where x evaluates its trust on y through formulation of trust and trust-update model. There is possibility of node x moving away from node y during sometimes, in that case no link is considered in between them, which hopefully requires no trust evaluation because the proposed trust-update model takes direct trust evaluation into account.

3.1.1 Formulation of Trust

In MANET, it can be expressed as ratio of successful packet forwarding to the total number of packet forwarded.

$$T_{x,y} = P_s/P_T \tag{2}$$

where P_s represents successful forwarded packet and P_T represents total number of packet to be forwarded. The packets are categorized as data packet and control packet. Therefore accordingly, this forwarding packet ratio is classified into data packet (DT) and control packet forwarding (CT). Propagated trust generated through each interaction between two nodes is formulated as:

$$rT_{x,y} = \alpha * DT_{x,y} + \gamma CT_{x,y} \tag{3}$$

where $\alpha, \gamma = (1 - \alpha) \geq 0$ are corresponding weights assigned to data packet and control packet trust ratio, respectively.

3.1.2 Fitness Function

Fitness function is mapped according to the application requirement as the quality of generated trial vector is judged on the basis of fitness function value. In case of MANET to search secure path, not-to-trust factor is mapped in the fitness function. The fitness function is given as follows in Eq. (4):

$$rT'_{x,y} = 1 - rT_{x,y} \tag{4}$$

Here, $rT'_{x,y}$ is not-to-trust factor of node x on node y. In trusted-DE algorithm, trust is mapped in terms of not-to-trust factor since minimization of fitness function is considered for the evaluation. In other words, not-to-trust factor is more effective when clubbed with minimization. Minimum value of not-to-trust factor results in having maximum certainty on trust. Now, the defined fitness function is as follows:

$$f = \sum_{\substack{s = i, \\ j = i+1 \\ i \in R}}^{D-1} rT'_{i,j} + \delta * h \tag{5}$$

Here, δ is trust threshold to add influence of malicious node, h is hop count, R indicates a set of member intermediate node between source (S) and destination (D), and $rT'_{i,j}$ is not-to-trust factor for nodes i and j.

3.2 Trust-Update Model

After every interaction, the trust level of entity varies; this trust-update model encodes these variations of each candidate nodes. In other words, trust-update model effectively inculcates the dynamics of trust. This model functions on the aggregation of previous and current trust with considering the effect of malicious node in terms of packet dropping ratio. In Eqs. (6) and (7), aggregation of previous and current is formulated as follows:

Case 1: In complete trust situation when $T^{cur}_{AB} = 1$ and no malicious activity of packet dropping

$$T^{new}_{AB} = \varphi T^{pre}_{AB} + (1 - \varphi)T^{cur}_{AB} \text{ if } T^{cur}_{AB} = 1 \tag{6}$$

Case 2: When current trust fluctuates due to the presence of packet dropping situation

$$T^{new}_{AB} = \begin{cases} \varphi T^{pre}_{AB} + (1 - \varphi)T^{cur}_{AB} * (1 - \beta), & \text{if } (P_f \leq P_s) \\ \varphi T^{pre}_{AB} + (1 - \varphi)T^{cur}_{AB} * \left(\frac{1-\beta}{1+\beta}\right), & \text{if } (P_f > P_s) \\ \varphi T^{pre}_{AB}, & \text{if } (T^{cur}_{AB} = 0) \wedge (P_s = 0) \end{cases} \tag{7}$$

Here, T^{new}_{AB} is the updated trust of node A on B. Similarly T^{cur}_{AB} and T^{pre}_{AB} are current and previous earned trust. Symbol "φ" is a trust balancer whose values are assigned from 0 to 1. The total number of successful packets and unsuccessful packets delivered are P_s and P_f, respectively. Punishment factor β is a ratio of unsuccessful packets and successful packet delivered which is expressed as P_f/P_s. This factor reduces the updated trust by including the influence of malicious packet dropping activity, and if the trust value falls below the threshold, node actively blacklists that neighbor node in its list. All nodes maintain the trust value in its own trust table. Each node calculates and updates its trust on other node by the given Eqs. (6) and (7).

3.3 Trusted-Differential Evolution Algorithm

Algorithm 1: Trusted- Differential evolution (TDE)
1.   Initialize: nodes of MANET as Population
2.   Initialize value of control parameters
             F=1 and Cr=0.8, , $\delta > 0.4$
3.   Encode the path
4.   Initialize every feasible solution for each candidate
5.   ITR=1
6.   While path_fitness < path_max_fitness
7.   Mutation: To explore feasible alternative path and produce donor vector through equation (8)
8.   Crossover: To search every possible partial path and Generate trial vector through (9)
9.   Selection: Comparison trial vector with target vector to select the best-suited vector among them by equation (10)
10.   ITR=ITR+1
11.   End
12.   Data transmission through path R
13.   If $(T_{AB}^{cur} = 1)$
14.   Update trust of each participant node by equation (6)
15.   Update trust by equation (7)

The member nodes of MANET are assumed as the population for Trusted-Differential Evolution (TDE) algorithm. The population is expressed by $G_p = [\overrightarrow{x_{1,G}}, \overrightarrow{x_{2,G}}, \ldots \overrightarrow{x_{N,g}}]$, where G_p represents population at G_{th} generation and $x_{i,G}(i = 1, 2, 3, 4, \ldots n)$ is an n-dimensional vector. Biologically, sudden variation in genetics is termed as mutation. In DE, mutation process is mathematically modeled to produce new genetic material known as donor vector. The donor vector $\overrightarrow{D_{i,G}}$ is produced against every target vector $\overrightarrow{x_{a,G}}$ selected from the current generation. The mutation strategy adopted for TDE is "DE/rand/1", referred in Eq. (8) as follows:

$$\overrightarrow{D}_{i,G} = \vec{x}_{r1,G} + F * \left(\vec{X}_{r2,G} - \vec{X}_{r3,G} \right) \tag{8}$$

where r1, r2, and r3 are randomly selected distinct values from population. Physically, mutation process generates all alternative feasible paths from mutant node to the destination node. The next step is crossover; this step is executed after mutation. The few genes of donor vector $\overrightarrow{D_{i,G}}$ are interchanged with the component of target vector to produce offspring vector $\overrightarrow{U}_{i,j,G}$. The crossover operation is performed by two methods binomial and exponential for TDE binomial crossover executed by the following Eq. (9).

$$\vec{U}_{i,,G} = \begin{cases} \overrightarrow{D_{i,,G}}, & \text{if } rand\,(0,\,1) \leq Cr \\ \overrightarrow{x_{i,G}}, & \text{otherwise} \end{cases} \tag{9}$$

$$\overrightarrow{x}_{i,G+1} = \begin{cases} \vec{U}_{i,G} & \text{if } f\left(\vec{u}_{i,g}\right) \leq f\left(\vec{X}_{i,g}\right) \\ \vec{x}_{i,G}, & \text{otherwise} \end{cases} \tag{10}$$

where rand is uniformly distributed random values selected from the range (0 1), and Cr is the crossover rate. Now, the generated vector is compared with target vector in selection phase and best one is chosen. The whole algorithm executes in iterative manner to search secure route, and for reference, TDE is explained in algorithm 1.

4   Simulation Setup and Result Analysis

In this section, the Trusted-DE algorithm is compared with genetic algorithm [5] and dynamic source routing [6] through computer simulations. All the simulations were executed on MATLAB 2013(a) on Microsoft Windows 7 machine with configuration 4 GB RAM CORE i5 and 2.2 GHz processor. The simulation parameters are listed in Table 3.

For this application, $\alpha = 0.5$, $\gamma = 0.5$, $\varphi = 0.4$, and $\delta = 0.4$. Initially, the trust level of each node is set to 0.5. The network size of 50 nodes is investigated with varying number of malicious nodes 2, 4, 6, 8, 10, and 12. The number of nodes is considered as the population of the proposed TDE, and almost 500 possible topologies are considered for each case.

Table 3 Simulation parameters

Parameter	Values
Simulation time	100 s
Map size	1000 m × 1000 m
Mobility model	Random waypoint
Traffic type	Constant bit rate (CBR)
Pause time	5 s
Transmission radius	250 m
Connection rate	4 pkts/s
Packet size	512 bytes

Fig. 1 Computation time (s) versus malicious node

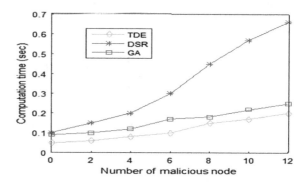

4.1 Result Analysis

A. Computation time: Time to search the eligible route for transmission. Simulation executes for 50 numbers of nodes in mobile ad hoc networks with varying number of malicious nodes 2, 4, 6, 8, 10, and 12. Simulation runs until search does not find most trusted path to destination. It works for every possible feasible path and compares to get best-suited path (chromosome). From experimental results shown in Fig. 1, TDE requires less computational time and does not deteriorate with increased number of malicious node. Although computation time should increase with number of increasing malicious nodes, the mechanism behind TDE is different from any traditional protocol, actually in iterative manner with various interpretations of chromosome suitable routes are found. The average computation time for TDE is about 0.134 s, and for other algorithm, it is approximately 0.347 s for dynamic source routing (DSR) and 0.161 s for GA (depicted in Fig. 1).

B. Packet delivery ratio: Percentage of total number of data packets delivered successfully to the destination. With no malicious node, the packet delivery loss ratio is about 3% and it gradually increases with increased number of malicious node up to 22%. As shown in Fig. 2, the packet delivery ratio degrades gently for TDE, whereas for DSR it degrades sharply. The base component not-to-trust factor directs the algorithm to search secure path without any malicious node; therefore, TDE has less packet delivery ratio. Still, packet drops about 22% with 12 number of malicious node; actually the reason behind this behavior is participation of low-trust node. The packet delivery ratio is about 97–78% for TDE.

C. Simulation for random topologies of mobile ad hoc networks: In general, the performances of any soft-computing algorithm are judged on the basis of convergence speed and quality of solution with maintaining diversity. Here, with varying number of nodes (20–80), we justify that TDE owns good scalability, through quality of solution. The numbers of nodes are considered as population size of each execution of TDE, and almost on 500 possible topologies are considered for each case. In Fig. 3, the quality of solutions for GA is compared with

Fig. 2 Packet delivery ratio versus number of malicious node

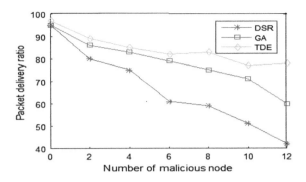

Fig. 3 Comparison of quality of solution

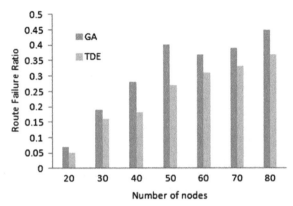

Table 4 Quality of solution to compare performance

Performance metric	Algorithms	Genetic algorithm	TDE
Route failure (Ratio)	Average	0.307	0.23
	Standard deviation	0.135	0.112

TDE algorithm and the results in the figure prove that TDE perceives better quality of solution. In the case of 40 nodes, TDE outperforms GA with approximate probability of 0.2. Furthermore, for network with 80 nodes, TDE outperforms with probability 0.08 (GA). The recorded statistics shows that route failure ratio (route optimality $= 77\%$) of TDE is about 0.23 in Table 4. Moreover, the standard deviation for TDE is 0.11 which is obliviously less than standard deviation of GA. That means with dynamically changing topologies, TDE retains its robustness to attain similar quality of solutions.

TDE outperforms GA due to the difference vector scheme in mutation strategy which enhances the search capability.

5   Conclusion

It is already known that MANETs are deployed without any central management and require full cooperation to achieve effective performance. In this paper, TDE ensures the full cooperation of nodes for successful data delivery through secure path by inhibiting malicious node; it shows that packet delivery ratio is better than GA and DSR in Fig. 2. The not-to-trust component guides to have secure and shortest route (with minimum hop count), and trust-updation model with the assistance to punishment factor lowers the node's current trust. Results discussed in Table 4 show that TDE is also scalable, as with increasing network density (20–80) have low standard deviation is 0.11. The trust is main building block of TDE, and through empirical results, it shows more effectiveness than traditional routing DSR and soft-computing algorithm GA.

References

1. Storn, R., & Price, K. (1995). *Differential evolution—A simple and efficient adaptive scheme for global optimization over continuous spaces*. Technical Report TR-95-012, ICSI.
2. Chakraborty, U., Das, S., & Abbott, U. (2011). Clustering in mobile ad hoc networks with differential evolution. In *IEEE Congress on Evolutionary Computation (CEC)* (pp. 2223–2228).
3. Govindan, K., & Mohapatra, P. (2012). Trust computation and trust dynamics in mobile ad hoc networks: A survey. *IEEE Communication Surveys & Tutorials, 14*(2). Second Quarter (2012).
4. Xia, H., Jia, Z., Li, X., Ju, L., & Sha, E. H.-M. (2013). Trust prediction and trust-based source routing in mobile ad hoc networks. *Ad Hoc Networks, 11*, 2096–2114 (2013).
5. Ahn, C. W., & Ramakrishna, R. S. (2002). A genetic algorithm for shortest path routing problem and the sizing of populations. *IEEE Transactions on Evolutionary Computation, 1*(1), 511–579.
6. Johnson, D., & Maltz, D. (1996). Dynamic source routing in ad-hoc mobile wireless networks. In I. Tomasz, K. Hank (Eds.), *Mobile computing* (pp. 153–181) (1st ed.). Kluwer Academic Press.
7. Gundry, S., Zou, J., Kusyk, J., Uyar, M. U., Sahin, C. S. (2012). Fault tolerance bio-inspired topology control mechanism for autonomous mobile node distribution in MANETS. In *Military Communications Conference, 2012 Milcom*. https://doi.org/10.1109/milcom.2012.6415743.
8. Gundry, S., Kusyk, J., Zou, J., Sahin, C. S., Uyar, M. U. (2013). Differential evolution based fault tolerant topology control in MANETs. In *Military Communications Conference, MILCOM 2013*.
9. Ren, J., Wang, J., Xu, Y., Cao, & L. (2015). Applying differential evolution algorithm to deal with optimal path issues in wireless sensor networks. In *IEEE International Conference on Mechatronics and Automation (ICMA)*. https://doi.org/10.1109/icma.2015.7237748.
10. Anjum, A., & Mohammed, G. N. (2012). Optimal routing in ad-hoc network using genetic algorithm. *International Journal of Advanced Networking and Applications, 03*(05), 1323–1328.
11. Yetgin, H., Cheung, K. T. K., & Hanzo, L. (2012). Multi-objective routing optimization using evolutionary algorithms. In *IEEE Wireless Communication and Networking Conference (WCNC)*. https://doi.org/10.1109/wcnc.2012.6214324.
12. Subramaniam, S., Saravanam, R., & Prakash, P. K. (2013). Trust based routing to improve network lifetime of mobile ad-hoc networks. *Journal of Computing and Information Technology CIT, 21*(3), 149–160.
13. Wei, Z., Tang, H., Yu, R. F., Wang, M., & Mason, P. (2014). Security enhancements for mobile ad hoc networks with trust management using uncertain reasoning. *IEEE Transaction on Vehicular Technology, 63*(9) (2014).

14. Sun Y. L., Yu, W., Han, Z., & Liu, K. J. R. (2006). Information theoretic framework of trust modeling and evaluation for ad hoc networks. *IEEE Journal on Selected Areas in Communications, 24*(2) (2006).
15. Xia, H., Wang, G.-D., & Pan, Z.-K. (2016). *Node trust prediction framework in mobile ad hoc networks* (pp. 50–56). IEEE: Trustcom/BigDataSE/ISPA.

Redundancy Elimination During Data Aggregation in Wireless Sensor Networks for IoT Systems

Sarika Yadav and Rama Shankar Yadav

Abstract Internet of things (IoT) has emerged as a natural evolution of environmental sensing systems such as wireless sensor networks (WSNs). Wireless sensor nodes being resource constrained in terms of limited energy supply through batteries, the communication overhead and power consumption are the most important issues for WSNs design. The sensor nodes have non-replaceable battery with limited amount of energy, which determines the sensor node's life as well as network lifetime. A key challenge in the design as well as during the operation of WSNs is the extension of the network lifetime even in harsh environment. This paper presents a review on various data aggregation techniques to reduce redundancy and proposes an approach for redundancy elimination during data aggregation exploiting support vector machine (SVM) so that network lifetime can be prolonged during communication of information from sensor nodes to base station. The proposed approach is simulated, and the results are analyzed in terms of average packet delivery ratio, average residual energy, accuracy, and aggregation gain ratio with state-of-the-art techniques.

Keywords Internet of things (IoT) · Wireless sensor networks (WSN) · Redundancy · Data aggregation · Support vector machine (SVM) · Aggregation head

1 Introduction

Internets of things (IoT) refer to the interconnection of uniquely identifiable embedded computing-like devices within the existing Internet infrastructure [1]. In future,

S. Yadav (✉) · R. S. Yadav
Department of Computer Science and Engineering, Motilal Nehru
National Institute of Technology, Allahabad 211004, India
e-mail: sarikayash18@gmail.com

R. S. Yadav
e-mail: rsy@mnnit.ac.in

© Springer Nature Singapore Pte Ltd. 2019
A. Khare et al. (eds.), *Recent Trends in Communication, Computing,
and Electronics*, Lecture Notes in Electrical Engineering 524,
https://doi.org/10.1007/978-981-13-2685-1_20

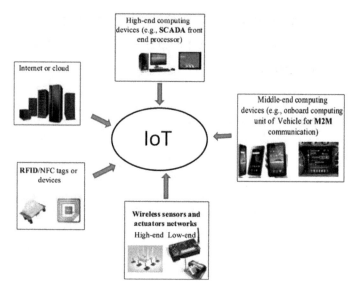

Fig. 1 Essential components of IoT [6]

WSNs are expected to be integrated into the "Internet of things," where sensor nodes join the Internet dynamically and use it to collaborate and accomplish their tasks. The fully functional IoT integrates four essential components [2]: wireless sensor networks (WSN), machine-to-machine (M2M) communications, radio frequency identification (RFID), and supervisory control and data acquisition (SCADA) as shown in Fig. 1. IoT, featuring a powerful sensing ability and wide coverage of detecting area, can make a real-time monitoring over the physical surroundings exploiting WSN integrated with M2M devices to receive round-the-clock information.

A wireless sensor network (WSN) consists of a large number of tiny and low power sensor nodes, which are arbitrarily or manually deployed through an unattended target region [3]. WSNs have diverse applications in environment monitoring, disaster warning systems, healthcare, defense exploration, and surveillance systems [4]. Majority of the applications deploy unattended wireless sensor nodes with non-replenished limited battery for energy supply to remain operational for a longer period of time.

The performance of WSN relies on three factors, i.e., data collection and aggregation, clustering, and routing [5] throughout the deliverance of sensed data from wireless sensor nodes to BS. The wireless sensor nodes disburse most of their energy in aggregation and communication of sensed data which reduce the network lifetime [5]. Since the wireless sensor nodes are distance and energy constrained, it is inefficient for all the wireless sensor nodes to transmit the data directly to the base station. Data collection and aggregation is employed for regular compilation of sensed data from multiple sensor nodes to be transmitted to the base station for processing ultimately. However, if multiple wireless sensor nodes are sensing same event, the data will be redundant and massive. Data redundancy means the duplicate data sensed by numerous nodes deployed in the event detected region. Eliminating redundancy is a major challenge in WSNs as too much redundancy means all nodes send the data to the BS, energy will be wasted, and node energy will deplete early. Hence, we need such aggregation techniques for combining redundant data into valuable information at the wireless sensor nodes which can reduce the total number of packets transmitted to the base station as estimated reduction in energy consumption and bandwidth.

In this paper, we are concern with the comparative study of various data collection and aggregation techniques that are proposed in the literature. We propose an approach to reduce redundancy during data collection and aggregation by classifying the multiple data copies using support vector machine (SVM). The rest of the paper is organized as follows: In Sect. 2, we confer the comprehensive literature review of state-of-the-art techniques and their comparative study. Section 3 presents the problem description and the proposed approach. The performance evaluation of our proposed approach and competing algorithms is given in Sect. 4. Finally, conclusion and future work is discussed in Sect. 5.

2 Literature Review

Data aggregation techniques make the sensor nodes of the network to be resource-aware, i.e., adapt their communication and computation to handle and transmit huge amount of sensed data irrespective of resource constraints [7].

Figure 2 presents a classification of data aggregation techniques as per their implementation objective. From the classification, it is observed that various aggregation techniques are developed for specific application and explicit objective such as enhancing reliability, reduction in data size, maintaining precision, reducing transmission energy.

Data redundancy in WSN is identified as of two types: spatial redundancy and temporal redundancy [8]. Spatial redundancy means the prospect to acquire data from a specific place/location from different sensor nodes. It is based on the deployment

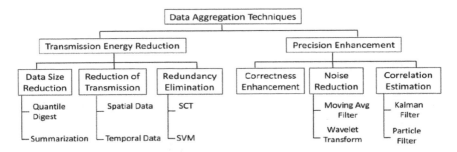

Fig. 2 Classification of data aggregation protocols for redundancy elimination

position of sensor nodes in the sensing region. Since WSNs are deployed densely, there is significant amount of spatial redundancy. Temporal redundancy means the prospect to acquire data from a specific action performed more than once, indefinite in time, followed in order to verify the results for reliability. It is also known as time redundancy and is used to improve the exactness of sensor nodes readings.

The data aggregation techniques are designed [9] as per the need whether spatial redundancy is required or temporal or both and are application-specific. The aggregation techniques are developed for specific application and can be categorized as having objective of transmission energy conservation and accuracy improvement. Transmission energy conservation data aggregation techniques encompass algorithms that reduce incoming data and conserve energy consumption during transmission. These algorithms are further classified in techniques with objective of data size reduction, number of transmission minimization, and redundancy elimination.

Data size reduction algorithms are further divided into three types: summarization [10], quantile digest, and representative data item. Summarization attempts to find out one single quantity extracted from the entire set of sensor readings by applying functions such as average, sum, count, min, max. Quantile digest [11] includes the partitioning of ordered data into equal-sized data subsets dynamically, e.g., finding median of readings. Representative data item technique is used to find a relatively small number of parameters or variables that represent a particular set of data.

Redundancy elimination algorithms exploit the fact that the sensor nodes in close proximity produce sensed data in high correlation. Hence, transmission energy is saved by identifying and discarding redundant data either spatially or temporally or both, e.g., duplicate suppression [12], other functions like count, sum, and average [13] for computation of final result.

The various proposed protocols have been summarized in Table 1 along with their merits and demerits.

From the above literature review, research issues identified in redundancy elimination during data aggregation for wireless sensor IoT systems are as follows:

- To develop a technique for sensibly selecting the information to be transmitted to aggregation head for reducing energy consumption during communication.
- To develop technique for combining data into high-quality information at the aggregation head by eliminating redundancy and identifying false alarms or outliers to conserve energy and bandwidth.

3 System Model and Terminology

In the proposed work, we consider a WSN with a large number of sensors distributed randomly in a certain deployment area. The AH is selected when the event is detected using aggregation head selection protocol [19], and the participating sensor nodes will collect physical environment data and send it to its AH. The AHs are responsible for

Table 1 Summary of related work in redundancy elimination

Authors: algorithm	Characteristics	Merits	Demerits
Sampoornam et al. [14]: Efficient data redundancy reduction (EDRR)	Conjugative sleep scheduling scheme with connected dominating set (CDS) for parent selection and DPCM	• Aggregation head on basis of highest residual energy • Use of DPCM for determining residual power levels	• Loss of accuracy while eliminating redundancy • Delay overhead in determining residual power levels
Kumar et al. [15]: Linguistic fuzzy trust based data aggregation and transmission (LDAT)	Selects trusted data aggregator using linguistic fuzzy trust mechanism	• Uses context-aware system for validation	• Transmission overhead • Loss of accuracy while eliminating redundancy
Manoj et al. [16]: Semantic correlation tree (SCT) based adaptive data aggregation (SCTADA)	Divides WSN into ring-like structure, further divided into sectors; each sector has aggregation head	• Minimizes redundancy • Eliminates false data using doorway algorithm	• Delay while determining false alarms
Patil et al. [17] : Support vector machine (SVM) based Data Redundancy elimination for data aggregation (SDRE)	Makes use of SVM for redundancy elimination	• Minimizes redundancy • Eliminates false data	• Transmission overhead • Loss in accuracy as only one node selected for value
Khedo et al. [18]: Redundancy elimination for accurate data aggregation (READA)	Applies a grouping and compression mechanism and uses a prediction model derived from cached values	• Redundancy eliminated at local as well as aggregation head level • Outlier detection	• False alarms may cause delay and loss of data • Reduced accuracy

aggregating and transferring the data to the BS wirelessly. A general block diagram of a data aggregation mechanism is given in Fig. 3.

The sensor nodes *Snode1*, *Snode2*, and *Snode3* collect data *Data1*, *Data2*, and *Data3* from the environment. Rather than each sensor node sending the data to BS, they send the data to an aggregation head AH node. The AH creates a single internal representation of the environment from its inputs. The single representation is then forwarded to the BS.

There is a lot of data redundancy in WSN as we have discussed in earlier section. In order to get solution for this problem, we propose a data aggregation technique that exploits support vector machine (SVM) of supervised learning model to eliminate redundancy. SVM performs two functions: One is classification, and another is correlated data elimination. It uses linear classifier method to represent the classification of local participating sensor nodes (LNs) and aggregation head (AH) nodes using

Fig. 3 Data aggregation mechanism

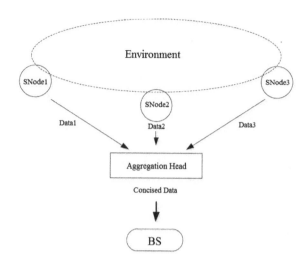

Fig. 4 Maximum-margin hyperplane and margins for an SVM [20]

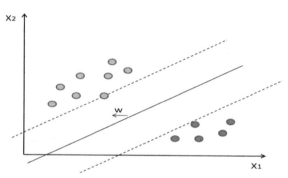

a hyperplane which divides it into two classes, redundant (no), and not redundant (yes).

The SVM model is shown in Fig. 4. The region bounded by these two hyperplanes is called the "margin," and the maximum-margin hyperplane is the hyperplane that lies halfway between them [20]. These hyperplanes can be described by the equations $\vec{w_i} \cdot \vec{x_i} + b = 1$ and $\vec{w_i} \cdot \vec{x_i} + b = -1$.

Geometrically, the distance between these two hyperplanes is $\frac{b}{||\vec{w}||}$, so to maximize the distance between the planes we want to minimize $||\vec{w}||$. As we also have to prevent data points from falling into the margin, we add the following constraint:

For each i, either $\vec{w_i} \cdot \vec{x_i} + b \geq 1$ if $y_i = 1$ or $\vec{w_i} \cdot \vec{x_i} + b \leq 1$, if $y_i = -1$.

These constraints state that each data point must lie on the correct side of the margin. This can be rewritten as $y_i (\vec{w_i} \cdot \vec{x_i} + b \geq 1)$, for all $1 \leq i \leq n$.

We can put this together to get the optimization problem:

"Minimize $||\vec{w}||$ subject to $y_i (\vec{w_i} \cdot \vec{x_i} + b \geq 1)$, for all $1 \leq i \leq n$".

The aggregation session is initiated by the AH on event detection, where the local participating sensor nodes (LNs) agree to participate in the session. In the first hyperplane, the LNs sense the data and compare it with threshold value and assemble the data in data sets. Next, LNs compress the assembled packets as per their set_ids. Before sending the compressed sets, LNs take decision and discard the redundant sets and then transmit it to AH. Similarly, in the second hyperplane, the AH takes decision and discards the duplicate sets and transmits the concise data to the BS.

4   Simulation and Results

The network simulation set up of the data aggregation model is designed and implemented using AADL with OSATE. The energy model is adopted from [19], sensor radio model [21], and simulation parameters [22].

It is considered that the wireless network loses its connectivity and the coverage of the sensed area when more than 40% of the sensor nodes are dead. The wireless sensor nodes are randomly deployed in a 100 m × 100 m field with a base station located at center (0, 0). The energy consumption power [23] for sending (24.75 mW) is quite close to receiving (13.5 mW), and the initial energy for each sensor node is taken as full (2.8 J).

We perform the comparative analysis of simulation results obtained for the current state-of-the-art protocols EDRR [14], LDAT [15], SCT [16], SDRE [17], and READA [18]. The evaluation parameters are average packet delivery ratio (APDR), average residual energy (ARE), accuracy and aggregation gain ratio (AGR).

A. **Performance metrics**: The performance evaluation of the proposed and current state-of-the-art protocols is measured using the following metrics as shown in Table 2.

B. **Results and Analysis**: In this simulation, initially 10 aggregation head nodes and later with step of 10, maximum 40 aggregation head nodes are selected for aggregation of data.

- **Effect of Aggregation Gain Ratio (AGR)**: Figure 5 represents the aggregation gain ratio of proposed and under consideration protocols. The proposed model depicts aggregation gain in terms of data collected and aggregated in between the range 0.55–0.95, whereas the READA shows this gain in range of 0.5–0.8,

Table 2 Performance evaluation metrics

Metric	Computation
Aggregation gain ratio (AGR)	$AGR = \dfrac{\sum \text{number of packets delivered after aggregation}}{\sum \text{number of packets transmitted before aggregation}}$
Average residual energy (ARE)	$E_{TOTAL} = \sum_{i=1}^{x} (E_{INIT} + E_{AH} + E_{NAH})$ $ARE = (E_{TOTAL})/\text{No. of Events}$
Accuracy	$Accuracy = \dfrac{\sum (Estimated\ Mean - Actual\ Mean)}{\sum Actual\ Mean} * 100$

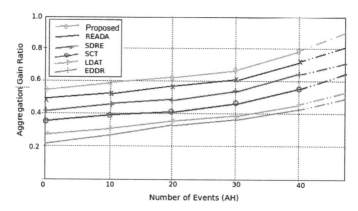

Fig. 5 Effect of aggregation gain ratio (AGR) over number of events

SDRE shows this gain in range of 0.45–0.7, SCT shows this gain in range of 0.38–0.62, LDAT shows this gain in range of 0.27–0.48, and EDRR shows this gain in range of 0.5–0.8. This is due to the fact that as the number of aggregation head increases due to the increase in the number of events, there is more increase in information collection and eventually more aggregation gain.

- **Effect of Average Residual Energy (ARE)**: Figure 6 shows the effect of average residual energy over number of events during the simulation. The proposed model shows the highest amount of average residual energy left with aggregation heads in range of 2.9–1.6 j as compared to simulated model of READA shows in range of 2.8–1.4, SDRE shows in range of 2.6–1.2, SCT shows in range of 2.4–1.0, LDAT shows in range of 1.9–0.6, and EDRR shows in range of 1.7–0.5. This is due to the fact that as the number of events increases, the number of aggregation heads increase which consumes energy in determination of redundancy and transfers of aggregated data at base station.

- **Effect of Accuracy Over Number of Aggregation Heads**: Figure 7 shows the effect of accuracy over number of aggregation heads generated during the simulation. The proposed model shows the highest ratio of accuracy in range of 0.5–0.9 as compared to simulated model of READA which shows in range of 0.45–0.8, SDRE shows in range of 0.4–0.7, SCT shows in range of 0.3–0.65, LDAT shows in range of 0.3–0.6, and EDRR shows in range of 0.23–0.5. This is due to the fact that as the number of events increases, the aggregation heads eliminate redundant data at local nodes before transmitting to base station. From the results, it is concluded that the proposed model and READA outperform the others protocols in terms of aggregation gain, average residual energy, and accuracy.

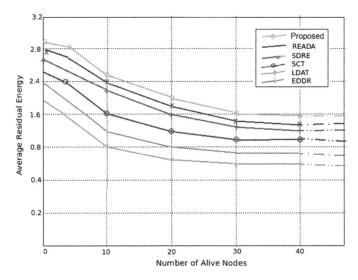

Fig. 6 Effect of average residual energy (ARE) over number of events

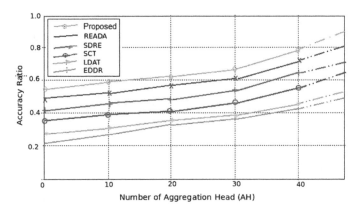

Fig. 7 Effect of accuracy over number of aggregation heads

5 Conclusion

Data collection and aggregation has been increasingly used in WSNs as an effective way of improving network lifetime. This paper presents the comparative study of redundancy elimination techniques and proposes an approach for redundancy elimination using the linear classifier of SVM for identifying the redundant data. We believe that the proposed technique for classification and comparative study of these data aggregation techniques with respect to type of improvements they offer will assist in energy conservation and increase the network lifetime. This work can further be enhanced by associating a correlation coefficient with sensed data so that the level of redundancy can be reduced without data loss. This is an ongoing work, and preparation of a more comprehensive and energy-efficient data aggregation technique in the form of a journal paper is underway.

References

1. Akyildiz, I. F., Su, W., Sankarasubramaniam, Y., & Cayirci, E. (2002). A survey on sensor network. *IEEE Communications Magazine, 40*(8), 102–114.
2. T. Koskiahde, J. Kujala, T. Norolampi, & Oy, F. (2008). A sensor network architecture for military and crisis management. In *IPCS International IEEE Symposium on Precision Clock Synchronization for Measurement, Control and Communication, IEEE* (pp. 110–114). Michigan: IEEE Xplorer, September 2008.
3. White paper. http://www.iec.ch/whitepaper/pdf/iecWP-internetofthings-LR-en.pdf; July 31, 2014.
4. Razzaque, M. A., Milojevic, M., Palade, A., & Clarke, S. (2016). Middleware for internet of things: a survey. *IEEE Internet of Things Journal, 3*(1), 70–96.
5. Yadav, S., & Yadav, R. S. (2016). A review on energy efficient protocols in wireless sensor networks. *Wireless Networks, Springer, 22*(1), 335–350.
6. Madamkam, S., Ramaswamy, R., & Tripathi, S. (2015). Internet of Things (IoT): A literature review. *Journal of Computer and Communications, 3*, 164–173.
7. Nakamura, E.F. et al. (2005). Using information fusion to assist data dissemination in wireless sensor networks. *Telecommunication Systems, 30*(1), 237–254.
8. Fasolo, E., Rossi, M., Widmer, J., & Zorzi, M. (2007). In-network aggregation techniques for wireless sensor networks: A survey. *IEEE Wireless Communications, 14*(2), 70–87.
9. Abdelgawad, A., & Bayoumi, M. (2012). Data fusion in WSN" In *Resource-Aware Data Fusion Algorithms for Wireless Sensor Networks*; *Lecture Notes in Electrical Engineering Series* (Vol. 118, pp. 17–35). Springer.
10. Qurat-ul-Ain, I. T., Ahmed, S., & Zia, H. (2012). An objective based classification of aggregation techniques for wireless sensor networks. In *Springer International Multi Topic Conference, IMTIC*.
11. He, T., Blum, B. M., Stankovic, J. A., & Abdelzaher, T. (2004). AIDA: Adaptive application independent data aggregation in wireless sensor networks. *ACM Transactions on Embedded Computing System, 3*(2), 426–457.
12. Pandey, V., Kaur, A., & Chand, N. (2010). A review on data aggregation techniques in wireless sensor network. *Journal of Electronics and Electrical Engineering, 1*, 01–08.
13. Xu, H., Huang, L., Zhang, Y., Huang, H, Jiang, S., Liu G. (2010). Energy-efficient cooperative data aggregation for wireless sensor networks. *Elsevier Parallel Distributed Computing, 70*(9), 953–961.
14. Sampoornam, K. P., & Rameshwaran, K. (2013). An efficient data redundancy reduction technique with conjugative sleep scheduling for sensed data aggregators in sensor networks. *WSEAS TRANSACTIONS on COMMUNICATIONS, 12*(9), 2224–2864. E-ISSN.
15. Kumar, M., & Dutta, K. (2016). LDAT: LFTM based data aggregation and transmission protocol for wireless sensor networks. *Journal of Trust Management*, 2–20.
16. Kumar, S. M., Rajkumar, N. (2014). SCT based adaptive data aggregation for wireless sensor networks. *Springer Wireless Personal Communication, 75*, 2121–2133.
17. Patil, P., & Kulkarni, U. (2013). SVM based data redundancy elimination for data aggregation in wireless sensor networks. *IEEE*, 1309–1316. 978-1-4673-6217-7/13.
18. Khedo, K., Doomun, R., & Aucharuz, S. (2010). READA: redundancy elimination for accurate data aggregation in wireless sensor networks. *Elsevier Computer Networks, 38*(4), 393–422.

19. Yadav, S., Yadav, R. S. (2015). Energy efficient protocol for aggregation head selection in wireless sensor network. *International Journal of Future Computer and Communication, 4*(5), 311–315.
20. Manevitz, L. M., & Yousef, M. (2001). One-class SVMs for document classification. *Journal of Machine Learning Research, 2,* 139–154.
21. Tripathi, R. K., Dhuli, S., Singh, Y. N., & Verma, N. K. (2014). Analysis of weights for optimal positioning of base station in a wireless sensor network. *IEEE Transaction,* 978-1-4799-2361-8/14.
22. Heinzelman, W. R., Chandrakasan, A., & Balakrishnan, H. (2002). An application specific protocol architecture for wireless microsensor networks. *IEEE Transactions on Wireless Communications, 1*(4), 660–670.
23. Wang, X. F., Xiang, J., & Hu, B. J. (2009). Evaluation and improvement of an energy model for wireless sensor networks. *Chinese Journal of Sensors and Actuators, 22*(9), 1319–1333.

A Lightweight and Secure IoT Communication Framework in Content-Centric Network Using Elliptic Curve Cryptography

Sharmistha Adhikari and Sangram Ray

Abstract In the recent era, content-centric network (CCN) is emerging as a future Internet paradigm to leverage scalable content distribution. Similarly, the Internet of things (IoT) is another upcoming technology which integrates as well as manages heterogeneous connected devices over the Internet. Recent literature have shown that IoT architecture can efficiently perform if it is implemented in CCN environment. In addition, considering the openness of the Internet used in IoT communication and limited capacity of IoT devices, security becomes a serious challenge which demands attention of the research community. In this paper, our main objective is to design a secure IoT communication framework that operates in CCN. A certificateless public key infrastructure is designed for our resource-constrained IoT communication framework in CCN. We have incorporated elliptic curve cryptography (ECC), a state-of-the-art lightweight cryptosystem, to ensure security of the proposed scheme. Finally, an in-depth security analysis confirms that the proposed scheme is resistant to various relevant cryptographic attacks.

Keywords Content-centric network (CCN) · Internet of things (IoT)
Elliptic curve cryptography (ECC)

1 Introduction

CCN and its importance: Since 2009, content-centric network (CCN) is envisaged as a future Internet architecture to cope up with the ever-increasing need for information exchange in current technological era. CCN is designed to leverage scalable content distribution over the Internet. In CCN, information/content gets more importance

S. Adhikari (✉) · S. Ray
Department of Computer Science and Engineering,
National Institute of Technology Sikkim, Sikkim 737139, India
e-mail: sharmistha.adhikari@gmail.com

S. Ray
e-mail: sangram.ism@gmail.com

© Springer Nature Singapore Pte Ltd. 2019
A. Khare et al. (eds.), *Recent Trends in Communication, Computing,
and Electronics*, Lecture Notes in Electrical Engineering 524,
https://doi.org/10.1007/978-981-13-2685-1_21

than the host where it is available and security is provided separately on the piece of content [1–4]. CCN uses human-readable hierarchical content naming approach to uniquely identify any content over the network [3]. Thus, CCN replaces the IP-based routing with name-based routing and uses content caching mechanism in intermediate routers to reduce network latency time. CCN router uses a limited buffer called content source (CS) to cache popular content for future use. Routers also maintain a forwarding information base (FIB) and a pending Interest table (PIT) to facilitate name-based routing. The major entities of CCN are content provider, publisher, consumer, and network router. CCN consumer sends Interest packet as a request for content. Any CCN router which receives the Interest checks the availability of the requested content in its CS. If it is available, the router consumes the Interest and sends back the requested content to the respective consumer. Otherwise, the Interest is forwarded to the content provider. Content provider supplies the content to its interface publisher for publishing in CCN. Finally, the requested content is forwarded to the requesting consumer following the reverse Interest path. As CCN enhances smooth delivery of content over the Internet, it can easily support the huge need for data/content exchange in Internet of things (IoT) framework.

IoT and its importance: With the advancement in consumer electronics, wireless communication, and intelligent sensors, IoT is becoming a reality where several heterogeneous things are interconnected through the Internet for information exchange and operated remotely without any human intervention. One of the most significant challenges in IoT framework is security which includes infrastructure-level security, application-level security, general system security, and communication security [5–11]. Now, a brief discussion on IoT communication framework is given here. IoT framework involves three significant entities, namely IoT devices, gateway server, and IoT end users. IoT device includes intelligent sensor devices and actuators which are deployed in the field such as a smart factory, smart home, hospital (in case of smart health care). In IoT network, intelligent sensor devices gather data from the environment and feed them to the gateway server. IoT gateway server manages security of the data communication and gives access of the gathered data to the authorized users who in return may give some commands to the IoT actuators via gateway server.

Background study: CCN as a future Internet architecture is ideally designed to address the emerging technology like IoT and its challenges at its design level [5–11]. In this regard, different researchers [6–10] have focused on IoT as a challenging Internet technology that can be significantly benefitted from CCN. Later, Suarez et al. [11] proposed an IoT management architecture based on information-centric network. Though the authors have addressed the overall IoT management architecture and its different aspects, the paper [11] lacks the cryptographic details of security measures used. In this paper, we have outlined a CCN-based IoT framework and designed a detailed communication security architecture using ECC [12–16].

Our contribution: In this paper, we have addressed an IoT communication framework , which operates in CCN environment. As security of IoT communication is an important challenge, we have proposed an ECC-based efficient and secure architecture for IoT communication. Considering the limited resource of IoT devices, a certificateless

ECC-based security architecture is designed that uses identity-based private keys for users and IoT devices. Moreover, a thorough security analysis is done to ensure that the proposed scheme is well protected from relevant cryptographic attacks.

Paper organization: The rest of the paper is organized as follows. In Sect. 2, the proposed scheme is presented and its security analysis is discussed in Sect. 3. Finally, Sect. 4 concludes the paper.

2 Proposed Scheme

In this section, we have designed a lightweight and secure IoT communication framework in CCN considering security of IoT communication as a major concern. The proposed IoT communication framework is depicted in Fig. 1 where the IoT gateway server is connected with the IoT devices and IoT end users using CCN framework. The connection is usually achieved through wireless connectivity where gateway server acts as a trusted authentication server (AS) and handles the security measures of the IoT communication system.

In the proposed scheme, we have used Interest, Content, and Manifest packets for communication among user, AS, and IoT device. The general structure of these packet types is shown in Fig. 2. In addition, general structures of two databases maintained by AS, namely IoT device database and IoT user database, are given in Fig. 3.

"Name prefix space" column of both the databases can have values for all the CCN namespaces involved in the IoT framework. The name prefix spaces are IoT_dev, IoT_user, IoT_admin, and IoT_as. The "commands" column of IoT device database has values like increase, decrease, print. The "device access permission" column stores the IoT device names for which the user has access. Similarly, "commands permitted" column lists the permitted commands which are used by a user to a particular IoT device. In our scheme, hierarchical CCN name is used for the IoT devices and end users. The example of CCN naming approach for the proposed IoT framework is shown in Fig. 4.

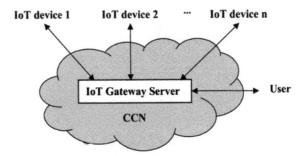

Fig. 1 Proposed IoT communication framework in CCN

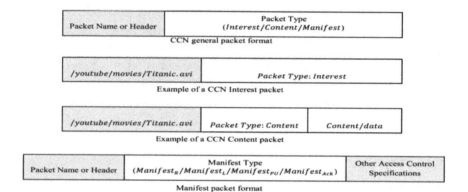

Fig. 2 General structure of different CCN packets

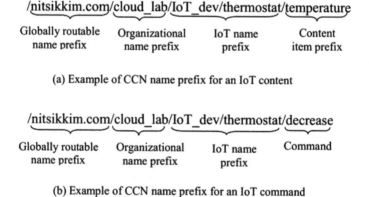

Fig. 3 General structure of IoT device database and IoT user database

/nitsikkim.com/cloud_lab/IoT_dev/thermostat/temperature

| Globally routable | Organizational | IoT name | Content |
| name prefix | name prefix | prefix | item prefix |

(a) Example of CCN name prefix for an IoT content

/nitsikkim.com/cloud_lab/IoT_dev/thermostat/decrease

| Globally routable | Organizational | IoT name | Command |
| name prefix | name prefix | prefix | |

(b) Example of CCN name prefix for an IoT command

Fig. 4 Example of CCN naming approach used for proposed IoT framework

Multiple end users may access an IoT framework but depending on their authorization, specific access is allowed by the gateway/AS. We have used Manifest packet

[4] to share access control information between the user and the AS. Manifest is also used by user to send commands to the IoT devices via AS. The proposed security framework is divided into three phases, namely (1) system initialization phase, (2) registration phase, and (3) authentication and session key negotiation phase. The details of these three phases are briefly described in the following subsections.

2.1 System Initialization

In this phase, the AS selects all the security parameters and publishes them in the respective CCN where the IoT framework operates. AS selects an elliptic curve E with prime order p and generator G and two secure one-way hash function h and h_1 where $h_1: \{0, 1\}^* \rightarrow \mathbb{Z}_p^*$. AS also chooses a random secret $s \in \mathbb{Z}_p^*$ and calculates its public key $PU_{AS} = s \cdot G$. Finally, AS publishes the security parameters as $\{E, p, G, h, h_1, PU_{AS}\}$.

2.2 Registration Phase

Initially, all the users and the IoT devices need to register to the AS in order to be included in the IoT framework. The registration phase is conducted through a secure channel where the user and IoT device get their private keys from AS depending on their CCN name prefix which is unique and treated as identity. The identity of any user or IoT device is verified by the AS at the time of registration. Any IoT device is registered at the time of its deployment in the network, and its private key gets hardcoded in the respective device through the registration procedure. The private key of the IoT device is used for symmetric encryption of the transmitted data between the IoT device and AS. In the registration process, user provides its identity ID_U to the AS and gets its private key $s \cdot h_1(ID_U)$. Similarly, an IoT device with identity ID_D gets its private key $s \cdot h_1(ID_D)$.

2.3 Authentication and Session Key Negotiation Phase

The user needs to login to AS in order to access IoT framework for which he/she is registered. After successful mutual authentication between the user and AS, a contributory secret session key is negotiated. The session key is changed in each session/user login to ensure security. Now, the detailed procedure is depicted in Fig. 5 and discussed stepwise where $X \rightarrow Y : M$ means sender X sends message M to receiver Y. Here, E/D_X means symmetric encryption/decryption using key X. In addition, dot operator (.) and concatenation operator (‖) are used for ECC point multiplication and message concatenation, respectively.

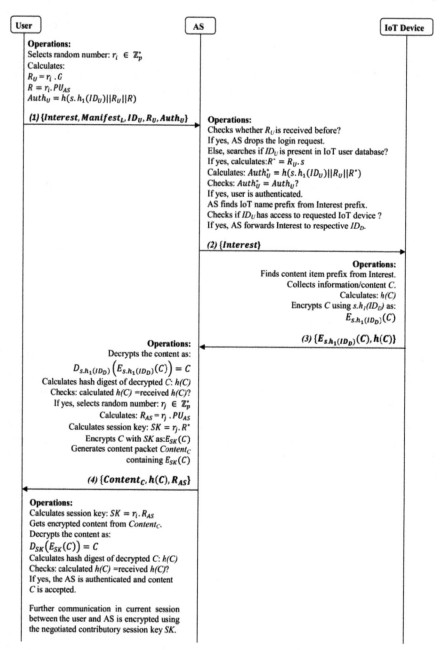

Fig. 5 Mutual authentication and session key negotiation between user and AS

Step 1: User ➔AS: $\{Interest, Manifest_L, ID_U, R_U, Auth_U\}$

Initially, the user selects a random number $r_i \in \mathbb{Z}_p^*$ and calculates $R_U = r_i \cdot G$, $R = r_i \cdot PU_{AS}$, and $Auth_U = h(s \cdot h_1(ID_U)||R_U||R)$. Finally, the user sends an Interest, requesting the content, a $Manifest_L$ for sending access control information regarding login process, along with identity ID_U, R_U, and authentication parameter $Auth_U$ to the AS. Here, Interest name prefix signifies the requested content name as shown in Fig. 4a.

Step 2: AS ➔ IoT Device: $\{Interest\}$

After receiving the authentication request from *step 1*, the AS initially checks whether R_U is received before. If yes, AS drops the login request. Else, searches if ID_U is present in IoT user database. If yes, AS calculates $R^* = R_U \cdot s$, $Auth_U^* = h(s \cdot h_1(ID_U)||R_U||R^*)$ and checks $Auth_U^* = Auth_U$? If yes, the user is authenticated; otherwise, AS drops the session. After successful authentication of the user, AS finds IoT name prefix from received Interest prefix and checks if ID_U has access to requested IoT device. If yes, AS forwards the Interest to the respective IoT device (ID_D).

Step 3: IoT Device ➔AS:$\{E_{s \cdot h_1(ID_D)}(C), h(C)\}$

After receiving the Interest from *step 2*, the IoT device finds content item prefix from Interest name prefix. Then, it collects information/content C from the environment. It calculates $h(C)$ and also encrypts C using its private key $s \cdot h_1(ID_D)$ as $E_{s \cdot h_1(ID_D)}(C)$. Finally, the IoT device sends the encrypted content to AS along with the content hash digest $h(C)$.

Step 4: AS ➔User:$\{Content_C, h(C), R_{AS}\}$

After receiving in *step 3*, AS decrypts the content as $D_{s \cdot h_1(ID_D)}\big(E_{s \cdot h_1(ID_D)}(C)\big)$ and gets requested content C. Then, AS calculates hash digest of decrypted C as $h(C)$, and checks if calculated $h(C) =$ received $h(C)$. If yes, selects a random number $r_j \in \mathbb{Z}_p^*$ and calculates $R_{AS} = r_j \cdot PU_{AS}$ and contributory session key $SK = r_j \cdot R^*$. Finally, AS encrypts C using SK as $E_{SK}(C)$, generates content packet $Content_C$ with the encrypted content, and sends $Content_C$, $h(C)$ and key part R_{AS} to the user.

After receiving the message in *step 4*, the user calculates contributory session key $SK = r_i \cdot R_{AS}$. User gets encrypted content from $Content_C$ packet and decrypts as $D_{SK}(E_{SK}(C))$ to get C. The user also calculates hash digest of decrypted C as $h(C)$ and checks if calculated $h(C) =$ received $h(C)$. If yes, the AS is authenticated, SK is negotiated, and the received content C is accepted. Any further communication in the current session between the user and AS is encrypted using the negotiated contributory session key SK.

3 Security Analysis

In this section, several relevant cryptographic attacks are analyzed to show that our scheme is well protected.

3.1 Mutual Authentication

Mutual authentication is an important parameter of any security framework that is being taken care of in our scheme. In the proposed scheme, the user sends $Auth_U$ which is dependent on user's secret key and random value r_i and can be verified only by the AS. Accordingly, AS authenticates the user. Similarly, AS sends encrypted content, content hash digest $h(C)$, and key part R_{AS}. Upon calculating SK using R_{AS}, the user decrypts the content, calculates its hash digest, and verifies received $h(C)$ with calculated $h(C)$. If verified, the user authenticates AS. Thus, the mutual authentication is completed.

3.2 Confidentiality

Confidentiality property is maintained in our scheme as none of the secret keys, session key SK, requested content, or authentication parameter $Auth_U$ travels openly rather either they are hashed or encrypted.

3.3 Replay Attack Resilience

In our scheme, replay attack by an intruder is successfully prevented as in *step 2* of the authentication phase, AS checks whether R_U is received before. If yes, AS drops the login request. Here, value of R_U is dependent on a random value r_i which is changed in each session. Hence, any repetition in the value of R_U is detected by the AS.

3.4 Man-in-the-Middle Attack Resilience

As mutual authentication and confidentiality are maintained in the proposed scheme, any attempt of message modification or replay by an intruder will be identified by the user or AS. Hence, man-in-the-middle attack is prevented.

3.5 Perfect Forward Secrecy

Perfect forward secrecy is a property which ensures that even if the long-term keys are compromised at a point in time, the sessions before that time are still secure. As the proposed scheme uses random values r_i and r_j to calculate SK in each session, even if user's long-term key $(s \cdot h_1(ID_U))$ becomes compromised nobody will be able to compute SKs before that time.

3.6 Known Session Key Attack Resilience

Known session key attack means the knowledge of one session key reveals other session keys of different sessions. This is successfully prevented in our scheme because SK is calculated by using random values r_i and r_j which are changed in every session.

3.7 Brute Force Attack Resilience

Our scheme is resilient to brute force attack as an intruder cannot guess SK. Strength of SK is dependent on three secret random numbers r_i, r_j, and s from $\mathbb{Z}_p^*$. Moreover, SK is a point on the elliptic curve. Hence, based on the security strength of elliptic curve discrete logarithmic problem, it is impossible to guess SK in polynomial time.

4 Conclusion

In this paper, a CCN-based IoT communication framework is proposed. Considering the openness of wireless connectivity and limited capacity of resource-constrained IoT devices, we have presented a lightweight security framework for CCN-based IoT communication using ECC. As ECC uses efficient point multiplication operation and smaller key size (160-bits) than other public key cryptosystems such as RSA (1024-bits) to provide same level of security, our scheme incurs low computation and communication overheads. Moreover, as identity-based private keys are used for user and IoT devices, the overheads of public key certificate generation, verification, management, etc., are eliminated. Finally, an in-depth security analysis ensures that the proposed scheme is resilient against relevant cryptographic attacks.

Acknowledgments This work is financially supported by Visvesvaraya Ph.D. Scheme, Ministry of Electronics and Information Technology, Government of India.

References

1. Jacobson, V., Smetters, D. K., Thornton, J. D., Plass, M. F., Briggs, N. H., Braynard, R. L. (2009). Networking named content. In *Proceedings of the 5th International Conference on Emerging Networking Experiments and Technologies* (pp. 1–12). ACM.
2. Smetters, D., Jacobson, V. (2009). *Securing network content* (pp. 2009–01). Technical report, PARC.
3. Mahadevan, P. (2014). *CCNx 1.0 Tutorial*. Technical report, PARC.
4. Kuriharay, J., Uzun, E., Wood, C. A. (2015). An encryption-based access control framework for content-centric networking. In *2015 IFIP Networking Conference (IFIP Networking)* (pp. 1–9). IEEE.
5. Li, S., Da Xu, L. (2017). *Securing the internet of things*. Syngress.
6. Corujo, D., Aguiar, R. L., Vidal, I., Garcia-Reinoso, J. (2012). A named data networking flexible framework for management communications. *IEEE Communications Magazine, 50*(12).
7. François, J., Cholez, T., Engel, T. (2013). CCN traffic optimization for IoT. In *2013 Fourth International Conference on the Network of the Future (NOF)* (pp. 1–5). IEEE..
8. Baccelli, E., Mehlis, C., Hahm, O., Schmidt, T. C., Wählisch, M. (2014). Information centric networking in the IoT: experiments with NDN in the wild. In *Proceedings of the 1st ACM Conference on Information-Centric Networking* (pp. 77–86). ACM.
9. Quevedo, J., Corujo, D., Aguiar, R. (2014). A case for ICN usage in IoT environments. In *Global Communications Conference (GLOBECOM)* (pp. 2770–2775). IEEE.
10. Ahlgren, B., Lindgren, A., Wu, Y. (2016). Experimental Feasibility Study of CCN-lite on Contiki Motes for IoT Data Streams. In *Proceedings of the 3rd ACM Conference on Information-Centric Networking* (pp. 221–222). ACM.
11. Suarez, J., Quevedo, J., Vidal, I., Corujo, D., Garcia-Reinoso, J., & Aguiar, R. L. (2016). A secure IoT management architecture based on Information-Centric Networking. *Journal of Network and Computer Applications, 63,* 190–204.
12. Hankerson, D., Menezes, A. J., Vanstone, S. (2006). *Guide to elliptic curve cryptography*. Springer Science and Business Media.
13. Stallings, W. (2006). *Cryptography and network security: principles and practices*. Pearson Education India.
14. Miller, V. S. (1985). Use of elliptic curves in cryptography. In: *Conference on the Theory and Application of Cryptographic Techniques* (pp. 417–426). Berlin, Heidelberg: Springer.
15. Koblitz, N. (1987). *Elliptic Curve Cryptosystems. Mathematics of Computation, 48*(177), 203–209.
16. Lauter, K. (2004). The advantages of elliptic curve cryptography for wireless security. *IEEE Wireless Communications, 11*(1), 62–67.

Enhancement of Security in the Internet of Things (IoT) by Using X.509 Authentication Mechanism

S. Karthikeyan, Rizwan Patan and B. Balamurugan

Abstract Internet of Things (IoT) is the interconnection of physical entities to be combined with embedded devices like sensors, activators connected to the Internet which can be used to communicate from human to things for the betterment of the life. Information exchanged among the entities or objects, intruders can attack and change the sensitive data. The authentication is the essential requirement for security giving them access to the system or the devices in IoT for the transmission of the messages. IoT security can be achieved by giving access to authorized and blocking the unauthorized people from the internet. When using traditional methods, it is not guaranteed to say the interaction is secure while communicating. Digital certificates are used for the identification and integrity of devices. Public key infrastructure uses certificates for making the communication between the IoT devices to secure the data. Though there are mechanisms for the authentication of the devices or the humans, it is more reliable by making the authentication mechanism from X.509 digital certificates that have a significant impact on IoT security. By using X.509 digital certificates, this authentication mechanism can enhance the security of the IoT. The digital certificates have the ability to perform hashing, encryption and then signed digital certificate can be obtained that assures the security of the IoT devices. When IoT devices are integrated with X.509 authentication mechanism, intruders or attackers will not be able to access the system, that ensures the security of the devices.

Keywords Internet of Things · Authentication · Security · Digital certificates
X.509 digital certificate · Authentication

S. Karthikeyan (✉) · R. Patan · B. Balamurugan
School of Computing Science and Engineering, Galgotias University,
Greater Noida 201310, Uttar Pradesh, India
e-mail: link2karthikcse@gmail.com

R. Patan
e-mail: prizwan5@gmail.com

B. Balamurugan
e-mail: kadavulai@gmail.com

© Springer Nature Singapore Pte Ltd. 2019
A. Khare et al. (eds.), *Recent Trends in Communication, Computing,
and Electronics*, Lecture Notes in Electrical Engineering 524,
https://doi.org/10.1007/978-981-13-2685-1_22

1 Introduction

The Internet of Things (IoT) is the grouping of devices like sensors, actuators are planted in physical objects (vehicles, mobiles, houses) that are connected by means of both wired and wireless networks that are forming patterns. The term IoT was first coined by the Auto-ID center [1] in Kevin 1999. As the technologies emerged drastically, many innovations came into a picture of the Internet, which reduced many activities that been done by humans for 24/7. The importance of Internet of Things (IoT) is realized by scientific and business communities and it started to emerge in enhancing everyday lifestyle. Consider an example [2], a motion sensor is also known as motion detector can detect if any anybody who reach any places where there is no authorization for the intruder or the stranger. A motion sensor uses one or multiple technologies for finding any changes in a location. If any intruder doing the illegal activities, a warning signal will be sent to the security system, which can be connected with the monitoring center, thus by alerting the concerned individuals as well as the monitoring center.

There are millions of devices are connected by the IoT, and those devices are secured by using the security mechanism called as X.509 authentication mechanism which do both hashing and encryption [3], after that the data are stored in the cloud for the retrieval or the manipulation of the information which is explained in Fig. 1 [4].

The privacy and the security of the information are more important in the IoT, as the data transferred from the one end of the local site to another end of the Internet site, and it will be easy for the intruder as the data exchange is happening through the cloud. Now the intruder will have the control throughout the office or home results in loss of the assets either by money or any other means. In Fig. 1, it signifies the importance of security of the connection between devices and Azure IoT over encryption technologies that is TLS using X.509-based certificates, consistent one. It also offers robustness of communication among the devices and the cloud by acknowledgments [4]. Messages are stored in the Hub of IoT. There are two protocols

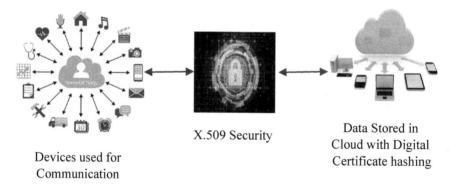

Devices used for
Communication

X.509 Security

Data Stored in
Cloud with Digital
Certificate hashing

Fig. 1 Securing the IoT communication with X.509

that are supported by Azure IoT such as the industry standard https protocol and the advanced message queuing protocol. Scalability can be increased by securing connection to both IP-enabled and non-IP-enabled devices. Now, the authentication is the primary security measures that should be implemented to reduce the attacks happening from the intruder [5].

A digital certificate is nothing but an electronic password which permits an individual, devices, system, an organization to transfer information safely to the cloud by the way of the public key infrastructure (PKI). The public key infrastructure comprises digital certificates and encryption as an ecosystem. The ecosystem made of not by software and hardware alone, but all of the individuals involved with the digital certificates. PKI certificates are received from a certificate authority (CA), responsible for creating and managing them. X.509 certificates are used to define authentication protocols and it also uses public key crypto and digital signatures [6].

The entities involved in X.509 certificate management are subjects and end entities, certification authority (CA), and registration authority (RA) [4]. The applications of the IoT are wide. The applications are not just limited to smart home, automated houses, theft detection, manufacturing, military, media, infrastructure management, agriculture, energy saving, environment monitoring, building, and construction.

2 Related Work

The security in IoT plays a major role, but still authentication helps to achieve it more. In [7] from an authentication perspective, X.509 certificates undeniably make a solution more secure. However, using X.509 certificate authentication greatly complicates a design if client-side certificates are required. It describes the security [8] issues of the IoT [9] and designed architecture with three-layer system structure such as application layer, network layer, and perception layer. In this paper [10], it confesses the challenges of the IoT like context awareness for privacy, digital device in a physical ambient, identification in the IoT environment, authenticating devices, data combination, scalability in IoT, secure setup and configuration, CI and IoT, conflicting market interest, considering IoT in an evolving Internet, human IoT trust relationship, data management, lifespan of every IoT's entities [11].

In this paper [8], the difficulties of the IoT and the ample opportunities are available for the challenges to be converted into problems. The IPv6 and Web services [1] are the basic components for IoT applications that ensures to give the following things like [12] different appliances with simplified development, homogeneous protocol ecosystem that connects with Internet hosts. Internet protocol version 6 (IPv6) [1] gives the quality of services, reliability with enhancement and high security for the improvement of IoT standard [13]. Thus, IoT is an open standard, and then IPv6 is the best protocol. The approaches to the security bootstrapping are three such as key pre-distribution-based, key distribution center (KDC)-based, and the certificate-based bootstrapping [14]. This paper describes Internet Engineering Task Force (IETF) to enhance security solutions for the IoT [15].

The X.509-based public key certificate implements either RSA or DSA signatures, where the entities are signed by certificate authority (RSA or DSA private key) [6, 16, 17]. In IoT applications, the digital certificate will be used in two different zones [14]. The first one is used for the end-to-end security which is based on the Datagram TLS (DTLS) [18] in the middle of a sensor device and an Internet host [14]. Two-factor authentication or 2FA [19] is an additional way to login which a user should give information more than the password for confirming the identity. As the usage of mobiles increasing every day by the customers such as banking, e-shopping [20], the requirement for security concerns has arisen that results in involvement of multi-factor authentication. It supports end-to-end session encryption key generation and mutual authentication [18]. Even though username and password have been received from the user, still we will send additional information to the mobile devices for verification; still, these factors are not applicable to the IoT. It uses a novel approach to create an authentication system provides transaction identification code (TIC) and SMS to give additional security other than normal login [17]. X.509 certificates are used in numerous situations for unique purposes [6, 21]. The standard certificate extensions are defined in the standard documents and ANSI X9.55 in various ways.

3   Problem Statement

Everyday new IoT devices are built to reduce the workload of human, IoT turning out to be a major emerging technology that has big impact on the globe, even though it's been used in many ways still the challenges are getting increased as the IoT are connected to the internet, it is open to all whoever needs to access the cloud, thus the security is the one of the major concern that affects the sensitive data stored in the cloud, any intruder will be able to hack or crack it and get the access to the cloud who can manipulate the data by any means. Thus, by the confidentiality, integrity, availability (CIA triad) can be compromised by the intruders.

4   Theoretical Deployment Process

As was formerly noted, a hacker who manages to extract the X.509 certificate/private key pair from a device can use the certificate/key for either eavesdropping on the communication or for exploiting an IoT solution. A unique certificate per device makes it possible to disable the exploited certificate/key by using a revocation list; however, this requires that the IoT solution can detect the exploitation in the first place.

Hashing converts the digital certificate which is not signed into hashed digital certificate and then followed by the encryption changes the hashed digital certificate into encrypted digital certificate, and then in the final phase, signed digital certificate

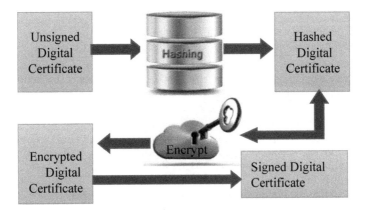

Fig. 2 Flow representation of communication

can be obtained which is explained in Fig. 2. In the event that it does as such, it will be a hacking endeavor.

1. **Authentication**. To confirm the trade [6], peers D and C register a validation label $\alpha D = \Gamma[KCD, (CD, D, CC, C, \rho D, \rho C)]\alpha C = \Gamma[KCD, (CC, C, CD, D, \rho C, \rho D)]$ where the $\Gamma[k, s]$ administrator alludes to a nonspecific symmetric confirmation calculation (e.g., an HMAC) taking a shot at the bit stream s by utilizing key k [6, 22].

2. **Key Derivation**. At long last, each associate figures the genuine session key PK (or different keys, to recognize encryption key from uprightness enter as generally done in standard security conventions) through a customary/standard key derivation function (KDF) χ [6], as: $PK = \chi(KCD, \rho D, \rho C)$—subtle elements in segment IV [6].

5 Practical Evaluation Model

X.509 certificates are the digital certificates uses X.509 public key infrastructure standard to subordinate a public key which possesses the identity contained in a certificate [23]. Certification Authority(CA) are the authority which distributes the X.509 certificates to the both party of the authorized IoT devices [21]. The CA is responsible for maintaining more than one special certificate called CA certificates which used to issue X.509 certificates [23]. Only the certification authority has access to CA certificates.

To solve this, Azure IoT hub is now added to supported X.509 certificates. Here, the communication between IoT and device can be inserted with a certificate and the required authorization can be ensured before establishing the connection.

___**Algorithm 1** for the process of fixing X.509 certification for IoT___

Input: IoT device Communication and Transmission
Output: Secure authentication for IoT Technology with X.509 Certificate

Step 1: Start
Step 2: Create new device protocols for authentication
Step 3: Verify the Authentication mechanism, data encryption
Step 4: Thumbprint creation for primary
Step 5: Thumbprint creation for secondary
Step 6: label the devices primary device $\alpha D = \Gamma[KCD,(CD, D, CC, C, \rho D, \rho C)]$
Step 7: label the devices secondary device $\alpha C = \Gamma[KCD,(CC, C, CD, D, \rho C, \rho D)]$
Step 8: Creating X.509 Key driven approach $PK = \chi(KCD, \rho D, \rho C)$
Step 9: Identify the key field
- serial number (unique within CA) identifying the certificate
- issuer unique identifier (v2+)
- subject unique identifier (v2+)
- extension fields (v3)

Step 10: signature (ofhash of all fields in the certificate)
Step 11: End

The unsigned digital certificate go through the process of hashing and then it will be converted into hashed digital certificate followed by the encryption which is the process of converting the plain text into cipher text, after the encryption done in the certificates, the hashed digital certificate transform into encrypted digital certificate; thus, the final signed digital certificate can be obtained from the encrypted digital certificate.

For creating an identity with Azure IoT hub, we would need a certificate which can be embedded in the device. X.509 Certificate will be sent as bytes from the header to the endpoint where the actual part of creating an identity is done using Device Client SDK. The performance of the proposed approach is accurate while comparing with the existing approach; it includes these performance parameters like identity, impacts, risks, safety, efficiency, and robustness [24].

6 Results and Discussion

The cryptographic and correspondence functionalities' effect has on every single nuclear operation. Initially, the 78 bytes verifiable endorsements can be sent inside only two link layer bundles, restricted to the 13 parcels required for the 725 bytes of the express X.509 declaration in the PEM arrange [6], or the 9 parcels required to send the 495 bytes of the unequivocal X.509 testament in the DER design. Second, cryptographic operations [6] of ECQV require a lower computational load concerning that expected to figure an ECDSA signature.

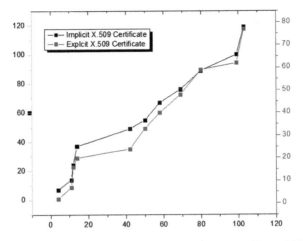

Fig. 3 IoT device authentication with X.509 certificates running from 77.1 to 86.7%

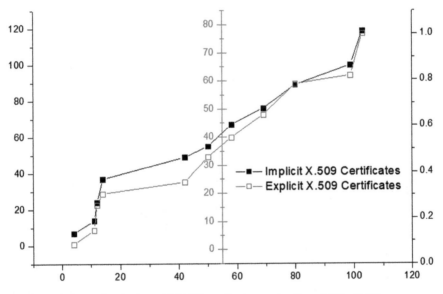

Fig. 4 IoT device authentication with X.509 certificates running from 50.9 to 84.7%

Figure 3 reports (with 95% certainty interims) the time required to finish the KMP [6] convention as a component of the number of jumps that isolates the included hubs. The Output from the X.509 mechanism achieves high security with maximal broadcast appointment sparsing. X.509 PEM endorsements running from 77.1 to 86.7% and from 50.9 to 84.7% with regard to express X.509 authentication in the DER organize as shown in Figs. 3 and 4 [6].

7   Conclusion and Future Work

Our paper proposed a high secure authentication mechanism for IoT secure and accurate communication. IoT is evolving rapidly with the innovation of technologies that are emerging every day by any means, which leads to usage of applications in the real-time environment. Even though IoT applied in many areas, still there are plenty of issues that are associated with the security. In this chapter, it described the modes of authentication mechanism which can be used for the prevention and detection of attacks that are produced by the intruders. Digital certificates such as X.509 certificate can also be used as one of the authentication mechanisms among the devices for ensuring the security and the data have not been breached by any possible means. Thus by using the X.509 digital certificates, it can produce the device authentication with the accuracy of more than 84.7%, whereas the previous results show the accuracy of 50.9% in the IoT Devices.

References

1. Heer, T., Garcia-Morchon, O., Hummen, R., Keoh, S. L., Kumar, S. S., & Wehrle, K. (2011). Security Challenges in the IP-based Internet of Things. *Wireless Personal Communications, 61*(3), 527–542.
2. Liu, J., Yang, X., & Philip Chen, C. L. (2012). Authentication and access control in the internet of things. In *2012 32nd International Conference on Distributed Computing Systems Workshops (ICDCSW)*, (pp. 588–592). IEEE.
3. Suo, H., Wan, J., Zou, C., Liu, J. (2012). Security in the internet of things: A review. In *2012 International Conference on Computer Science and Electronics Engineering*.
4. Alrawais, A., Abdulrahman A., & Xiuzhen C. (2015). X. 509 Check: A Tool to Check the Safety and Security of Digital Certificates. In *2015 International Conference on Identification, Information, and Knowledge in the Internet of Things (IIKI)* (pp. 130–133). IEEE.
5. Ranjan, A. K., Kumar, V., & Hussain, M. (2014). Security analysis of TLS authentication. In *Proceedings of the International Conference on Contemporary Computing and Informatics (IC3I)* (pp. 1356–1360), November 2014.
6. Sciancalepore, S., Student Member, IEEE, Piro, G., Member, IEEE, Boggia, G., Senior Member, IEEE, & Bianchi, G., Public Key Authentication and Key agreement in IoT devices with minimal airtime consumption.
7. Zhao, K., & Lina G. (2013). A survey on the internet of things security. In *2013 9th International Conference on Computational Intelligence and Security (CIS)* (pp. 663–667). IEEE.
8. Zhang, Z. K., Cho, M. C. Y., Wang, C. W., Hsu, C. W., Chen, C. K. & Shieh, S. (2014, November) IoT security: ongoing challenges and research opportunities. In *2014 IEEE 7th International Conference on Service-Oriented Computing and Applications (SOCA)*, (pp. 230–234). IEEE.
9. Xu, X. (2013). Study on security problems and key technologies of the internet of things. In *2013 Fifth International Conference on Computational and Information Sciences (ICCIS)* (pp. 407–410), 21–23 June 2013. https://doi.org/10.1109/iccis.2013.114.
10. Zolanvari, M., & Jain, R. (2015). IoT Security: A Survey.
11. Gurpreet Singh, M., Upadhyay, P., & Chaudhary, L. (2014). The internet of things: challenges & security issues. In *2014 International Conference on Emerging Technologies (ICET)* (pp. 54–59). IEEE.
12. Kim, E., Kaspar, D., Chevrollier, N., & Vasseur, J. P. (2011). Design and application spaces for 6LoWPANs draft-ietf-6lowpan-usecases-09, January 2011.

13. Perrig, A., Szewczyk, R., Wen, V., Culler, D., & Tygar, J. D. (2002). Spins: Security protocols for sensor networks. *Wireless Networks Journal* (2002).
14. Park, Chang-Seop. (2017). A secure and efficient ECQV implicit certificate issuance protocol for the internet of things applications. *IEEE Sensors Journal, 17*(7), 2215–2223.
15. Roman, R., Najera, P., & Lopez, J. (2011). Securing the internet of things. *Computer, 44*(9), 51–58.
16. https://www.ipa.go.jp/security/rfc/RFC3280-04EN.html.
17. Sanyal, S., Tiwari, A., Sanyal, S. (2010). A multifactor secure authentication system for wireless payment. In *Emergent web intelligence: Advanced information retrieval*, C. Richard et al. (Ed.) 1st ed. Chapter 13, (Vol. XVI, pp. 341–369). Springer Verlag London Limited. https://doi.org/10.1007/978-1-84996-074-8_13.
18. Granlund, D., Åhlund, C., Holmlund, P. (2015). EAP-Swift: An efficient authentication and key generation mechanism for resource constrained WSNs. *International Journal of Distributed Sensor Networks, 2015* Article ID 460914.
19. Borgohain, T., Amardeep, B., Kumar, U., & Sanyal, S. (2015). Authentication systems in Internet of Things. arXiv:1502.00870.
20. Acharya, S., Polawar, A., Pawar, P. (2013). Two factor authentication using smartphone generated one time password. *IOSR Journal of Computer Engineering (IOSR-JCE), 11*(2), 85–90.
21. Chau, S.Y., Omar, C., Endadul, H., Huangyi, G., Aniket, K., Cristina, N.-R., et al. (2017). Sym-Certs: practical symbolic execution for exposing noncompliance in X. 509 certificate validation implementations. In 2017 IEEE Symposium on Security and Privacy (SP) (pp. 503–520). IEEE.
22. Blake, I., Seroussi, G., & Smart, N. (1999). Elliptic curves in cryptography. Cambridge University Press.
23. https://aws.amazon.com/documentation/.
24. Suresh, K., Rizwan, P. & RajasekharaBabu, M. (2016). EEIoT: Energy efficient mechanism to leverage the internet of things (IoT). In *IEEE International Conference on Emerging Technological Trends*, Kollam, India (pp. 14–22).

Part IV
Signal Processing

Effect of Secondary Path Lengths on the Performance of FxLMS and Virtual Sensing Technique Based ANC System

Manoj Kumar Sharma, Renu Vig and Gagandeep Sahib

Abstract With the urbanization, exposure of mankind to the noise is increasing day by day leading to many health issues. For low-frequency noise reduction, active noise control system is widely applied in many applications. In the present paper, the effect of different secondary path lengths in FxLMS and virtual sensing technique based ANC system is studied. The virtual sensing technique based ANC is applied in the cases where it is not feasible to place the error microphone physically at the desired location. The filter coefficients of secondary path of ANC system are measured experimentally for the three different filter lengths, k (i.e., k = 64, 128, and 256) using Texas Instruments TMS320C6713 processor in the semi-anechoic chamber. The performance of ANC system with different filter lengths is analyzed in the terms of residual noise, signal-to-noise ratio, computational load, and error plots. The comparison suggests that secondary path of different filter lengths suits for FxLMS and virtual sensing technique based ANC systems.

Keywords Active noise control · Virtual sensing technique · FxLMS
Secondary path filter length · Coefficients

M. K. Sharma (✉)
Electrical and Electronics Engineering Department, UIET,
Panjab University, Chandigarh 160014, India
e-mail: mks_uiet@pu.ac.in

R. Vig · G. Sahib
Electronics and Communication Engineering Department, UIET,
Panjab University, Chandigarh 160014, India
e-mail: renuvig@hotmail.com

G. Sahib
e-mail: gagan007sahib@gmail.com

© Springer Nature Singapore Pte Ltd. 2019
A. Khare et al. (eds.), *Recent Trends in Communication, Computing,
and Electronics*, Lecture Notes in Electrical Engineering 524,
https://doi.org/10.1007/978-981-13-2685-1_23

1 Introduction

Active noise control (ANC) system has been implemented successfully in many real-time applications. This system is based on electro-acoustic technique in which noise signal is acquired and its anti-noise signal is produced using algorithm and it is superimposed to suppress unwanted noise. The ANC system applies adaptive algorithms and most commonly used is filtered reference least mean square (FxLMS) algorithm. However, for the proper implementation of ANC system, it requires estimate of the secondary path.

The basic principle of active noise control is superposition which requires acoustic combination of primary disturbance $d(n)$ and canceling signal $y'(n)$ (Fig. 1) [1].

However, the secondary path $S(z)$ from canceling loudspeaker $y(n)$ to error microphone $e(n)$ is added to compensate for acoustic superposition. $S(z)$ includes the digital-to-analog (D/A) converter, reconstruction filter, power amplifier, loudspeaker, acoustic path from loudspeaker to error microphone, error microphone, preamplifier, anti-aliasing filter, and analog-to-digital (A/D) converter [1].

Active noise control (ANC) technique involves the use of conventional algorithms such as LMS, FxLMS, leaky FxLMS, filtered-U recursive [2–5]. This paper focuses on use virtual sensing technique and FxLMS algorithm for ANC system. The weight updation is done using filtered normalized least mean square (FNLMS) algorithm.

1.1 Paper Outline

Section 2 gives the brief idea about virtual sensors. Section 3 describes virtual sensing technique, and secondary path modeling and path estimation are explained in Sect. 4. The experimental setup, simulations, and results are shown in Sect. 5. Section 6 deals with the computational load analysis. Finally, conclusions are drawn in Sect. 7.

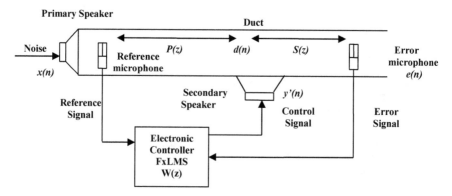

Fig. 1 Schematic diagram of ANC system

2 Virtual Sensors

A device which measures a physical quantity and converts it to a signal which can be observed and processed by the observer on a suitable instrument is known as a sensor. For example, piezoelectric and pyroelectric sensors convert pressure and heat into electrical signals. Sensors can be broadly classified into two categories: physical sensors (PS) and virtual sensors (VS). Physical sensors usually take direct measurements of a physical quantity at the place to be sensed. Virtual sensors, also called soft sensors or estimators, measure physical quantity for physical sensors at a virtual place, i.e., away from the place where actual physical quantity is to be sensed and calculate output using complex models. Virtual sensors are used in many applications such as intelligent transportation systems, computer science applications, and active noise cancelation [6].

In the present work, virtual sensor is applied in active noise control at the desired location called a zone of silence (ZoS), away from the physical sensor and the effect on the performance is observed by considering the different secondary path lengths.

3 Virtual Sensing Technique

Virtual sensing technique involves the shifting of the zone of silence to the desired location. This technique finds its application in practical situations where error microphones cannot be placed at the desired location due to feasibility problems such as medical issues. In this technique, a zone of silence (ZoS) is created away from error microphone and that is the most challenging aspect of this technique [7, 8]. A microphone is placed at the virtual location for the purpose of system identification which is removed later when noise cancelation setup is running [9].

Thus, three acoustic paths are formed as shown in Fig. 2, i.e., between secondary loudspeaker and virtual microphone location $S_v(z)$, between secondary loudspeaker and physical microphone location $S_p(z)$ and between physical and virtual microphone location $H(z)$. In Fig. 2 , $\hat{S}_v(z)$, $\hat{S}_p(z)$, and $\hat{H}(z)$ are used which are the estimate of $S_v(z)$, $S_p(z)$ and $H(z)$ and are found during the process of system identification. Figure 3 [8] shows the block diagram of ANC using virtual sensing feed-forward architecture depicting the control system approach which is validated using basic control system theory. $P(z)$ represents the primary path, and $W(z)$ represents the coefficients of control filter.

As seen from Fig. 3 desired signal, $d(n)$ is given as

$$d(n) = p(n) * x(n) \tag{1}$$

Output at canceling loudspeaker, $e(n)$ is calculated as

$$e(n) = d(n) - y(n) \tag{2}$$

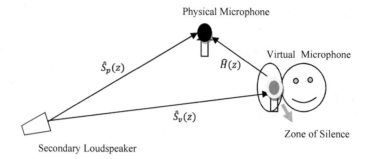

Fig. 2 ANC using virtual sensing technique

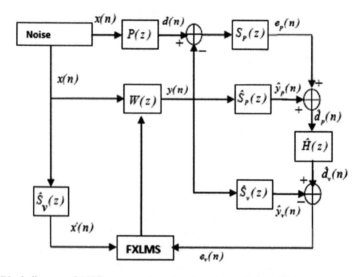

Fig. 3 Block diagram of ANC system using virtual sensing technique [8]

Error signal at physical, $e_p(n)$ location is given as

$$e_p(n) = s_p(n) * e(n) \tag{3}$$

Estimate of the primary disturbance $d_p(n)$ is calculated as

$$\hat{d}_p(n) = e_p(n) + \hat{s}_p(n) * y(n) \tag{4}$$

Estimate of primary disturbance at virtual location $\hat{d}_v(n)$ is calculated using Eq. (5).

$$\hat{d}_v(n) = h(n) * \hat{d}_p(n) \tag{5}$$

Error signal at virtual location is given as Eq. (6)

$$e_v(n) = \hat{d}_v(n) - \hat{s}_v(n) * y(n) \tag{6}$$

Weights of control filter is updated using Eq. (7) as

$$w_i(n+1) = w_i(n) + \frac{\alpha}{(N+M)P_{x'}(n)} * x'(n-i) * e_v(n) \tag{7}$$

where

N   is the length of secondary path.
M   is the length of the primary path that exists between the noise source and canceling loud speaker which is kept equal to N in proposed work.

$$P_{x'}(n) = \beta * P_{x'}(n) + [1 - \beta] * x'(n) * x'(n) \tag{8}$$

4   Secondary Path Modeling and Path Estimation

The secondary path is the path existing between canceling loudspeaker and error microphone which includes both acoustic and electrical paths as shown in Fig. 4. The acoustic path being the path from which the sound wave travels and the electrical path counts for ADC, DAC, pre-amplifiers, anti-aliasing filters, etc. The success of ANC system in real-time applications depends upon the accuracy of secondary path identification. Several techniques have been proposed and compared for secondary path modeling [10]. Offline secondary path modeling is done prior to the real-time evaluation of the system. In the present work, system identification has been done using offline secondary path modeling technique.

In the proposed work, DSP kit TMS320C6713 is used for the purpose of system identification. Code Composer Studio version 3.1 (CCSv3.1) was used to interface DSP kit with the system. To determine $\hat{S}_v(z)$, $\hat{S}_p(z)$ and $\hat{H}(z)$, FIR filters were implemented whose weights are updated using LMS algorithm. White noise was generated through the speakers and was captured using error microphone which in turn was connected to DSK6713. The final weights of the FIR filter after updating process were obtained in the memory of CCS. These final weights are the coefficients of secondary path. After the identification of these three paths, virtual microphone is removed. The above-mentioned procedure is followed for different filter lengths (i.e., 64, 128, and 256). The different path lengths are then used in ANC system and performance comparison is done.

Figure 5 shows the complete hardware setup used for real-time system identification. It consists of speakers for the purpose of generating white noise, a microphone for capturing the white noise, mixed signal oscilloscope (MSO) for viewing the FIR

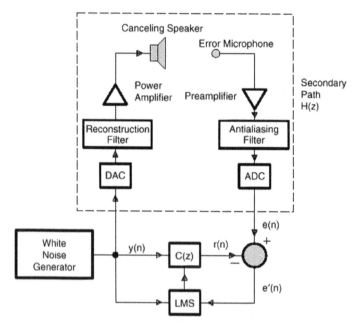

Fig. 4 Modeling of the secondary path of an ANC system

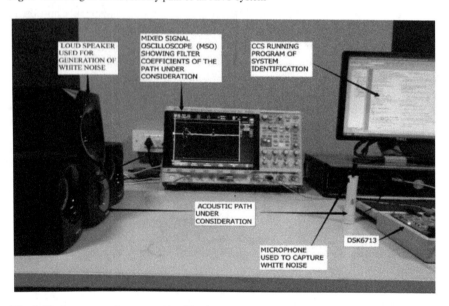

Fig. 5 Hardware setup for system identification

filter weights or secondary path, DSP TMS320C6713 processor which implements FIR filter. The experiments are carried out in semi-anechoic chamber to nullify the effect of external noise.

5 Simulations and Results

In this section, the filter coefficients of secondary paths with three different filter lengths, i.e., 64, 128, and 256, obtained using hardware setup are used for simulating ANC system using FxLMS algorithm with and without virtual sensing technique. The results are compared for FxLMS and virtual sensing techniques. The different filter lengths have been considered to observe its effect on the performance of ANC system. For these different filter lengths, parameters such as signal-to-noise ratio (SNR), smoothened ensemble average squared error (SEASE), have been analyzed and compared. Furthermore, error plots are also analyzed to show the effect to filter length on each of these techniques. For simulation purpose, step size (α) is taken as 0.05, length of the input signal is taken as 10^5 samples, and the sampling frequency is 16 kHz.

5.1 Reference Noise Signal

The analysis has been done for the random noise signal which is generated using MATLAB through *rand* command. The length of input for analysis is taken as 100,000 samples. Figure 6 shows the first 1000 samples of the input signal used for analysis. Power density spectrum (PSD) of complete 100,000 samples is shown in Fig. 7. From PSD plot, it is evident that there is no dominating frequency component and power is distributed almost equally throughout all frequencies. Thus, input signal closely resembles the white noise.

5.2 Signal-to-Noise Ratio (SNR)

For observing to what extent noise is suppressed signal-to-noise ratio (SNR) is calculated using Eq. (9).

$$SNR = 10 \log_{10}\left(\Sigma d^2(n)/\Sigma e^2(n)\right) \tag{9}$$

where

$e(n) = e_p(n); d(n) = d_p(n)$ for calculating SNR at physical location for FxLMS technique.

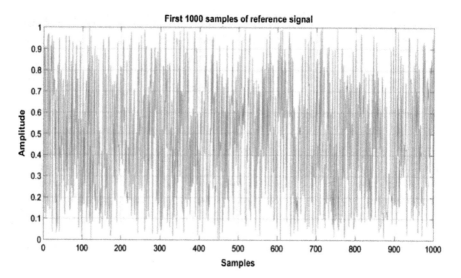

Fig. 6 Reference (random noise) signal (first 1000 samples)

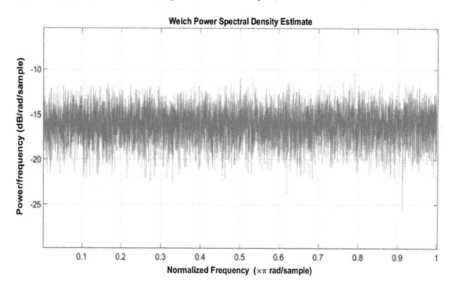

Fig. 7 Power spectrum estimate of input signal

$e(n) = e_v(n); d(n) = d_v(n)$ for calculating SNR at virtual location for virtual sensing technique.

$e(n) = e_v(n); d(n) = d_p(n)$ for calculating SNR at virtual location for FxLMS technique.

$e(n) = e_p(n); d(n) = d_v(n)$ for calculating SNR at physical location for virtual sensing technique.

Table 1 SNR values for different filter lengths for ANC system using FxLMS and virtual sensing technique

Filter length (K)	SNR (in dB) physical location		SNR (in dB) virtual location	
	FxLMS	Virtual sensing	FxLMS	Virtual sensing
64	19.4954	2.1068	0.9415	21.0156
128	21.5484	2.6731	0.7129	20.0663
256	12.7705	−2.374	−1.5985	12.3047

Above-mentioned three acoustic paths are determined for three different filter lengths viz. 64, 128, and 256. Corresponding to each, SNR at the physical location and the virtual location is calculated and results are shown in Table 1.

From Table 1, it has been concluded that for filter length (K) of 256, SNR for virtual sensing technique at the virtual location and SNR for FxLMS technique at the physical location have the lowest value among all three. Furthermore, for $K = 256$, the negative value of SNR for FxLMS and virtual sensing technique at virtual and physical location, respectively, indicates that noise level at that location is more than the signal level. The reason for the low performance of ANC system for $K = 256$ is that higher filter length involves more computational load, due to which it takes more time for the processor to generate the anti-noise signal and hence proper overlapping of input noise and anti-noise signal does not take place due to which performance of ANC system degrades.

For FxLMS technique, $K = 128$ gives the highest SNR which is 21.5 dB, and in case of virtual sensing technique, $K = 64$ gives the best SNR which is 21.0 dB. If filter length is kept below 64, then it also affects the performance of ANC system as for K less than 64, and acoustic path estimation is not proper. During the process of system identification, minimum filter length needs to be maintained so that proper estimation of the path is obtained.

5.3 Smoothened Ensemble Average Squared Error (SEASE)

It is observed that for different values of step size (α), error plot has slightly different shapes. So, to generalize and account for the effect of α, a smoothened ensemble average squared error (SEASE) plot was analyzed. While computing SEASE plots, different values of alpha were considered. Corresponding to each value of alpha, the error signals were computed and processed. SEASE plots for FxLMS and virtual sensing technique using different filter lengths are shown in Fig. 8 and Fig. 9, respectively.

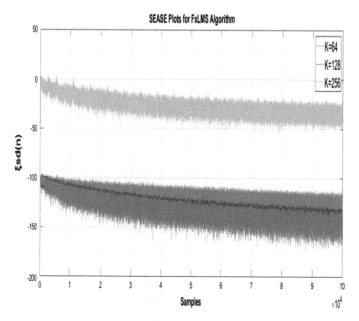

Fig. 8 SEASE plots for FxLMS algorithm-based ANC

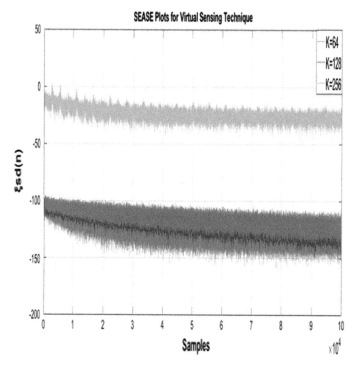

Fig. 9 SEASE plots for virtual sensing technique

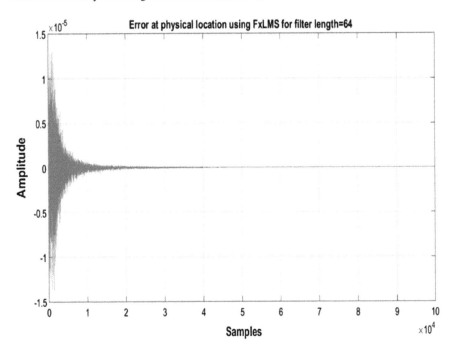

Fig. 10 Error plot for k = 64 using FxLMS at physical location

5.4 Error Plots

Error plots give the measure of error between input noise signal and the anti-noise signal. Our goal is to reduce this error as low as possible. From error, one can also observe how effectively and quickly an algorithm works to cancel out the noise.

From Figs. 10 and 11, it is evident that virtual sensing technique gives better results from FxLMS for K = 64 which is in accordance with Table 1. Adaptation is more rapid and steady-state value is achieved faster in case of virtual sensing technique as compared to FxLMS.

From Figs. 12 and 13, it is evident that FxLMS technique performs better for K = 128. Steady-state value is much larger in case of virtual sensing, and thus, low SNR value is achieved which is in accordance with Table 1. Thus, for filter length equal to 128 FxLMS outperforms virtual sensing technique.

From Figs. 14 and 15, it is evident that FxLMS and virtual sensing technique perform equally for K = 256. It takes a longer time to reach the steady-state value for both, and hence, low SNR values are obtained for both, which is the case according to Table 1.

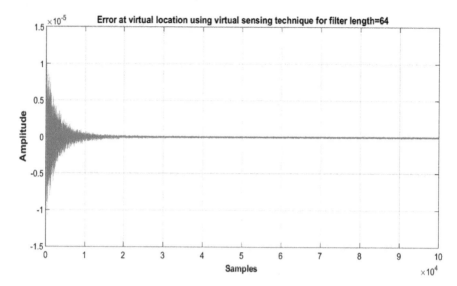

Fig. 11 Error plot for k = 64 using virtual sensing technique at virtual location

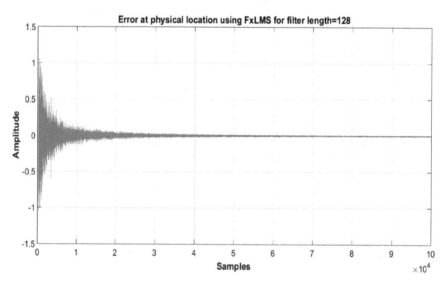

Fig. 12 Error plot for k = 128 using FxLMS at physical location

6 Computational Load

While computing computational load, the whole algorithm was divided into three independent parts, i.e., system identification, filtering and estimation, weight upda-tion. A number of additions, subtractions, multiplications, and divisions are computed

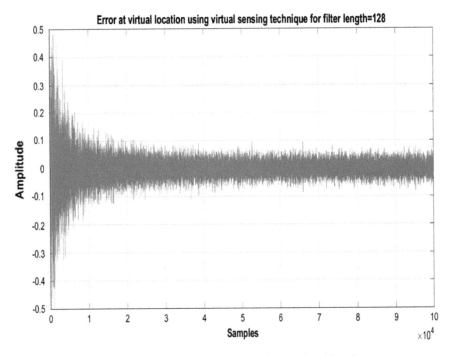

Fig. 13 Error plot for k = 128 using virtual sensing technique at virtual location

Table 2 Comparison of computational load (in general) for FxLMS and virtual sensing technique

Calculations	Virtual sensing				FxLMS			
	Additions	Subtractions	Multiplications	Divisions	Additions	Subtractions	Multiplications	Divisions
System identification	12 K − 3	3	9 K	Nil	4 K − 1	1	3 K	Nil
Filtering and estimation	(7 K + 2 h − 2)L	2L	(5 K + h)L	Nil	(7 K − 1)L	L	5 KL	Nil
Weight updation	(5K − 1)L	L	(4 K + 3)L	L	(5 K − 1)L	L	(4 K + 3)L	L
Total	(12 K + 2 h − 3)L + 12 K − 3	3L + 3	(9 K + h + 3)L + 9 K	L	(12 K − B2)L + 4 K − 1	2L	(9 K + 3)L + 3 K	L

for each of three above-mentioned parts. Their values are shown in Table 2. K is filter length, L is the no. of samples or signal length, and h is the length of $\hat{H}(z)$.

From Table 2, it is evident as the filter length (K) increases, number of additions and multiplications increases. Furthermore, the total number of subtractions and divisions in both the techniques are constant w.r.t. filter length (k).

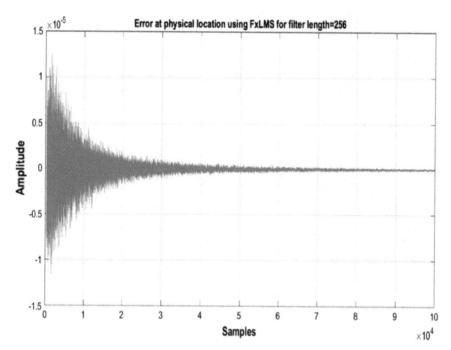

Fig. 14 Error plot for k = 256 using FxLMS at physical location

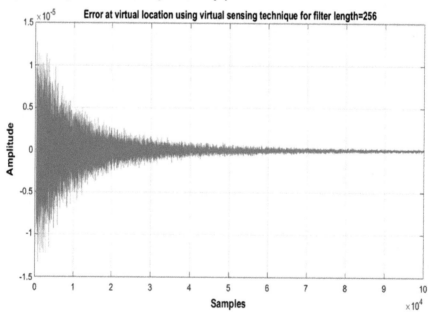

Fig. 15 Error plot for k = 256 using virtual sensing technique at virtual location

7 Conclusion

The filter coefficients for the secondary path of ANC system are obtained experimentally for filter lengths of 64, 128, and 256. Their effect on the performance of ANC system is analyzed for different parameters. Signal-to-noise ratio (SNR) of 21.5 dB is best achieved for filter length of 128 for the FxLMS algorithm, and for virtual sensing technique based ANC system, the filter length of 64 gives best SNR of 21 dB. The error (SEASE) plots were obtained and analyzed which indicate that filter length K of 128 for FxLMS and of 64 for virtual sensing technique based ANC system gives best result. The error plots for FxLMS and virtual sensing ANC for different filter lengths are compared. The computational load of different techniques is calculated and its comparison for different filter lengths is done. It is concluded that selection of appropriate secondary path length and its filter coefficient is vital for good performance and real-time implementation of ANC system.

Acknowledgements The work in this article is supported by Special Assistance Programme (SAP) of University Grants Commission (UGC), New Delhi [grant no. F.3.32, 2012].

References

1. Kuo, S. M., & Morgan, D. (1996). *Active noise control systems-algorithms and dsp implementation.* New York: Wiley.
2. Bjarnason, E. (1995). Analysis of the filtered-x LMS algorithm. *IEEE Transactions on Speech and Audio Processing, 3*(6), 504–514. https://doi.org/10.1109/89.482218.
3. Omour, A., Zidouri, A., Iqbal, N., & Zerguine, A. (2016). Filtered-X least mean fourth (FXLMF) and leaky FXLMF adaptive algorithms. *EURASIP Journal on Advances in Signal Processing, 39.* https://doi.org/10.1186/s13634-016-0337-z.
4. Zeb, A., Mirza, A., & Sheikh, S. A. (2015). Filtered-x RLS algorithm based active noise control of impulsive noise. In *7th International Conference on Modelling, Identification and Control (ICMIC).* IEEE. https://doi.org/10.1109/icmic.2015.7409414.
5. Kuo, S. M., & Morgan, D. (1999). Active noise control: a tutorial review. *Proceedings of the IEEE, 87*(6), 943–973. https://doi.org/10.1109/5.763310.
6. Liu, L., Kuo, S. M., & Zhou, M. (2009). Virtual sensing techniques and their applications. In *International Conference on Networking, Sensing and Control, ICNSC '09.* https://doi.org/10.1109/icnsc.2009.4919241.
7. Garcia-Banito, J., Elliott, S. J., & Boucher, C. C. (1997). Generation of zones of quiet using a virtual microphone arrangement. *Journal of the Acoustical Society of America, 101*(6), 3498–3516. https://doi.org/10.1121/1.418357.
8. Sharma, M. K., Vig, R., Pal, R. & Veena, S. (2018). Quiet zone for the patient in an ambulance: Active noise control technology for siren noise reduction. Archives of Acoustics (in press).
9. Kestell, C. D., Cazzolato, B. S., & Hansen, C. H. (2001). Active noise control in a free field with virtual sensors. *Journal of the Acoustical Society of America, 109*(1), 232–243. https://doi.org/10.1121/1.1326950.

10. Gupta, P., Sharma, M. K, & Thangjam, S. (2015). Ambulance siren noise reduction using noise power scheduling based online secondary path modeling for ANC System. In *International Conference on Signal Processing, Computing and Control (ISPCC)*. IEEE Explore (pp. 63–67). https://doi.org/10.1109/ispcc.2015.7374999.

FPGA Realization of Scale-Free CORDIC Algorithm-Based Window Functions

Shalini Rai and Rajeev Srivastava

Abstract Filtering is an immense process in spectral analysis of signals. In designing of filters, window functions are usually used. In this paper, we present the variety of window functions based on the scale-free COordinate Rotation DIgital Computer (CORDIC) algorithm for the target angle which covers the complete coordinate space. To overcome the problem of more occupied area and speed, we present a study of a different design that is scale-free CORDIC algorithm-based window function architectures. The current paper presents the simulation and synthesis results of two designs which are coded in very high speed integrated circuit hardware description language (VHDL). The Xilinx 13.1 software is used for the simulation and synthesis of coded design, and also these designs are mapped into Virtex-5(XC5VLX20T-FF323) field-programmable gate array (FPGA) device.

Keywords Window functions · Field-programmable gate array (FPGA) Scale-free CORDIC processor

1 Introduction

Window filtering techniques [1] are widely used and important methods in signal processing. These functions are used for both time and frequency-based signal processing. For finite impulse response filters (FIR), several window functions are developed depending upon the requirements like reduction of sidelobe, dynamic range. In the past few decades, hardware efficient VLSI architectures for window function generator were designed using lookup tables but these architectures occupied large space and have more latency and word length. By this technique, the length of window could not vary. The fixed length window functions are inefficient. Therefore, a flexi-

S. Rai · R. Srivastava (✉)
Department of Electronics and Communication, University of Allahabad, Allahabad, India
e-mail: rajeev_jk@rediffmail.com

S. Rai
e-mail: shaliniece_04@yahoo.com

© Springer Nature Singapore Pte Ltd. 2019
A. Khare et al. (eds.), *Recent Trends in Communication, Computing, and Electronics*, Lecture Notes in Electrical Engineering 524,
https://doi.org/10.1007/978-981-13-2685-1_24

ble size window-based functions which are reconfigurable were implemented based on CORDIC algorithm [2, 3]. The CORDIC algorithm has various applications like design of digital filters [4], FFTs [5] and several window functions. Also, CORDIC is useful in the calculation of many transcendental algebraic functions, which can be used in various applications, VLSI design such as multiplication, division, hyperbolic tangent and sigmoid function. CORDIC is an efficient hardware and has simplicity as well as low computational complexity property. The major drawback of the CORDIC implementation is that it results in high latency or large expense of hardware of scale factor compensation network design. For the minimization of latency and reduction of numbers of iterations, parallel CORDIC architectures [6, 7] have been proposed. In parallel CORDIC architectures, the latency is reduced but the cost of hardware and time to implement the scale factor compensation is increased. The savings of hardware obtained by employing variable scale factor [8, 9] compensation network, but these methods increase the area or otherwise affect throughput or latency. Scale-free CORDIC processor [10–12] has been suggested for optimum solution for hardware area savings and less complex hardware as well as high throughput and low latency. Our proposed work deals with the study of scale-free CORDIC processor design with two different window functions of variable length. That is, in the proposed study, reconfigurable architectures have been used.

Rest of the paper is organized as follows: Sect. 2 presents overview of CORDIC algorithm. In Sect. 3, we present design aspects of the scale-free CORDIC processor. The design of scale-free CORDIC processor-based architecture of different window functions is given in Sect. 4. Section 5 presents simulation results along with the performance measures and finally the conclusions are given in Sect. 6.

2 CORDIC Algorithm

(A) Basic CORDIC Algorithm:

A great scientist Jack. E. Volder is an inventor of an original CORDIC algorithm [13, 14]. It converts the rectangular coordinate s(x, y) to polar coordinates (R, θ). It is a shift and adds steps to perform the vector rotation. The basic CORDIC algorithm equations are:

$$x' = x \cos\theta - y \sin\theta = \cos\theta(x - y\tan\theta) \tag{1}$$

$$y' = y \cos\theta + x \sin\theta = \cos\theta(y + x\tan\theta) \tag{2}$$

If the rotation angle θ is divided into a set of small angles for rotation in a set of steps θ can be approximated by $\theta = \sum_{i=0}^{n} \delta i\theta i$, where $\delta_i = \{1, -1\}$, δi is the sign of rotation (+ve for counterclockwise and −ve for clockwise rotation). There are several admissible values that may be chosen for the rotation steps. If the iteration is chosen as $\theta_i = \tan^{-1} 2^{-i}$. This value is selected because it is easier to implement in hardware; therefore, the new coordinates after each rotation (x_{i+1}, y_{i+1}) can be expressed as

$$x_{i+1} = \cos\theta i(xi - yi . \delta i . 2^{-i}) \tag{3}$$

$$y_{i+1} = \cos\theta i(yi + xi . \delta i . 2^{-i}) \tag{4}$$

$$K_i = \cos\theta_i = \frac{1}{\sqrt{1 + 2^{-2i}}} \tag{5}$$

$$K = \prod_0^N K i \text{ is} \tag{6}$$

Defined as the scale factor.

(B) Unified CORDIC Algorithm:

In 1971, J. S. Walther reinvented the generalized CORDIC algorithm [15, 16] having three different trajectories like circular ($m = 1$), linear ($m = 0$) and hyperbolic ($m = -1$). For each trajectory, two rotation directions are included (vectoring and rotation). For vectoring, a vector with starting coordinates (x_0, y_0) is rotated in such a way that the vector finally lies on the abscissa by iteratively converging y_n to zero. For a rotation, a vector with starting coordinates (x_0, y_0) is rotated by an angle θ_0 in such a way that the final value of the angle register converges to zero. The unified CORDIC algorithm is defined as follows:

$$\begin{bmatrix} x_{i+1} \\ y_{i+1} \end{bmatrix} = K_i \begin{bmatrix} 1 & -m\delta_i 2^{-i} \\ \delta_i 2^{-i} & 1 \end{bmatrix} \begin{bmatrix} x_i \\ y_i \end{bmatrix} \tag{7}$$

$$K_i = \frac{1}{\sqrt{1 + m . 2^{-2i}}} \tag{8}$$

$$\theta = \sum_{i=0}^n \delta_i \theta_i \tag{9}$$

$$\theta_i = \frac{1}{\sqrt{m}} \tan^{-1} \sqrt{m} . 2^{-i} \tag{10}$$

where $m = \begin{cases} -1 & \textit{for hyperbolic} \\ 0 & \textit{for linear} \\ 1 & \textit{for circular} \end{cases}$

There are six operational modes exist by using different combination of three trajectories and two modes, and they are summarized in Table 1.

Direct computation—multiplication—$x \times y$, Division—$\frac{y}{x}$.

Trigonometric functions-$\sin z$, $\cos z$, $\tan z$, $\sinh z$, $\cosh z$, $\tanh z$, $\tan^{-1} z$, $\tanh^{-1} z$.

Additional function may be computed by choosing appropriate combination of multiples modes of operation and appropriate initialization.

Table 1 Modes m of operation for the CORDIC algorithm

m	$z_n \to 0$	$y_n \to 0$	θi
1	$X_n = k(x_0 \cos z_0 - y_0 \sin z_0)$ $Y_n = k(y_0 \cos z_0 + x_0 \sin z_0)$ $k = 1.647$	$X_n = \sqrt{x_0^2 + y_0^2}$ $Z_n = z_0 + \arctan\left(\frac{y_0}{x_0}\right)$	$\arctan 2^{-i}$
0	$X_n = x_0$ $Y_n = y_0 + x_0 z_0$	$X_n = x_0$ $Z_n = z_0 + \frac{y_0}{x_0}$	2^{-i}
-1	$X_n = k'(x_0 \cosh z_0 + \sinh z_0)$ $Y_n = k'(x_0 \cosh z_0 + \sinh z_0)$ $k' = 0.828$	$X_n = \sqrt{x_0^2 + y_0^2}$ $Z_n = z_0 + \text{arctanh}\left(\frac{y_0}{x_0}\right)$	$\text{arctanh} 2^{-i}$

$$\tanh z = \frac{\sinh z}{\cosh z} \qquad \text{modes: } m = -1, 0$$
$$e^z = \sinh z + \cosh z \quad \text{modes: } m = -1$$
$$y = f(z) = 1/1 + e^{-z} \text{ modes: } m = -1, 0$$

(C) Scale-free CORDIC Algorithm

There are several improvements in CORDIC algorithm. For improvements of architecture performance and reduction of cost, an abundance of development has been
established in the area of algorithm design and advancement of architecture. For
enhancement of throughput the parallel and pipelined CORDIC architectures are preferred. The pipelined scaling-free CORDIC [10–12] is a very enormous development
in the research area of upgradation of the CORDIC algorithm. The Taylor series-
based scale-free CORDIC algorithm is a great invention in the field of CORDIC
algorithm improvements [17]. The rotation matrix for scaling-free CORDIC is given
as:

$$Rp = \begin{bmatrix} 1 - 2^{-(2i+1)} & -2^{-i} \\ 2^{-i} & 1 - 2^{-(2i+1)} \end{bmatrix} \qquad (11)$$

The sine and cosine approximated to

$$\sin \alpha_i = 2^{-i} \qquad (12)$$
$$cos\alpha_i = 1 - 2^{-(2i+1)} \qquad (13)$$

The conventional CORDIC processor gives two direction rotations but the Taylor
series-based scale-free CORDIC processor gives only one direction rotation. The
Taylor series of sine and cosine terms defined as

$$\cos \alpha_i = \sum_{n=0}^{\infty} (-1)^n \frac{\alpha i^{2n}}{!2n} = 1 - \frac{\alpha i^2}{!2} + \frac{\alpha i^4}{!4} - \qquad (14)$$

$$\sin \alpha_i = \sum_{n=0}^{\infty} (-1)^n \frac{\alpha i^{2n+1}}{!2n+1} = \alpha i - \frac{\alpha i^3}{!3} + \frac{\alpha i^5}{!5} - \qquad (15)$$

(D) Window Filtering Techniques

During spectral analysis, the input signals are to be truncated to fit a finite observation window according to the length of FFT processor. In frequency domain, there are several phenomena occur like picket fence effect and spectral leakage due to the direct truncation by using rectangular window. There are some different window functions by which we reduce these effects. Window filtering is a popular process for limiting any signals to small-time segments in a desired fields. The most common accessible windowing techniques are rectangular, Gaussian, Hamming, Hanning, Blackman-Harris and Kaiser. The assortment of the available windows depends on the spectral characteristics desired by the applications. As given below, the equations explain the Hanning, Hamming and the Blackman window functions [18]

$$W_{Hann}(n) = 0.5 - 0.5 \cos\left(\frac{2\pi n}{(N-1)}\right) \qquad (16)$$

where N is the window length

$$W_{Hamm}(n) = \alpha - \beta \cos\left(\frac{2\pi n}{(N-1)}\right) \qquad (17)$$

where $\alpha + \beta = 1$.

To maximize sidelobe cancellation, the values of α and β are determined. For Hamming window, the coefficients are calculated as $\alpha = 25/46$ and $\beta = 21/46$.

$$W_{Blackman}(n) = \alpha_0 + \alpha_1 \cos\left(\frac{2\pi n}{N}\right) + \alpha_2 \cos\left(\frac{4\pi n}{N}\right) \qquad (18)$$

where $\alpha_0 + \alpha_1 + \alpha_2 = 1$. The Blackman window with coefficients $\alpha_0 = 0.42$, $\alpha_1 = 0.5$ and $\alpha_2 = 0.08$.

3 Design Aspects of Scale-Free CORDIC Processor

Decomposition of angle of rotation into micro-rotations in conventional CORDIC, the angle of rotation is used as follows: (i) the elementary angles are defined according to the 2^{-i} where i is the no iteration and ROM is used as a storage circuit for the elementary angles, (ii) the micro-rotation corresponding to all the elementary angles are performed in clockwise or anti-clockwise and (iii) each elementary angle is non-repeated, but in scale-free CORDIC processor, the micro-rotations are rotated in only one direction with multiple times corresponding to the initial shifts, and for other shifts, non-repeated iterations are included.

(a) For elimination of the ROM which is used for storage of elementary angles and for simplification of the hardware define the elementary angles [19] as: $\alpha i = 2^{-s_i}$ where s_i is the number of shifts for ith iteration.

(b) The most significant one location represents the bit position of the one (1) in an input string of bits starting from most significant bit (MSB). The MSO location identifier (MSO-LI) generates an n-bit output for a 2^n bit input string. It is used for finding the shift index. $s_i = N - M$, N is the word length of the input data and M is the location of the most significant bit (one) in N input string.

(c) The order of approximation of Taylor series decides the largest elementary angle. The basic shift and the largest elementary angle for third order of approximation are to be:

$$S_b = \left\lfloor \frac{l - \log 2(4!)}{4} \right\rfloor \tag{19}$$

$$\alpha_{max} = 2^{-S_b} \tag{20}$$

where l is the word length. For 16-bit word length, $s_b = \lfloor 2.854 \rfloor$. Depending upon the desired accuracy, one can either select $s_b = 2$ or $s_b = 3$. Any rotation angle θ is expressed as:

$$\theta = n_1 . \alpha_{max} + n_2 . \sum \alpha_{si} \tag{21}$$

where $s_i \geq s_b$ and $n = n_1 + n_2$, n is the total number of iterations 'n' is a constant. The number of frequentness for third-order Taylor series approximation is seven.

For designing of scale-free CORDIC processor and the micro-rotation sequence generation, we take input angle to be rotated θ_i and most significant ones bit location is represented by ML (location identifier). If ML = 15, then elementary angle $\alpha = 0.25$ radians, shift si = 2, $\theta_{i+1} = \theta_i - \alpha$. If ML is other than 15, then shift $si = 16 - ML$ and $\theta_{i+1} = \theta_i$ with $\theta_i[ML] = '0'$.

Table 2 shows that the elementary angles corresponding to the basic shift values.

Table 2 Elementary angles versus corresponding shifts [19]	Shifts (si)	Elementary	Angle (α_i)
		Decimal	16-bit hexadecimal
	2	0.25	4000H
	3	0.125	2000H
	4	0.0625	1000H

Fig. 1 Design of coordinate
calculation unit

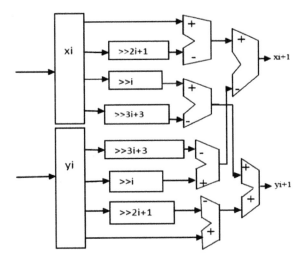

The percentage error for the sin and cos is indistinguishable from the range $(0, \pi/4)$. So the maximum angle of rotation handled by micro-rotation sequence generation lies in the range $(0, \pi/4)$.

The following points are important for the designing of micro-rotation sequence generator.

(i) For $(N - MSOB_{location} < s_b)$. Then the shift index would be used corresponding the highest elementary angle $\alpha_{max} = 0.25$ radians with shift index $= 2$.

(ii) For $(N - MSOB_{location} \geq s_b)$. The highest elementary angle (α_{si}) would be employed for the CORDIC iteration corresponding to the $s_i = 16 - M$.

The third-order Taylor series augmentation of sine–cosine functions gives the revolving matrix for the proposed architecture. In complete scale-free CORDIC algorithm for simplification of equations of rotation matrix, the Taylor series coefficient!3 is shifted by 2^3.

$$Ri = \begin{bmatrix} 1 - 2^{-(2si+1)} & -(2^{-si} - 2^{-(3si+3)}) \\ 2^{-si} - 2^{-(3si+3)} & 1 - 2^{-(2si+1)} \end{bmatrix} \tag{22}$$

Figure 1 shows the coordinate calculation unit by which calculate the x_i and y_i value. Shift index calculation s_i unit shown in Fig. 2. Shift index calculation depends on the elementary angles. The elementary angles' calculation or micro-rotation sequence generator unit is shown in Fig. 3.

Fig. 2 Shift index si
calculation unit

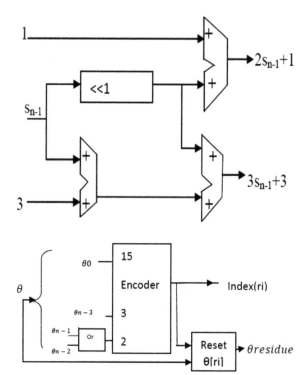

Fig. 3 Micro-rotation
sequence generator

4 Architecture for Window Functions

For implementation of windows functions, we use the pipelined architecture. We
have designed the window architecture for 16-bit output width. Here we designed
the window functions by using circular CORDIC processor, linear CORDIC processor and angle generator circuit and these window functions are also designed by
using circular CORDIC processor, window coefficient multiplier which is designed
by using booth multiplier. Figure 4 shows the block diagram of Angle Generator Unit.
Figures 5 and 6 show the block diagram for generating different window functions.
The different window functions are depending on the window select pins as the Hanning (ws0=0, ws1=0), Hamming (ws0=1, ws1=0), Blackman (ws0=0, ws1=
1) [3, 18] window families. The circuit is the combination of blocks of angle generator unit (AGU), window coefficient multiplier (WCM), circular CORDIC processor
(CCP) and first input first output register. Angle generator unit generates two angles
$\theta = \frac{2\pi n}{N}$ and $2\theta = \frac{4\pi n}{N}$. For multiplication of the window coefficient used a linear
CORDIC which is based on conventional CORDIC algorithm or optimized shift-add
network which is designed using booth multiplier. CORDIC processor which is in
rotation mode and circular trajectory is employed for producing the cosine terms,
and it is used in the window functions equations.

Angle generator unit produces two angles for evaluating the window functions.

Fig. 4 Angle generator unit

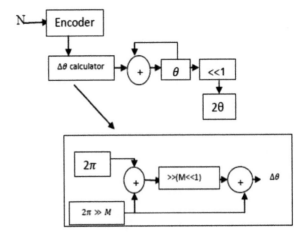

Fig. 5 Block diagram for generating window functions using LCP and CCP

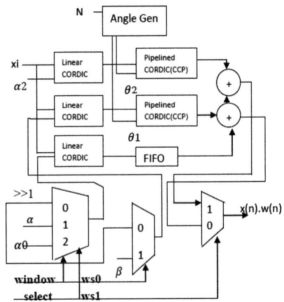

$$\theta = \frac{2\pi n}{N} \text{ and } 2\theta = \frac{4\pi n}{N} \tag{23}$$

where N is a multiple of 2 such that $N = 2^M$

The difference between the consecutive values of θ is given by

$$\Delta\theta = \theta_{n+1} - \theta_n, \quad \Delta\theta = \frac{2\pi}{(N-1)} \tag{24}$$

Fig. 6 Block diagram for
generating window functions
using shift-add n/w and CCP

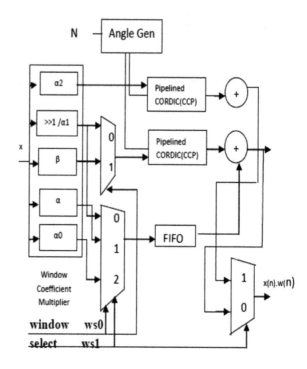

For $N = 2^M$

$$\Delta\theta = \frac{2\pi}{(2^M - 1)} = \frac{2\pi \left(1 - 2^{-M}\right)^{-1}}{2^M} \tag{25}$$

Using binomial theorem (BT), we simplify to

$$\Delta\theta = \frac{2\pi}{2^M} + \frac{2\pi}{2^{2M}} + \frac{2\pi}{2^{3M}} \tag{26}$$

The CCP unit is designed for the target angle range $[0, \pi/4]$. The range of target angle is enhanced by using the octant symmetry which is shown in Fig. 7 and Table 3 shows the initial coordinate values for enhancement of the angle of target angle.

5 FPGA Implementation Results

Scale-free CORDIC algorithm-based window functions architectures designed using linear CORDIC and circular CORDIC processor and also by using add-shift network and circular CORDIC processor. These are designed by using Xilinx13.1 VHDL module and are mapped into Virtex-5(XC5VLX20T-FF323) device. Table 4 shows

Fig. 7 Octant symmetry

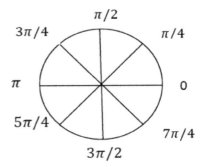

Table 3 Initial coordinate values for octant symmetry

Angle	Coordinate values
$[0, \pi/4]$	(x, y)
$[\pi/4, \pi/2]$	(y, x)
$[\pi/2, 3\pi/4]$	(−y, x)
$[3\pi/4, \pi]$	(−x, y)
$[\pi, 5\pi/4]$	(−x, −y)
$[5\pi/4, 3\pi/2]$	(−y, -x)
$[3\pi/2, 7\pi/4]$	(y, −x)
$[7\pi/4, 2\pi]$	(x, −y)

Table 4 Complexity comparison: window function generator

ROC (radians)	[−0.785, 0.785]	[−0.785, 0.785]
Parameters	Window function generator-Design1	Window function generator-Design2
	LCP+CCP	Shift-add N/W + CCP
BELs	8053	9349
Flip flops	2153	3177
Max. Freq. MHz (A)	70.598	70.961
No. of iterations (B)	26	10
Delay nsec. (B/A)	368	140
Total delay	50.786 ns	2.826 ns
Logical delay	8.850 ns, 17.4%	2.540 ns, 89.9%
Route delay	41.936 ns, 82.6%	0.286 ns, 10.1%

that for 16-bit implementation, the first design consumes 1128 slices and 7674 4-input LUTs, with a maximum operating frequency 70.598 MHz. The total delay is 50.786 ns. In this design, logical delay is 8.850 ns and route delay is 41.936 ns. The second design consumes 1098 slices and 8119 4-input LUTs, with a maximum operating frequency 70.961 MHz. The total delay is 2.862 ns in which 2.54 ns is for logical delay and 0.286 ns for route delay.

5.1   Area

In the design with linear and circular CORDIC processor, seven, 16-bit adder/subtractor and 261 registers were used. The number of latches, comparators and multiplexers used is 94,803 and 150, respectively. The number of XOR logic gate used is 12177.

Device usage summary: Selected Device: 5vlx20tff323-2
Slice Logic Utilization: (a) Number of Slice Registers: 2137 out of 12480-17% (b) Number of Slice LUTs: 7703 out of 12480-61% (c) Number used as Logic: 7703 out of 12480-61%.

In the design with add-shift network and circular CORDIC processor, 106, 16-bit adder/subtractor and 98, 32-bit adder/subtractor were used. The numbers of registers and latches used are 1232 and 142, respectively. The comparators are 803. The number of XOR logic gate is 4752.

Device usage summary: Selected Device: 5vlx20tff323-2. *Slice Logic Utilization*: (a) Number of Slice Registers: 3146 out of 12480-25% (b) Number of Slice LUTs: 8011 out of 12480-64% (c) Number used as Logic: 8011 out of 12480-64%.

5.2   Latency and Delay

The throughput of all the architecture is equal; it is one data/clock cycle. Latency is the number of iteration of the pipelined CORDIC processor. So it is different for both the architecture. In the first design, there are two circular and three linear CORDIC processors. So the total pipeline stages for first design are 26, while in the second design the total pipeline stages are only 10. This shows that the latency is low in second architecture as compared to the first architecture, and also, the total delay in the second architecture is less as compared to the first architecture.

6   Conclusion

In this paper, we performed comparative study of the two different types of window function generator one is designed using circular CORDIC processor and linear CORDIC processor. Another one is designed using circular CORDIC processor and add-shift network. Add-shift network is designed using booth multiplier. We observe that the total delay is comparatively small in the proposed architecture, i.e. the design with CORDIC processor and add-shift network. This is due to the use of add-shift network reduces the number of pipelining stages which results in the number of iteration and latency. Further, we observe that all the operations of multiplications could be performed directly by the use of add-shift network with booth multiplier.

References

1. Parhi, K. K. (1999). *VLSI digital signal processing systems*. Wiley.
2. Ray, K. C., & Dhar, A. S. (2006). CORDIC–based unified VLSI architecture for implementing window functions for real time spectral analysis. *IEEE Proceedings: Circuits, Devices and Systems, 153*(6), 539–544.
3. Ray, K. C., & Dhar, A. S. (2008). High throughput VLSI architecture for Blackman windowing in real spectral analysis. *Journal of Computers, 3*(5), 54–59.
4. Vaidyanathan, P. P. (1985). A unified approach to orthogonal digital filters and wave digital filters based on the LBR two- pair extraction. *IEEE Transactions on Circuits and Systems I, CAS-32*, 673–686.
5. Banerjee, A., Dhar, A. S., & Banerjee, S. (2001). FPGA realization of a CORDIC based FFT processor for biomedical signal processing. *Microprocessors and Micro System, 25*(3), 131–142.
6. Gisuthan, B., & Srikanthan, T. (2000). FLAT CORDIC: A unified architecture for high speed generation of trigonometric and hyperbolic functions. In *Proceedings of the 43rd IEEE Midwest Symposium on Circuits and Systems*, Lansing MI (pp. 1414–1417).
7. Juang, T. B., Hsiao, S. F., & Tsai, M. Y. (2004). Para-CORDIC: Parallel CORDIC rotation algorithm. *IEEE Transactions on Circuits and Systems I, 51*(8), 1515–1524.
8. Lin, C. H., & Wu, A. Y. (2005). Mixed-scaling-rotation-CORDIC (MSR-CORDIC) algorithm and architecture for high-performance vector rotational DSP applications. *IEEE Transactions on Circuits and Systems I, 52*(11), 2385–2396.
9. Sumanasen, M. G. B. (2008). A scale factor correction scheme for the CORDIC algorithm. *IEEE Transactions on Computers, 57*(8), 1148–1152.
10. Maharatna, K., Troya, A., Banerjee, S., & Grass, E. (2004). Virtually scaling free adaptive CORDIC rotator. *IEEE Proceedings Computers and Digital Techniques, 151*(6), 448–456.
11. Maharatna, K., & Banerjee, S. (2005). Modified virtually scaling free adaptive CORDIC rotator algorithm and architecture. *IEEE Transactions on Circuits and Systems for Video Technology, 15*(11), 1463–1474.
12. Jaime, F. J., Sanchez, M. A., Hormigo, J., Villalba, J., & Zapata, E. L. (2010). Enhanced scaling free CORDIC. *IEEE Transactions on Circuits and Systems Video Technology, 57*(7), 1654–1662.
13. Volder, J. E. (1959). The CORDIC trigonometric Computing technique. *IRE Transactions on Electronic Computers, 8*(3), 330–334.
14. Volder, J. E. (2000). The birth of CORDIC. *Journal of VLSI Signal Processing, 25*(2), 101–105.
15. Walther, J. S. (1971). A unified algorithm for elementary functions. In *Proceedings of AFIPS Spring Joint Computer Conference* (pp. 379–385).
16. Walther, J. S. (2000). The story of unified CORDIC. *Journal of VLSI Signal Processing, 25*(2), 107–112.
17. Meher, P. K., Valls, J., Juang, T. B., Sridhara, K., & Maharatna, K. (2009). 50 years of CORDIC algorithms, architectures and applications. *IEEE Transactions Circuits and Systems I, 56*(9), 1893–1907.
18. Aggarwal, S., Khare. K. (2012). Redesigned-scale-free CORDIC algorithm based FPGA implementation of window functions to minimize area and latency. *International Journal of Reconfigurable Computing, 2012*(185784), 1–8.
19. Aggarwal, S., Meher, P. K., & Khare, K. (2013). Scale-free hyperbolic CORDIC processor and its application to waveform generation. *IEEE Transactions on Circuits and System-I Regular Papers, 60*(2), 314–326.

Part V
Image Processing and Computer Vision

An Acceleration of Improved Segmentation Methods for Dermoscopy Images Using GPU

Pawan Kumar Updhyay and Satish Chandra

Abstract Medical images have made a great influence on medicine, diagnosis, and treatment. The essential element of medical image processing is segmentation for identifying the region of interest. The fundamental methods of image segmentation are unable to process the large dataset of images and require scaling to make them more interactive. In order to address this issue, an exponential entropy-based image segmentation methods are proposed which are based on boundary demarcation, contour- and learning-based approaches. To accelerate these methods on graphical processing unit, a well-defined concept of memory preallocation and vectorization are incorporated in the novel approach. Results have been investigated on 240 gold standard dermoscopy images. These results reveal that the optimized methods of segmentation are computationally benefited from GPU processing in terms of speed and accuracy for skin lesion detection.

Keywords Adaptive thresholding · Active snake · Pulse-coupled neural network
Graphical processing unit · Exponential entropy · Dermoscopy images

1 Introduction

Image segmentation has been the most useful technique in image processing. It is an essential part of image analysis, object representation, and visualization used for extracting region of interest. Image segmentation is computationally expensive for high dimension dataset of satellite images and high modality of medical images. GPU

P. K. Updhyay (✉) · S. Chandra
Department of CSE & IT, JIIT University, Noida, UP, India
e-mail: pawan.upadhyay@jiit.ac.in
URL: http://jiit.ac.in

S. Chandra
e-mail: satish.chandra@jiit.ac.in

© Springer Nature Singapore Pte Ltd. 2019
A. Khare et al. (eds.), *Recent Trends in Communication, Computing,
and Electronics*, Lecture Notes in Electrical Engineering 524,
https://doi.org/10.1007/978-981-13-2685-1_25

provides more efficient computation for high data parallelism and shared memory. After the inception of CUDA architecture by NVIDIA in 2007, GPU has shown rapid growth in computer vision and medical image analysis. GPU commences concurrent visualization and interactive segmentation and generates better results as compared to general processing system. GPU running on parallel threads as compared to a few concurrent thread of CPU and it helps to achieve higher level of concurrency for the system. Diversified computing gives heterogeneity among the model of CPU and GPU; i.e., the sequential part of the application runs on the CPU, and computationally intensive part is accelerated by the GPU [1]. However, factors that affect GPU computing are data parallelism, thread count, branch divergence memory usage, and synchronization [2]. Segmentation techniques like level set and adaptive snake were drastically improved in recent years.

Thresholding is one of the simplest methods of image segmentation in which pixels are partitioned according to their intensity values. In this paper, author describes the optimization of local adaptive thresholding with or without neural network techniques by GPU processing. In addition to this, we adopted exponential entropy for better divergence unlike intensity. The rest of the paper is organized as follows. Section 2 describes the various existing application of image segmentation with their applications on GPU. In Sect. 3, we present three approaches with their optimization by exponential entropy function and accelerated with parallel concept performed on graphical processing unit for in-depth analysis of these algorithms. Section 4 describes about the dataset of dermoscopy skin lesion, and Sect. 5 reports about many interesting results indicating the potential of GPU on image segmentation algorithms.

2 Related Works

One of the most significant discussions in the current era of noninvasive diagnostic is segmentation of medical images and its performance on graphical processing unit. Life cycle of image processing includes filtering, segmentation, detection, and recognition. Preprocessing requires image denoising and segmentation algorithm for medical images having various transforms including perfect contourlet transform [3]. In addition, the numerical results were analyzed with threshold concept adopted by contourlet coefficients and gave higher peak signal-to-noise ratio (PSNR) as compared to wavelet-based methods. In a decade, many algorithms have been proposed to identify maximum entropy of skin lesion in dermoscopic images using image segmentation [4]. Moreover, there are various types of entropies like Shannon, Renyi, and Tsallis employed by many researchers in various methods of image segmentation [5]. Out of these methods, Shannon and Renyi entropy-based methods produced better segmentation results. In 2015, one scientist proposed an improved version of adaptive segmentation for 3D model of medical images [6]. In addition to this, a quick shift algorithm on color images is used for improving the performance of segmentation [7]. Further, the quick shift was applied on GPU for further improvement of results, i.e., 10–50 times. In 2006, an algorithm based on Potts model is proposed for image segmentation [1]. In addition to this, we have also studied the effect of nonparametric image segmentation. Further, they have used metropolis algorithm for image segmentation on GPU for smaller to larger dimensional images. Since only color information is considered, it shows better performance in smaller segments for image segmentation [1]. Annealing iterations were computed on GPU with higher cost in terms of execution time [2]. In 2013, Erik Smistad discussed recent advances of GPU in the field of medical image segmentation [8]. In 2013, Zafer Glerand Ahmetnar presents image segmentation on GPU by using scaling approach similar to level set method [9]. Although it did not use partial differentiation method for convergence, it reduces the computational time drastically. This signifies that the large operations are required for GPU processing. Author was posited the functioning of GPU through linear and integral programming and architecture of CUDA, respectively [10]. In 2011, T. Romen Singh et al. emphasized that the binarization done through local adaptive thresholding for grayscale text documents is an efficient way [11]. In 2009, Ying Zhuge et al. identified a parallel fuzzy connected image segmentation algorithm on GPU with help of CUDA programming language for interactive speed and compared with CPU.

3 Comparison of GPU Accelerated Image Segmentation Methods

Image segmentation is the process of partitioning an image into multiple regions (set of pixels). The results of image segment method are set of segments that collectively form an entire image. Our aim is to improve the performance of the image segmentation methods by the proposed algorithms described in the subsection given below. In addition to this, these segmentation methods after profiling with exponential function are accelerating with parallel concept on GPU [Geforce 940 processor]. The computationally improved results in pulse-coupled neural network, adaptive thresholding, active snake methods are efficiently used as noninvasive methods of abnormality detection. However, the fundamental concept is at regular partitioning of an image and this process can be performed with parallel threads. Finally, submission of these parallel threads as a global single thread, and it signifies for complete image on graphic processing platform. Due to this reason, these three methods are compared in results section.

3.1 Learning-Based Image Segmentation

Learning techniques are implied in imaging problem, especially in image segmentation methods [12]. The taxonomy combines unsupervised and supervised method which includes clustering-based techniques, artificial neural networks, and fuzzy-based algorithms. There are various trade-off of learning algorithm, and its related model is selected on the basis of data used in it. There are various types of model network such as dynamic, static, and memory-based neural networks, and out of them, we are selecting self-learning model of neural network known as pulse-coupled neural network, and it is used for real-time applications. PCNN follows the concept of self-learning, having number of inputs which is equal to numbers of pixel in the input image and it significantly used for the applications of computer vision [13–15]. Our aim is to identify the objective function for calculating the threshold field in certain region with minimum numbers of iteration. These iterations generate best results during segmentation and generate binary output. The adaptive pulse-coupled neural network is as follows:

Step 1: The adaptive pulse coupled neural network follows those neurons (pixels) having maximum energy and entropy in the network (image space) and these neurons are considered to be a firing neuron.

Step 2: The concept of unsupervised learning is govern by exponential entropy measure, and it minimizes the number of neurons for the segmentation of abnormality in skin images.

Step 3: This model removes the time lag between the correct output frames and increases the computation speed. This technique identifies the most probable pixels having maximum energy and entropy used to extract the object from the input image. The adaptive pulse-coupled neural network partitions the input image into four regions, and each of the regions is computed through parallel threads. Execution time is reduced to one-fourth for identifying maximum entropy pixels as lesion region in dermoscopy image space. The concept of vectorization and preallocation of variables enhances the computation speed for this algorithm on CUDA platform.

3.2 Boundary Demarcation-Based Image Segmentation

Boundary demarcation segmentation methods are primarily described by adaptive thresholding. It is a histogram based technique used to calculate the threshold value to decide the decision boundary. There can be multiple thresholds according to the range of values assigned to the histogram. In order to separate two objects, only one threshold value is required, and it grows based on number of objects to segment. In image processing, we deal with pixels and their intensities, if the intensity of the pixel lies below the threshold most likely to be skin region and if greater than boundary value, it represents an abnormality region as lesion. Mathematical form of thresholding is given in Eq. [1] where f(x, y) is a function defined as binary for the input image pixel I(x, y), and the threshold T(x, y) of image pixel is defined by histogram techniques. In addition to this, image pixel is classified as lesion if the value of it is greater than T(x, y). The optimization in proposed adaptive thresholding is given below with the following steps:

Step 1: Blue color band is having minimum discrimination of intensities which help to calculate the exponential entropy at every pixel of image, and it is divided by the image matrix into 25 bins of length 10.

Step 2: Calculating probability density function of each histogram and identifies the first local maxima which decides its threshold for image segmentation.

Step 3: Histograms that are not considered in the process of segmentation are offset. So, the total histogram for the lesion is given below in Eq. [1].

$$T(max) = T(i) + offset(i) \qquad (1)$$

Step 4: Dynamic allocation in GPU is not possible since space for every operation should not exceed the predefined limit.

Step 5: Memory preallocation and vectorization help to improve the space and time complexity.

Step 6: Vectored code more promptly decide the region boundaries with minimum error rate and increases the computational speed as well.

3.3 Contour-Based Image Segmentation

Active contours are often attracted by fine edges which do not belong to the object boundary. They generally appear in dermoscopic images due to artifacts such as hair, specular reflections, or even from variations in skin region. Therefore, it requires an efficient method for artifacts removal at outer edges. The adaptive snake is more appropriate for this issue [16]. It is partitioned in two phases, primarily it detects contour by stroking during segmentation and evaluate the edge linking and then approximates a subset of them using entropy point estimation algorithm based on the expectation maximization (EM) method. The proposed method of optimized active snake is described as given below:

Step 1: Exponential entropy minimizing the number of iterations for edge detection which are based on radial analysis, and this process is called stroke detection.

Step 2: The strokes detection is bifurcated into two steps. First, we detect entropy transitions points for its maximum entropic value of pixels, and the edge linking is performed by inhomogeneity concept.

Step 3: User intervention is required to select the regions, and this process helps to perform profiling using preallocation concept of GPU.

First, let us briefly describe the robust estimation of the contour. Only a subset of the detected strokes is valid and should be considered using estimation maximization method. The EM method recursively updates the elastic contour as skin lesion in the entitled image.

4 Dataset

The database of dermoscopic images is provided by Edinburgh Research and Innovation for Dermofit Image Library 2.00 under the Academic License Agreement (reference number: TEC2813) [16]. There are 1300 images obtained from high focal length camera having ten classes of lesion; out of them six are considered for evaluating perfect results as describe in the next section. The lesion classes belong to the type precancerous/cancerous samples are (MN: malignant melanoma, SCC: squamous cell carcinoma, BCC: basal cell carcinoma, AK: actinic keratosis, MEL: melanocytic nevi, IEC: intraepithelial carcinoma) and noncancerous samples are (PYO: pyogenic granuloma, DF: dermatofibroma, SK: seborrheic keratosis).

5   Experiment and Results

The experiments are performed on ten classes of dermoscopic images, and results are calculated using graphics processor of NVIDIA [GeForce 940M] having memory of 2 GB DDR5. Apart from GPU, Intel i5 Ivy Bridge processor is used at 2.5 GHz and 8 GB RAM. Comparisons of various types of segmentation methods after optimization and acceleration with existing one in terms of accuracy is illustrated in Table [1–3], and corresponding computation time is described in Figure [2].

$$Precision: T_p/T_p + F_p \tag{2}$$

$$Recall: T_p/T_p + F_n \tag{3}$$

$$Accuracy = T_p + T_n/P + N \tag{4}$$

where true positive (Tp), true negative (Tn), false positive (Fp), false negative (Fn), number of positive pixels (P: Tp + Fp), negative pixels of image (N: Tn + Fn).

The results shown in Figure [1] reveal that the adaptive pulse-coupled neural network performance is much better than other two of the methods for precancerous (AK, IEC) class samples. The APCNN method is having better segmentation accuracy for complex class of cancer such as BCC and SCC is illustrated in Table [1–3]. However, the computation time is independent to the skin lesion class type

Lesion classes	Precision	Recall	Accuracy
AK	1.00	0.97	0.97
BCC	1.00	0.97	0.97
DF	0.99	1.00	0.99
HAEM	1.00	1.00	1.00
IEC	1.00	0.88	0.88
ME	0.97	1.00	0.97
MN	0.95	1.00	0.95
PYO	0.98	1.00	0.98
SCC	0.97	1.00	0.97
SK	1.00	0.80	0.81

Table 1 Performance of adaptive pulse-coupled neural network for image segmentation

Table 2 Performance of improved adaptive thresholding for image segmentation

Lesion classes	Precision	Recall	Accuracy
AK	0.01	1.00	0.02
BCC	0.92	1.00	0.93
DF	1.00	0.69	0.98
HAEM	0.98	1.00	1.00
IEC	1.00	0.98	0.98
ME	0.98	1.00	0.98
MN	1.00	0.96	0.97
PYO	1.00	0.94	0.94
SCC	0.64	1.00	0.66
SK	0.99	1.00	0.99

Table 3 Performance of improved active snake for image segmentation

Lesion classes	Precision	Recall	Accuracy
AK	0.02	0.01	0.02
BCC	0.97	0.97	0.97
DF	0.09	0.07	0.09
HAEM	0.98	0.98	0.98
IEC	0.98	1.00	0.98
ME	0.98	0.98	0.98
MN	0.97	1.00	0.97
PYO	0.94	1.00	0.94
SCC	0.66	0.65	0.66
SK	0.99	0.99	0.99

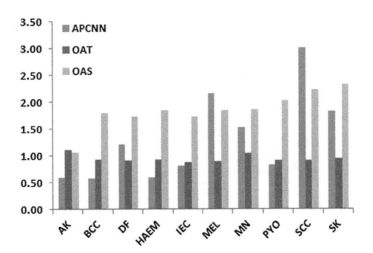

Fig. 1 Computation time on GPU

as cancerous or noncancerous. They completely depend upon the compactness of structural region in a skin lesion. The more compact lesion classes are AK, BCC, HAEM, IEC, PYO and due to this, it consume less computation time as compared to the time taken by scatter lesion patterns such as DF0, MEL, SCC, SK (Fig. 1 and Tables 1, 2, 3).

6 Conclusion

This paper presents an improved version of image segmentation methods on GPU platform. GPU performs vectorization and preallocation of data for data parallelism and reduces the artifacts of signal inhomogeneity at edges during the image segmentation. It performs this activity in two steps, initially by optimizing the segmentation methods by exponential entropy measure and then profiling the code by memory allocation concepts of GPU. Results reveal that the proposed method speed up the segmentation process approximately to half second per image. In future, proposed algorithms are going to implement on various other modalities of medical images.

References

1. Abramov, A., & Kulvicius, T. (2010). Real-time image segmentation on a GPU. In *Facing the Multicore-Challenge* (pp. 131–142).
2. Boyer, V., & El Baz, D. (2013). Recent advances on GPU computing in operations research. In *2013 IEEE International Symposium on Parallel and Distributed Processing Workshops and PhD Forum* (pp.1778–1787).
3. Engineering, C. (2011). Medical image denoising using aaptive. *2*(2), 52–58.
4. Khattak, S. S., Saman, G., Khan, I., & Salam, A. (2015). Maximum entropy based image segmentation of human skin lesion. *9*(5), 1060–1064.
5. Silveira, M., Nascimento, J. C., Marques, J. S., Marçal, A. R. S., Mendonça, T., Yamauchi, S., et al. (2009). Comparison of segmentation methods for melanoma diagnosis in dermoscopy images. *3*(1), 35–45.
6. Kim, C. H., & Lee, Y. J. (2015). Medical image segmentation by improved 3D adaptive thresholding (pp. 263–265).
7. Fulkerson, B., & Soatto, S. (2012). Really quick shift: Image segmentation on a GPU. Lecture Notes in Computer Science (including Subseries Lecture Notes in Artificial Intelligence and Lecture Notes in Bioinformatics) (Vol. 6554, pp. 350–358) LNCS.
8. Smistad, E., Falch, T. L., Bozorgi, M., Elster, A. C., & Lindseth, F. (2015). Medical image segmentation on GPUs—a comprehensive review.
9. Güler, Z., & Cinar, A. (2013). GPU-based image segmentation using level set method with scaling approach. *Computer Science & Information Technology*, 81–92.
10. Ghorpade, J., Parande, J., Kulkarni, M., & Bawaskar, A. (2012). Gpgpu processing in cuda architecture. *Advanced Computing: An International Journal, 3*(1), 105–120.
11. Singh, T. R., Roy, S., Singh, O. I., Sinam, T., & Singh, K. M. (2011). A new local adaptive thresholding technique in binarization. *8*(6), 271–277.
12. Srivastava, N., Hinton, G. E., Krizhevsky, A., Sutskever, I., & Salakhutdinov, R. (2014). Dropout: A simple way to prevent neural networks from overfitting. *The Journal of Machine Learning Research, 15*, 1929–1958.

13. Johnson, J. L., & Padgett, M. L. (1999). PCNN models and applications. *IEEE Transactions on Neural Networks, 10*(3), 480–498.
14. Wang, H., Ji, C., Gu, B., & Tian, G. (2010). A simplified pulse-coupled neural network for cucumber image segmentation. In *Proceedings of 2010 International Conference on Computational and Information Sciences ICCIS* (pp. 1053–1057).
15. Li, J., Zou, B., Ding, L., & Gao, X. (2013). Image segmentation with PCNN model and immune algorithm. *Journal of Computers, 8*(9), 2429–2436.
16. Emre Celebi, M., & Mishra, N. K. (2016). An overview of melanoma detection in dermoscopy images using processing and machine learning. Arxiv Statistics—Machine Learning, 1–15.

Dense Flow-Based Video Object Segmentation in Dynamic Scenario

Arati Kushwaha, Om Prakash, Rajneesh Kumar Srivastava and Ashish Khare

Abstract Segmenting object from a moving camera is a challenging task due to varying background. When camera and object both are moving, then object segmentation becomes more difficult and challenging in video segmentation. In this paper, we introduce an efficient approach to segment object in moving camera scenario. In this work, first step is to stabilize the consecutive frame changes by the global camera motion and then to model the background, non-panoramic background modeling technique is used. For moving pixel identification of object, a motion-based approach is used to resolve the problem of wrong classification of motionless background pixel as foreground pixel. Motion vector has been constructed using dense flow to detect moving pixels. The quantitative performance of the proposed method has been calculated and compared with the other state-of-the-art methods using four measures, such as average difference (AD), structural content (SC), Jaccard coefficients (JC), and mean squared error (MSE).

Keywords Object segmentation · Moving camera · Dense flow
Non-panoramic model · Background modeling

A. Kushwaha · R. K. Srivastava · A. Khare (✉)
Department of Electronics and Communication,
University of Allahabad, Allahabad, Uttar Pradesh, India
e-mail: ashishkhare@hotmail.com

A. Kushwaha
e-mail: aratikushwaha.jk@gmail.com

R. K. Srivastava
e-mail: rkumarsau@gmail.com

O. Prakash
Department of Computer Science and Engineering,
Nirma University, Ahmedabad, India
e-mail: au.omprakash@gmail.com

A. Khare et al. (eds.), *Recent Trends in Communication, Computing, and Electronics*, Lecture Notes in Electrical Engineering 524,
https://doi.org/10.1007/978-981-13-2685-1_26

1 Introduction

Segmenting object from a moving camera and detection is an important and challenging area of research in the field of computer vision from last few decades due to increasing need of automated video analysis. It is the first step in the most of the automated visual event detection applications, such as automated surveillance, vehicular traffic analysis, moving object analysis, human activity recognition, human–computer interaction (HCI), behavior description and robot vision. The main goal of object segmentation is to extract target object. Object segmentation is a complex process due to 3D world projection into 2D scene, the presence of noise, complex object motion, object occlusion, varying object shape, and dynamic background. Based on the motion of camera, the object segmentation can be categorized as static and moving camera segmentation [1, 2].

In last few decades, a lot of researches have been conducted to extract moving objects using background subtraction techniques [3–6]. These techniques work well in static camera only. For the moving cameras such as camera present in vehicles, the segmentation task becomes more difficult and challenging due to the relative motion of the object and camera. In case of moving camera, two independent motions are present—background and object motion. Changing background yields in inappropriate segmentation using background subtraction. To handle this problem, first we need to estimate camera motion as well as object motion.

The basic algorithm of moving camera segmentation is done by computing absolute difference between consecutive frames followed by the predetermined threshold, to extract moving object of interest. It is not possible to model background using background subtraction techniques since background varies with time. To detect moving object, in varying background with moving camera, both camera motion and object motion are separately estimated. Then, object is detected using temporal difference of two consecutive frames at two consecutive time instances. The superiority of temporal difference method is due to its speed of computation and easy to implement. The major disadvantage of temporal difference method is its false object segmentation. Another most commonly used methods of object segmentation in dynamic scenarios are based on panoramic model [7–12], and these are extension of background subtraction. Due to varying background in dynamic scenarios, backgrounds are modeled by applying registration techniques in between consecutive frames. To obtain the panoramic model from input frames, many registration techniques are used such as tonal mosaic alignment and Lucas–Kanade Tracking (LKT) [19]. In these methods, background is modeled by panoramic images that are stitched from all possible views of input frames. The moving objects are segmented by comparing the current frame with the corresponding region in the panoramic model, i.e., then follow segmentation rules as in static camera segmentation techniques [7]. However, extraction of moving object with moving camera deals with problem like registration error, parallax effect, slow initialization, large computation memory due to the construction of the large possible field of view and time problem [14]. Some algorithms based on non-panoramic background model for moving camera segmentation

[13–18] have been studied to resolve these problems but still suffer from wrong classification of motionless background pixels as a foreground pixel. In the proposed work, the problem of false labeling of background pixel as foreground pixel was resolved by calculating the motion vector of each pixel with the assumption that background pixels have negligible magnitude compared to moving pixels. Motion vectors have been calculated using dense optical flow. In the proposed work, no offline training is needed, which yields the fast and robust, object segmentation in real scenes.

The rest of the paper is organized as follows: in Sect. 2, we briefly discuss the proposed algorithm, and in Sect. 3, the experimental results are presented, and finally, the conclusion is given in Sect. 4.

2 The Proposed Method

The overall framework of the proposed approach has been shown in Fig. 1. In case of moving camera scenes, two independent motions are present, viz. camera motion and object motion. To detect moving object, firstly, we estimated camera motion, i.e. to stabilize camera motion first task is to extract feature vectors in background frames and track corresponding features in current frame by Lucas–Kanade Tracking (LKT) [19] which results in homography matrix.

A Homography is a 3 × 3 projective transformation matrix that gives point-to-point mapping from one view to another. Using this homography matrix, overlapped region between current frame and background frame is projected followed by spatio–temporal Gaussian background modeling technique [4, 13, 14]. In this work, we have calculated motion vectors of each pixel. Since moving pixels have some magnitudes, background pixels have negligible values; therefore for identification of background and foreground pixels, labeling first spatio–temporal Gaussian back-

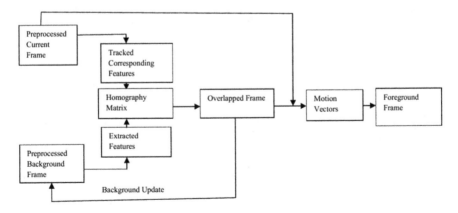

Fig. 1 Block diagram of the proposed algorithm

ground modeling technique is used to model background and followed by threshold on motion vectors for identification of moving pixel to avoid false identification of moving foreground pixels [14].

3 Experimental Results and Discussion

The proposed work has been implemented in C++ using OpenCV 3.0, Microsoft visual studio 2010. The written code has been tested in Intel Core i3 1.9 GHz PC, and it is tested in several real-time moving camera videos. More than 30 frames were processed in each second, and algorithm performs well in both static and moving camera case. In this paper, we have presented the results with a movie clip of 1000 frames of size 640×360. Three methods have been taken for comparative studies [14, 16, 17]. For representative purpose, the experimental results for frame numbers 1–1000 at an interval of 200 have been shown in Fig. 2.

From Fig. 2, we observe that the proposed algorithm performs more accurately compared to the methods used in comparison. The proposed algorithm works well in dynamic background with moving camera for real-time applications.

The performance has been evaluated based on four quantitative measures given in [20, 21]. They are average difference (AD), mean squared error (MSE), structural content (SC), and Jaccard coefficients (JC). Their computational equations are given below.

Let $G_{i,j}$ is ground truth frame and $S_{i,j}^{x}$ Segmented frame and m, n are dimensions of the frame.

The smaller value of average difference represents better segmentation. Average difference can be calculated as follows:

$$AD = \frac{\sum_{i=1}^{m} \sum_{j=1}^{n} \left(G_{i,j} - S_{i,j}^{'} \right)}{mn} \tag{1}$$

The lower value of mean squared error shows better segmentation. Mean squared error can be calculated as follows:

$$MSE = \frac{\sum_{i=1}^{m} \sum_{j=1}^{n} \left(G_{i,j} - S_{i,j}^{'} \right)^2}{mn} \tag{2}$$

The higher value of structural content shows better segmentation. Structural content can be calculated as follows:

$$SC = \frac{\sum_{i=1}^{m} \sum_{j=1}^{n} (G_{i,j})^2}{\sum_{i=1}^{m} \sum_{j=1}^{n} (S_{i,j}^{'})^2} \tag{3}$$

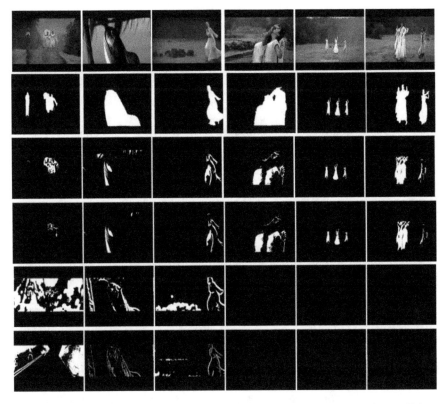

Fig. 2 Frames are arranged from left to right 1–1000 on each 200 regular intervals (row 1) input frames, (row 2) ground truth, (row 3) proposed method, (row 4) SGM method [14], (row 5) dense flow-based method [16], (row 6) Wronskian model [17]

The higher value of Jaccard coefficients shows better segmentation. Jaccard coefficients (JC) can be defined as [21]:

$$JC = \frac{TP}{TP + FP + FN} \qquad (4)$$

where TP is number of true positive pixels, FP is number of false positives pixels, and FN is number of false negative pixels.

Table 1 Performance measure in terms of AD

Frame no./methods	Proposed method	SGM method [14]	Dense flow based [16]	Wronskian method [17]
1	**0.0395**	0.0406	0.2989	0.2209
200	**0.1933**	0.2050	0.3509	0.3312
400	**0.1290**	0.1689	0.1772	0.1844
600	**0.0379**	0.0425	0.3443	0.3277
800	**0.1860**	0.1929	0.1652	0.1660
1000	**0.0080**	0.0092	0.0249	0.0262

Table 2 Performance measure in terms of MSE

Frame no./methods	Proposed method	SGM method [14]	Dense flow based [16]	Wronskian method [17]
1	**0.0395**	0.0406	0.2989	0.2209
200	0.2129	**0.2050**	0.3509	0.3312
400	0.1820	**0.1689**	0.1772	0.1844
600	**0.3015**	0.3054	0.3443	0.3277
800	**0.1250**	0.1307	0.1652	0.1660
1000	**0.0087**	0.0102	0.0262	0.0262

Table 3 Performance measure in terms of SC

Frame no./methods	Proposed method	SGM method [14]	Dense flow based [16]	Wronskian method [17]
1	**0.4701**	0.3853	5.1461	3.163
200	0.3981	**0.4211**	0.1128	0.0821
400	**0.2006**	0.1174	0.1088	0.0456
600	**0.1456**	0.1342	0.0939	0.0789
800	**0.2719**	0.2377	0.0378	0.0686
1000	**0.6802**	0.6406	0	0

Tables 1, 2, 3, 4 show the quantitative performance measure values of the results given in Fig. 2. The best values have been shown in bold for each method. From Tables 1, 2, 3, 4 and Fig. 2, we observe that in most of the cases our proposed method gives improved result than the other methods used in comparison.

Table 4 Performance measure in terms of JC

Frame no./methods	Proposed method	SGM method [14]	Dense flow based [16]	Wronskian method [17]
1	**0.2576**	0.2352	0.0612	0.0214
200	0.2775	**0.2893**	0.0394	0.0530
400	0.0967	**0.0979**	0.0748	0.0308
600	**0.1269**	0.1182	0.0538	0.0694
800	**0.2132**	0.1916	0.0353	0.0466
1000	**0.4010**	0.3805	0	0

4 Conclusions

In this work, we have presented a method for video object segmentation in moving camera case. Background motion and object motion of a video object were estimated and used in object segmentation. In this work, first camera motion was stabilized then spatio–temporal Gaussian model was used for background modeling. For detection of foreground and background pixels, motion vectors were used based on dense optical flow technique. The proposed method of object segmentation was compared with others state-of-the-art methods. The statistical performance of the results demonstrates that the proposed method performs better than the methods used in our comparison. The proposed approach is suitable for the segmentation of object sizes ranging from very small to that of large object. The method is robust, and it works well in simple and cluttered background.

References

1. Vrigks, M., Nikou, C. & Kakadiaries, L. A. (2015). A review of human activity recognition methods. *Frontiers in Robotics and AI*, 2(28).
2. Ke, S.-R., Uyen Thuc, H.-L, Lee, Y.-L., Hwang, J.-N., Yoo, J.-H., & Choi, K.-H. (2013). A review on video-based human activity recognition. *Computers*, 2, 88–131.
3. Ko, T., Soatto, S., & Estrin, D. (2010). Warping background subtraction. In *2010 IEEE Conference on Computer Vision and Pattern Recognition (CVPR)* (pp. 1331–1338).
4. Stauffer, C., & Grimson, W. (1999). Adaptive background mixture models for real-time tracking. In *IEEE Computer Society Conference on Computer Vision and Pattern Recognition* (Vol. 2, pp. 246–252).
5. Tavakkoli, A., Nicolescu, M., Bebis, G., & Nicolescu, M. (2009). Nonparametric statistical background modeling for efficient forackground modelling from n detection. *Machine Vision Applications, 20*, 395–409.
6. Bouwmans, T., El Baf, F., & Vachon, B. (2008). Background modeling using mixture of gaussians for foreground detection—a survey. *Recent Patents on Computer Science, 1*(3), 219–237.
7. Cho, S. H., & Hang, B. K. (2011). Panoramic background generation using mean-shift in moving camera environment. In *Proceedings of the International Conference on Image Processing, Computer Vision, and Pattern Recognition*.

8. Guillot, C., Taron M., Sayd, P., Pha, Q. C., Tilmant, C., & Lavest, J. M. (2010). Background subtraction adapted to pan tilt zoom cameras by key point density estimation. In *Computer Vision–ACCV Workshops* (pp. 33–42), Springer: Berlin Heidelberg.

9. Xue, K., Liu, Y., Ogunmakin, G., Chen, J., & Zhang, J. (2013). Panoramic Gaussian Mixture Model and large-scale range background substraction method for PTZ camera-based surveillance systems. *Machine Vision and Applications, 24*(3), 477–492.

10. Robinault, L., Bres, S., & Miguet, S. (2009). Real time foreground object detection using pan tilt zoom camera. *VISSAPP, 1*, 609–614.

11. Murray, D., & Basu, A. (1994). Motion tracking with an active camera. *IEEE Transactions on Pattern Analysis and Machine Intelligence, 16*, 449–459.

12. Xue, K., Liu, Y., Ogunmakin, G., Chen, J., & Zhang, J (2013). Panoramic Gaussian Mixture Model and large-scale range background substraction method for PTZ camera-based surveillance systems. *Machine Vision and Applications, 24*, 477–492.

13. Viswanath, A., Kumari Beherab, R., Senthamilarasub, V., & Kutty, K. (2015). Background modelling from a moving camera. *Procedia Computer Science, 58*, 289–296.

14. Kim, S. K., Yun. K., Yi, K. M., Kim, S. J., & Choi, H. Y. (2013). Detection of moving objects with a moving camera using non-panoramic background model. *Machine Vision and Applications, 24*, 1015–1028.

15. Yi, K., Yun., K., Kim, S., Chang, H., & Choi, J. (2013). Detection of moving objects with non-stationary cameras in 5.8 ms: Bringing motion detection to your mobile device. In *Proceedings of the IEEE Conference on Computer Vision and Pattern Recognition Workshops* (pp. 27–34).

16. Kurnianggoro, L., Shahbaz, A., & Jo, K.-H. (2016). Dense optical flow in stabilized scenes for moving object detection from a moving camera. In *16th International Conference on Control, Automation and Systems (ICCAS 2016)*, October 16–19. HICO, Gyeongju, Korea.

17. Kadim, Z., Daud, M. M, Radzi, S. S. M., Samudin, N., & Woon, H. H. (2013). Method to detect and track moving object in non-static PTZ camera. In *Proceedings of the International MultiConference of Engineers and Computer Scientists*, 3 (Vol. I) IMECS 2013, March 13–15, Hong Kong.

18. Hu, W.-C., Chen, C.-H., Chen,T.-Y., & Huang, D.-Y., & Wu, Z.-C. (2015). Moving object detection and tracking from video captured by moving camera. *Journal of Visual Communication and Image Representation, 30*, 164–180.

19. Lucas, B. D., & Kanade, T. (1981). An iterative image registration technique with an application to stereo vision. In *International Joint Conference on Artificial Intelligence* (pp. 674–679).

20. Khare, M., & Srivastava, R. K. (2012). Level set method for segmentation of medical images without reinitialization. *Journal of Medical Imaging and Health Informatics, 2*(2), 158–167.

21. Rosin, P., & Ioannidis, E. (2003). Evaluation of global image thresholding for change detection. *Pattern Recognition Letters, 24*(14), 2345–2356.

Image Quality Assessment: A Review to Full Reference Indexes

Mahdi Khosravy, Nilesh Patel, Neeraj Gupta and Ishwar K. Sethi

Abstract An image quality index plays an increasingly vital role in image processing applications for dynamic monitoring and quality adjustment, optimization and parameter setting of the imaging systems, and finally benchmarking the image processing techniques. All the above goals highly require a sustainable quantitative measure of image quality. This manuscript analytically reviews the popular reference-based metrics of image quality which have been employed for the evaluation of image enhancement techniques. The efficiency and sustainability of eleven indexes are evaluated and compared in the assessment of image enhancement after the cancellation of speckle, salt and pepper, and Gaussian noises from MRI images separately by a linear filter and three varieties of morphological filters. The results indicate more clarity and sustainability of similarity-based indexes. The direction of designing a universal similarity-based index based on information content of the image is suggested as a future research direction.

Keywords Image quality measurement · Error measurement
Similarity measurement · Noise cancellation · Image enhancement

The correspondent author is Neeraj Gupta (neeraj.gupta@oakland.edu). M. Khosravy also jointly collaborates with Electrical Engineering Department, University of the Ryukyus, Okinawa, Japan, and School of Computer and Engineering Science, Oakland University, MI, USA.

M. Khosravy
Electrical Engineering Department, Federal University of Juiz de Fora, Juiz de Fora, Brazil
e-mail: dr.mahdi.khosravy@ieee.org

N. Patel · N. Gupta (✉) · I. K. Sethi
School of Computer and Engineering Science, Oakland University, Rochester, MI, USA
e-mail: neerajgupta@oakland.edu

© Springer Nature Singapore Pte Ltd. 2019
A. Khare et al. (eds.), *Recent Trends in Communication, Computing, and Electronics*, Lecture Notes in Electrical Engineering 524,
https://doi.org/10.1007/978-981-13-2685-1_27

1 Introduction

A valid image quality index plays a vital role in the efficiency of a variety of image processing applications, e.g., image enhancement [1], image adaptation [2], image compression, medical imaging, image-based medical diagnosis. Thereof, for the sake of automation and optimization, it highly demands a consistent image quality metric. For instance, in medical applications, the higher enhancement of an image leads to the earlier and more proper diagnosis of a disease, as well as more successful dealing with epidemics like breast cancer which even in developed countries, one in ten women faces its risk of mortality [3]. In most of the medical images, the indicative features are very small, the detection and interpretation are a difficult task even for an expert physician, and an efficient image enhancement technique helps a lot to avoid improper diagnosis by being used in computer-aided analysis programs. Clearly, the efficiency, optimization, and automation of an image enhancement technique are highly dependent on the deployed metric of image quality.

In recent research works on image quality analysis, there is a major emphasis on a numerical understanding of the human visual system (HVS) to incorporate HVS preferences in image quality indexes. However, it is extremely complex to comprehend HVS preference direction by the current psychophysical tools, but as it is reported even implementation of a simple model of HVS in the quality indexes leads to a higher match with the human observer image quality check.

This manuscript reviews the common reference-based image quality indexes as it analytically evaluates the indexes through assessment their clarity in indicating the image enhancement level after the cancellation of different types of noise from MRI images. Classically, image quality measures were in use for the evaluation of the image compression techniques effect on image quality, wherein the image after decompressing was compared with the original image before compression.

2 Reference-Based Image Quality Indexes

The classic measures of image quality are all bivariate functions exploiting the differences or similarities between corresponding pixels in the image and the reference image, and they are, respectively, called error-based and similarity-based criteria.

2.1 Error-Based Image Quality Indexes

The error-based indexes obtain a measure of the difference between the two images by deploying the statistical distribution differences of their pixel values. These image quality criteria have been listed [4] and analyzed [5] in the literature. Table 1 gives a brief list of reference-based image quality indexes, wherein $F(j, k)$ and $\hat{F}(j, k)$

Table 1 Error-based image quality indexes

Mean absolute error $= \sum_{j=1}^{M} \sum_{k=1}^{N} \frac{\lvert F(j,k)-\hat{F}(j,k)\rvert}{MN}$	Mean square error $= \sum_{j=1}^{M} \sum_{k=1}^{N} \frac{(F(j,k)-\hat{F}(j,k))^2}{MN}$
Average difference $= \sum_{j=1}^{M} \sum_{k=1}^{N} \frac{(F(j,k)-\hat{F}(j,k))}{MN}$	Structural content $= \dfrac{\sum_{j=1}^{M}\sum_{k=1}^{N} F(j,k)^2}{\sum_{j=1}^{M}\sum_{k=1}^{N} \hat{F}(j,k)^2}$
N. cross-correlation $= \dfrac{\sum_{j=1}^{M}\sum_{k=1}^{N} F(j,k)\hat{F}(j,k)}{\sum_{j=1}^{M}\sum_{k=1}^{N} F(j,k)^2}$	Correlation quality $= \dfrac{\sum_{j=1}^{M}\sum_{k=1}^{N} F(j,k)\hat{F}(j,k)}{\sum_{j=1}^{M}\sum_{k=1}^{N} F(j,k)}$
Image fidelity $= 1 - \dfrac{\sum_{j=1}^{M}\sum_{k=1}^{N}\left(F(j,k)-\hat{F}(j,k)\right)^2}{\sum_{j=1}^{M}\sum_{k=1}^{N}\left(F(j,k)\right)^2}$	Maximum difference $= \max\{\lvert F(j,k)-\hat{F}(j,k)\rvert\}$
Laplacian MSE $= \dfrac{\sum_{j=1}^{M}\sum_{k=2}^{N}\left(O(F(j,k))-O(\hat{F}(j,k))\right)^2}{\sum_{j=1}^{M}\sum_{k=2}^{N}\left(O(F(j,k))\right)^2}$	Peak MSE $= \dfrac{1}{MN}\dfrac{\sum_{j=1}^{M}\sum_{k=1}^{N}\left(F(j,k)-\hat{F}(j,k)\right)^2}{\max\left(F(j,k)\right)^2}$
N. absolute error $= \dfrac{\sum_{j=1}^{M}\sum_{k=1}^{N}\lvert O(F(j,k))-O(\hat{F}(j,k))\rvert}{\sum_{j=1}^{M}\sum_{k=1}^{N}\lvert O(F(j,k))\rvert}$	NMSE $= \dfrac{\sum_{j=1}^{M}\sum_{k=1}^{N}\left(O(F(j,k))-O(\hat{F}(j,k))\right)^2}{\sum_{j=1}^{M}\sum_{k=1}^{N}\left(O(F(j,k))\right)^2}$
L_p norm $= \left(\frac{1}{MN}\sum_{j=1}^{M}\sum_{k=1}^{N}\lvert F(j,k)- \hat{F}(j,k)\rvert^p\right)^{\frac{1}{p}}$, $p=1,2,3$.	Peak signal-to-noise ratio $= 10\log_{10}(255^2/MSE(F,\hat{F}))$

denote the samples of the reference image and the image undergone quality measurement, for Laplacian MSE $O(F(j,K)) = F(i+1,k) + F(j-1,k) + F(i,k+1) + F(j,k-1) - 4F(j,k)$ and for N. Absolute error, NMSE, and L_2-norm there are three definitions of $O(F(j,k))$: (i) $O(F(j,k)) = F(j,k)$, (ii) $O(F(j,k)) = F(j,k)^{\frac{1}{3}}$, and (iii) $O(F(u,v)) = H(\sqrt{(u^2+v^2)})F(u,v)$; i.e., $F(u,v)$ is cosine transform, and $H(.)$ is the model of rotationally symmetric spatial frequency response of HVS [6] as $H(r) = (0.2 + 0.45r)e^{-0.18r}$.

2.2 Similarity-Based Image Quality Index

The image quality assessment based on error does not take into account the structural comparison of the image with the reference image. Reference [7] establishes a structure for image quality analysis based on the assumption of adaptability of the human visual system to derive structural information from the vision. In the same context, the change in structural information is measured as an approximation of distortion in perceived vision [8, 9]. The similarity assessment is comprising three comparisons of the image with the reference image in luminance, contrast, as well as the structure. The luminance of each image is estimated as the average intensity:

$$\mu_F = \frac{1}{MN}\sum_{j=1}^{M}\sum_{k=1}^{N} F(i,j), \tag{1}$$

and the luminance comparison is obtained by the bivariate function $l(\mu_F, \mu_{\hat{F}})$. The intensity standard deviation is used as the contrast of each image,

$$\sigma_F = \left(\frac{1}{MN-1} \sum_{j=1}^{M} \sum_{k=1}^{N} (F(i, j) - \mu_F)^2\right)^{\frac{1}{2}} \qquad (2)$$

and thereof the contrast comparison function is obtained $c(\sigma_F, \sigma_{\hat{F}})$. The structural comparison $s(F, \hat{F})$ is by comparison of normalized images $\frac{F-\mu_F}{\sigma_F}$ and $\frac{\hat{F}-\mu_{\hat{F}}}{\sigma_{\hat{F}}}$, and finally the structural similarity index measure (SSIM) is as follows

$$SSIM(F, \hat{F}) = l(\mu_F, \mu_{\hat{F}})^\alpha c(\sigma_F, \sigma_{\hat{F}})^\beta s(F, \hat{F})^\gamma \qquad (3)$$

where parameters α, β, and γ are set considering their relevant importance which for simplification all can be considered equal to one. However, a deep study on SSIM shows it fails in the assessment of severely blurred images. To overcome this drawback and having a more consistency with HVS, especially for blurred images, the gradient-based structural similarity index measure (GSSIM) has been presented [10]. GSSIM compares the edge areas information between the image and the reference one and deploys the resultant gradient contrast comparison $c_g(.)$ and gradient structure comparison $s_g(.)$ instead of the conventional ones. Another similarity-based index presented in the literature [11] is the complex wavelet structural similarity (CW-SSIM), which utilizes the fact that certain image distortions result in constant changes in the phase of local wavelet coefficients, but not affects the structural content of the image.

Maximum cross-correlation (MCC) and mutual information (MI) are two other very popular indexes for comparison of images/signals. Both indexes have been widely used, especially in the evaluation of blind source separation [12–15], while the distance between separated images from each other is measured by MI and the similarity between the original source images and the recovered ones by is evaluated by MCC.

2.3 Image Degradation Indexes

There are image quality assessments based on measuring the image degradation by noise and other interferences in light of consideration of a linear model of HVS frequency response [16–18]. Reference [19] forms a visual distortion function (VDF) by using contrast threshold function (CTF) and contrast sensitivity function (CSF) and thereof introduces an image distortion index by computing the area under VDF:

$$\text{Distortion measure} = \int_0^{f_{max}} \left[1 - \text{CTF}(\frac{f_r}{f_N})\right]\text{CSF}(f_r)df_r \tag{4}$$

where CTF is a nonlinear response of HVS to a single frequency which is the minimum amplitude necessary to just detect a sine wave of a given frequency. Indeed, CTF helps to obtain a frequency response of HVS known as contrast sensitivity function CSF. $f_r = \sqrt{f_x^2 + f_y^2}$ denotes the radial frequency, wherein f_x and f_y are, respectively, horizontal and vertical frequencies. f_N and f_{max} are, respectively, the Nyquist and the maximum frequency involved in distortion measure. A universal image quality index with capability of applicable to various image types models the image distortion as combination of (i) correlation loss, (ii) luminance distortion, and (iii) contrast distortion [20] and gives an image quality index in the range of $[-1, 1]$ where 1 is exactly as the reference image and -1 is as $\hat{F}(j, k) = 2\bar{F} - F(j, k)$ for all (j, k) spatial indexes. The universal quality index is defined as follows:

$$\text{UQI} = \frac{4\sigma_{F\hat{F}}\bar{F}\bar{\hat{F}}}{(\sigma_F^2 + \sigma_{\hat{F}}^2)[(\bar{F})^2 + (\bar{\hat{F}})^2]}, \tag{5}$$

$$\bar{F} = \frac{1}{MN}\sum_{j=1}^{M}\sum_{k=1}^{N}F(j,k), \; \bar{\hat{F}} = \frac{1}{NN}\sum_{j=1}^{M}\sum_{k=1}^{N}\hat{F}(j,k), \sigma_F^2 = \frac{1}{(M-1)(N-1)}\sum_{j=1}^{M}\sum_{k=1}^{N}(F(j,k) - \bar{F})^2$$

$$\sigma_{\hat{F}}^2 = \frac{1}{(M-1)(N-1)}\sum_{j=1}^{M}\sum_{k=1}^{N}(\hat{F}(j,k) - \bar{\hat{F}})^2, \sigma_{F\hat{F}} = \frac{1}{(M-1)(N-1)}\sum_{j=1}^{M}\sum_{k=1}^{N}(F(j,k) - \bar{F})(\hat{F}(j,k) - \bar{\hat{F}}),$$

where M and N are dimensions of the image. Another distortion measure has been presented based on singular value distance (SVD) of two images [21] wherein for ith sub-blocks of both images the singular value distances are computed as follows:

$$D_i = \sqrt{\sum_{i=1}^{n}(s_i - \hat{s}_i)^2} \tag{6}$$

where s_i and $\hat{s}_i$ are the singular values of ith sub-blocks of the image and the reference image, respectively. Then, the SVD distortion measure is expressed as follows:

$$\text{SVD-distortion} = \frac{\sum_{i=1}^{(\frac{k}{n})\times(\frac{k}{n})}|D_i - D_{mid}|}{(\frac{k}{n})(\frac{k}{n})} \tag{7}$$

where D_{mid} denotes the median of D_i values, and k and n are, respectively, the image and the sub-block sizes.

3   Analytical Results

We have selected eleven dominant and popular reference-based image quality indexes for analytical evaluation. The evaluation is over a set of MRI image contaminated by three types of noise as speckle, salt and pepper, and Gaussian and then filtered by four techniques comprising a simple linear filtering (LF), morphological filtering (MF) [22, 23], weighted morphological filtering (WMF) [24], and mediated morphological filtering (MMF) [25–27]. The power of these four operators in noise cancellation and enhancing the image quality has been evaluated and compared in the literature; thereof, we have used them for the evaluation of the sustainability of the indexes, while the result of the comparisons is expected to be

$$Q(LF) < Q(MF) < Q(WMF) < Q(MMF) \tag{8}$$

For benchmarking, a MRI image has been deployed in Fig. 1 wherein the very left column is the noise contaminated by different types of noise and the other columns from left to right are, respectively, LF, MF, WMF, and MMF results. The source image has the courtesy of the authors as it is from their data set (https://sites.google. com/view/sip-digital-data).

This image set we have provided it the value of the eleven indexes applied to the contaminated images by each type of noise and the filtered images by the above techniques for each type of noise has been demonstrated in Table. 2. The conclusion

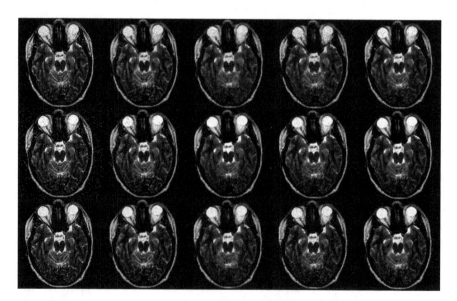

Fig. 1 Benchmark MRI images for the evaluation of reference-based image quality indexes, each row from left to right: noisy and its corresponding recovered image by linear filtering, morphological filtering, weighted morphological filtering, and mediated morphological filtering. From two to down, speckle, salt and pepper, and Gaussian noise

Table 2 Values of reference-based image quality indexes in the evaluation of cancellation of three types of noise (speckle, salt and pepper, and Gaussian) by four methods (linear filtering (LF), morphological filtering (MF), weighted MF, and mediated MF)

	MAE	MSE	PSNR	L_2	L_3	SC	IF	SSIM	MCC	UIQ	MI
Speckle noisy image	3.4742	31.38	33.16	0.0215	0.0021	0.9971	0.7588	0.9073	0.8885	0.9165	0.4718
Linear filtering	3.9068	35.17	32.67	0.0228	0.0023	0.9289	0.7297	0.8641	0.8966	0.8445	0.4849
Morphological filtering (MF)	3.1853	26.55	33.89	0.0198	0.0021	0.9743	0.7960	0.7979	0.8563	0.7798	0.4572
Weighted MF	2.6344	22.52	34.61	0.0182	0.0020	0.9776	0.8270	0.8286	0.8745	0.8160	0.4722
Mediated MF	2.7425	24.12	34.31	0.0189	0.0020	0.9802	0.8146	0.8298	0.8807	0.8154	0.4752
Salt and pepper noisy image	2.9025	3.5208	42.66	0.0072	0.0009	0.9997	0.9729	0.6551	0.9502	0.6623	0.5174
Linear filtering	6.1691	53.021	30.89	0.0280	0.0025	0.8472	0.5925	0.6702	0.8941	0.6119	0.4819
Morphological filtering (MF)	3.3679	26.238	33.94	0.0197	0.0021	0.9675	0.7984	0.8146	0.8709	0.8018	0.4677
Weighted MF	2.6348	21.316	34.84	.0177	0.0019	0.9712	0.8362	0.8556	0.8917	0.8363	0.4854
Mediated MF	2.6814	22.745	34.56	0.0183	0.0020	0.9782	0.8252	0.8571	0.9024	0.8448	0.4930
Gaussian noisy image	9.8330	84.048	28.89	0.0352	0.0028	0.7933	0.3541	0.4510	0.8968	0.5137	0.4740
Linear filtering	8.7507	86.962	28.74	0.0358	0.0032	0.6983	0.3317	0.4989	0.8984	0.5379	0.4854
Morphological filtering (MF)	8.5424	84.534	28.86	0.0353	0.0032	0.6973	0.3503	0.4452	0.8600	0.4702	0.4590
Weighted MF	7.6004	75.366	29.36	0.0333	0.0031	0.7167	0.4208	0.4797	0.8811	0.5079	0.4759
Mediated MF	5.5912	49.947	31.14	0.0271	0.0027	0.8582	0.6161	0.6201	0.8895	0.5259	0.4796

from the observations presented in thirty-three blocks of Table 2 (eleven indexes for three types of noises) are summarized as follows: (i) MAE, MSE and PSNR, L_2 norm, L_3 norm and Image Fidelity (IF) in short words the error-based indexes indicate the improvement but not clearly and very close to each other, especially in the case Salt & Pepper noise where even shows slightly opposite to the expectation. (ii) The (similarity-based indexes; mutual information (MI), universal image quality (UIQ), maximum cross-correlation (MCC), structural similarity (SSIM), and structural content (SC) show a good correlation with our expectation of qualities. (iii) Structural content that classically has been categorized as an error-based index has good correlation with similarity-based indexes, and it shows well the quality difference and matches with inequality 8.

4 Conclusion

The reference-based indexes have been traditionally in use for digital image quality analysis before and after processes. However, these indexes are well known for the lack of reasonable correlation with the human observer's response known as subjective evaluation. This paper has reviewed and analyzed the popular reference-based image quality indexes. The evaluation of eleven indexes in quality of noise cancellation by four different linear and nonlinear filtering indicates: (i) The error-based indexes suffer lack of clarity in comparative analysis of image enhancement techniques; thereof, they are expected for the lack of efficiency for other autonomous and optimization applications, and (ii) the similarity-based indexes are more sustainable for the evaluation of the efficiency of the techniques as they show higher correlation to subjective evaluation by human observers. As a conclusion, the evaluation indexes more than being relied on value differences, they should be based on informative measures of similarity to the reference image. As a future work, besides a comprehensive review of indexes, a preference-based index will be designed by using the informative content of the image.

References

1. Khosravy, M., Gupta, N., Marina, N., Sethi, I. K., & Asharif, M. R. (2017). Brain action inspired morphological image enhancement. In Nature-Inspired Computing and Optimization (pp. 381–407). Springer: Cham.
2. Khosravy, M., Gupta, N., Marina, N., Sethi, I. K., & Asharif, M. R. (2017). Perceptual Adaptation of Image Based on ChevreulMach Bands Visual Phenomenon. *IEEE Signal Processing Letters, 24*(5), 594–598.
3. Mencattini, A., Salmeri, M., Lojacono, R., Frigerio, M., & Caselli, F. (2008). Mammographic images enhancement and denoising for breast cancer detection using dyadic wavelet processing. *IEEE transactions on instrumentation and measurement, 57*(7), 1422–1430.

4. Eskicioglu, A. M., & Fisher, P. S. (1993). A survey of quality measures for gray scale image compression. In *Proceedings of the 1993 Space and Earth Science Data Compression Workshop (NASA Conference Publication 3191)*, Snowbird, Utah, April 2 (pp. 49–61).

5. Eskicioglu, A. M., & Fisher, P. S. (1995). Image quality measures and their performance. *IEEE Transactions on Communications, 43*(12), 2959–2965.

6. Nill, N. (1985). A visual model weighted cosine transform for image compression and quality assessment. *IEEE Transactions on communications, 33*(6), 551–557.

7. Wang, Z., Bovik, A. C., & Lu, L. (2002). Why is image quality assessment so difficult? In *IEEE International Conference on Acoustics, Speech, and Signal Processing* (pp. 3313–3316).

8. Wang, Z., & Bovik, A. C. (2003). Multi-scale structural similarity for image quality assessment. In *Proceedings of IEEE Asilomar Conference on Signals* (vol. 2, no. Ki L, pp. 1398–1402).

9. Wang, Z., Bovik, A. C., Sheikh, H. R., & Simoncelli, E. P. (2004). Image quality assessment: From error visibility to structural similarity. *IEEE Transactions on Image Processing, 13*(4), 600–612.

10. Chen, G., Yang, C., & Xie, S. L. (2006). Gradient-based structural similarity for image quality assessment. In *International Conference on Image Processing* (No. 1, pp. 2929–2932).

11. Sampat, M. P., Wang, Z., Gupta, S., Bovik, A. C., & Markey, M. K. (2009). Complex wavelet structural similarity: A new image similarity index. *IEEE Transactions on Image Processing, 18*(11), 2385–2401.

12. Khosravy, M., Asharif, M. R., & Yamashita, K. (2009). A PDF-matched short-term linear predictability approach to blind source separation. *International Journal of Innovative Computing, Information and Control (IJICIC), 5*(11), 3677–3690.

13. Khosravy, M., Asharif, M. R., & Yamashita, K. (2011). A theoretical discussion on the foundation of Stones blind source separation. *Signal, Image and Video Processing, 5*(3), 379–388.

14. Khosravy, M., Asharif, M., & Yamashita, K. (2008). A probabilistic short-length linear predictability approach to blind surce sparation. In *23rd International Technical Conference on Circuits/Systems, Computers and Communications (ITC-CSCC 2008)*, Yamaguchi, Japan (pp. 381–384).

15. Khosravy, M., Alsharif, M. R., & Yamashita, K. (2009). A PDF-matched modification to stones measure of predictability for blind source separation. In *International Symposium on Neural Networks Springer* (pp. 219–228). Heidelberg: Berlin.

16. Lin, Q. (1993). Halftone image quality analysis based on a human vision model. *Proceedings of SPIE, 1913*, 378–389.

17. Mitsa, T., Varkur, K. L., & Alford, J. R. (1993). Frequency channel based visual models as quantitative quality measures in halftoning. *Proceedings of SPIE, 1913*, 390–401.

18. Mitsa, T., & Varkur, K. L. (1992). Evaluation of contrast sensitivity functions for the formulation of quality measures incorporated in halftoning algorithms. In *Proceedings of the IEEE International Conference on Acoustics, Speech, and Signal Processing*, (Vol. 3, pp. 313–316).

19. Damera-Venkata, N., Kite, T. D., Geisler, W. S., Evans, B. L., & Bovik, A. C. (2000). Image quality assessment based on a degradation model. *IEEE Transactions on Image Processing, 9*(4), 636–650.

20. Zhou, W., & Bovik, A. C. (2002). A universal image quality index. *IEEE Signal Processing Letters, 9*(3), 81–84.

21. Shnayderman, A., Gusev, A., & Eskicioglu, A. M. (2006). An SVD-based grayscale image quality measure for local and global assessment. *IEEE Transactions on Image Processing, 15*(2), 422–429.

22. Khosravy, M., Gupta, N., Marina, N., Sethi, I. K., & Asharif, M. R. (2017). Morphological filters: An inspiration from natural geometrical erosion and dilation. In Nature-Inspired Computing and Optimization (pp. 349–379). Springer: Cham.

23. Sedaaghi, M. H. & Khosravy, M. (2003). Morphological ECG signal preprocessing with more efficient baseline drift removal. In *Proceedings of the 7th IASTED International Conference*, ASC (pp. 205–209).

24. Sedaaghi, M. H., & Wu, Q. H. (1998). *Weighted Morphological Filter Electronics Letters, 34*(16), 1566–1567.

25. Sedaaghi, M. H., Daj, R., & Khosravi, M. (2001). Mediated morphological filters. In *IEEE International Conference on Image Processing*, ICIP2001 (Vol. 3, pp. 692–695).
26. Khosravy, M, Asharif, M. R., & Sedaaghi, M. H. (2008). Medical image noise suppression using mediated morphology. In *IEICE Technical Report* (Vol. 107, no. 461, pp. 265–270), MI2007–111, Okinawa, Japan.
27. Khosravy, M., Asharif, M. R. & Sedaaghi, M. H. (2008). Morphological adult and fetal ECG preprocessing employing mediated morphology. In *IEICE Technical Report* (Vol. 107, no. 461, pp. 363–369), MI2007–129, Bunka, Tenbusu, Okinawa, Japan.

Copy-Move Forgery Detection Using Shift-Invariant SWT and Block Division Mean Features

Ankit Kumar Jaiswal and Rajeev Srivastava

Abstract Digital images are used in courtrooms as evidence. We cannot predict nativity of the image without forensic analysis. Tampering with the image is common nowadays with a lot of online and offline tools. To hide an object in an image, regions of the same image are copied and pasted on that object, and this is known as copy-move forgery. In this paper, we have introduced a technique to detect such type of forgery, known as CMFD. In this technique, the image is pre-processed by converting RGB into YCbCr and then Y channel is decomposed into four components of translation-invariant stationary wavelet transform (SWT). Its LL (approximation) component is then divided into 8×8 blocks. Further, from each block, we have taken six mean features which are calculated by dividing each block into four squares and two triangular blocks and put them into feature vector with block location. After sorting these feature vectors into lexicographical order, we get the location of forged regions.

Keywords Copy-move forgery detection · Digital image forgery
Feature extraction · Translation-invariant · Stationary wavelet transform

1 Introduction

Due to the advancement of smart handheld devices, low cost and higher bandwidth internet facilities, use of multimedia and video/image editing applications are increasing day by day. Using these applications, one can easily manipulate, tamper or synthesize image without any trace. The integrity and authentication of the image may be no longer preserved.

A. K. Jaiswal (✉) · R. Srivastava
Computing and Vision Lab, Department of Computer Science and Engineering,
Indian Institute of Technology (BHU) Varanasi, Varanasi, India
e-mail: akjiitbhu@gmail.com

R. Srivastava
e-mail: rajeev.cse@iitbhu.com

© Springer Nature Singapore Pte Ltd. 2019
A. Khare et al. (eds.), *Recent Trends in Communication, Computing,
and Electronics*, Lecture Notes in Electrical Engineering 524,
https://doi.org/10.1007/978-981-13-2685-1_28

Two forensics schemes, active and passive, are widely used for the forgery detection in an image. The tampered region can be extracted using a pre-embedded watermark in active schemes [1]. However, source files must be there to embed watermark first for this scheme. On the other hand, the passive scheme finds some intrinsic fingerprint clues of images to detect manipulated areas. Passive attacks are imperially used when prior knowledge about the image is unavailable. Therefore, development of image investigation schemes to verify the authenticity and integrity is very important nowadays. Professionals or non-professionals tamper the image by hiding the region using copy and move operation of the object from the same image or adding a new object on the region that they want to tamper. Copy and move technique is often used to manipulate or hide the information of the image, where some objects or pixels of the image are being copied and pasted to other region of the different image or same image. Here in Fig. 1, in an original image, there is a pigeon on the floor and all the pixels covering pigeon are copied and pasted to another location which seems that there are two pigeon in an image (forged).

An article of news came into picture, "Hurricane Harvey: Viral photograph of a shark swimming in Houston flood is a hoax". A blogger from Dublin posted an image in which a shark was swimming on the highway of Houston. After some time, it was cleared in an investigation that this was a forged image. We read this type of article in the newspaper daily. Courtrooms hearing and police investigations are also used in digital images as an evidence. How can one ensure an integrity and authenticity of a digital image? Now, detection of digital image forgery detection comes into the picture. How can forged area in a digital image be detected?

An image is processed through various camera components before the final image is produced in the form of pipelining after capturing it from any digital camera. The captured image is modified by a processing algorithm each time when it passes through any component. These processing algorithms may leave some intrinsic fingerprint clues to detect the tempered area. In 2003, J. Fridrich worked on it and proposed a method to detect the tampered area in an image. A lot of researchers were also done from 2005 to 2017 to detect copy-move forgery (CMF). Common workflow of this copy-move forgery detection is as in Fig. 2.

Fig. 1 Forgery based on copy and move

Original Image   Forged Image

Fig. 2 Workflow of copy-move forgery detection (CMFD)

Pre-processing (Optional): One would like to improve the image data by removing the noise, suppressing the undesired distortion or enhancing the image features. To reduce the dimensionality of an image, colour conversion of an RGB image to grey scale is also a method of pre-processing so that complexity can be reduced and speed of processing will be increased.

Feature Extraction: Selection of relevant information of an image—discrete cosine transform, discrete wavelet transform, log-polar transform, invariant, key-points, texture and intensity are common methods given by researchers.

Matching: Finding similarity is done by either block-based or key-point-based, and it depends on the extracted features.

Visualization (Optional): This step is used to visualize the result of processed copy-move forgery detection (CMFD) method. The tampered region will be displayed in the forged image. The forged region will be displayed either by colouring or mapping forged region with original image when block-based detection method will be used. On the other hand, the key-point-based method will be visualized by line transformation between each matching point.

Distinguishing between the tampered image and a native image is very difficult. Tampering of the image can be done with a single image or more than one image also. Copying of region from an image and pasted it into the same is a type of copy-move forgery (CMF). Detection of such forged region is not possible with naked eyes. In this paper, the same approach of block-based CMFD scheme has been proposed. Translation-invariant stationary wavelet transform (SWT) with block division mean features has been used to detect and localize the forged regions in the image. In this approach, the first image is pre-processed and converted from RGB into YCbCr; then, SWT of Y channel has been used to divide the image into 8×8 overlapped blocks. Each block is used to take six mean features as four squared mean features and two upper and lower triangular mean features. After feature extraction, features are sorted in lexicographical order and matched using shift vector.

This section was the basic introduction to digital image forgeries, detection of forgeries and basic idea about the workflow of the CMFD scheme. The next section is all about the previous works done by a researcher in this particular area. In the third section, a proposed method is given with algorithm and flowchart. Experimental work and result analysis are explained in the fourth section. The final section concludes and explains the future scope in this area.

2   Related Works

According to the Oxford Dictionary [2], the definition of the tamper is, "Interfere with (something) in order to cause damage or make unauthorized alterations". In digital images, a portion of the image is copied and moved to another region of the same image or another image. This is known as tampering of the digital image. In block-based approach, an image of size $M \times N$ splits into square or circle blocks in overlapped or non-overlapped manner for feature extraction from those blocks. Extracted features of each block are then compared with other blocks of the image to find the similarities between blocks. In this way, forged regions in the image are detected. These extracted features are frequency transform, texture and intensity, moment-invariant, log-polar transform, dimension reduction, and others. From the year 2008, we get more than ten research papers from different accesses on frequency transform feature. Instead of splitting the image into blocks, key-point-based approach extracts distinctive local features presented in the image such as corners, edges and blobs.

Fridrich et al. [3] proposed the first method for copy-move forgery detection of an image. This method identifies matched (exact and robust) segments in the image. The method is based on block-based approach, and extracted feature in this robust matching method is discrete cosine transform (DCT). DCT feature of each block is compared with DCT feature of all other blocks in the image. Based on the comparison, matched regions of the image are detected. Generally, images are JPEG compressed images in which forgeries are done.

Doctored JPEG image detection [4] method is based on block artefact grid (BAG) mismatch algorithm. BAG is a grid which is presented in an image where block artefacts appear, and DCT feature of a block and BAG are matched together in a non-forged image. Usually, an attacker does not care about the BAG while copying the slice of an image, so the BAG contains in slice also moved from one place to another place, which causes a mismatch. A portion of grids which contains by slice mismatches when portion moved to another portion, the author picked up the mistake done by the attacker. This detection approach is not limited to copied region from the same image but also works for region copied from another image. Here, problem is that method can be applied to JPEG compressed image only, not on other data sets. Another problem with this method is its complexity.

Popescu et al. [5] substituted DCT feature with principal component analysis (PCA), rest the technique is same as [3]. In this method, PCA is applied on each fixed-sized block of the image and then sorted in lexicographical order. An improved DCT feature-based CMFD method was proposed in [6] with better performance.

The proposed method by Huang et al. [6] gives better result in terms of performance. In this method, author truncated the high frequency of the DCT coefficient, which is called an improved DCT coefficient. After truncating, coefficient row vectors are reshaped in zigzag order in the method. This method is robust to JPEG compression, but when copied regions are too small, this method may not give the better result. Though this method detects the copied region with better performance, still, there are problems with this method. During forgery, attackers attack image such as intermediate/geometric transform like rotation, scaling, mirror, reflection and translation, which provides a spatial synchronization and homogeneity between the copied region and its neighbours and post-processing attacks like JPEG compression, Gaussian noise and blurring of an image. Post-processing operations eliminate any visual findings from a forged image like sharp edges. Such problems are resolved in the proposed method given by [7].

Cao et al. presented a method [7] for CM forgery detection using DCT; the first image is divided into square blocks, and then DCT is applied on each block to calculate quantized coefficient as mentioned in the above-presented method. After applying DCT on each block, dimensions are reduced to four appropriate features by representing each block into the circular block and extracting features from that circular block. Blocks are then sorted in lexicographical order to find the matched block.

Mahmood et al. [8] proposed a method in which image is first transformed into SWT, and then, the image is divided into 4×4 or 8×8 overlapped blocks; from each block, four triangular sub-blocks' mean features are extracted. These mean features are sorted and matched to localize and detect the forgery.

Alkawaz et al. [9] proposed an approach to detect and localize the forged region in the image using DCT coefficient of a greyscale image. In this approach, first RGB image is converted into the greyscale image; then, the greyscale image is divided into 8×8 or 4×4 overlapped blocks, and its 2D DCT transform coefficient is used as feature vector using zigzag scanning. The feature vectors are sorted into lexicographical order, and duplicate regions are found using Euclidean distance.

Michael et al. [10] presented a method to detect the forged region in an image using discrete wavelet transform (DWT) and principal component analysis–eigenvalued decomposition (PCA-EVD). Similar to all other block-based methods, in this technique overlapped blocks are used and named as a window. PCA-EVD of each block row has been formed, and then, all rows are sorted into lexicographical order.

3 Proposed Method

The purpose of the proposed technique is to detect and localize the translated forged regions with higher accuracy. To overcome the problem of translation which was existing with DCT and DWT, we used shift-invariant stationary wavelet transform (SWT). To get higher accuracy, we extracted block division mean feature vector. We get these mean features by dividing 8×8 blocks into four squares and two triangular blocks. We put these means into feature vector with the location of its 8×8 block. Then, we sorted these feature vectors into lexicographical order and found the matched vectors location. Details of each step are given:

3.1 Pre-processing

Image colour conversion operations like RGB to grey scale or RGB to YCbCr are the pre-processing operations. In the proposed technique, the RGB image has been converted into YCbCr colour space. Mathematically, conversion process can be defined as (Fig. 3):

$$Y = \left(\frac{77}{256}\right)R + \left(\frac{150}{256}\right)G + \left(\frac{29}{256}\right)B$$

$$Cb = -\left(\frac{44}{256}\right)R - \left(\frac{87}{256}\right)G + \left(\frac{131}{256}\right)B + 128$$

$$Cr = \left(\frac{131}{256}\right)R - \left(\frac{110}{256}\right)G - \left(\frac{21}{256}\right)B + 128$$

This YCbCr colour space is used in the standard process of digital video encoding. In above formulas, R, G and B are red, green and blue—three channels of the image.

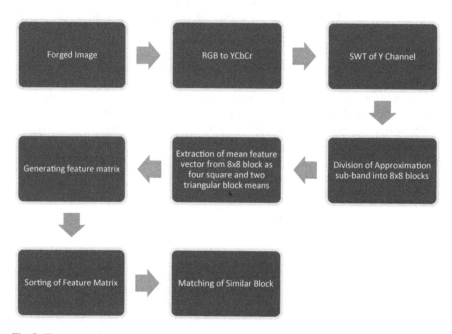

Fig. 3 Flowchart of proposed algorithm

3.2 Feature Extraction

From the pre-processed image, features are extracted. In this approach, pre-processed image is first decomposed into various sub-bands of shift-invariant SWT. An approximation and detailed coefficient can be obtained by convolving a 2D image with low- and high-pass filters (refer Fig. 4). Four obtained sub-bands, LL, HL, LH and HH, are approximation, vertical, horizontal and diagonal sub-bands, respectively. In the presented technique, approximation sub-band LL is divided into 8×8 overlapped blocks.

If given image size is $m \times n$, then the number of overlapping blocks will be:

$$N_{blocks} = (m - b + 1) \times (n - b + 1)$$

To extract mean features from the 8×8 size block, four squared mean features and two triangular mean features are calculated.

$$M_i = \frac{\sum f(x, y)}{n} \quad where \, i = 1, 2, 3, 4, 5, 6$$

Extracted mean features with the location (x, y) [upper left corner position of pixel] of the block are stored in the feature vector.

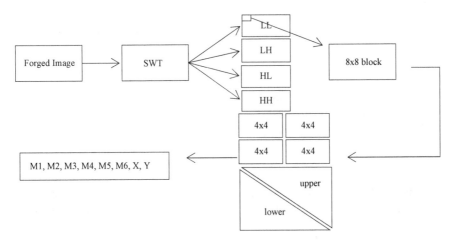

Fig. 4 Feature extraction from forged image

Fig. 5 Forged images used for experiment and result analysis

3.3 Matching

A feature matrix is generated using mean feature vectors. This matrix has mean features of each block with the location of the block. Matching of blocks is done by using these feature vectors excluding the location of the block. Feature vectors excluding location are sorted into lexicographical order. The features closer to each other will be close to the sorted matrix. It may be possible that the location of similar feature vectors is closer; in such case, they are not copied and moved; instead, they have same features. A shift vector [3] has been used to avoid such blocks. Let two matching block positions are (i1, i2) and (j1, j2) A shift vector can be calculated as:

$$s = (s1, s2) = (i1 - j1, i2 - j2)$$

Because $-s$ and s both are the same shift, it can be normalized by multiplying -1 to the shift vector. A shift vector counter is incremented for each matching pair block by one.

$$C(s1, s2) = C(s1, s2) + 1$$

A threshold for comparing shift vector counter, if the occurrence of shift vector is more than the threshold value, then the region is copied. Finally, similar blocks are highlighted.

4 Experiment and Result Analysis

In this section, experiment and result analysis is performed on standard dataset of copy-move forgery detection (Fig 5). Thirty-five forged and ground truth images of size 512×512 from CoMoFoD dataset [11] were taken. Our implementation was

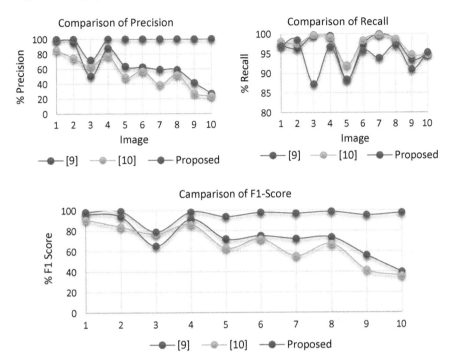

Fig. 6 Comparison of precision, recall and F1 score

Table 1 Comparison of the proposed technique with prevision techniques

Technique	Precision (%)	Recall (%)	F1 score (%)
Proposed technique	97.17	94.08	95.47
DCT [9]	64.52	96.57	75.16
DWT-PCA [10]	55.60	97.49	68.76

run on MATLAB 2017R server version. The configuration of the system was Linux server (CentOS) with 16 GB RAM and Xeon processor. We performed the proposed solution and matched output with ground truth. Here is a table of compared precision, recall and F1 score value with DCT [9] and DWT-PCA [10] (Fig. 6 and Table 1).

Out of thirty-five images, ten images have been taken to compare the results of precision, recall and F1 score graphically with DCT [9] and DWT-PCA [10]. In the above visual graph, it can be seen that for each image, precision value of the proposed method is approximately one. The calculation of precision, recall and F1 score is based on the false matched and true matched pixel values of the proposed technique with ground truth image. Mathematically, formulas can be derived as:

$$P = \frac{tp}{tp + fp}, \quad R = \frac{tp}{tp + fn}, \quad f1 - score = 2 \times \frac{P \times R}{P + R}$$

P = Precision, R = Recall, tp = True Positive
f = False Positive, fn = False Negative.

The overall performance of the proposed technique can be visualized graphically. In most of the cases, the F1 score percentage is more than 90% which is comparatively more than DCT and DWT-PCA.

5 Conclusion

Distinguishing between the tampered image and a native image is very difficult. Tampering of the image can be done with a single image or more than one image also. Copying of region from an image and pasting it into the same is a type of copy-move forgery (CMF). Detection of such forged region is not possible with naked eyes. The purpose of the proposed technique is to detect and localize the translated forged regions with higher accuracy. The proposed technique is translation- and shift-invariant. It gives a better result than DCT and DWT-PCA techniques. In this technique, overlapping blocks are used to extract and match the feature vector. In case of a high-resolution image, the proposed technique may take much time, but the result will be better.

References

1. Ng, T. T., Chang, S. F., Lin, C. Y., & Sun, Q. (2006). Passive-blind Image Forensics. *Multimedia Security Technologies* for *Digital Rights Management* (pp. 383–412). https://doi.org/10.1016/b978-012369476-8/50017-8.
2. English Oxford Living Dictionaries. Retrieved Oct 5, 2017 from https://en.oxforddictionaries.com/definition/tamper.
3. Fridrich, J. S., & David Lukáš, J. (2003). Detection of copy-move forgery in digital images. *Procedure Digital Forensic Research Workshop, 3,* 652–663. https://doi.org/10.1109/PACIIA.2008.240.
4. Weihai, L., & Yu, N., Yuan, Y. (2008). Doctored JPEG image detection. In *Proceedings of IEEE International Conference on Multimedia and Expo* (pp. 253–256). https://doi.org/10.1109/icme.2008.4607419.
5. Popescu, A. C., & Farid, H. (2004) Exposing digital forgeries by detecting duplicated image regions. Department of Computer Science Dartmouth College Technical Report TR2004-515 1–11. https://doi.org/10.1109/tsp.2004.839932.
6. Huang, Y., Lu, W., Sun, W., & Long, D. (2011). Improved DCT-based detection of copy-move forgery in images. *Forensic Science International, 206,* 178–184. https://doi.org/10.1016/j.forsciint.2010.08.001.
7. Cao, Y., Gao, T., Fan, L., & Yang, Q. (2012). A robust detection algorithm for copy-move forgery in digital images. *Forensic Science International, 214,* 33–43. https://doi.org/10.1016/j.forsciint.2011.07.015.
8. Mahmood, T., Mehmood, Z., Shah, M., & Khan, Z. (2017). An efficient forensic technique for exposing region duplication forgery in digital images. *Applied Intelligence,* 1–11. https://doi.org/10.1007/s10489-017-1038-5.

9. Alkawaz, M. H., Sulong, G., Saba, T., Rehman, A. (2016). Detection of copy-move image forgery based on discrete cosine transform. *Neural Computing & Applications*, 1–10. https://doi.org/10.1007/s00521-016-2663-3.

10. Michael, Z., Xingming, S. (2011). DWT-PCA(EVD) based copy-move image forgery detection. *International Journal of Digital Content Technology and its Applications*, 5.

11. Tralic, D., Zupancic, I., Grgic, S. (2017). GM CoMoFoD—New database for copy-move forgery detection. In *Proceedings of 55th* International Symposium *ELMAR-2013*. Retrieved Oct 11, 2017 form http://www.vcl.fer.hr/comofod/download.html.

Dual Discrete Wavelet Transform Based Image Fusion Using Averaging Principal Component

Ujjawala Yati and Mantosh Biswas

Abstract Image fusion is a process of combining two or more images into one, in order to obtain more relevant data or information from it. In this paper, we have proposed dual discrete wavelet transform (DDWT)-based image fusion method in frequency domain with averaging the principal component analysis that overcomes the spatial distortion, blocking artifact, and shift variance of the fusion methods. The results of the proposed method have been promising for qualitative and quantitative evaluations that are performed on subjective and objective criteria, respectively, which are shared in the experimental results for considered test images over fusion methods.

Keywords Image enhancement · Image fusion · DDWT
Principal component analysis

1 Introduction

Fusion is a process of combining two or more distinct entities, similar is in the case of image fusion. Say, there are two images image A and image B containing some information, both images when analyzed individually possess no or less information but, after we fuse them together in resultant fused image; we find that resultant image gives some useful information. Image fusion is a technique which can be used in many fields such as military, security and surveillances, computer vision, astronomical images, medical imaging, navigation. Image fusion can be done from images with same modality or images from different modality as well; e.g., CT and MRI can also be fused together to obtain relevant and useful information [1]. If we try to understand

U. Yati (✉) · M. Biswas
Department of Computer Engineering Kurukshetra, National Institute
of Technology Kurukshetra, Kurukshetra, India
e-mail: ujjawala.yati093@gmail.com

M. Biswas
e-mail: mantoshb@gmail.com

© Springer Nature Singapore Pte Ltd. 2019
A. Khare et al. (eds.), *Recent Trends in Communication, Computing,
and Electronics*, Lecture Notes in Electrical Engineering 524,
https://doi.org/10.1007/978-981-13-2685-1_29

in simple terms, addition of best pixels from two different images into third image could give relevant information and thus be called image fusion. This could be a simple technique of image fusion but, in order to get more relevant information from them, we have to work little harder, so we have to find better ways to fuse images to get more relevant information. Therefore, there are number of methods proposed in the field of image fusion in both the spatial and frequency or transformation domains, but the methods of image fusion should have the properties; that is, it transfers the advantageous information from input images to the integrated image, while making no loss of information during processing, and also should not add any artifacts in the resultant fused image. The spatial domain methods combine information in spatial space according to certain fusion rules like minimum-maximum, average. In contradiction, the frequency domain methods conduct multi-scale/multi-resolution decompositions of image before combining information from them and then the fusion rules are designed on basis of frequency context. Basically, it includes three stages, i.e., decomposition, coefficients fusion, and reconstruction [2]. The image fusion method of the spatial domain in the initial stages ordinarily acquires an image fusion scheme that is based on block, where each pair of blocks is combined with an intended activity level measurement such as summation of modified pixel values by applying Laplacian function [3, 4]. In the last few years, pixel-based image fusion methods have been proposed in [5, 6], deploying gradient information. However, spatial domain methods were not able to achieve all the properties stated above because spatial domain methods suffer from spatial distortion, and also the problem of blocking artifacts arises from the block-based spatial domain methods.

Therefore, to meet the requirement stated above for the image fusion, transformation domain methods were proposed in which images are transformed to frequency domain, i.e., wavelet transforms and pyramid methods [2, 7]. These methods firstly decompose the input images into approximation coefficients having variations in large scale and seriate detail coefficient having variations on small scale, after that fusion rules are applied on them for integrating them into a single fused resultant image. An image pyramid is a sequence of images in which the image at each level is constructed by applying low-pass filtering and subsampling on its predecessor in such a way that at each level size of image gets halved, such as gradient, Laplacian, and morphological pyramid, but these methods suffer from halo effect [4–8]. Further, transform provides better time-frequency characterization over pyramid. Consequently, it gained popularity in the field of image fusion methods. There are many fusion methods developed in the frequency or transformation domain. Elias et al. [9] proposed a method for image fusion using discrete cosine transform with Laplacian pyramid approach. In this method, Laplacian pyramid is constructed for input images, and at each level the size of image is reduced by half to obtain next level where both spatial density and resolution are reduced. This method suffers from the spatial distortion and blocking artifact. Kaur et al. [10] proposed pixel level-based image fusion method using discrete wavelet transform with enhanced Laplacian pyramid by mapping the local binarized pixels of images. This method has drawbacks of blocking artifact due to the use of pyramid-based method and lack k of shift invariance. Pradnya et al. [11] proposed the wavelet transform-based image fusion in the transformation domain using stationary wavelet transform and form fusion. They used spatial frequency measurement but have limitation, i.e., spatial distortion, and its results are just satisfactory. Aymaz et al. [12] have proposed image fusion method in frequency domain based on stationary wavelet transform in combination with PCA. This method overcomes the problem mentioned in other wavelet transforms, but it does not provide better results comparatively. There are many areas, like military, surveillance, astronomy, remote sensing, medical, where images are captured at different frames, but sometimes these images become less informative when seen individually, and we need to extract more information from them and to resolve the problem that arises in above-stated methods; we have proposed an image fusion method based on dual discrete wavelet transform (DDWT) which is an enhanced version of DWT. In our proposed method, we have in collaborated DDWT with principal component analysis and averaging. In our method, the input images are decomposed by DDWT to generate coefficients, after that principal component is applied to the coefficients and average of the principal component is calculated for all the coefficients that will form the weights to the fusion rule for resultant fused image.

The paper is compiled as follows: Proposed method is given in Sect. 2, and in Sect. 3, the experimental results are provided for the proposed method over the considered image fusion method on the test images. Lastly, the paper is concluded in the conclusion section.

2   Proposed Method

In our proposed image fusion method, we have used dual discrete wavelet transform (DDWT) with averaging PCA. The block diagram of the proposed method is shown in Fig. 1. Our proposed method can be divided into four steps, firstly in step 1, we decompose the input image using DDWT to obtain coefficients of subbands, and secondly in step 2, we apply PCA on coefficients to produce principle components that will provide weight to the subbands of the input image. Thirdly, in step 3, weights of the subbands obtained in step 2 are used for assigning weights to the input images using averaging, and finally, in step 4, we apply the fusion rule by utilizing the weights of input images from step 3 for the fused image. These four steps of our proposed method are stated and explained below.

Step 1: The DWT encapsulates both the frequency and temporal content. The DWT is a multi-resolution method which provides a time to frequency representation of the any signal which means that the different resolutions are analyzed on different frequencies [2, 10]. The DWT decomposition of the input images can be represented as the binary tree of approximation and detail coefficients [1]. In our proposed method, we have employed DWT with two different types of mother wavelets, namely Daubechies (DWT^D) and Biorthogonal (DWT^B), forming two binary trees of approximation and detail coefficients for each of the input images. The subbands

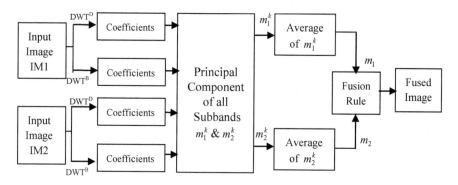

Fig. 1 Block diagram of proposed fusion method

generated after applying the DWT with one of the other wavelets, i.e., Daubechies arc LL_1^D, LH_1^D, HL_1^D, HH_1^D, and with another mother wavelet, i.e., Biorthogonal are LL_1^B, LH_1^B, HL_1^B, and HH_1^B on the input image IM1. Similarly, for the another input image IM2, the subbands generated after applying the DWT with one of the mother wavelets, i.e., Daubechies are LL_2^D, LH_2^D, HL_2^D, HH_2^D, and with another mother wavelet, i.e., Biorthogonal are LL_2^B, LH_2^B, HL_2^B, and HH_2^B. These subbands are passed for the PCA where weights are assigned to the subbands by computing the principal component for the subbands in step 2.

Step 2: Principal component analysis (PCA) is a statistical tool including mathematical processing that is used to convert correlated variables to uncorrelated ones known as principal components. We have thought of using PCA in our work because it is not biased against the variables that have more number of distinct values, and also, it is easy and simple to compute [13]. The input images IM1 and IM2 are decomposed using DDWT generating the subbands. Those subbands are passed to the PCA, where the principal components calculate by selecting the maximum value of eigenvector for each of the subbands. The two computed principal components are $m_1(LL_1^D)$ and $m_2(LL_2^D)$ provide weights to the subbands LL_1^D and LL_2^D of the input images IM1 and IM2, respectively. Similarly, the weights to other subbands are also computed and represented as $m_1(LH_1^D)$, $m_1(HL_1^D)$, $m_1(HH_1^D)$, $m_1(LL_1^B)$, $m_1(LH_1^B)$, $m_1(HL_1^B)$, $m_1(HH_1^B)$, $m_2(LH_2^D)$, $m_2(HL_2^D)$, $m_2(HH_2^D)$, $m_2(LL_2^B)$, $m_2(LH_2^B)$, $m_2(HL_2^B)$, and $m_2(HH_2^B)$. After calculating the weights of all the subbands, it is used to compute the weight for the input images IM1 and IM2 in the next step.

Step3: The weights to the input images IM1 and IM2 are assigned by using weights gained for each of the subbands that are obtained from the principal component analysis above step 2. The weights of the subbands are combined together by taking the summation of them and dividing it by decomposition level, N, namely m_1 and m_2 for IM1 and IM2, respectively, as given below:

$$m_1 = (m_1(LL^D) + m_1(LH^D) + m_1(HL^D) + m_1(HH^D)$$
$$+ m_1(LL^B) + m_1(LH^B) + m_1(HL^B) + m_1(HH^B))/N \qquad (1)$$

$$m_2 = (m_2(LL^D) + m_2(LH^D) + m_2(HL^D) + m_2(HH^D)$$
$$+ m_2(LL^B) + m_2(LH^B) + m_2(HL^B) + m_2(HH^B))/N \qquad (2)$$

Step 4: Finally in this step, image fusion rule is applied after and on obtaining the weights for each of the input images. To get the resultant fused image, each of the input images is multiplied with their weights and then summed together to form the resultant image as given below:

$$F = m_1 * IM_1 + m_2 * IM_2 \qquad (3)$$

3   Experimental Results

The characteristic and visual upshots for the proposed image fusion method are carried out experimentally for the evaluation of the fused results with respect to the test input images, namely book, clock, flower, motor, and saras, are provided in Fig. 2, of size 512 * 512 in comparison with the fusion methods: DWT, DWT+LPA, DCT+LPA, SWT, PCA, and SWT+PCA. We have implemented the fusion methods on MATLAB 2017a with RAM 4 GB and i5 processor of the PC. In our proposed method, we have used two types of mother wavelets, namely Daubechies and Biorthogonal with level 1 decomposition level. To verify the preserved spectral characteristics and the improvement of the spatial resolution, we have employed both the subjective and objective performance evaluation parameters for the evaluation of the proposed fusion method with considered fusion methods for the visualization and quantitative comparison, respectively.

Subjective evaluation is a method of human visual analysis for fused image, which is simple and instinctual [1]; therefore, we have used subjective evaluation provided in Figs. 3, 4, 5, 6, and 7 for the test images. It is dependent totally on human visualization and interpretation; so its backdrops cannot be underestimated, and therefore, quantitative analysis is also performed. So, to do quantitative analysis of the results, number of objective parameters have been developed by the researchers, three of them are used in this paper, namely feature mutual information (FMI), that is non-reference image fusion metric; it computes the degree of dependency between fused image with input images [14], fused peak signal-to-noise ratio (FPSNR) which is based on the PSNR [14], and structural similarity index measure (SSIM) that is the measure of the amount of structural information in that image [15]. A higher value indicates more information is transferred.

The resultant fused image of book in Fig. 3, it can be seen that the texts in the image of the book are clearer as it is readable more easily of our proposed method

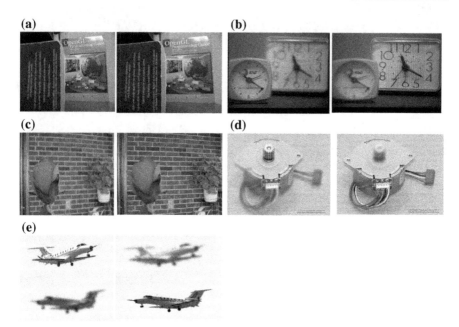

Fig. 2 Test images **a** book, **b** clock, **c** flower, **d** motor, and **e** saras

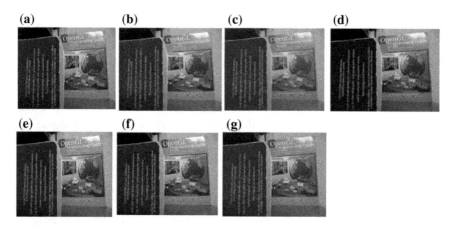

Fig. 3 Fused images of book for methods: **a** DWT, **b** DWT+LPA, **c** DCT+LPA, **d** SWT, **e** PCA, **f** SWT+PCA, and **g** proposed

(Fig. 3g) over the fusion methods DWT, DWT+LPA, DCT+LPA, SWT, PCA, and SWT+PCA. The resultant fused image of the clock for the proposed method (Fig. 4g) is better than the considered fusion methods that are considered in Fig. 4. From Fig. 5, it can be realized that the resultant fused image of proposed method (Fig. 5g) for flower can be better visualized over the heeded image fusion methods. It can be

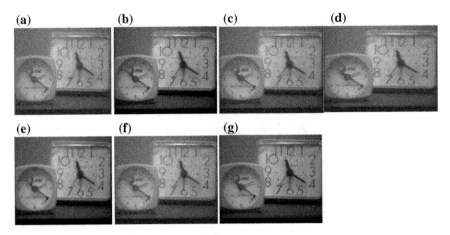

Fig. 4 Fused images of clock for methods: **a** DWT, **b** DWT+LPA, **c** DCT+LPA, **d** SWT, **e** PCA, **f** SWT+PCA, and **g** proposed

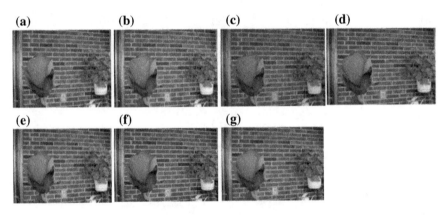

Fig. 5 Fused images of flower for methods: **a** DWT, **b** DWT+LPA, **c** DCT+LPA, **d** SWT, **e** PCA, **f** SWT+PCA, and **g** proposed

seen from Fig. 6 that the resultant fused image of motor of the proposed method (Fig. 6g) gives good result; especially, the rings of rotor and boundary of the wires are clearly differentiable over considered fusion methods that are examined. In Fig. 7 of saras fused image, it can be seen that the edges in the fused image of the proposed method (Fig. 7g) for the saras are clearer as it can be visualized more easily over the considered fusion methods.

The FMI results of the proposed method in comparison with considered fusion methods are given in Table 1. It can be realized from Table 1 that the proposed method gives better performance over the heeded image fusion methods: DWT, DWT+LPA,

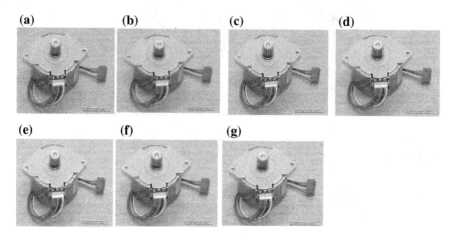

Fig. 6 Fused images of motor on methods: **a** DWT, **b** DWT+LPA, **c** DCT+LPA, **d** SWT, **e** PCA, **f** SWT+PCA, and **g** proposed

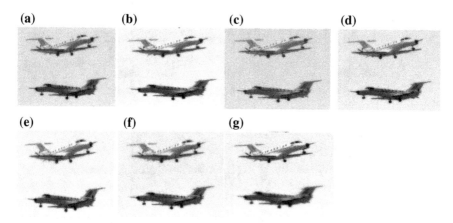

Fig. 7 Fused images of saras on methods: **a** DWT, **b** DWT+LPA, **c** DCT+LPA, **d** SWT, **e** PCA, **f** SWT+PCA, and **g** proposed

SWT, PCA, and SWT+PCA on the images: Book and clock but gives comparative results for the flower, and saras and the image motor, the proposed method gives good results over the methods: DWT, DWT+LPA, SWT, and PCA, whereas comparative results of SWT+PCA. The results on FPSNR and SSIM parameter for the proposed method and considered fusion methods are stated in Tables 2 and 3, respectively, and graphically in Fig. 8. From the table ,it can be seen that the proposed method gives better performance for the test images, namely book and clock, flower, motor, and saras, over the considered methods: DWT, DWT+LPA, SWT, PCA, and SWT+PCA.

Table 1 FMI results for fusion methods: DWT, DWT+LPA, DCT+LPA, SWT, PCA, SWT+PCA and proposed on images: book, clock, flower, motor, and saras

Image name	DWT	DWT+LPA	DCT+LPA	SWT	PCA	SWT+PCA	Proposed
Book	0.8647	0.8547	0.8654	0.8671	0.8660	0.8668	**0.8827**
Clock	0.903	0.8839	0.9068	0.9104	0.9007	0.9099	**0.9210**
Flower	0.8584	0.8129	0.8759	0.8735	0.8712	0.8744	**0.8840**
Motor	0.9284	0.8910	0.9454	0.9437	0.9464	**0.9473**	0.9464
Saras	0.9418	0.9208	0.9456	0.9463	0.9488	0.9459	**0.9495**

Table 2 FPSNR results for fusion methods: DWT, DWT+LPA, DCT+LPA, SWT, PCA, SWT+PCA and proposed on images: book, clock, flower, motor, and saras

Image name	DWT	DWT+LPA	DCT+LPA	SWT	PCA	SWT+PCA	Proposed
Book	29.631	23.791	31.354	32.144	31.010	33.577	37.817
Clock	30.197	28.787	30.065	30.415	30.347	30.650	38.563
Flower	29.497	21.846	33.580	32.002	30.964	32.157	39.989
Motor	30.762	23.305	31.271	31.779	31.856	32.204	33.378
Saras	26.933	24.765	26.352	27.175	27.389	27.374	30.715

Table 3 SSIM results for fusion methods: DWT, DWT+LPA, DCT+LPA, SWT, PCA, SWT+PCA and proposed on images: book, clock, flower, motor, and saras

Image name	DWT	DWT+LPA	DCT+LPA	SWT	PCA	SWT+PCA	Proposed
Book	0.9343	0.8000	0.9424	0.9583	0.9571	0.9766	0.9901
Clock	0.9275	0.9027	0.9080	0.9336	0.9335	0.9410	0.9880
Flower	0.9161	0.6718	0.9609	0.9581	0.9459	0.9649	0.9931
Motor	0.9235	0.7413	0.9507	0.9575	0.9513	0.9591	0.9636
Saras	0.9311	0.9070	0.9076	0.9351	0.9391	0.9388	0.9636

From the graphical representation for the fusion results of objectives evaluation metrics, namely FMI, FPSNR, and SSIM separately as (a)–(c), respectively, in Fig. 8 for the examined fusion methods together with the proposed method on the given images on horizontal axis numbered from 1 to 5 and range for each metrics are labeled on vertical axis. In Fig. 8a showing the graphical representation of the FMI, it can be seen that the height of bar is high for the proposed method over the heeded methods for the test images, namely book, clock, flower, and saras. While the motor fused image, the proposed method height of bar is high over DWT, DWT+LPA, DCT+LPA, SWT, and PCA methods but relatively lower than SWT+PCA. The height of bar is high for proposed method over considered methods for FPSNR and SSIM metric on all the test images in Fig. 8b–c, respectively.

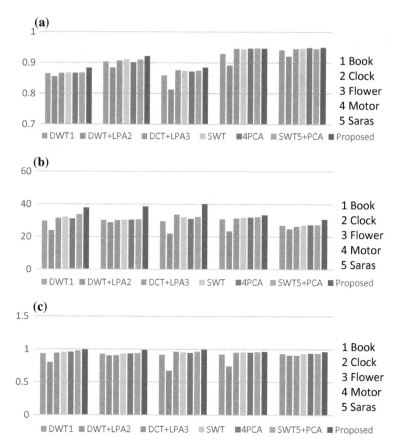

Fig. 8 Comparative performance graph of test images with considered fused methods for parameters of **a** FMI, **b** FPSNR, and **c** SSIM

4 Conclusion

Image fusion is a method of reducing redundancy and integrating the complementary information from many images which are obtained from different sensors, into a single resultant image which is specific to the area of application. We have presented, in this paper, dual discrete wavelet transform which resolves the problem of shift variance, spatial distortion, and blocking artifacts in collaboration with averaging of PCA to fuse the image and for extraction of more information from them. Experimentally, we have tested the working of proposed fusion method with test images, and results have seen to be impressive and much more informative in our proposed method over DWT, DWT+LPA, DCT+LPA, SWT, PCA, and SWT+PCA on both subjective and objective evaluations.

References

1. Pajares, G., & Cruz, J. M. D. (2004). A wavelet-based image fusion tutorial. *Pattern Recognition Journal, 37*(9), 1855–1872.
2. Shi, W., Zhu, C., Tian, Y., & Nichol, J. (2005). Wavelet-based image fusion and quality assessment. *International Journal of Applied Earth Observation and Geoinformation, 6*, 241–251.
3. Huang, W., & Jing, Z. (2007). Evaluation of focus measures in multi-focus image fusion. *Pattern Recognition Letter, 28*(4), 493–500.
4. Li, S., Kang, X., & Hu, J. (2013). Image fusion with guided filtering. *IEEE Transaction on Image Processing, 22*(7), 2864–2875.
5. Zhou, Z., Li, S., & Wang, B. (2014). Multi-scale weighted gradient-based fusion for multi-focus images. *Information Fusion, 20*(1), 60–72.
6. Burt, P. J. (1992). A gradient pyramid basis for pattern selective image fusion. *International Symposium Digest of Technical Papers, 5*(2), 467–470.
7. Burt, P., & Adelson, E. (1983). The Laplacian pyramid as a compact image code. *IEEE Transaction on Communication, 31*(4), 532–540.
8. Toet, A. (1989). A morphological pyramidal image decomposition. *Pattern Recognition Letters, 9*(4), 255–261.
9. Naidu, V. P. S., & Elias, B. (2013). A novel image fusion technique using DCT based laplacian pyramid. *International Journal of Inventive Engineering and Sciences, 1*(2), 1–9.
10. Kaur, H., & Rani, J. (2015). Image fusion on digital images using laplacian pyramid with DWT. In *Third International Conference on Image Information Processing* (pp. 393–398).
11. Pradnya, P. M., & Ruikar, S. D. (2013). Image fusion based on stationary wavelet transform. *International Journal of Advanced Engineering Research and Studies, 2*(4), 99–101.
12. Aymaz, S., & Kös, C. (2017). Multi-focus image fusion using stationary wavelet transform with PCA. In *10th International Conference on Electrical and Electronics Engineering* (pp. 1–5).
13. Patil, U., & Mudengudi, U. (2011). Image fusion using hierarchical PCA. In *International Conference on Image Information Processing* (pp. 1–6).
14. Haghighat, M., & Razian, M. (2014). Fast-FMI: Non-reference image fusion metric. In *8th International Conference on Application of Information and Communication Technologies* (pp. 1–3).
15. Wang, Z., Bovik, A. C., Sheikh, H. R., & Simoncelli, E. P. (2004). Image quality assessment: From error visibility to structural similarity. *IEEE Transactions on Image Processing, 13*(4), 1–14.

Comparative Study on Person Re-identification Using Color and Shape Features-Based Body Part Matching

Sejeong Lee, Jeonghwan Gwak, Om Prakash, Manish Khare, Ashish Khare and Moongu Jeon

Abstract Researches on person re-identification have become more prevalent as a core technique in visual surveillance systems. While a large volume of person re-identification (Re-ID) methods adopt experimental settings with a large patch size, relatively small ones are not much considered. Thus, this work focuses on investigating comparison of person Re-ID performance when color and shape features-based are adopted for small patch size images with low resolution. First, the deep decompositional network is used to divide the person into upper and lower body parts. Then, color and shape features are extracted. Finally, using single or combination of features, similarity-based ranked matching scores are computed. The person Re-ID performance is evaluated based on VIPeR dataset. From the experiment results, we found that color-based feature is better than shape-based features, and the combination of color and shape features-based can be meaningful.

S. Lee · M. Jeon (✉)
School of Electrical Engineering and Computer Science, Gwangju Institute
of Science and Technology, Gwangju 61005, Korea
e-mail: mgjeon@gist.ac.kr

J. Gwak (✉)
Biomedical Research Institute, Seoul National
University Hospital (SNUH), Seoul 03080, Korea
e-mail: james.han.gwak@gmail.com

J. Gwak
Department of Radiology, Seoul National
University Hospital (SNUH), Seoul 03080, Korea

O. Prakash
Centre of Computer Education, Institute of Professional Studies,
University of Allahabad, Allahabad, India

M. Khare
Dhirubhai Ambani Institute of Information and Communication
Technology, Gandhinagar, Gujarat, India

A. Khare
Department of Electronics and Communication,
University of Allahabad, Allahabad, Uttar Pradesh, India

© Springer Nature Singapore Pte Ltd. 2019
A. Khare et al. (eds.), *Recent Trends in Communication, Computing,
and Electronics*, Lecture Notes in Electrical Engineering 524,
https://doi.org/10.1007/978-981-13-2685-1_30

313

Keywords Person re-identification · Deep decompositional network
Body part matching · Combination of features · Color feature · Shape feature

1 Introduction

Person re-identification (Re-ID) deals with the issue of matching pedestrian image patches (or sometimes video volumes) observed in disjoint multiple cameras and finding the same pedestrians [1]. In order to assist automated visual surveillance systems (e.g., [2–5]), it is known as a fundamental research topic. As an example of its application, it can be utilized in finding lost persons (e.g., children, old persons with disease) or criminals in broadcasting using the persons' information and publicly available multiple cameras. Despite hectic and long-term research, person Re-ID is still difficult and yet challenging. This task tends to be very difficult because it is vulnerable to occlusion, similar color clothes among different persons, the background clutter, variation of viewpoint, illumination, pose and resolution. Under these conditions, it will be very difficult to extract robust information to discriminate two pedestrian images obtained from different cameras, and as a result, it tends to produce very poor performance even for the same pedestrian images. Also, existing person Re-ID methods require very high computation power due to the complexity of its similarity computation, but in practical person Re-ID system applications, real-time processing capability is the essential requirement.

Person Re-ID can be divided into appearance-based Re-ID and salience-based Re-ID. The former method deals with appearance information of persons as the feature. As we discussed, since appearance information is very vulnerable to environmental conditions, it is difficult to use it solely. Salience-based Re-ID extracts characteristic or obvious parts (e.g., a backpack of the pedestrians or color parts distinguishable with others). However, this is weak at viewpoint variations (e.g., the backpack may not be visible from the front view of the pedestrian). While person Re-ID using deep learning is getting increased attention, the main disadvantage of deep learning-based person Re-ID is the need of retraining when a new person appears. Person Re-ID in surveillance applications is very challenging because in realistic situations, pedestrians are captured at low resolution.

In this work, we propose a framework to evaluate color and shape features for person Re-ID. For the color appearance matching, a simple but robust body part model was used based on the deep decompositional network (DDN) [6]. In detail, instead of using the whole image of a person, we extract its torso and leg parts without its background. By eliminating the background, it tends to be robust to background clutter. Moreover, extracting torso and leg parts makes it robust to pose variations. However, by removing the background, the edge information of the body parts can be lost, even though the edge can be a very important shape feature. Therefore, we do not perform the background removal process for extraction of the shape features including histogram of oriented gradients (HOG) [7], scale invariant feature transform (SIFT) [8] and speeded up robust feature (SURF) [9]. Figure 1 shows the proposed system.

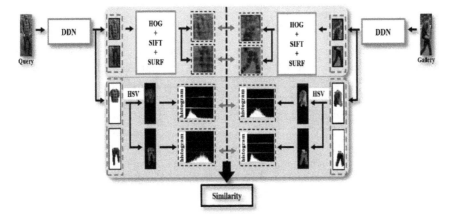

Fig. 1 Proposed person Re-ID system

We divide a body part into the upper body and the lower body; set the color feature and shape features of the query image (probe) and gallery images; and compare each corresponding parts between them.

2   Overview of the Proposed Person Re-ID System

In the proposed method, we separate the meaningful body parts (e.g., torso and leg) using DDN to extract the color information of pedestrians. In this process, the background is removed and as the result, the body of the pedestrian is remained. Then, we create an HSV color histogram for each part and calculate the distance between the probe and gallery images. Using this information, we can obtain the color similarity by comparing the upper body part image of the probe with the upper body image of the gallery and comparing the lower body part image of the probe with the lower body part image of the gallery. To obtain shape information of the pedestrian, we separate the patch image into the upper body and lower body based on the location of the waist. In this process, the background is remained unlike the color feature extraction process. Next, we extract the HOG, SIFT and SURF from each body part image. Shape similarity between the probe and gallery images can be calculated with the feature(s). Finally, we can determine whether the probe and the gallery images are the same using the color and shape information.

2.1   Division of Body Parts

In general, when the resolution is low, e.g., as shown in Fig. 2 from the viewpoint invariant pedestrian recognition (VIPeR) dataset [1], the usage of face information in person Re-ID is difficult because details of faces cannot have distinguishable information. In the VIPeR dataset, it is very difficult to recognize whether they are the same person by using the face information. Therefore, following the previous studies [6, 10–13], we exclude face part and extract features from the upper body and the lower body. Then, features of the upper body part of the probe are compared with those of the upper body part of the gallery, and features of the lower body part of the probe are compared with those of the lower body part of the gallery. By separating the upper body part and the lower body part, we can extract more reliable features. For example, suppose that there are two men dressed differently. One is dressed in a white shirt and black jeans, and another man dressed in black shirts and white pants. When we compare a man with another man, by partitioning the upper and lower body parts, the differences are easily noticed. In contrast, this may not be easy without such partitioning. In addition, if we divide the body part using DDN, the background is removed and thus the color feature is less affected by the background clutter.

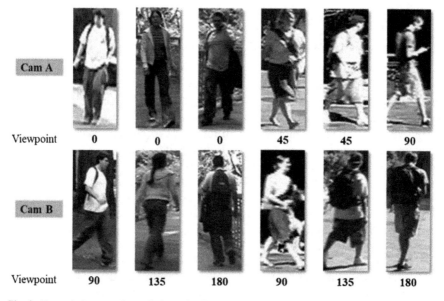

Fig. 2 Example image pairs and viewpoint from the VIPeR

Fig. 3 CMC curves on VIPeR dataset when using only HSV color histogram

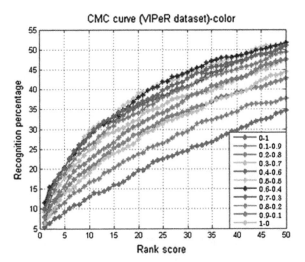

2.2 Color Feature

Among several visual appearance features, color is the most important feature for the Re-ID task. This is because under surveillance conditions with very low resolutions, the other information except color can be very susceptible to varying conditions. The HSV color space is more robust to changes in illumination. Thus, we create the histogram from the upper body part and the lower body part, and then calculate the distance between two HSV histograms of the probe and gallery images for each corresponding body part. Figure 3 shows the results of cumulated matching characteristic (CMC) curve for 11 weight combinations between the upper body part and the lower body part. The best performance was drawn from the proportion 0.6 (the upper body part): 0.4 (the lower body part), which was almost similar to the DDN-based body part division results.

2.3 HOG Feature

To obtain the HOG features, the image is divided into cells of a certain size. Then, a histogram is obtained for each cell about the edge direction. HOG is a vector expressing the histogram which contains the shape information of the image. For extracting edge information around the boundary of body parts, the background information is also important, and thus we remain the background for this feature.

Table 1 Comparison of SIFT and SURF

Features	Time	Scale invariance	Rotational invariance	Affine invariance
SIFT	Normal	Good (scale space search)	Good (orientational DoG)	Good
SURF	Good	Good (scale space search)	Normal	Good

2.4 SIFT and SURF Features

To extract SIFT and SURF features, similar to the extraction process of the HOG feature, we used images of the upper body part and the lower body part without removing the background. Next, when we get the SIFT, we created a Gaussian pyramid to obtain difference of Gaussian (DoG) and then obtained the feature points. Since DoG is obtained according to each scale, there is an advantage that it is robust against the size change. It is also robust to rotation because it uses orientation. The SURF complements the slow speed of the SIFT algorithm as summarized in Table 1.

From Fig. 4a–c, it is shown that there is no certain proportion to separate the upper body part and the lower body part.

3 Experimental Results

The experiments were carried out using the color feature and three shape features (HOG, SIFT and SURF). We compare the accuracy and find the appropriate the upper and lower body ratios for each feature. Using this ratio, we compare the three methods (only color, color+HOG, color+HOG+SIFT+SUFR) with the CMC curve.

3.1 Viewpoint Invariant Pedestrian Recognition (VIPeR) Dataset

We evaluated our approach on the VIPeR dataset. The VIPeR dataset consists of 632 person image pairs from two different surveillance cameras. Each image pair includes two images of the same person obtained from different cameras, Cam A and Cam B. As shown in Fig. 2, Cam A represents the viewpoint between 0° and 90°, and Cam B between 90° and 180°. 0° means the front perspective image, and thus Cam A and Cam B have the average angle different of 90°. This person Re-ID dataset is challenging because of the viewpoint, poses and illumination change between the two cameras.

Fig. 4 CMC curves on
VIPeR dataset using **a** HOG,
b SIFT and **c** SURF

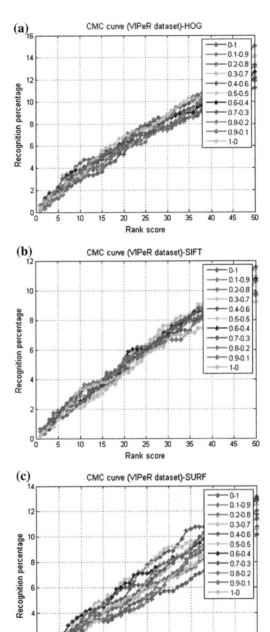

3.2 Re-ID Results and Discussion

We calculate the similarity between the two image patches using the Bhattacharyya distance. The most accurate ratio of the upper body to lower body was 6:4 when using the color feature only. In general, the accuracy of HOG, SIFT and SURF was good when the upper body weight was higher than the lower body weight, which indicates that the lower body part of the pedestrian is less discriminative than the upper body part, and the upper body is important for person Re-ID. From Table 2 and Fig. 5, it can be seen that the only color method is much less accurate than the other two methods (especially when rank is 1). However, color+HOG+SIFT+SURF slightly outperformed color+HOG. In other words, SIFT and SURF are not much important for person Re-ID if HOG is adopted. Therefore, if the high accuracy is the top priority, the adoption of all the features will be the better option, but if is it important to reduce the amount of computation for faster execution time, it is better to choose the color feature or at most color+HOG features.

Table 2 Comparing color, color+HOG and color+HOG+SIFT+SURF on the VIPeR dataset

Methods	$r=1$	$r=5$	$r=10$	$r=20$	$r=50$
Only color	11.6	21.0	28.4	37.3	52.6
Color+HOG	14.5	23.9	31.1	41.1	56.0
Color+HOG+SIFT+SURF	15.2	24.4	32.0	42.6	57.2

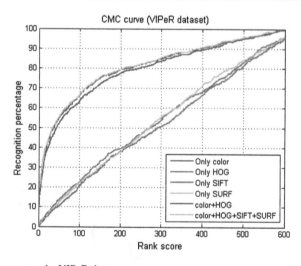

Fig. 5 CMC curves on the VIPeR dataset

4   Concluding Remarks

In this work, we implemented the framework of utilizing color and shape features for person Re-ID. The results of the three cases—when using the only color, using the color and HOG and using the color, HOG, SIFT and SURF—were compared on the VIPeR dataset by the CMC curve. The results showed that the accuracy was higher when using both the color and the shape features than when using the color solely. Further, there was the accuracy gain of approximately 1% when using HOG, SIFT and SURF than when using only HOG as the shape feature, which indicates that SIFT and SURF may not significantly affect for the person Re-ID. From this quantitative study, we could conclude that depending on the situations that require person Re-ID, we can conditionally choose whether to use the shape feature(s) or what types of shape features can be combined in addition to the color feature.

Acknowledgements This work was supported by the ICT R&D program of MSIP/IITP [B0101-16-0525, Development of global multi-target tracking and event prediction techniques based on real-time large-scale video analysis], and the Basic Science Research Program through the National Research Foundation of Korea (NRF) funded by the Ministry of Education (NRF-2017R1D1A1B03036423) and the Brain Research Program through the NRF funded by the Ministry of Science, ICT & Future Planning (NRF-2016M3C7A1905477).

References

1. Gray, D., & Tao, H. (2008). Viewpoint invariant pedestrian recognition with an ensemble of localized features. In *Proceedings of the ECCV* (pp. 262–275).
2. Gwak, J., Park, G., & Jeon, M. (2017). Viewpoint invariant person re-identification for global multi-object tracking with non-overlapping cameras. *KSII Transactions on Internet and Information Systems, 11*(4), 2075–2092.
3. Gwak, J. (2017). Multi-object tracking through learning relational appearance features and motion patterns. *Computer Vision and Image Understanding, 162,* 103–115.
4. Yang, E., Gwak, J., & Jeon, M. (2017). CRF-boosting: Constructing a robust online hybrid boosting multiple object trackers facilitated by CRF learning. *Sensors, 17*(3), 1–617.
5. Yang, E., Gwak, J., & Jeon, M. (2017). Multi-human tracking using part-based appearance modelling and grouping-based tracklet association for visual surveillance applications. *Multimedia Tools and Applications, 76*(5), 6731–6754.
6. Luo, P., Wang, X., & Tang, X. (2013). Pedestrian parsing via deep decompositional network. In *Proceedings of the ICCV* (pp. 2648–2655).
7. Dalal, N., & Triggs, B. (2005). Histograms of oriented gradients for human detection. In *Proceedings of the CVPR* (pp. 886–893).
8. Lowe, D. G. (2004). Distinctive image features from scale-invariant keypoints. *International Journal of Computer Vision, 60*(2), 91–110.
9. Bay, H., Ess, A., Tuytelaars, T., & Van Gool, L. (2008). Speeded-up robust features (SURF). *Computer Vision and Image Understanding, 110*(3), 346–359.
10. Cheng, D. S., Cristani, M., Stoppa, M., Bazzani, L., & Murino, V. (2011). Custom pictorial structures for re-identification. In *Proceedings of the BMVC* (p. 1).

11. Bazzani, L., Cristani, M., & Murino, V. (2014). SDALF: Modeling human appearance with symmetry-driven accumulation of local features. In *Person re-identification* (pp. 43–69). London: Springer.
12. Bazzani, L., Cristani, M., & Murino, V. (2013). Symmetry-driven accumulation of local features for human characterization and re-identification. *Computer Vision and Image Understanding, 117*(2), 130–144.
13. Corvee, E., Bremond, F., & Thonnat, M. (2010). Person re-identification using spatial covariance regions of human body parts. In *Proceedings of the AVSS* (pp. 435–440).

Performance Comparison of KLT and CAMSHIFT Algorithms for Video Object Tracking

Prateek Sharma, Pranjali M. Kokare and Maheshkumar H. Kolekar

Abstract Human detection and tracking is one of the most crucial tasks in video analysis. We can find its applications in areas like video surveillance, augmented reality, traffic supervision. KLT and CAMSHIFT are two popular algorithms for this task. In this paper, we present a comparison of their performance in different scenarios. As a result, this paper provides concrete statistics to choose an appropriate algorithm for tracking, given the nature of the objects and surrounding. Our experiments show that KLT algorithm is advantageous for crowded scenes, whereas CAMSHIFT performs better for tracking a specific target. Based on our analysis, we conclude that KLT algorithm performs more efficiently than CAMSHIFT algorithm for video object tracking.

Keywords Video object tracking · KLT · CAMSHIFT
Video surveillance system

1 Introduction

1.1 Motivation

With progress in the areas of computer vision, artificial intelligence and security, video surveillance techniques have recently been developing at a rapid rate. The ability to monitor areas of interest remotely has intrigued generations. One such notable technique which plays a key role in automating video surveillance is tracking of objects. Tracking of objects plays a key role in several scenarios like abnormal activity detection [1, 2], augmented reality, traffic supervision, human computer

P. Sharma (✉) · P. M. Kokare · M. H. Kolekar (✉)
Indian Institute of Technology Patna, Bihta, Patna 801103, Bihar, India
e-mail: prateek.cs14@iitp.ac.in

M. H. Kolekar
e-mail: mahesh@iitp.ac.in

© Springer Nature Singapore Pte Ltd. 2019
A. Khare et al. (eds.), *Recent Trends in Communication, Computing, and Electronics*, Lecture Notes in Electrical Engineering 524,
https://doi.org/10.1007/978-981-13-2685-1_31

interaction, vehicle navigation. Various machine learning techniques can also be used for human activity recognition such as Hidden Markov Model [HMM] [3], Bayesian Belief Network [4, 5]. If we detect any person performing abnormal activity, surveillance system can track such suspicious person such as Kanade–Lucas–Tomasi (KLT) and continuously adaptive mean shift (CAMSHIFT) tracking algorithms.

Video object tracking helps us to generate the trajectory of an object over time by locating its position in every frame of the video. Object tracker may also provide the complete region in the image that is occupied by the object at every time instant [6]. Based on what information we hope to retrieve from the video, the nature of objects and surroundings and the distance of the target from the camera, we hope to choose a video tracking algorithm that is more suitable for our case.

Nowadays, more and more public places are being secured through video cameras. With limited manpower, manual analysis of these videos for detection of anomalous activities, traffic supervision, vehicle tracking, etc. becomes difficult. There are immediate requirements for fully automated video surveillance systems in law enforcement, commercial and defense applications [1, 7, 8].

To develop such intelligent systems, we need to understand the advantages and disadvantages of the existing automated video analysis approaches. And object tracking technique is an important parameter to analyze this. By analyzing different object tracking algorithms, we will be able to understand the scenarios where a particular algorithm is most efficient. As a result, we will be able to deploy suitable object tracking algorithms for different scenarios, improving the accuracy and efficiency of our systems. Furthermore, it will also provide us the opportunity to experiment by combining different algorithms on the basis of their use cases and properties.

1.2 Related Work

There are multitudinous algorithms available for object tracking. These algorithms can be classified into different categories based on the approach they use, and the purpose they serve. The trackers we predominantly investigate in this paper are kernel-based trackers. Kernel-based trackers consider object shape and appearance as cues. Objects are tracked by computing the motion of the kernel in consecutive frames. This motion is usually in the form of a parametric transformation such as translation, rotation, and affine [9].

One of the common approaches for object tracking is the optical flow-based approach. Considering that the apparent velocity of the brightness pattern varies smoothly almost everywhere in the image, Horn and Shunk [10] propose an approach which computes the flow velocity of each pixel in the frame to locate dense flow fields. Followed by that, Tomasi and Shi [11] proposed the KLT tracker which uses feature points to track the object. A detailed description of this algorithm is given in Sect. 2.

In 2003, Comaniciu and Meer [12] used circular region as the kernel. They employ a metric derived from the Bhattacharyya coefficient as a similarity measure and use the mean shift procedure to perform the optimization. The mean shift procedure tries to add the maximum density region by iterative calculation of mean of the pixels inside the assumed region. An extension to the mean shift algorithm is the CAMSHIFT algorithm [13], which adjusts its windows size adaptively. A detailed description of CAMSHIFT algorithm is given in Sect. 3.

For performance evaluation, [14] provides a rich set of metrics to assess different aspects of performance of motion tracking. They use six different video sequences that represent a variety of challenges to illustrate the practical value of the proposed metrics by evaluating and comparing two motion tracking algorithms.

1.3   Contributions

In this paper, we analyze and compare the performance of two popular object tracking algorithms, CAMSHIFT algorithm and KLT algorithm. These algorithms are compared based on various performance parameters, and the study has been summarized. For our experiments, we have used standard database [15]. In certain scenarios, one is more advantageous compared to the other. Our paper presents an analysis on the basis of parameters like time required to analyze a video, number of times the correct frame is missed, and the number of people in the video to provide a proper outline to choose an appropriate algorithm for the given scene. Apart from that, this analysis done on standard dataset will be helpful for experiments like combination of tracking algorithm, designing efficient surveillance systems.

The contribution of our work is that it allows the identification of specific weakness and strengths of the two trackers. We have compared the robustness of the two tracking algorithms in different crowd density scenarios with the help of relevant metrics to assess the performance.

2   KLT Tracker

KLT algorithm is a feature-based tracker which uses feature points to track an object. To find the location of our target in current frame, we apply temporal constraints on the previous state. Based on the input frame and a predefined template T(x), the tracker aims to align the pixels from the template to the input image. If the image and template do not converge, method searches for another point which will be the best t. Now this state is used as the initial state for the next frame. Figure 1 shows the flow diagram for KLT algorithm.

For calculation of the alignment

$$\sum_x \left[I(W(x; p)) - T(x) \right]2 \tag{1}$$

Fig. 1 KLT tracker
algorithm

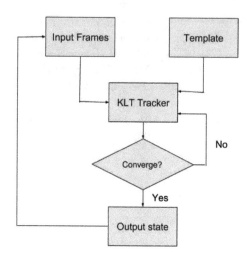

where p is the displacement parameter. Assuming we know initial value of p, we find Δp.

$$\sum_x \left[I(W(x; p + \Delta p)) - T(x) \right]2 \tag{2}$$

We can calculate p by using Taylor's expansion series and differentiating w.r.t p.

$$\Delta p = H^{-1} \sum_x \left[\nabla I \frac{\partial W}{\partial p} \right]^T \cdot [T(x) - I(W(x; p))] \tag{3}$$

where H is the Hessian Matrix. Using Δp, we can calculate the next best t point.

3 Continuously Adaptive Mean Shift Tracking

Continuous adaptive mean shift, popularly known as CAMSHIFT algorithm, uses the density appearance to track objects. It aims to find the maximum density area with the help of a tracking window. CAMSHIFT is an extension to mean shift algorithm which uses iterative searching to find the extreme value of probability distribution. It has an advantage that it adjusts the search window adaptively. CAMSHIFT uses adaptive continuous probability, and hence, we consider calculate probability for each frame [13].

Algorithm 1: CAMSHIFT Algorithm

1. Choose the initial location of the search window
2. Mean Shift (one or many iterations); store the zeroth moment.
3. Set the search window size equal to a function of the zeroth moment found in Step 2
4. Repeat Steps 2 and 3 until convergence (mean location moves less than a preset threshold).

In the above algorithm, we can calculate the zeroth moment as follows,

$$M_{00} = \sum_{x} \sum_{y} [I(x, y)] \tag{4}$$

The rst moment for x and y will be

$$M_{10} = \sum_{x} \sum_{y} x[I(x, y)] \tag{5}$$

$$M_{01} = \sum_{x} \sum_{y} y[I(x, y)] \tag{6}$$

Then, the mean search window location (centroid) will be

$$x_c = \frac{M_{10}}{M_{00}} \tag{7}$$

$$y_c = \frac{M_{01}}{M_{00}} \tag{8}$$

where $I(x, y)$ is the pixel (probability) value at position (x, y) in the image, and x and y range over the search window.

4 Results

We compared KLT and CAMSHIFT tracker based on two parameters, run-time and track completeness [14]. Track completeness is the time span that the system's track (ST) overlapped with the ground truth (GT) track divided by the total time span of GT track. The spatial overlap is defined as the overlapping level $A(GT_{ik}; ST_{jk})$ between GT_{ik} and ST_{jk} tracks in a specific frame

$$A(GT_{ik}; ST_{jk}) = \frac{Area(GT_{ik} \cap ST_{jk})}{Area(GT_{ik} \cup ST_{jk})} \tag{9}$$

We also define the binary variable, $O(GT_{ik}; ST_{jk})$ based on a threshold T_{ov}

Table 1 Experimental results showing comparison of KLT and CAMSHIFT for single-human video and multi-human videos

Dataset	KLT tracker		CAMSHIFT tracker	
	Average time (s)	TCM	Average time (s)	TCM
Single-human videos	23.4	0.41	47.4	0.36
Multi-human videos	39.2	0.35	8.9	0.21

$$O\left(GT_{ik}, ST_{jk}\right) = \begin{cases} 1 \ if \ A\left(GT_{ik}; ST_{jk}\right) > T_{ov} \\ 0 \ if \ A\left(GT_{ik}; ST_{jk}\right) < T_{ov} \end{cases} \tag{10}$$

$$TC = \frac{\sum_{k=1}^{N} O\left(GT_{ik}, ST_{jk}\right)}{number \ of \ GT_i} \tag{11}$$

Average track completeness of a video sequence is defined as

$$TCM = \frac{\sum_{k=1}^{N} \max(TC_{GT})}{N} \tag{12}$$

where N is the total number of GT tracks.

For CAMSHIFT, we used a predefined search window. For KLT tracker, we used Harris Corners as good features. All the tests were run on a i5-6200U processor. For experiments, we used two datasets from SPEVI datasets, single face dataset [16] and multiple faces dataset [15]. Table 1 shows the comparison results for both the above categories for KLT and CAMSHIFT tracker.

We can see from the result KLT took very less time in comparison with CAMSHIFT for single-human videos, but the ratio of time taken by KLT to CAMSHIFT increased considerably for multi-human videos. This happened due to increase in the tracking points in KLT as number of targets increased. Figure 2 shows the output of KLT tracker, and Fig. 3 shows the output of CAMSHIFT tracker for single-human video. Figure 4 shows the output of CAMSHIFT and KLT tracker for multi-human videos.

CAMSHIFT algorithm performs better for single-human video compared to multi-human video. Crowd density has very less impact over the performance of KLT tracker. To distinguish objects of same color, we observe that KLT algorithm performs better than CAMSHIFT. This happens due to the fact that KLT uses special feature points to track objects. But as the distance of object increases from camera, KLT is unable to locate the features and hence is less efficient in such cases. CAMSHIFT performs better in such scenarios as window adapts its size as the video progresses.

Table 1 provides a comparison of KLT and CAMSHIFT trackers based on parameters discussed above.

(a) **(b)** **(c)** **(d)**

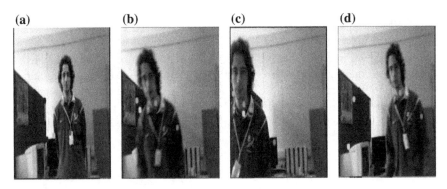

Fig. 2 Output of KLT tracker for single-human videos

(a) **(b)** **(c)** **(d)**

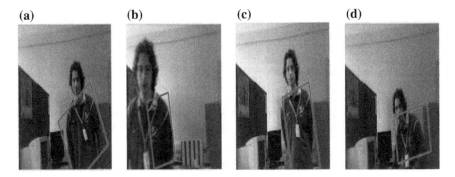

Fig. 3 Output of CAMSHIFT tracker for single-human videos

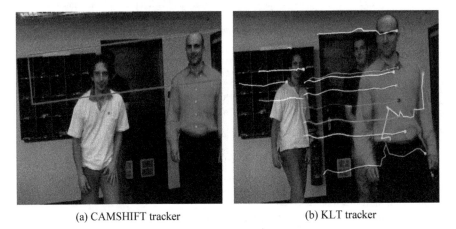

(a) CAMSHIFT tracker   (b) KLT tracker

Fig. 4 Output of CAMSHIFT and KLT tracker for multiple human video

5 Conclusion

In this paper, we implemented KLT and CAMSHIFT algorithm and presented a comparison of their performance. Experimental results show that KLT algorithm is computationally faster than CAMSHIFT algorithm. Our analysis shows that KLT algorithm performs better than CAMSHIFT. We also observed that KLT performs poorly in the cases where target is far from the point of view, whereas CAMSHIFT does not have a significant impact on accuracy in such cases.

References

1. Kolekar, M. H., Bharti, N., & Patil, P. N. (2016). Detection of fence climbing using activity recognition by support vector machine classifier. In *2016 IEEE Region 10 Conference (TEN-CON)* (pp. 398–402). Singapore.
2. Kushwaha, A. K. S., Prakash, O., Khare, A., & Kolekar, M. H. (2012). Rule based human activity recognition for surveillance system. In *Proceedings of the IEEE 4th International conference on Intelligent Human Computer Interaction*, 27–29 December 2012, (pp. 466–471). IIT Kharagpur, India.
3. Kolekar, M. H., & Dash, D. P. (2016). Hidden Markov model based human activity recognition using shape and optical ow based features. In *Proceedings of the IEEE Region Conference (TENCON)*, November 2016, (pp. 393–397).
4. Kolekar, M. H. (2011). Bayesian belief network based broadcast sports video indexing. *An International Springer Journal of Multimedia Tools and Applications, 54*, 27–54.
5. Kolekar, M. H., & Sengupta, S. (2015). Bayesian network-based customized highlight generation for broadcast soccer videos. *IEEE Transactions on Broadcasting, 61*(2), 195–209.
6. Rai, H., Kolekar, M. H., Keshav, N., & Mukherjee, J. K. (2015). Trajectory based unusual human movement identification for video surveillance system. In H. Selvaraj, D. Zydek, & Chmaj, G. (Eds.), *Progress in systems engineering. Advances in intelligent systems and computing* (Vol. 366). Springer, Cham.
7. Kolekar, M., Palaniappan, K., Sengupta, S., & Seetharaman, G. (2009). Semantic concept mining based on hierarchical event detection for soccer video indexing. *Journal of Multimedia, 4*, 298–312. https://doi.org/10.4304/jmm.4.5.
8. Kolekar, M. H., & Sengupta, S. (2006). Event-importance based customized and automatic cricket highlight generation. In *2006 IEEE International Conference on Multimedia and Expo*, Toronto, Ont, (pp. 1617–1620).
9. Yilmaz, A., Javed, O., & Shah, M. (2006). Object tracking: A survey. *ACM Computing Surveys, 38*(4), 145.
10. Horn, B. K. P., & Schunck, B. G. (1981). Determining optical flow. *Artificial Intelligence, 17*(1–3), 185–203.
11. Tomasi, C., & Shi, J. (1994). Good features to track. In *IEEE Conference on Computer Vision and Pattern Recognition (CVPR)* (pp. 593–600).
12. Comaniciu, D., Ramesh, V., & Meer, P. (2003). Kernel-based object tracking. *IEEE Transactions on Pattern Analysis and Machine Intelligence, 25*, 564–575.
13. Bradski, G. R. (1998). Computer vision face tracking for use in a perceptual user interface. In *IEEE Workshop on Applications of Computer Vision*, Princeton, NJ, (pp. 214–219).
14. Yin, F., Makris, D., & Velastin, S. (2007). Performance evaluation of object tracking algorithms. In *10th IEEE International Workshop on Performance Evaluation of Tracking and Surveillance (PETS2007)*.

15. Maggio, E., Piccardo, E., Regazzoni, C., Cavallaro, A. (2006). Particle PHD filter for multi-target visual tracking. In *Proceedings of the IEEE International Conference on Acoustics Speech and Signal Processing (ICASSP 2007)*, Honolulu (USA), April 15–20.
16. Maggio, E., Cavallaro, A. (2005). Hybrid particle filter and mean shift tracker with adaptive transition model. In *Proceedings of the IEEE International Conference on Acoustics, Speech and Signal Processing (ICASSP 2005)*, Philadelphia, 19–23 March 2005, (pp. 221–224).

Delving Deeper with Dual-Stream CNN for Activity Recognition

Chandni, Rajat Khurana and Alok Kumar Singh Kushwaha

Abstract Video-based human activity recognition has fascinated researchers of computer vision community due to its critical challenges and wide variety of applications in surveillance domain. Thus, the development of techniques related to human activity recognition has accelerated. There is now a trend towards implementing deep learning-based activity recognition systems because of performance improvement and automatic feature learning capabilities. This paper implements fusion-based dual-stream deep model for activity recognition with emphasis on minimizing amount of pre-processing required along with fine-tuning of pre-trained model. The architecture is trained and evaluated using standard video actions benchmarks of UCF101. The proposed approach not only provides results comparable with state-of-the-art methods but is also better at exploiting pre-trained model and image data.

Keywords Activity recognition · Deep learning · Spatio-temporal features
Convolution neural network

1 Introduction

An activity is a collection of human/object movements with particular semantic meaning. Activity recognition is thus task of labelling video streams with corresponding activity class. From the computer vision perspective, the recognition of an activity involves parsing the video sequence to learn about activity representation, and in turn the learned knowledge is used to identify similar activities in future. There are numerous applications of activity analysis including human–computer interac-

Chandni (✉) · R. Khurana · A. K. S. Kushwaha
Department of CSE, IKGPTU, Kapurthala, India
e-mail: chandnikathuria5@gmail.com

R. Khurana
e-mail: khuranarajat19@gmail.com

A. K. S. Kushwaha
e-mail: dr.alokkushwaha@ptu.ac.in

© Springer Nature Singapore Pte Ltd. 2019
A. Khare et al. (eds.), *Recent Trends in Communication, Computing,
and Electronics*, Lecture Notes in Electrical Engineering 524,
https://doi.org/10.1007/978-981-13-2685-1_32

Sports Analysis   SurveillanceTraffic monitoring

Fig. 1 Application exemplars for activity recognition

tion, health monitoring and sports analysis, as shown in Fig. 1. In surveillance, it is used to monitor activities in smart environments to detect suspicious activities so as to alert concerned authorities. In nursing homes, it helps to monitor activities of patients. Due to this paramount importance of activity recognition, it has become active research area and numerous methods have been proposed.

Identification of activities from videos basically involves feature extraction followed by classification. Feature extraction has proven to be a particularly difficult task when referring to real-world data. There are several challenges involved with this task [1], such as intra- and inter-class variations, dynamic or cluttered background and viewpoint variations. Further, it involves handling large video datasets. Earlier approaches to activity recognition include local and global feature representations. The representatives include motion energy image (MEI), motion history image (MHI) introduced in Bobick et al. [2], spatio-temporal interest points used by Ivan Laptev [3]. But handcrafted features are quite labour-intensive and demand for the domain knowledge; hence, deep learning has recently drawn the attention of the researchers. Deep learning is a particular kind of machine learning that has great power and flexibility to represent the world as nested hierarchy of concepts. Convolution neural network (CNN) is deep artificial neural network designed specifically for analysis for visual imagery with minimal effort. This means that the CNN learns the filters that in traditional methods were hand-engineered. This independence from prior domain knowledge and human effort in feature design is a major benefit of deep learning-based solutions.

The contributions of our work are twofold. We demonstrate (i) the use combination of deep 2D and 3D residual networks in dual-stream model to get comparable results on UCF101 dataset [4] and (ii) fine-tuning of pre-trained model in one of network streams to better exploit available video data. Rest of the paper is organized as follows: Sect. 2 discusses the related work. In Sect. 3, we present our methodology along with implementation details. Section 4 presents experimentation with dataset and comparison with previous work. Section 5 concludes overall work.

2   Related Work

The decoupled approaches to activity recognition involve handcrafted features, which in turn are classified using separate classification model. Second class of activity recognition involves neural network architecture that jointly learns features and classifier. The proposed work relates to second class. Among 2D CNN-based approaches, Dobhal et al. [5] proposed human activity recognition model based on 2D representation of action and reported to be invariant to holes, shadows, partial occlusions and speed of actions performed using BMI. To cope with the problems posed by application of CNN to small datasets, Wang et al. [6] explore information offered by depth maps that contain information relating to the distance of the surfaces of scene objects from a viewpoint. Most of recent approaches involve using multiple network streams with different input modalities to get improved performance. Karpathy et al. [7] introduced idea of slow fusion to increase the temporal awareness of convolution network. Simonyan et al. [8] proposed popular two-stream model based on idea of visual processing by humans. One of the network streams learns spatial feature from RGB input images, whereas the other stream is focused on learning temporal features from optical flow inputs. Final activity class results are computed as an average of the outputs of two network streams. Feichtenhofer et al. [9] explored early fusion within CNN to fuse the separate streams, with idea to retain pixel-wise correspondence between spatial and temporal features. Mohammadreza et al. [10] proposed chained multiple stream architecture comprising three streams trained separately with pose information, optical flow and RGB frames. Finally, the joint probability over all input streams factorizes into the conditional probabilities over the separate input streams and final class label is obtained at the end of the chain. Action recognition from video needs to capture motion information encoded in sequence of video frames. CNNs with 3D spatio-temporal convolution filters serve as natural extension of 2D CNNs to video and hence can address this issue well. 3D CNNs have been investigated for activity recognition in [7, 11–15]. In this work, we investigated combination of 2D and 3D CNNs with deep residual network architecture in dual-stream model. However, for the sake of achieving comparable results with minimal pre-processing, we have explored only RGB frames as input modality.

3   Deep Dual-Stream CNN for Activity Recognition

3.1   System Overview

Motivation behind work comes from the crux of the activity recognition problem that it involves costly training of deep networks. Hence, our work focuses on fine-tuning of pre-trained model. Also idea is to get results comparable to the state of the art using simplified multi-stream approach that involves inputting raw video frames to both streams. The simplest approach in the direction to apply CNN for activity recog-

Fig. 2 Overview of the
proposed system for activity
recognition (ours)

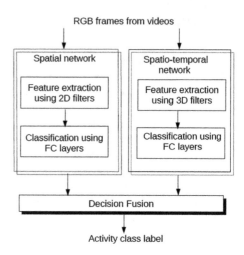

nition from videos is to treat video frames as still images as video streams consist of
sequence of video frames across temporal dimension. Convolutional networks can
then be used to identify activities at individual frame level. This simple approach,
however, does not take into account motion encoded in sequence of frames, and
thus, it simply ignores temporal information in activity recognition. Identification of
some activities is possible from appearance only; however, other activities may need
motion/temporal information. There are different ways to take account of temporal
information, like optical flow, motion history images. But it needs additional processing of video data. Going in the direction of multi-stream models [8–10] to take
advantage of ensemble learning and considering above-mentioned points, we implement dual-stream model for activity recognition. The basic architecture comprises
two network streams named "spatial stream" and "spatio-temporal" stream. Figure 2
shows overview of our work.

Both network streams accept still images/frames extracted from video dataset as
input. Spatial stream is aimed at learning activity representation using 2D convolutions, and spatio-temporal stream learns spatio-temporal features using 3D convolutions. Training network using still video frames provides opportunity to exploit large
image datasets such as ImageNet [16]. The final activity class is calculated based on
fusion of predictions of spatial and spatio-temporal network streams. Architecture
of both network streams is illustrated and described in Sect. 3.2.

3.2 Implementation Details

In neural network design, network architecture is one important factor. For both
network streams, we use residual network architecture [17], which is one of the
most successful architectures which bagged all the ImageNet challenges including

Fig. 3 Residual learning
[17]

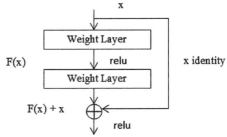

Fig. 4 Block structure for
ResNet and ResNext

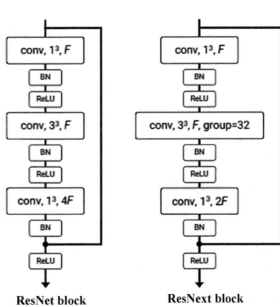

ResNet block   ResNext block

classification, detection and localization. The residual learning framework eases the training of networks and enables them to be substantially deeper. It provides shortcut connections that allow signal to bypass one layer and hop to the next layer in the sequence as illustrated in Fig. 3. Network block structure is shown in Fig. 4.

Here, conv, x3, F represent kernel size and group as the number of groups of group convolutions. Convolution layers are followed by batch normalization (BN) and ReLU activation.

Spatial Network Stream

Previous works by different authors have shown that deeper structures improve performance of computer vision tasks [18]. This is the reason we chose ResNet 101 architecture; however, the original two-stream model [8] used a relatively shallow network structure. Layers of spatial stream are summarized in column "ResNet 101(Spatial net)" of Table 1. We also initialize the ResNet with ImageNet [16] weights as pre-

Table 1 Network architecture of dual-stream model (Ours)

Layer	ResNet 101 (Spatial Net)	ResNext 101 (Spatio-temporal Net)
Conv1_x	7 * 7, 64, stride = 2	7 * 7 * 7, 64, temporal stride = 1 spatial stride = 2
Conv2_x	3 ResNet blocks	3 ResNext blocks
Conv3_x	4 ResNet blocks	24 ResNext blocks
Conv4_x	23 ResNet blocks	36 ResNext blocks
Conv5_x	3 ResNet blocks	3 ResNext blocks
Fully connected	FC (fully connected) layer	FC layer

training has proven to be successful way to initialize deep networks when the target dataset does not have enough training samples.

Spatio-temporal Network Stream

Convolution networks cannot naturally preserve temporal information, and hence, it needs different input modality like optical flow as used in [8, 9], motion history images or modifications to CNN such as 3D convolutions which are able to learn spatio-temporal features [12]. Our spatio-temporal stream is implemented using ResNext 101 architecture. ResNext block uses concept of cardinality as different dimensions which refer to the number of middle convolutional layer groups in the bottleneck block. For our spatio-temporal stream, we set cardinality of 32. Layers of spatio-temporal stream are summarized in Table 1. As we are more focused on fine-tuning of pre-trained model for problem at hand, we consider kinetics pre-trained model for transfer learning as it has sufficient data for CNN. Fine-tuning of conv5_x and the FC layers for UCF101 achieved the best results.

Decision Fusion

For final activity prediction, we simply averaged the softmax class scores of both streams that generate final activity class scores, whereas class with highest score is the final activity class label for the given video stream.

4 Experiments

We conduct experiments on widely used and challenging benchmarks for action recognition: UCF101 is described in Sect. 4.1. Section 4.2 gives training and testing settings. In Sect. 5, we compare our work with state-of-the-art methods.

| Apply Eye Makeup | Drumming | Horse riding |

Fig. 5 Example snapshots of RGB frames extracted from videos in different activity classes from UCF101

4.1 Dataset and Performance Measure

UCF101 [16] is a common video dataset for human activity recognition. It comprises total 13320 videos of 101 action categories, and Fig. 5 shows some of these activity classes. The reason we chose this dataset is that it has high diversity and variation in camera motion, object appearance, pose and scale, viewpoint, cluttered background, illumination conditions, etc., which allow the recognition algorithms to test and verify their robustness and effectiveness in the realistic situation. The videos in 101 action categories are grouped into 25 groups, where each group can consist of 4–7 videos of an action. The videos from the same group may share some common features such as similar background, similar viewpoint. The action categories include five types: (1) human–object interaction, (2) body-motion only, (3) human–human interaction, (4) playing musical instruments and (5) sports.

For evaluation, we rely on video accuracy which is standard evaluation protocol.

4.2 Network Training

UCF101 comprises three train–test splits, and we consider split-1 for our experiments. Mini-batch training is used with size of batch adjusted to GPU memory. For spatial stream, we randomly selected three frames from each video in mini-batch and 224×224 sub-image is randomly cropped from each selected frame. For calculating loss, video-level prediction is obtained by consensus among these frames. For spatio-temporal stream, the size of each sample is 3 channels × 16 frames × 112 pixels × 112 pixels. We also horizontally flip each sample by probability of 50%. In our model, stochastic gradient descent (SGD) with momentum is used to train the network. We use cross-entropy losses and backpropagation of gradients. Weight decay and momentum are 0.001 and 0.9, respectively.

Table 2 Results on UCF split-01

Stream	UCF101 split-01 (%)
DD $_{spatial}$	82
DD $_{spatio-temp}$	88
DD $_{avg-fusion}$	90

Table 3 Comparison of our deep dual-stream (DD) model with state of the art. (*mean accuracy over three UCF splits)

Method	Input modality	Accuracy (%)
[7] Slow fusion	RGB	65.4*
[8] Spatial stream	RGB	72.7
[8] Temporal stream	Optical flow	81.2
[8] Fusion by Avg	RGB + optical flow	86.2
[8] Fusion by SVM	RGB + optical flow	87.0
[9] Spatial stream	RGB	74.2 (VGGM-2048) 82.61 (VGG-16)
[9] Temporal stream	Optical flow	82.34 (VGGM-2048) 86.25 (VGG-16)
[9] Late fusion	RGB + optical flow	85.94 VGGM-2048), 90.62 (VGG-16)
DD $_{spatial}$(ours)	RGB	**82**
DD $_{spatio-temp}$(ours)	RGB	**88**
DD $_{avg-fusion}$(ours)	RGB	**90**

5 Results and Discussion

Table 2 presents test results on UCF split-01 for spatial (DD $_{spatial}$) and spatio-temporal (DD $_{spatio-temp}$) streams. Using 3D convolution on RGB frames outperforms 2D convolutions in spatial stream approximately by 7%. Softmax class scores of both streams are then combined by averaging, and class with maximum scores gives activity class label, resulting in accuracy of 90% as depicted by DD avg-fusion.

Comparison with State of the art

We compare performance of our work to the state of the art on UCF101 datasets as given by Table 3. Our focus is on using only RGB frames for activity recognition in multi-stream model, and hence, we consider some of RGB-based and multi-streams methods on UCF101 for comparison. Our DD $_{spatial}$ clearly outperforms spatial stream in [8, 9] (VGGM-2048) by good margin. DD $_{spatio-temp}$ is able to improve by ~ 7% (81.2 vs. 88) when compared with the temporal stream of [8] and by ~ 2% (86.25 versus 88) for [9]. Combined result of both streams DD $_{avg-fusion}$ outperforms combined results of [8] by ~ 3%, and results are also comparable to late fusion (VGG-16) in [9].

6 Conclusion

We proposed the dual-stream deep model activity recognition with competitive performance, which comprises two network streams, implemented using CNN. For spatio-temporal network stream, we examined CNNs with spatio-temporal 3D convolutional kernels and fine-tuning of kinetics pre-trained model for UCF101 which gave comparable results with minimal training and only RGB frames as input modality. Experimentation with residual networks with 3D kernels also emphasizes that deeper 3D CNNs have the potential to contribute to significant progress in fields related to various video analysis tasks such as action detection, video summarization, object segmentation. In our future work, we will investigate the use of some other modality with residual networks for activity recognition and consider other standard video action benchmarks.

References

1. Poppe, R. 2010). A survey on vision-based human action recognition. *Image and Vision Computing, 28*(6), 976–990.
2. Bobick, A. F., Davis, J. W. (2001). The recognition of human movement using temporal templates. *IEEE Transactions on Pattern Analysis and Machine Intelligence, 23*(3), 257–267.
3. Laptev, I. (2005). On space-time interest points. *Int. Journal of Computer Vision, 64*(2), 107–123.
4. Soomro, K., Roshan Zamir, A., & Shah, M. (2012). UCF101: A dataset of 101 human action classes from videos in the wild CRCV-TR-12-01, 1, 2, 3, 5.
5. Dobhal, T., et al. (2015). Human activity recognition using binary motion image and deep learning. *Procedia Computer Science, 58,* 178–185.
6. Wang, P., Zhang, J., & Ogunbona, P. O. (2015). Action recognition from depth maps using deep convolutional neural networks. *IEEE Transactions on Human-Machine Systems.*
7. Karpathy, A., Toderici, G., Shetty, S., Leung, T., Sukthankar, R., Fei-Fei, L., 2014. Large-scale video classification with convolutional neural networks. In: Proc. IEEE Conference on Computer Vision and Pattern Recognition (CVPR). pp. 1725–1732.
8. Simonyan, K., & Zisserman, A.. (2014). Two-stream convolutional networks for action recognition in videos. In *Proceedings of the Advances in Neural Information Processing Systems (NIPS)* (pp. 568–576).
9. Feichtenhofer, C., Pinz, A., & Zisserman, A.. (2016). Convolutional two-stream network fusion for video action recognition. In *Proceedings of the IEEE Conference on Computer Vision and Pattern Recognition (CVPR)* (pp. 1933–1941).
10. Zolfaghari, M., Oliveira, G. L., Sedaghat, N., & Brox, T. Chained Multi-stream Networks Exploiting Pose, Motion, and Appearance for Action Classification and Detection. https://arxiv.org/abs/1704.00616.
11. Tran, D., Bourdev, L., Fergus, R., Torresani, L., & Paluri, M. (2015) .Learning spatiotemporal features with 3D convolutional networks. In *ICCV*.
12. Ji, S., Xu, W., Yang, M., & Yu, K. (2010). 3D convolutional neural networks for human action recognition. In *ICML*.
13. Taylor, G. W., Fergus, R., LeCun, Y., & Bregler, C. (2010). Convolutional learning of spatio-temporal features. In *ECCV*.
14. Baccouche, M., Mamalet, F., Wolf, C., Garcia, C., & Baskurt A.. (2011). Sequential deep learning for human action recognition, A.. A. Salah & B. Lepri (Eds.) HBU, LNCS 7065 (pp. 29–39).

15. Varol, G., Laptev, I., & Schmid, C. (2016). Long-term Temporal Convolutions for Action Recognition. arXiv:1604.04494.

16. Deng, J., Dong, W., Socher, R., Li, L., Li, K., & Li, F. (2009). ImageNet: a large-scale hierarchical image database. In *CVPR* (pp. 248–255).

17. He, K., Zhang, X., Ren, S., & Sun, J. (2016). Deep residual learning for image recognition. In *Proceedings of the IEEE Conference on Computer Vision and Pattern Recognition (CVPR)* (pp. 770–778), 1, 2, 3, 4, 5.

18. Szegedy, C., Liu, W., Jia, Y., Sermanet, P., Reed, S., Anguelov, D., et al. (2015). Going deeper with convolutions. In *CVPR* (pp. 1–9).

Part VI
Multimedia Analytics

Enabling More Accurate Bounding Boxes for Deep Learning-Based Real-Time Human Detection

Hyunsu Jeong, Jeonghwan Gwak, Cheolbin Park, Manish Khare, Om Prakash and Jong-In Song

Abstract While human detection has been significantly recognized and widely used in many areas, the importance of human detection for behavioral analysis in medical research has been rarely reported. Recently, however, efforts have been actively made to recognize behavior diseases by measuring gait variability using pattern analysis of human detection results from videos taken by cameras. For this purpose, it is very crucial to establish robust human detection algorithms. In this work, we modified deep learning models by changing multi-detection into human detection. Also, we improved the localization of human detection by adjusting the input image according to the ratio of objects in an image and improving the results of several bounding boxes by interpolation. Experimental results demonstrated that by adopting the proposals, the accuracy of human detection could be increased significantly.

Keywords Human detection · Deep learning · Bounding box regression
Localization · Real-time analysis

H. Jeong · C. Park · J.-I. Song (✉)
Department of Electrical Engineering and Computer Science, Gwangju Institute
of Science and Technology, 123, Cheomdangwagi-ro, Buk-gu, 61005 Gwangju, Korea
e-mail: jisong@gist.ac.kr

J. Gwak (✉)
Biomedical Research Institute, Seoul National
University Hospital (SNUH), Seoul 03080, Korea
e-mail: james.han.gwak@gmail.com

J. Gwak
Department of Radiology, Seoul National
University Hospital (SNUH), Seoul 03080, Korea

M. Khare
Dhirubhai Ambani Institute of Information
and Communication Technology, Gandhinagar, Gujarat, India

O. Prakash
Centre of Computer Education, Institute of Professional Studies,
University of Allahabad, Allahabad, India

© Springer Nature Singapore Pte Ltd. 2019
A. Khare et al. (eds.), *Recent Trends in Communication, Computing,
and Electronics*, Lecture Notes in Electrical Engineering 524,
https://doi.org/10.1007/978-981-13-2685-1_33

1 Introduction

Human detection in this work deals with the problem to find people in a given frame. In this work, from our video dataset taken for gait variability analysis, it will enable subsequent annotation and search of the person who is our analysis target. The problem of human detection is approached in two ways. The first is to classify people, and the second is to localize human as bounding box information. As traditional methods, the histogram of gradient (HOG) feature, deformable part model (DPM) and support vector machine (SVM) classifier are commonly used for human detection [1, 2]. In recent years, however, it has become possible to classify and localize objects more accurately using deep learning models, for example, based on convolutional neural network (CNN) in order to detect multiples objects from an image [3–7].

Human detection methods are required in several areas such as medical and computer vision field [8–11] to enable practical applications [12]. In autonomous vehicles, it must be possible to alert the presence of pedestrians on streets. In addition, by measuring gait variability through human detection, the gait variability is used to figure out behavioral diseases [13]. In this paper, to get a new gait variability in terms of human center point, we used deep learning-based human detection algorithms and devised a method of allocating more robust bounding box information to enable more accurate localization.

In this paper, to utilize the benefit of hierarchical layer-wise representation learning, we adopt deep learning models that are able to model data at more abstract representations [3]. Compared to low-level handcrafted features such as scale-invariant feature transform (SIFT) and HOG, they will improve the accuracy of classification and localization of human objects. In addition, since recent deep learning models have a bounding box regression module unlike conventional machine learning approaches (such as the combination of handcrafted features, DPM and SVM), deep learning-based detection models tend to obtain higher bounding box accuracy. However, while there are several deep learning-based object detection models, most of them are optimized for multiple objects. Therefore, for human detection without consideration of the other objects, we need to confine the detection to persons and improve low bounding box localization performance of the existing models. To this end, this work focuses on devising the functionalities to improve the accuracy.

The rest of this work is organized as follows: Some closely related work is given in Sect. 2. Section 3 shows the overview of the proposed method. Section 4 describes the detailed bounding box localization method. Experimental results are given in Sect. 5. Finally, conclusions and future work are in Sect. 6.

2 Related Work

Object detection model has been developed in the order of RCNN, Fast RCNN, You Only Look Once (YOLO), Faster RCNN and Single-Shot Multi-Box Detector (SSD) [4–7, 14]. RCNN and Fast RCNN use selective search algorithms as a region proposal method, but it makes object detection slower. YOLO improved the speed by converting selective search into grid as region proposal. Although Faster RCNN uses predefined anchor boxes for region proposal and region proposal network (RPN) making better mean average precision (mAP) than YOLO, RPN causes the entire network to be split into two networks making quite slower than YOLO. YOLO and SSD have brought a lot of speed improvements by integrating a region proposal part into one network instead of using it as a separate network, and thus, SSD and YOLO are widely used when considering execution time without significant mAP degradation. Therefore, in this work, for real-time behavior modeling we tested models based on YOLO and SSD.

There is a lot of research to assist the patients with Alzheimer's disease (AD), using behavior detection and deep learning technologies. Monitoring systems can make them possible to live more independently in their house with flexibility. This is because the activity data of a patient captured from a vision sensor system can be further transmitted to a guardian or related persons and then can provide assistance when necessary [15]. In addition, compared with normal persons, the AD patients are well known to have more agitation, dysphoria and aberrant motor behaviors [16]. According to the studies given above, we found that there are very few deep learning-based approaches that can detect persons solely with high accuracy and robustness in video (i.e., across a series of consecutive frames) and this led us to propose the new functionalities.

3 Overview of the Proposed Method

The most widely used deep learning-based detection models are Faster RCNN, SSD and YOLO. Faster RCNN has slower detection speed than SSD or YOLO [6, 7, 14], but it has relatively higher detection accuracy. Since the goal of this work is to establish real-time detection with high accuracy, we adopted YOLO V2 (viz. YOLO 9000) [17] and the SSD with inception V2 (SSD V2) which is one of Google Object Detection models [18] for comparative study. The two models detect multiple objects, such as a dog, a person and a cup. From our comparative study, we found that the SSD-based model outperforms YOLO V2 in terms of bounding box localization accuracy, and thus, we selected it to our base model for improvement. Since we need to detect the person only because the object of interest is a patient, we modified the selected SSD-based multi-object detection model to become human detection. The selected model has 90 class IDs and returns detection results including class IDs, bounding boxes and confidence scores of multiple objects in an image. Therefore,

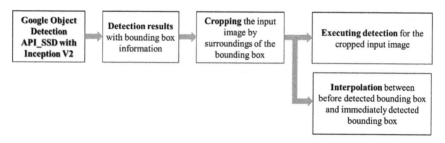

Fig. 1 Flowchart of the proposed SSD-based modified method

we need to filter out the other IDs except the person ID and then to remain bounding box information of the person ID on the image (Fig. 1).

Although the selected SSD-based model has better bounding box localization performance than the YOLO V2 model, the bounding box results sometimes are not always good in consecutive frames. For example, if the size of the detected object is fairly smaller than the size of the entire image, the locations and sizes of the bounding box results tend to be very different even for consecutive two frames. To solve this problem, we adjusted the input image of the deep learning model by enlarging the width and height. In addition, to solve the problem that the bounding box of the currently detected frame is very different compared to its previous frame, we suggest a method to correct the bounding box result by using an interpolation method for three consecutive frames.

3.1 Google Object Detection Model

To detect the person object in real time, we compared the two deep learning models YOLO V2 and SSD V2. From our repeated experiments for randomly selected 500 frames, we found that SSD V2 gives more accurate bounding box information, i.e., *Intersection over Union (IoU)* is higher than YOLO V2 in most of the cases. For example, as we can see from Fig. 2, the bounding box of SSD V2 is more precise than YOLO V2.

3.2 Human Detection

Original SSD V2 has the output parameters: `boxes` are the set of box coordinates, `scores` are the set of confidence scores, and `classes` are the set of classes of the detected objects.

```
Human_Detection(boxes, scores, class)
{
```

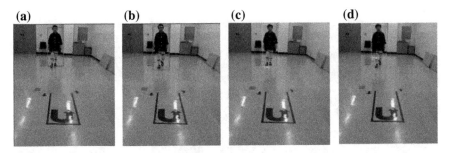

Fig. 2 Detection results of YOLO V2 at the 49th frame (**a**), at the 149th frame (**b**), the results of SSD V2 at the 49th frame (**c**), and at the 149th frame (**d**)

Fig. 3 **a** Detection results of original SSD V2, **b** detection results of modified SSD V2

```
If(classes[i]  == person_id){
    people_boxes  = boxes[i];
    people_scores  =  scores[i]
    draw_bounding_box(class[i])
    }
}
```

In Fig. 3a the two types of objects, persons and kites, are detected using original SSD V2 in which bounding boxes of person objects are drawn in the green color and those of kite objects in the yellow color. In contrast, from Fig. 3b, the modified SSD V2 detects persons only.

3.3 GIST Behavioral Variability Dataset

We first converted a video into 500 frames with the frame rate of 30 FPS. First, 250 frames are the frontal view data, 251th to 350th frames are the turning view (from left to right) data, and 351th to 500th frames are the rear view data. In our experiment, for ease of experimental simplicity, the frontal and the rear data are only used. Further, to

(a) (b)

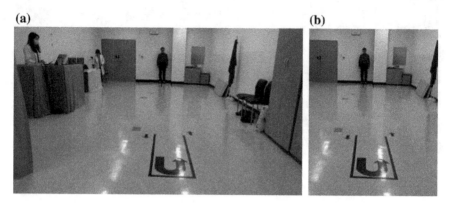

Fig. 4 **a** Original image frame. **b** Its cropped image of the region of interest

(a) (b) (c) (d)

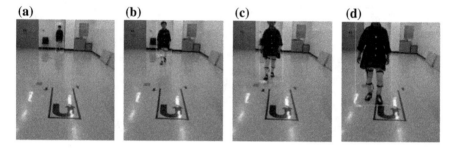

Fig. 5 Left two images are a small-size object at the 7th frame (**a**), at the 149th frame (**b**), the right two images are the big-size object at the 204th frame (**c**), and at the 249th frame (**d**)

get a gait variability easily, we cropped region of interest, as shown in Fig. 4, so that only a person can be captured to avoid inclusion of nontarget persons. By cropping the region of interest, we could reduce the search space 60–80% compared to the full size frame, which also resulted in speeding up the detection operations.

4 Proposed Accurate Detection Method

4.1 Resetting Input Image Region

As we mentioned above, we have 250 frames per a patient as the frontal data. When we compared small-size bounding boxes (from 1 to approximately 200th frames) with relatively big-size bounding boxes (from 201 to 250th frames), e.g., as shown in Fig. 5, we could see that bounding box localization performance of the former frames is worse than latter frames due to small person object sizes.

(a)   **(b)**   **(c)**   **(d)**

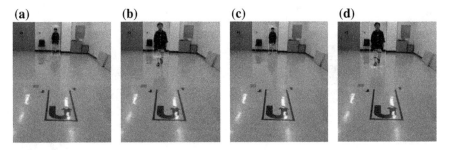

Fig. 6 Original SSD V2 results at the 7th frame (**a**), at the 149th frame (**b**), and adjusted results at the 7th frame (**c**) and the 149th frame (**d**)

To improve bounding box fitness for the small-size objects, we enlarged the input image size of SSD V2 in horizontal and vertical directions by 80 pixels each. As the results, as shown in Fig. 6, more accurate bounding box could be obtained for the small-size objects.

4.2 Interpolation Method

In general, detection results are not much different between the ith frame bounding box and the i+1th frame bounding box, but in some cases, we found that the two bounding boxes are quite different in the two consecutive frames. This was mainly because the lower body (i.e., leg) part of the bounding box sometimes is not well detected due to the rapid change while walking. Therefore, if the difference between the height of the previous bounding box and that of the current bounding box is more than a certain threshold level (e.g., 3.75% in this work), we performed the following simple interpolation method.

```
Interpolation (i-1th frame, ith frame, i + 1th frame)
{
    differ_height = |i-1th frame height - ith#160;frame height|
    if (differ_height > threshold)
    {
ith frame bottom = i-1th frame bottom - (((i-1th frame
bottom) - (i + 1th frame bottom)) / 2)
    }
}
```

As shown in Fig. 7e, if the difference between the detection results is higher than the threshold, the interpolation among the three frames is performed as the post-processing, and as a result, more accurate bounding box can be obtained.

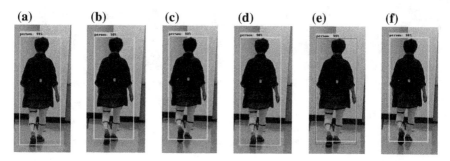

Fig. 7 Original detection results at the 457th frame (**a**), 458th frame (**b**), 459th frame (**c**), interpolation results at the 457th frame (**d**), 458th interpolated frame (**e**) and 459th frame (**f**)

5   Experiment Results and Discussion

5.1   Detection Accuracy

We fine-tuned the three models, SSD V2, YOLO V2 and modified SSD V2, using 400 frames excluding the turning view data. Then, for the test data consisting of 1200 frames, the three models were compared based on average IoU

$$IoU(GT, DBB) = \frac{GT \cap DBB}{GT \cup DBB} \tag{1}$$

where GT is a ground truth and DBB is the detected bounding box.

From the results in Table 1, we could see that average IoU of SSD V2 is slightly higher than YOLO V2. Further, the proposed model could improve its detection accuracy, compared to SSD V2, by approximately 7%.

5.2   Detection Speed

The experiments were run using TensorFlow 1.4, python 3.6 and Ubuntu 16.04 with i7-7700k CPU, 32 GB RAM and GeForce GTX 1070 GPU. As can be seen in

Table 1 Results of detection accuracy

Model	Measurement
	IoU
YOLO V2	0.88193
SSD V2	0.88517
Modified SSD V2	0.95364

Table 2 Results of detection average time

Model	Measurement
	Milliseconds
SSD V2	70
YOLO V2	45
Modified SSD V2	45

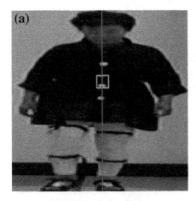 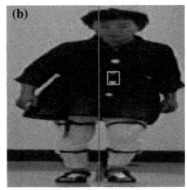

Fig. 8 Movement between the 45th frame's center point (**a**) and the 62th frame's center point (**b**)

Table 2, the detection speed of the proposed method showed real-time possibility and we believe that it can be further improved by adopting a better performing GPU.

5.3 Gait Variability

In this work, the new gait variability is a human center point as a new biomarker. In Fig. 8, the white boxes are the center point and we can see its movement between the 45th frame and the 62th frame. By measuring the gait variability as the center point movement across all frames, we can detect behavior differences simply.

As we can see from Fig. 9, because the object sizes are different between the 1st frame and the 147th frame, direct comparison between them is not feasible. To compensate the drawback, we resized the bounding boxes of all frames to be the same.

Figure 10 shows the difference of the positions of center points at the 3rd frame based on the fitness of the bounding boxes. For meaningful comparison across frames, correct and accurate bounding box localization is essential.

If we can make the correct bounding box and resize it to a standard size at every frame, we can measure the gait variability to analyze the difference of human center point across frames. It can be an important biomarker in finding behavior disease and can be used as a feature for deep learning or machine learning methods to classify different behavior patterns.

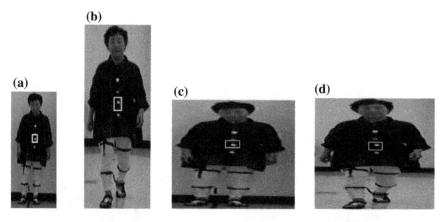

Fig. 9 Left two images are the first frame (**a**), 147th frame before resizing (**b**), right two images are the first frame (**c**) and the 147th frame after resizing (**d**)

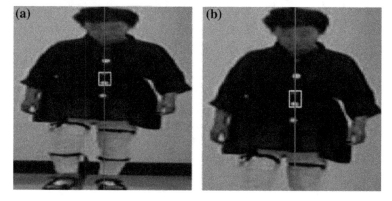

Fig. 10 Left image is correct bounding box at the 3rd frame (**a**) and the right image is incorrect bounding box at the 3rd frame (**b**)

6   Conclusions and Future Work

In this paper, we adopted a deep learning-based object detection model, which is generally better than machine learning-based detection models, to detect a person region called a bounding box in a video more accurately in real time. For human gait variability, we modified the conventional deep learning model to detect a person. In order to improve the bounding box result of the deep learning model, the input images are confined to the person object region of interest. In addition, interpolation is applied to improve the inaccurate bounding box localization. To get gait variability effectively, resizing is also performed at every frame. The future study has two directions. First, for more accurate gait variability analysis, we need to improve bounding box localization performance further, and thus, we are devising a further optimized detection model in terms of both accuracy and execution speed by adopting optimized predefined bounding box sizes and using parallel programming. Second, from a classification viewpoint, we are devising a deep learning-based automatic behavior analysis module to identify different behavior patterns of patients and classify them at different AD stages.

Acknowledgements This work was supported by the Brain Research Program through the National Research Foundation of Korea (NRF) funded by the Ministry of Science, ICT & Future Planning (NRF-2016M3C7A1905477, NRF-2014M3C7A1046050) and the Basic Science Research Program through the NRF funded by the Ministry of Education (NRF-2017R1D1A1B03036423). This study was approved by the Institutional Review Board of Gwangju Institute of Science and Technology (IRB no. 20180629-HR-36-07-04). All procedures performed in studies involving human participants were in accordance with the ethical standards of the institutional and/or national research committee and with the 1964 Helsinki Declaration and its later amendments or comparable ethical standards. This article does not contain any studies with animals performed by any of the authors.

References

1. Nguyen, D., & Li, W. (2016). Human detection from images and videos: a survey. *Pattern Recognition, 51,* 148–175. https://doi.org/10.1016/j.patcog.2015.08.027.
2. Felzenszwalb, P. (2008). A discriminatively trained, multiscale, deformable part model. In *10th IEEE International Symposium on High Performance Distributed Computing* (pp. 1–8). New York: IEEE Press. https://doi.org/10.1109/hipc.2008.4587597.
3. Krizhevsky, A., Sutskever, I., & Hinton, G. (2012). ImageNet classification with deep convolutional neural networks. *Advances in Neural Information Processing Systems 25* (pp. 1097–1105). New York: Curran Associates, Inc. https://doi.org/10.1145/3065386.
4. Girshick, R., Donahue, J., Darrell, T., & Malik, J. (2014). Rich feature hierarchies for accurate object detection and semantic segmentation. In *2014 IEEE Conference on Computer Vision and Pattern Recognition* (pp. 580–587). New York: IEEE Press. https://doi.org/10.1109/cvpr.2014.81.
5. Girshick, R. (2015). Fast R-CNN. In *IEEE International Conference on Computer Vision* (pp. 1440–1448). New York: IEEE press. https://doi.org/10.1109/iccv.2015.169.
6. Ren, S., He, K., & Girshick, R. (2016). Faster R-CNN: Towards real-time object detection with region proposal networks. *IEEE Transactions on Pattern Analysis and Machine Intelligence, 39,* 1137–1149. https://doi.org/10.1109/TPAMI.2016.2577031.
7. Redmon, J., Divvala, S., Girshick, R., & Farhadi, A. (2016). You only look once: Unified, real-time object detection. In *2016 IEEE Conference on Computer Vision and Pattern Recognition* (pp. 779–788). New York: IEEE Press. https://doi.org/10.1109/cvpr.2016.91.
8. Gwak, J., Park, G., & Jeon, M. (2017). Viewpoint invariant person re-identification for global multi-object tracking with non-overlapping cameras. *KSII Transactions on Internet and Information Systems, 11,* 2075–2092.
9. Gwak, J. (2017). Multi-object tracking through learning relational appearance features and motion patterns. *Computer Vision and Image Understanding, 162,* 103–115.
10. Yang, E., Gwak, J., & Jeon, M. (2017). CRF-boosting: Constructing a robust online hybrid boosting multiple object trackers facilitated by CRF learning. *Sensors, 17,* 617:1–617:18.
11. Yang, E., Gwak, J., & Jeon, M. (2017). Multi-human tracking using part-based appearance modelling and grouping-based tracklet association for visual surveillance applications. *Multimedia Tools and Applications, 76,* 6731–6754.

12. Prakash, O., Gwak, J., Khare, M., Khare, A., & Jeon, M. (2018). Human detection in complex real scenes based on combination of biorthogonal wavelet transform and Zernike moments. *Optik—International Journal for Light and Electron Optics, 157,* 1267–1281.
13. Yu, H., Riskowski, J., & Brower, R. (2009). Gait variability while walking with three different speeds. In: *2009 IEEE International Conference on Rehabilitation Robotics* (pp. 823–827). New York: IEEE Press. https://doi.org/10.1109/icorr.20095209486.
14. Liu, W., Anguelov, D., Erhan, D., Szegedy, C., Reed, S., Fu, C., et al. (2016). SSD: Single shot multibox detector. In *European Conference on Computer Vision* (pp. 21–37). Cham: Springer. https://doi.org/10.1007/978-3-319-46448-0_2.
15. Lam, K. Y., Tsang, N. W. H., & Han, S. (2017). Activity tracking and monitoring of patients with alzheimer's disease. *Multimedia Tools and Applications, 76,* 489–521. https://doi.org/10.1007/s11042-015-3047-x.
16. Mega, S., & Gornbein, F. (1996). The spectrum of behavioral changes in Alzheimer's disease. *Neurology, 46,* 130–135. https://doi.org/10.1212/WNL.46.1.130.
17. Redmon, J., Divvala, S., Girshick, R., & Farhadi, A. (2017). YOLO9000: Better, faster, stronger. In *2017 IEEE Conference on Computer Vision and Pattern Recognition* (pp. 6517–6525). New York: IEEE Press. https://doi.org/10.1109/cvpr.2017.690.
18. Huang, J., Rathod, V., Sun, C., & Zhu, M. (2017). Speed/accuracy trade-offs for modern convolutional object detectors. In *2017 IEEE Conference on Computer Vision and Pattern Recognition* (pp. 3296–3297). New York: IEEE Press. https://doi.org/10.1109/cvpr.2017.351.

Fusion of Zero-Normalized Pixel Correlation Coefficient and Higher-Order Color Moments for Keyframe Extraction

B. Reddy Mounika, Om Prakash and Ashish Khare

Abstract Keyframe extraction of videos is useful in many application areas such as video copy detection, retrieval, indexing, summarization. In this paper, we propose a novel shot-based keyframe extraction algorithm. The proposed algorithm is capable of detecting both shots and keyframes of any video efficiently. For extraction of keyframes, frames of video are clustered into shot transitions. These shot transitions of the video are obtained using higher-order color moments and zero-normalized pixel correlation coefficients. In each shot, all the frames are scanned to detect frame with highest standard deviation in that particular shot and chosen as keyframe to that shot. The proposed method is tested on videos of personal interviews with luminaries. Performance of the proposed method is evaluated on the basis of five parameters—recall, figure of merit , detection percentage, accuracy and missing factor. The proposed method is able to detect both abrupt and gradual shot transitions with comparatively less computational complexity. The exhaustive analysis of results shows the sound performance of the proposed method over the methods used in this study.

Keywords Zero-normalized pixel correlation coefficient · Color moments
Shot detection · Cut shot transition · Gradual shot transition
Keyframe extraction

1 Introduction

Managing multimedia applications such as video conferencing, e-learning videos, video-on-demand were always a challenging task and notified the need of proper storage and retrieval system. The first and foremost step toward storage and retrieval

B. Reddy Mounika (✉) · A. Khare (✉)
Department of Electronics and Communication, University of Allahabad, Allahabad,
Uttar Pradesh, India
e-mail: ashishkhare@hotmail.com

O. Prakash
Department of Computer Science and Engineering, Nirma University, Ahmedabad, India

© Springer Nature Singapore Pte Ltd. 2019
A. Khare et al. (eds.), *Recent Trends in Communication, Computing,
and Electronics*, Lecture Notes in Electrical Engineering 524,
https://doi.org/10.1007/978-981-13-2685-1_34

of video systems is keyframe extraction. Reduced storage space requirement, time and computational complexity made keyframe extraction technique attractive to all researchers. Keyframe extraction techniques are classified as sequential-based and cluster-based algorithms. In sequential-based approach [1], difference between visual features of frames is utilized to extract keyframes. Another possible approach toward keyframe extraction is clustering the similar frames into shots according to some similarity measures. A shot is defined as a sequence of frames recorded by a unique camera. The transition between shots can be of two types: abrupt or gradual. Abrupt shot transition is also known as cut or camera break that takes place within one frame, while a gradual shot transition takes place over short span of frames. A gradual transition can be fade-in, fade-out or dissolve resulted from different editing effects. The process of extraction of shots by detecting the boundaries of a shot is known as shot boundary detection (SBD). In general, any SBD algorithm consists of three steps: feature extraction for frames of the video, similarity measurement with the help of computed features and then classification of continuous frames into shots. Classification is done by designing a proper threshold. Threshold designing algorithm may be statistical based or learning. The challenges toward SBD are special effects like panning, zooming, illumination changes, object/camera motion. The entire shot can be mapped into a representative frame known as keyframe.

In this article, we propose a novel, fast and precise shot detection-based keyframe extraction algorithm. The proposed algorithm uses zero-normalized pixel correlation coefficient and higher-order color moments as feature set. Variation in zero-normalized pixel correlation coefficient between R, G, B channels of consecutive frames is calculated and subjected to the designed threshold to extract shot transition boundaries. The extracted shots based on SBD and the shots extracted by employing higher-order color moments are combined together resulting in overall shot transitions. Each shot is scanned to get the frame of highest standard deviation (STD) that frame is decided as the keyframe for that particular shot. The experiments have been performed on different videos of personal interviews with luminaries. The results are compared with the state-of-the-art methods. The performance analysis of the proposed methods and its comparison with other state-of-the-art methods shows the promising performance in both cases when transitions are abrupt or gradual.

Rest of the paper is organized as follows: In Sect. 2, we present the literature review; Sect. 3 presents the overview of zero-normalized pixel correlation coefficient and the proposed method. Section 4 presents the experimental results. Finally, the concluding remarks are given in Sect. 5.

2   Literature Review

An extraction of keyframe in any video is a challenging and difficult task. A large number of researchers are working in the area. A proper keyframes extraction with high accuracy is still a challenging task due to the varying lighting conditions and relative motion of the object and camera. Shot detection plays a vital role in keyframe extraction. A unified model to detect different types of video shot transitions with the help of frame transition parameters has been proposed by Mohanta et al. [2]. Mohanta et al. estimated current frame from previous and next frame using frame estimation scheme. Shot transitions are detected using frame transition parameters and frame estimation errors. A novel and robust algorithm for fast and accurate detection of shot transitions has been introduced by Birinci and kiranyaz [3]. Tavassolipour et al. [4] employed hue histogram difference and motion vector for feature extraction followed by the use of support vector machine to classify shot boundaries. Lu et al. [5] used adaptive thresholds to predict the positions of shot boundaries and lengths of gradual transitions. They employed singular value decomposition (SVD) on frame feature matrices of all candidate segments to reduce feature dimension. Thouraya Ayadi [6] presented a formal study on multilevel interior growing self-organizing maps (MIGSOM) to classify action and nonaction movie scenes. Spatiotemporal Gaussian mixture models and variation Bayesian algorithm have been used by Loukas et al. [7] to detect shot boundaries in endoscopic surgery videos. Debabrata et al. [8] detected shot transitions by computing similarity of frames with respect to a reference frame and employed postprocessing techniques to reduce false transitions. Poonam et al. [9] detected shots using higher-order color moments obtained from block-based image histogram; then, the frames with maximum mean and standard deviation of the shot were taken as key frames. Sheena et al. [10] also used histogram to detect key frames, in which key frames are extracted by thresholding the difference between histogram of consecutive frames of the video. Camera motion has been utilized by Ivan Gonzalez [11] in which a parametric model was built using motion vectors to achieve temporal segmentation and one representative frame of each motion pattern was taken as key frames. SIFT point distribution histogram has also been utilized to detect shot transitions by Hannane et al. [12] where they employed entropy-based singular values measure to detect keyframes. Thakre et al. [13] shots were detected by calculating distance between the wavelet-transformed blocks of consecutive frames of the video, and later, the frames with local minima and maxima of each shot were taken as keyframes. According to Dang and Radha [14] key frames were detected by decomposing input data into a set of sparse components and low-rank components both which were then combined into a single $\ell 1$-norm-based nonconvex optimization problem.

3   Proposed Method

This section presents the proposed method of keyframe extraction. The proposed method draws the motivation from the fact that zero-normalized pixel correlation coefficient measures the similarity between pixels of the two frames. Similarity is low in case of cut shot transition, whereas in case of gradual transition the value of similarity index does not change more. Higher-order color moments can help some extent to capture gradual transition. The proposed method mainly consists of five phases: feature extraction, similarity measurement, threshold computation and classification of continuous frames into shots and keyframe extraction.

Feature extraction—Firstly different frames of a video are extracted then for each frame, all the three color channels—red, green and blue—are extracted and stored separately into different matrices.

Similarity measurement—Zero-normalized pixel correlation coefficient (ZNCC) is used to measure similarity. For all frames, the zero-normalized pixel correlation coefficient between red, green and blue channels of consecutive frames is calculated separately and similarity scores are stored.

Threshold computation—The mean for the computed zero-normalized pixel correlation coefficient of three channels has been considered as threshold value.

Classification of frames into shots—Higher value of ZNCC means more similarity. A shot boundary is declared at the frames where the value of ZNCC is less than threshold. All the shot boundaries obtained by three different channels ZNCC matrices are combined along with the shot transitions obtained by the Poonam et al. [9] method to get overall shot transitions of the video.

Keyframe extraction—Each and every shot is scanned to get the frame with highest standard deviation which is selected as keyframe to that particular shot.

The proposed work consists of two main tasks—keyframe extraction and shot detection. For the keyframe extraction, we utilized zero-normalized pixel correlation coefficient and the work proposed by Poonam et al. [9].

Zero-normalized pixel correlation coefficient between consecutive frames of the video was computed using Eq. (1).

$$ZNCC = \frac{\sum_{i=1}^{N} (x_i - \overline{x})(y_i - \overline{y})}{\sqrt{\sum_{i=1}^{N} (x_i - \overline{x})^2 \sum_{i=1}^{N} (y_i - \overline{y})^2}} \tag{1}$$

where $\overline{x}$ and $\overline{y}$ are mean intensity values.

The mean of the above resultant matrix is calculated to determine threshold. A shot transition is declared by examining the correlation score of the consecutive frames.

3.1 The Algorithm: Shot Detection

Input: video clip

(i) *Extract the video frames.*
(ii) *For each frame red, green, blue channels are extracted separately.*
(iii) *For all frames, determine the zero-normalized correlation coefficient between red, green and blue channels of consecutive frames separately and stored in a matrix r1, g1 and b1, respectively.*
(iv) *Find mean of matrices r1, g1, b1 and three threshold values for three different matrices are set as*

$$T_i = mean\,(i); \quad where\,i = r1, g1, b1.$$

T_{r1}, T_{g1}, T_{b1} are the threshold values for $r1, g1, b1$ matrices, respectively.
(v)   Take three different matrices corresponding to threshold values.

$$L_i = find(i < T_i); for\,i = r1, g1, b1\,respectively.$$

(vi)   Take L_i as set of frames for which the correlation value is less than the threshold T_i.
(vii)   Detect the shots using Poonam et al. [9] method and store into matrix L.
(viii)   Take the union of L_i and L to determine the overall shot transitions as
Shot transition indices = union $(L, L_{r1}, L_{g1}, L_{b1})$

Output: Shot transition frames

3.2   Keyframe Extraction

In the proposed method key frames are extracted from the shots obtained by the algorithm discussed in Sect. 3.1. Each and every frame of shot is scanned to get *the frame with highest standard deviation, which is selected as keyframe to that* particular shot.

4   Experimental Results

The experiments are conducted on two video clips total having 101628 frames of personal interview with luminaries, and among them, one is downloaded from https://www.youtube.com/watch?v=mLi8tLkjqBM, and the other is downloaded from https://www.youtube.com/watch?v=kkCiCpyOlrA. In order to validate the results, the comparisons of performance for two videos of luminaries having frames of 77628 and 24000, respectively, are obtained by various methods as Poonam et al. [9], Sheena et al. [10], Thakre et al. [13] and the proposed hybrid method. These results are shown in Tables 1 and 2. Ground truth is build using publicly available tool Virtual Dub [15]. The measures are briefly described as below—

Recall (R): The fraction of relevant documents that are successfully retrieved is denoted by recall [8] and is given as

$$R = \frac{TP}{TP + FN} \tag{2}$$

where TP and FN stand for true positive and false negative. *Precision (P)* [8] describes the percentage of retrieved documents relevant to the query, and it is described as

Table 1 Performance calculations for Video 1 personal interview of luminary 1 having frames of 77628

Method	Shots detected	Key frames detected	No. key frames in ground truth	R	F	DP	A	*mf*
Sheena [9]	1910	2250	6654	0.0236	0.0367	2.3595	0.1180	41.3822
Poonam [10]	11058	22116		0.1413	0.1061	14.1268	0.4112	6.0787
Wavelet [13]	2468	4936		0.0310	0.0452	3.0959	0.2208	31.3010
Proposed	20448	22726		0.2857	0.2294	28.5693	0.4705	2.5003

Table 2 Performance calculations for Video 2 personal interview of luminary 2 having frames of 24000

Method	Shots detected	Key frames detected	No. key frames in ground truth	R	F	DP	A	*mf*
Sheena [9]	332	756	1500	0.0678	0.1107	6.7843	0.1713	13.7400
Poonam [10]	3199	6398		0.2137	0.1348	21.3704	0.4417	3.6794
Wavelet [13]	376	752		0.0631	0.1005	6.3094	0.1843	14.8495
Proposed	4207	4861		0.3114	0.1449	31.1398	0.4730	2.2113

$$P = \frac{TP}{TP + FP}; \quad \text{FP stands for False Positive} \tag{3}$$

F measure (*F*): A measure that combines both precision and recall is F measure [8] and is computed as

$$F = \frac{2 * (P) * (R)}{R + P} \tag{4}$$

Detection percentage (*DP*): Detection percentage is computed as given in Poornima et al. [16] as

$$DP = 100 \times \frac{TP}{TP + FN} \tag{5}$$

Accuracy (A): Accuracy can also be stated as percentage of corrected classification [17] and is defined as

$$\text{Accuracy} = \frac{\text{TP} + \text{TN}}{\text{TP} + \text{FP} + \text{TN} + \text{FN}} \tag{6}$$

where TN and FN denote true negative and false negative, respectively.

Missing factor (mf): According to Ibrahim et al. [18] missing factor (mf) is defined as

$$mf = \frac{FN}{TP} \tag{7}$$

The values of recall (R), precision (P), F measure and accuracy (A) are in the range 0–1, and the range of detection percentage and missing factor lies between 1 and 100.

It is observed from the tables that the values of boundary recall (R), accuracy (A) and figure of merit (F) are in the range 0–1. Higher values of boundary recall, figure of merit, accuracy are desired for an efficient method. Good segmentation will have higher value of accuracy. Maximum value of detection percentage (DP) is desirable for any method to be an efficient, and it lies in the range 1–100. The lower value of missing factor (*mf*) is better for performance of method. It is worth mentioning that the proposed method urges that the present results are highly superior to the values of the methods outlined; therefore, we believe that the comparison supports very well validity of the present results.

5   Conclusions

In the present article we proposed a new method for extracting keyframes based on shot transition detection using pixel correlation. Pixel correlation between two frames is the measure of similarity between first image and shifted copy of the second image as a function of the lag. Correlation has the advantage of property shift invariance, linear, easy to implement and less time complexity which lead to use correlation in the proposed method. We have implemented shot transition detection using pixel correlation, and key frames are extracted using a hybrid methodology. The proposed method is experimented on 10,1628 frames. The results obtained are compared with the methods of the literature [2, 5, 19], and from the results we can draw the conclusion the proposed method is showing good accuracy and the missing factor is low.

References

1. Guan, G., Wang, Z., Lu, S., Deng, J. D., & Femng, D. D. (2013). Keypoint-based keyframe selection. *IEEE Transactions on Circuit System Video Technology, 23*(4), 729–734.
2. Mohanta, Partha Pratim, Saha, Sanjoy Kumar, & Chanda, Bhabatosh. (2012). A model-based shot boundary detection using frame transition parameters. *IEEE Transactions on Multimedia, 14*(1), 223–233.
3. Birinci, M., Kiranyaz, S. (2014). A perceptual scheme for fully automatic video shot boundary detection. *29*(3), 410–423.
4. Tavassolipour, M., Karimian, M., & Kasaei, S. (2014). Event Detection and Summarization in Soccer Videos Using Bayesian Network and Copula. *IEEE Transactions on Circuits and Systems for Video Technology, 24*(2), 291–304.
5. Lu, Z. M., & Shi, Y. (2013). Fast video shot boundary detection based on SVD and pattern matching. *IEEE Transactions on Image Processing, 22*(12), 5136–5145.
6. Ayadi, T., Hamdani, M., Alimi, T. M., & Adel, M. (2013). Movie scenes detection with MIGSOM based on shots semisupervised clustering. *Neural Computing and Applications, 22*(7), 1387–1396.
7. Loukas, C., Nikiteas, N., Schizas, D., Georgiou, E. (2016). Shot boundary detection in endoscopic surgery videos using a variational Bayesian framework. *11*(11), 1937–1949.
8. Dutta, D., Saha, S. K., & Chanda, B. (2016). A shot detection technique using linear regression of shot transition patterns. *Multimedia Tools and Applications, 75*(1), 93–113.
9. Jadhava, P. S., & Jadhav, D. S. (2015). Video summarization using higher order color moments. In *Proceedings of the International Conference on Advanced Computing Technologies and Applications (ICACTA)* (Vol. 45, pp. 275–281).
10. Sheena, C. V., Narayanan, N. K. (2015). Key-frame extraction by analysis of histograms of video keyframes using statistical methods, In *Proceedings of the 4th International Conference on Eco-friendly Computing and Communication Systems* (Vol. 70, pp. 36–40).
11. Gonzalez-Diaz, I., Martinaz-Cortes, T., Gallardo-Antolin, A., & Diaz-de-Maria, F. (2015). Temporal segmentation and keyframe selection methods for user-generated video search-based annotation. *Expert Systems with Applications, 42*, 488–502.
12. Hannane, R., Elboushaki, A., Afdel, K., Naghabhushan, P., Javed, M. (2016). An efficient method for video shot boundary detection and keyframe extraction using SIFT-point distribution histogram. *International Journal of Multimedia Information Retrieval.* 10.1007%2Fs13735-016-0095-6.
13. Thakre, K. S., Rajurkar, A. M., Manthalkar, R. R. (2015). Video partitioning and secured keyframe extraction of MPEG video. In *Proceedings of the International Conference on Information Security & Privacy (ICISP2015), Nagpur, India, Procedia Computer Science,* (Vol. 45, pp. 275–281).
14. Dang, C., & Radha, H. (2015). RPCA_KFE: Key frame extraction for video using robust principal component analysis. *IEEE Transactions on Image Processing, 24*(11), 1–12.
15. Lee. Virtual Dub home page. http://www.virtualdub.org/index.html.
16. Poornima, K., & Kanchana, R. (2012). A method to align images using image segmentation. *International Journal of Soft Computing and Engineering, 2*(1), 294–298.
17. Khare, M., Srivastasava, R. K., Khare, A. (2015). Moving object segmentation in daubechies complex wavelet domain. *Journal of Signal, Image and Video Processing, 9*(3), 635–650.
18. Shaker, I. F., Abd-Elrahman, A., Abdel-Gawad, A. K., Sherief, M. A. (2011). Building extraction from high resolution space images in high density residential areas in the Great Cairo region. *Remote Sensing, 3*, 781–791.
19. Martn, R. V., & Bandera, A. (2013). Spatio-temporal feature-based keyframe detection from video shots using spectral clustering. *Pattern Recognition Letters, 34*(7), 770–779.

An Improved Active Contour Model for Salient Object Detection Using Edge Cues

Gargi Srivastava and Rajeev Srivastava

Abstract In this paper, we solve the problem of salient object detection by using an ensemble mechanism of edge detection and active contours. Edge cues are used to provide a solution to the initialisation problem of active contour. The active contour method suffers from the initialisation problem. If the initial contour lies in a region with low probability of salient object, the final salient object detection provides inaccurate results. In this paper, the problem is addressed by generating a binary mask using Sobel edge detection method which acts as the initial contour. The binary mask makes sure that the contour lies in the region with high probability of finding a salient object. This work is an improvement upon the active contour model. The method is simple and fast and follows basic human intuition to find salient objects. The proposed work is compared against seven recent works and gives better results in terms of precision, recall and false positive rates.

Keywords Salient object detection · Active contour · Edge

1 Introduction

Salient object detection is the process of extracting salient objects from the image. Humans have a natural capability to identify salient objects in a scene, referred to as the human attention mechanism [1]. Through salient object detection the goal is to automate this task using computers. Several works proceed in the direction following exactly the biological mechanism behind how humans differentiate between salient

G. Srivastava (✉) · R. Srivastava
Indian Institute of Technology (BHU) Varanasi, Varanasi 221005, UP, India
e-mail: gargis.rs.cse16@iitbhu.ac.in

R. Srivastava
e-mail: rajeev.cse@iitbhu.ac.in

© Springer Nature Singapore Pte Ltd. 2019
A. Khare et al. (eds.), *Recent Trends in Communication, Computing,*
and Electronics, Lecture Notes in Electrical Engineering 524,
https://doi.org/10.1007/978-981-13-2685-1_35

and nonsalient objects in a scene [2]. Other works focus on utilising image features like intensity and contrast, whereas some other are based on the perceptual grouping of pixels. Researchers have approached the problem from neurobiological, psychological and computer vision point of view [3]. The motivation behind salient object detection is that it is the first step in various other applications of image processing like: image tagging, image captioning, content-based image retrieval systems, object recognition, object detection, image and video compression and image segmentation. It is the basis for artificial intelligence and computer vision. Image tagging can further help in organising and maintaining large image databases and for the purpose of social media networking. Image captioning further utilises the result of salient object detection to generate appropriate captions of the images for a better understanding and for helping visually impaired people with several tasks.

The paper is organised as follows: Section 2 describes the related recent works in this field. Section 3 gives a description of the proposed work. Section 4 provides details of the experimental work performed. Section 5 discusses the results. Section 6 concludes the work giving possible future insights.

2   Related Work

Salient object detection is basically a two-step procedure. The first step is to identify salient regions or salient spots which specify areas where probability of finding an object is high. After identifying those areas of the image, the goal is to extract the object out from those regions. The step can involve trimming down or expanding the salient regions or spots identified in the first step. Borji in his paper [4] states that major works in this field proceed directly to the second step by assuming that the salient object is present in the centre of the image. In this paper, we focus on the first step significantly by proposing to find out salient regions using edge detection mechanism. Jian et al. in their paper [5] use a wavelet-based directional patch detection mechanism that detects patches from intensity and colour channels. Salient objects are then generated from feature maps obtained from both the channels. Wang et al. in their paper [6] use contrast information with respect to colour, space and texture information of the image. The paper also mentions that the visual principles of humans do not conform to the models prepared for interpreting saliency information.

3   Proposed Work

In this paper, the problem of saliency detection is approached using an active contour with level-set method. The reason behind using this particular approach is that the method is simple, helps us to detect clear object boundaries, overcomes problem of intensity homogeneity and is able to detect multiple salient objects. The output of an active contour method depends majorly on the initialisation of the contour.

Fig. 1 Flowchart for the
proposed algorithm

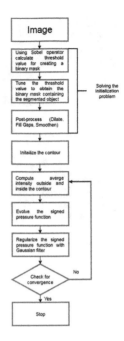

If the initial contour is chosen to be lying in probable salient region, the output is
better than when the initial contour lies outside the probable salient region. To solve
this initialisation problem, a basic edge detection operator—Sobel operator—is used
to solve the initialisation problem. With the help of Sobel operator, a rough image
segmentation is performed. The output is then used as the initial contour. This initial
contour is used to extract the complete object. The framework of this concept can be
given as in Fig. 1.

The first step is to calculate a threshold for generating a binary mask. The threshold
is found using Sobel operator. The threshold value is then tuned to generate the binary
mask containing the segmented object. Post-processing of the segmented object is
done by dilation, filling gaps and smoothening. Now this post-processed segmented
object is used as the initial contour for the active contour step. Intensity within and
outside the contour is calculated, and a signed pressure function (spf) is evolved
according to them. Each time the spf is regularised using a Gaussian filter. The
convergence is checked by the number of iterations. Figure 2 describes the procedure
visually by taking sample input images from three publicly available data sets.

The first column depicts the original input image. Columns 2 and 3 represent intermediate outputs corresponding to image segmentation and final contour obtained.
The contour is represented in blue colour. The second column images serve as the
initial contour for active contour algorithm. The fourth column represents the salient
object obtained, and the last column is the ground truth data corresponding to the
input images.

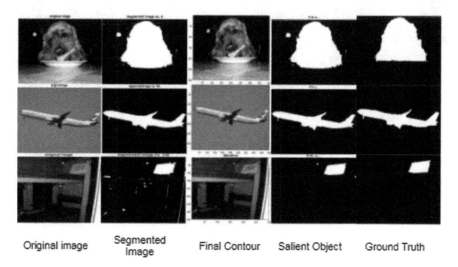

Original image   Segmented   Final Contour   Salient Object   Ground Truth
                     Image

Fig. 2 Visual Steps for the proposed algorithm

Fig. 3 Original image

3.1 Generating Initial Contour Using Sobel Operator

The basic aim of this step is to create a binary mask containing salient object which will serve the purpose of initialising the contour. The object of interest usually highly differs in contrast from the background. To utilise this contrast information, gradient of the image is calculated. Upon this gradient, a threshold can be applied to create the binary mask as shown in Figs. 3 and 4. After the binary gradient mask is obtained, the mask is post-processed using dilation, filling of gaps and smoothening. The results are shown in Figs. 5 and 6. The gradient mask obtained after filling holes is smoothened Fig. 7. This serves as an initial contour for the active contour model. After the active contour step is performed, the salient object is found, as shown in Fig. 8.

Fig. 4 Gradient mask

Fig. 5 Dilated gradient mask

Fig. 6 Gradient mask after filling holes

Fig. 7 Gradient mask after smoothing

Fig. 8 Final salient object

3.2 Active Contour Algorithm

In this section, the working methodology of the active contour algorithm is described. After the initial contour is defined, the first step is to get the contrast information of the image. The assumption is that the salient object differs from the background in terms of contrast. Average intensity within and outside the contour is calculated which helps in defining the spf function. The spf function is given as:

$$spf = I - \frac{c_1 + c_2}{2} \tag{1}$$

where I represents the input image, c_1 is the average intensity outside the contour and c_2 is the average intensity inside the contour. The evolution of the contour is given by:

$$\frac{\partial \phi}{\partial t} = spf(I(x, y)).\alpha |\nabla \phi| \quad x, y \in \Omega \tag{2}$$

where α is the balloon force and Ω is a bounded subset of $I(x, y \in \Re^2)$ in the given image.

At each iteration Gaussian filter is used to for the stabilisation of the contour and maintaining its sensitivity to noise. The two parameters balloon force α and standard deviation σ are highly image dependent. A high α may cause the contour to pass narrow edges and a low α may not allow the contour to pass weak edges. Setting σ high can cause increased blurring leading to edge leakage whereas a low value of σ can increase the sensitivity to noise. At the edges, the level-set function evaluates to zero which stops the evolution of contour.

4 Experimental Analysis

This section covers experimental details of the work done.

4.1 Data sets

Three publicly available benchmark data sets, namely MSRA10K, PASCAL-S and DUT-OMRON, are used for the evaluation of our model. The image data sets have around 10,000, 1000 and 5000 images, respectively. Ground truth of the images is available with the data sets. MSRA10K is a simple image data set with simple backgrounds and mostly only a single salient object in the centre. DUT-OMRON is a complex data set with cluttered background and multiple salient objects.

4.2   Execution Environment

The model has been tested on 64-bit MATLAB R2017a (9.2.0.556344) on a desktop with processor configuration Intel(R) Xeon(R) CPU E5-2630 v3 @ 2.40 Ghz.

4.3   Evaluation Parameters

The model has been evaluated for the following parameters:

1. Precision-recall (PR) Curve: If a large area is present under the PR-curve, it means the model is performing well.
2. Receiver operating characteristic (ROC) Curve: A large area under the ROC curve is reflection of a well-performing model.

4.4   Comparison with Existing Algorithms

On the line of the above-mentioned evaluation parameters, this work is compared along with seven other significant works done in this area on the same data sets. These are: PCAS [7], GC, UFO [8] and RC are biologically inspired methods, GS and MR are background prior-based methods and BMS is a psychology inspired method. In [9], i.e. GC, global cues are obtained from large-scale perceptually homogeneous elements. Information from different parts having same spatial distribution provides soft abstraction. Gathering information from both cues salient objects are obtained. In [10], i.e. RC contrast information and centre prior are used to detect salient regions. In [11], i.e. GS background and connectivity priors are used to subtract the background from images and then whatever is left is considered foreground. In [12], i.e. MR graph-based techniques and background prior methods are used in combination with the assumption that distance between two points lying in background region is smaller than when one point lies in salient region and the other in background region. In [13], i.e. BMS saliency maps are constructed by gaining information from topographical cues.

5   Results and Discussion

This section discusses the results obtained using the proposed model. Sobel operator was used against the choice of other edge detection mechanisms like Prewitt, Roberts, Laplacian of Gaussian, Canny and zero cross. The choice is justified by the accuracy obtained by using each of them. To get better results, threshold obtained using Sobel operator in first step is tuned. The tuning process is performed so as not to use the exact threshold provided automatically by MATLAB, instead use a multiplicative factor of it which serves the purpose of salient object detection. The factor used to tune the threshold is called the fudge factor. The value of fudge factor was chosen experimentally to be 0.7. Another factor σ is used for Gaussian regularisation. The value was selected by checking for the accuracy. The value is selected as 32. For convergence, the value of 100 was calculated experimentally. The value of balloon force was chosen to be 25. The comparison of other algorithms was done after setting these parameters. The results are shown below. From seeing the graphs, it is clear that our method is performing better than other methods. The data for comparison has been taken from [14]. For visual comparison of the result, a collection of sample images from all databases for all the compared algorithms is given. From the visual

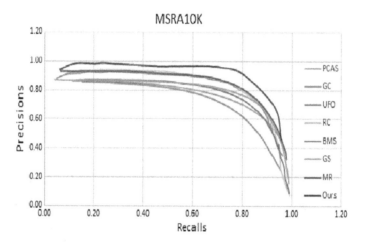

Fig. 9 MSRA10K precision-recall curve

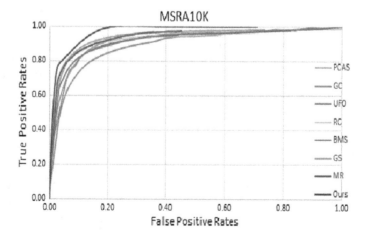

Fig. 10 MSRA10K ROC

result, it is seen that our algorithm gives better output than the other algorithms. No post-processing is performed on the final salient object obtained from the active contour algorithm. Figures 9, 10, 11, 12, 13 and 14. depict all the results obtained using the proposed model.

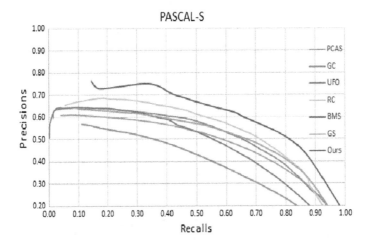

Fig. 11 PASCAL-S precision-recall curve

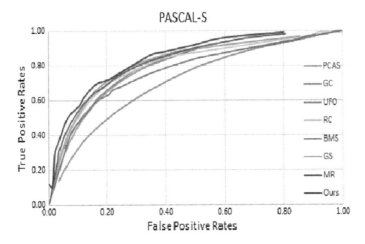

Fig. 12 PASCAl-S ROC

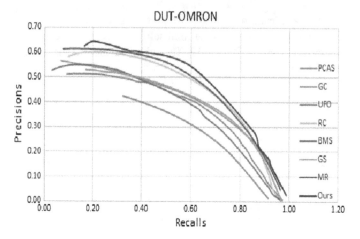

Fig. 13 DUT-OMRON precision-recall

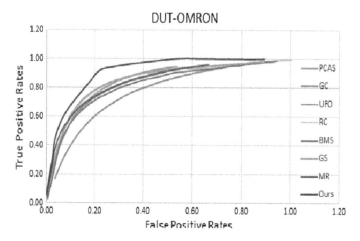

Fig. 14 DUT-OMRON ROC

6 Conclusion

In this paper, an attempt is made to extract salient objects from the images using an active contour algorithm by solving the initialisation problem using edge detection. The method is simple and fast to execute. The method is compared against seven state-of-the-art algorithms and is found to be better in performance. The method utilises contrast and intensity cues to find the salient object. The limitation is that the intensity of background should be either minimum or maximum for clear demarcation between foreground and background. The procedure fails when the intensity of background is neither minimum nor maximum. Salient object detection is an important research filed for artificial intelligence. Not only higher accuracy is desired, but simple techniques that can reduce the complexity of the procedure are also needed [13].

References

1. Jin-Gang, Y., Xia, G.-S., Gao, C., & Samal, A. (2016). A computational model for object-based visual saliency spreading attention along gestalt cues. *IEEE Transactions on Multimedia, 18,* 273–286. https://doi.org/10.1109/TMM.2015.2505908.
2. Zhu, L., Klein, D. A., Frintrop, S., Cao, Z., & Cremers, A. B. (2014). A multisize superpixel approach for salient object detection based on multivariate normal distribution estimation. *IEEE Transactions on Image Processing, 23,* 5094–5107. https://doi.org/10.1109/TIP.2014. 2361024.
3. Chen, T., Lin, L., Liu, L., Luo, X., & Li, X. (2016). DISC deep image saliency computing via progressive representation learning. *IEEE Transactions on Neural Networks and Learning Systems, 27,* 1135–1149. https://doi.org/10.1109/TNNLS.2015.2506664.
4. Borji, A. (2015). What is a salient object a dataset and a baseline model for salient object detection. *IEEE Transactions on Image Processing, 24,* 742–756. https://doi.org/10.1109/TIP. 2014.2383320.
5. Jian, M., Lam, K.-M., Dong, J., & Shen, L. (2015). Visual-patch-attention-aware saliency detection. *IEEE Transactions on Cybernetics, 45,* 1575–1586. https://doi.org/10.1109/tcyb. 2014.2356200.
6. Wang, Q., Yuan, Y., & Yan, P. (2013). Visual saliency by selective contrast. *IEEE Transactions on Circuits and Systems for Video Technology, 23,* 1150–1155. https://doi.org/10.1109/tcsvt. 2012.2226528.
7. Margolin, R., Tal, A., & Zelnik-Manor, L. (2013). What makes a patch distinct? *Computer Vision and Pattern Recognition,* 1139–1146. https://doi.org/10.1109/CVPR.2013.151.
8. Jiang, P., Ling, H., Yu, J., & Peng, J. (2013). Salient region detection by UFO uniqueness, focusness and objectness. *International Conference on Computer Vision,* 1976–1983. https:// doi.org/10.1109/ICCV.2013.248.
9. Cheng, M.-M., Warrell, J., Lin, W.-Y., Zheng, S., Vineet, V., & Crook, N. (2013). Efficient salient region detection with soft image abstraction. In *International Conference on Computer Vision* (pp. 1529–1536). https://doi.org/10.1109/ICCV.2013.193.
10. Cheng, M.-M., Mitra, N. J., Huang, X., Torr, P. H. S., & Hu, S.-M. (2015). Detection, global contrast, & region, based salient. *IEEE Transactions on Pattern Analysis and Machine Intelligence, 37,* 569–582. https://doi.org/10.1109/TPAMI.2014.2345401.
11. Wei, Y., Wen, F., Zhu, W., & Sun, J. (2012). Geodesic saliency using background priors. In *European Conference on Computer Vision* (pp. 29–42). https://doi.org/10.1007/978-3-642-33712-3_3.
12. Yang, C., Zhang, L., Lu, H., Ruan, X., & Yang, M.-H. (2013). Saliency detection via graph-based manifold ranking. *Computer Vision and Pattern Recognition,* 3166–3173. https://doi. org/10.1109/CVPR.2013.407.
13. Zhang, J., & Sclaroff, S. (2013). Saliency detection: A Boolean map approach. In *IEEE International Conference on Computer Vision* (pp. 153–160). https://doi.org/10.1109/ICCV.2013. 26.
14. Liu, Q., Hong, X., Zou, B., Chen, J., Chen, Z., & Zhao, G. (2017). Hierarchical contour closure-based holistic salient object detection. *IEEE Transactions on Image Processing, 26,* 4537–4552. https://doi.org/10.1109/TIP.2017.2703081.

Machine Learning-Based Classification of Good and Rotten Apple

Shiksha Singh and Nagendra Pratap Singh

Abstract An apple is one of the most cultivated and consumed fruits in the world and continuously being praised as a delicious and miracle food. It is a rich source of Vitamin A, Vitamin B1, Vitamin B2, Vitamin B6, Vitamin C, and folic acid etc, whereas the rotten fruits affect the health of human being as well as cause big economical loss in agriculture sectors and industries. Therefore, identification of rotten fruits has become a prominent research area. This paper focuses on the classification of rotten and good apple. For classification, first extract the texture features of apples such as discrete wavelet feature, histogram of oriented gradients (HOG), Law's Texture Energy (LTE), Gray level co-occurrence matrix (GLCM) and Tamura features. After that, classify the rotten and good apples by applying various classifiers such as SVM, k-NN, logistic regression, and Linear Discriminant. The performance of proposed approach by using SVM classifier is 98.9%, which is found better with respect to the other classifiers.

Keywords Apple images · Texture features · Machine learning · Classification

1 Introduction

The degradation in quality and quantity of the fruits and vegetables is the consequence of the disease present in the fruits. This results in great loss for economy of agriculture sectors and industries. Considering the example of soybean rust (rust is fungal disease in soybean plant), it is reported that by removing the disease in plant by just 20%, the profit of 11 million- dollar was incurred [1]. The infections or disease in tree leads to defect in tree's fruits, branches, leaves and even the twigs. If the illness in fruits is

S. Singh (✉) · N. P. Singh
Department of Computer Science and Engineering, MMM University of Technology,
Gorakhpur 273010, UP, India
e-mail: shikshait3075@gmail.com

N. P. Singh
e-mail: npscs@mmmut.ac.in

© Springer Nature Singapore Pte Ltd. 2019
A. Khare et al. (eds.), *Recent Trends in Communication, Computing, and Electronics*, Lecture Notes in Electrical Engineering 524,
https://doi.org/10.1007/978-981-13-2685-1_36

inspected at initial stage, the dispersion of the disease to other parts can be reduced or even it can be eliminated. The reduction in infection in fruits will directly increase the economy of the agriculture industries.

The disease present in plant evaluates the stability, quality and quantity in yield of fruits. Till date as per our best knowledge, no such techniques or sensor is developed which differentiate between the healthy and defected fruits. Only the traditional method for distinguishing the images are used, these are bare-eyes observation and scouting method [2]. Scouting method is the adaptive approach which is man driven, consumes lots of time and implementation is costly.

The method proposed in these papers can be implemented for developing an automated system for the detection of healthy and defected apples. There are n-number of application for image processing in agriculture industries. These include color scanners and cameras for taking images as input and then process it for getting the desired result. In our work, we have tried to elaborate the image processing and analyzing them through machine learning techniques [3].

In this paper, we have taken apple dataset for implementing the approach and thus verifying it. Common disease in apples fruits is caused due to infection or fungal activity; these diseases are apple rot, apple scab, apple blotch, dagger nematode etc.

In our work, we have given an analytical approach and verified the approach for the apple classification. This classification is performed by using the various features of images. Like we are using color and texture features. The approach goes like, the defected portion of the apple images is identified with k-means clustering segmentation technique, and then, features are computed over the segmented images on the color and texture features. Once the feature is extracted, we apply fivefold and hold-out validation technique. And in last by implementing various classifiers, we compute the accuracy [2].

Rest of the paper is organized as follows: Sect. 2 contains a deep literature survey. Section 3 explains the proposed method in detail. Section 4 contains the experimental results and discussion. Finally, the conclusion of the proposed work is given in Sect. 5.

2 Literature Survey

In this section, we have reviewed work done by various authors in feature extraction and classification. The authors have used the different features extraction technique and classifiers for attaining the maximum accuracy in result. Table 1 shows work done by the researchers in field of apple classification using different classifiers and feature extraction techniques [4].

Table 1 Literature survey of various proposed approach with their accuracy

References	Dataset	Preprocessing techniques	Feature extraction	Classifier	Accuracy
[2]	320 images of normal apple	K-means clustering	Color: GCH, CCV Texture: LBP, CLBP Shape: Zernike moments	Multiclass SVM	The accuracy of 95.94% was obtained on the combination of feature descriptor CCV, CLBP and ZM
[11]	100 images of Jonagold apples	Threshold segmentation	Texture: standard deviation Color: mean color index Shape: area major inertia moment	Quadratic discriminant Analysis	73% accuracy
[12]	500 images of Jonagold apples	Artificial neural network-based segmentation	Average standard deviation defected ratio	LDC, k-NN, fuzzy k-NN, SVM, Adaboost	SVM and Adaboost gives the accuracy of 90.3%
[13]	166 images of golden delicious apples	Gabor wavelet decomposition Gabor	Gabor feature vectors	Gabor-kernel PCA, PCA, SVM, Gabor PCA, kernel PCA	Gabor-kernel PCA gives the highest accuracy of 90.5%
[14]	Fuji apples	Vector median filter	Color: euclidean distance Shape: round variance, eclipse variance, tightness, ratio of perimeter and square area	SVM	Combination of color and shape feature gives the accuracy of 57.2%, accuracy given by color is 67.3% and shape is 69.2% separately
[15]	Jonagold apples	MLP-based segmentation.	Statistical: arithmetic mean, standard deviation median minimum maximum textural: contrast angular second moment sum-of-squares variance invariant moments geometric: defect ratio perimeter circularity fuzzy C-means algorithm	Statistical: LDC k-NN fuzzy k-NN SVM Syntactical: Decision tree C 4.5	Accuracy evaluated by the statistical classifier is 83.6% and syntactical classifier 85.6%

(continued)

Table 1 (continued)

References	Dataset	Preprocessing techniques	Feature extraction	Classifier	Accuracy
[16]	210 images of normal apple	1. Conversion into HSV color space and thresholding 2. transforming image into several windows or planes	Mean standard deviation	Nearest neighbor classifier	Accuracy 92% was detected
[17]	431 images of normal apple	1. Conversion into L*a*b* color space 2. K-means clustering	GCH, CCV, LBP, CLBP	Multiclass SVM	Highest accuracy 90% was given by CLBP
[18]	Normal apple images	Converting RGB to HSI color model	Color: histogram difference Morphology: erosion Texture: homogeneity	Back propagation neural network	Accuracy was 91.17% in case of color and morphology
[19]	65 images of normal apples	Global threshold segmentation in L*a*b* color space	Mean boundary gradient, mean intensity, fourier descriptors filler blank, invariant moments	Probabilistic neural network	The accuracy came out to be 83.33%

3 The Proposed Method

Image classification depends on integration of spatial, structure and statistical method [5]. In our research, we have used statistical approach for texture-based feature for classification of defected and healthy images. Statistical model evaluates the image on basis of entropy, energy, wavelet etc. Structural method evaluates the image on basis of the appearance of object like shape, spots etc., on the image. At last, spectral method uses spectral space representation like Fourier spectrum [5]. The proposed approach contains four main steps, namely the preprocessing, feature extraction, feature selection, and classification step as mentioned in Fig. 1.

3.1 Preprocessing of Images

Preprocessing of image involves refining the images. This refinement involves resizing the images and enhancing the image quality in terms of pixel brightness. The main

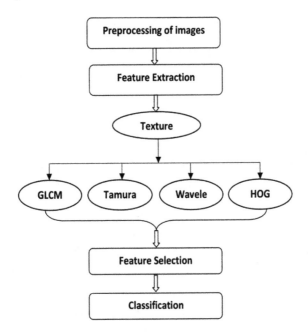

Fig. 1 Proposed approach

objective behind preprocessing of image is to remove the unnecessary distortion and enhance the quality of images.

In a preprocessing step of proposed approach, first convert the color image into grayscale image by using PCA-based color-to-grayscale conversion method because it preserves both color discriminability and the texture by using simple linear computations in subspaces with low computational complexity [6, 7] and then applying contrast limited adaptive histogram equalization (CLAHE) because CLAHE enhanced the quality of image and helpful for further image processing steps such as segmentation, feature extraction, restoration, registration etc. [6–8].

3.2 Feature Extraction and Selection

Feature extraction is a process of transforming the input data into set of features [9]. In our propose approach, texture features are extracted from the all healthy and defected apple images. Texture refers to the repetition of basic texture element known as "TEXEL" which contains several pixels that are placed randomly, periodic or quasi-periodic [9]. There are various feature descriptors for texture feature such as GLCM, Tamura, HOG, Wavelet transformation, Local-Binary pattern (LBP), Completed Local Binary Pattern (CLBP) and many more. In the proposed approach, we have taken the feature descriptor as GLCM, Wavelet transformation, Tamura, HOG and LTE for extracting the texture feature.

Gray Level Co-occurrence Matrix (GLCM): GLCM also known as co-occurrence distribution. This is second order statistical method for the analysis of texture feature of the image. The image is composed of pixels and they are associated with a specific Gray level, that is, intensity and the frequency of combination of different gray levels co-occurrence is tabulated. This tabulation is known as GLCM. GLCM tabulation is used for the calculation of texture feature. It gives the measure of fluctuation in intensity of pixels.

Discrete Wavelet Transform (DWT): DWT is a decomposition technique of any continuous signal with respect to time such as image in frequency domain. Several applications such as data compression, removing noise and features detection are based on this technique. In this technique, an image is first decomposed into series of wavelet coefficient corresponding to its four frequency sub-bands Low-Low, Low-High, High-Low, High-High. Most of the image energy/information is concentrated at its Low-Low frequency sub-band.

Histogram of Oriented Gradients (HOG): HOG is used to detect and extract the orientation features of an image. This technique is used to extract the distribution of gradient direction (orientated gradients). Gradients mean the X and Y derivatives of an image. The following flow is used to extract the HOG features.

Tamura's Features: These features are based on three fundamental texture features, contrast, directionality and coarseness. Variety of the texture pattern is the measurement of Contrast. The difference in luminance that's make an object is the Contrast. In an image direction of gray level values are measurement of Directionality and granularity of an image is the measurement of Coarseness.

Law's Texture Energy: Law's determined some masks which were used to discriminate the different texture. In this method, texture energy transformation is applied to the image to calculate the energy within the pass region of filters. The masks of technique are derived from 1D vectors of five pixel length. By convolving one vertical vector with one horizontal vector, a 2D filter was generated. All the 2D filters are convolved over the image to extract the features.

After extracting total 108 features, various machine learning-based classifiers such as support vector machine (SVM), k-nearest neighbors (k-NN), Logistic Regression, Linear Discriminant analysis are used for the classification of healthy and defected apple image. Short discussion about the various used machine learning-based classifiers is as follows:

Support Vector Machine: Support Vector Machine (SVM) is a supervised learning model. This algorithm is used to analyze the data for both classification and regression. It is a discriminative classifier defined by a separating hyperplane. The algorithm creates the optimal hyperplane as an output for the labeled training dataset. This hyperplane divides the dataset in two parts; these two parts are none other than the two different classes. The SVM does the separation of classes.

k- Nearest Neighbor: It is a supervised and non-parametric algorithm (does not perform any pre-assumption regarding the distribution) which is deployed for both classification and regression. The main principle behind the algorithm is to find an existing number of training dataset which is nearest in distance to the new point and then classifies these training datasets in various classes. The neighbors are determined by calculating the Euclidian and Hamming distance. The algorithm classifies the data input through the class membership of its k-neighbors and most of the neighbors represent the data classified as output.

Logistic Regression: It is used to analyze the relationship between a categorical dependent variable and categorical independent variables. It combines the independent variables to estimate the probability for the happening of event, that is, the object will belong to which class.

Linear Discriminant: Linear Discriminant Analysis (LDA) is a supervised learning model. LDA is used to reduce the dimensionality in the preprocessing for the classification. The main aim of implementing LDA is to avoid the problem of overfitting that is "curse of dimensionality". The output of LDA represents the feature space of dataset of dimensions n into the smaller subspace k, where $k <= (n - 1)$, by maintaining the difference in classes. It is a statistical tool that is required to figure out the dependent variable in various categories. It helps in analysis of accuracy of a classification.

4 Experimental Results and Discussion

For evaluating and validating our proposed approach for classification of healthy and rotten apple, we have collected the images of both rotten and healthy images. These images are collected from various sources [5, 10]. Some images are taken manually with the help of mobile phone camera while some are downloaded from Google Images [5, 10]. These are downloaded by entering the keywords "healthy apple dataset" for normal apple and "apple + disease name" for defected apple image on Google Images. We have taken 42 healthy apple images and 52 rotten apple images; total 94 images are present in dataset. Figures 2 and 3 show the sample images for the healthy and defected apple. After downloading the images, resize them because some of the downloaded image are in different sizes.

The proposed approach has been implemented on all 94 different apple images taken from the various sources [5, 10]. For experimental analysis, total 108 features are extracted from each 94 images. These 108 features are collection of 22 GLCM features, 15 LTE Features, 32 Wavelet features, 03 Tamura features, and 36 HOG Features. After extracting the features, apply the various machine learning-based approaches together with k-fold cross-validation technique and found that the per-

Fig. 2 Good/healthy apple images taken from [19, 20]

Fig. 3 Rotten apple images
taken from [19, 20]

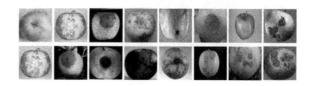

formance of the proposed approach is better with respect to other existing approaches in the literature. The performance of proposed approach is measured by evaluating an accuracy (ACC) using Eq. (1) as given below:

$$Accuracy = \frac{TP + TN}{TP + TN + FP + FN} \tag{1}$$

where TP, TN, FP, and FN represent true positive, true negative, false positive, and false negative values of confusion matrix, respectively.

The performance of classifiers is evaluated by applying two different scenarios. In first scenario, evaluate the performance of classifiers by selecting all extracted features and then apply k-fold validation by using thee different values of k such as fivefold, 10-fold, and 15-fold. The results are mentioned in Table 2. In second scenario, evaluate the performance of classifiers by selecting some suitable features by applying PCA-based feature selection technique and then apply k-fold validation by using thee different values of k such as fivefold, 10-fold, and 15-fold. The results are mentioned in Table 3.

Table 2 Performance of classifiers by selecting all extracted features and k-fold validation technique

	Accuracy		
	5-fold	10-fold	15-fold
SVM	98.9	97.9	98.9
K-NN	97.9	97.9	97.9
Logistic	78.7	81.6	86.2
Linear discriminant	78.7	92.6	92.6

Table 3 Performance of classifiers by applying PCA-based feature selection technique and k-fold validation technique

	Accuracy		
	5-fold	10-fold	15-fold
SVM	98.9	97.9	98.9
K-NN	97.9	97.9	97.9
Logistic	97.9	97.9	90.4
Linear discriminant	97.9	97.9	92.6

5 Conclusion

An apple is a rich source of important antioxidants, flavonoids, and dietary fiber and helps to reduce the risk of developing cancer, hypertension, diabetes, and heart disease, whereas the rotten fruits affect the human health as well as cause big economical loss in agriculture sectors. Therefore, identification of rotten fruits had become an important issue for research. This paper focuses on the classification of rotten and healthy apple by extracting the features and then classifying the rotten and healthy apples by applying various classifiers. Based on experimental observation, it was found that the performance of proposed approach by using SVM classifier is 98.9%, which was much better with respect to the other classifiers.

References

1. Roberts, M. J., Schimmelpfennig, D. E., Ashley, E., Livingston, M. J., Ash, M. S., Vasavada, U., et al. (2006). The value of plant disease early warning systems: a case study of usda's soybean rust coordinated framework. Technical report, United States Department of Agriculture, Economic Research Service.
2. Dubey, S. R., & Jalal, A. S. (2016). Apple disease classification using color, texture and shape features from images. *Signal, Image and Video Processing*, *10*(5), 819–826.
3. Dubey, S. R., & Jalal, A. S. (2014). Adapted approach for fruit disease identification using images. arXiv:1405.4930.
4. Sindhi, K., Pandya, J., & Vegad, S. (2016). Quality evaluation of apple fruit: A survey. *International Journal of Computer Applications* (0975–8887), *136*(1).
5. Healthy Apples Image. Retrieved January 15, 2018, from https://www.google.co.in/search?tbm=isch&q=apple+images&chips=q:apple+images,g_1:red,g_10:real&sa=X&ved=0ahUKEwiy3KXN55zXAhWMpY8KHTIzBbcQ4lYIOSgA&biw=1366&bih=588&dpr=1.
6. Singh, N. P., Srivastava, R. (2016). Segmentation of retinal blood vessels by using a matched filter based on second derivative of gaussian. *International Journal of Biomedical Engineering and Technology*, *21*(3), 229–246.
7. Seo, J. W., & Kim, S. D. (2013). Novel pca-based color-to-gray image conversion. In 2013 20th IEEE International Conference on Image Processing (ICIP), pp. 2279–2283. IEEE.
8. Singh Rajeev, N. P. (2018). Extraction of retinal blood vessels by using an extended matched filter based on second derivative of gaussian. In *Proceedings of the National Academy of Sciences, India Section A: Physical Sciences* 2016.
9. Jain, A. K. (1989). Fundamentals of digital image processing. In *Prentice-Hall information and system sciences series*. Prentice-Hall.
10. Ivars, D. J. B., & Garca, D. S. C. (2018). Image database: Apple golden'. Retrieved January 15, 2018, from http://www.cofilab.com/portfolio/goldendb/.
11. Leemans, V., Destain, M.-F. (2004). A real-time grading method of apples based on features extracted from defects. *Journal of Food Engineering*, *61*(1), 83–89.
12. Unay, D., & Gosselin, B. (2005). Artificial neural network-based segmentation and apple grading by machine vision. In 2005. IEEE International Conference on Image Processing, ICIP, (Vol. 2, p. II–630). IEEE.
13. Zhu, B., Jiang, L., Luo, Y., & Tao, Y. (2007). Gabor feature-based apple quality inspection using kernel principal component analysis. *Journal of Food Engineering*, *81*(4), 741–749.
14. Wang, J.-J., Zhao, D., Ji, W., Tu, J., & Zhang, Y. (2009). Application of support vector machine to apple recognition using in apple harvesting robot. In *2009 ICIA '09 International Conference on Information and Automation* (pp. 1110–1115). IEEE.

15. Unay, D., Gosselin, B., Kleynen, O., Leemans, V., Destain, M.-F., & Debeir, O. (2011). Automatic grading of bi-colored apples by multi-spectral machine vision. *Computers and Electronics in Agriculture, 75*(1), 204–212.
16. Arlimatti, S. R. (2012). Window based method for automatic classi_cation of apple fruit. *International Journal of Engineering Research and Applications, 2*(4), 1010–1013.
17. Dubey, S. R., & Jalal, A. S. (2012). Detection and classification of apple fruit diseases using complete local binary patterns. In *2012 Third International Conference on Computer and Communication Technology (ICCCT)*, (pp. 346–351). IEEE.
18. Jhuria, M., Kumar, A., & Borse, R. (2013). Image processing for smart farming: Detection of disease and fruit grading. In *2013 IEEE Second International Conference onImage Information Processing(ICIIP)*, (pp. 521–526). IEEE.
19. Ashok, V., & Vinod, D. S. (2014). Automatic quality evaluation of fruits using probabilistic neural network approach. In *2014 International Conference on Contemporary Computing and Informatics (IC3I)*, (pp. 308–311). IEEE.
20. Gonzalez, R. C., & Woods, R. E. (2002). *Digital image processing* (2nd ed.). Prentice Hall.

Classification of Normal and Abnormal Retinal Images by Using Feature-Based Machine Learning Approach

Pratima Yadav and Nagendra Pratap Singh

Abstract The human eye is one of the most beautiful and important sense organs of human body as it allows visual perception by reacting to light and pressure. Human eyes are capable of differentiating approximately 10 million colors. It contains more than 2 million tissues and cells. Along with these entire specialties, human eyes are the most delicate and sensitive organ. If not taken proper care, it may be infected with various diseases like glaucoma, myopia, hyper-myopia, diabetic retinopathy, age-related macular disease. Therefore, early-stage detection of these diseases could help in curing them completely and prevent from complete blindness. In this paper, we propose an approach to classify the normal (healthy) and abnormal (disease-infected) retinal images by using retinal image feature-based machine learning classification approach. The performance of proposed approach by using SVM classifier is 77.3%, which is found better with respect to the other classifiers like k-NN, linear discriminant, quadratic discriminant and decision tree classifiers.

Keywords Retina images · Texture features · Machine learning and classification

1 Introduction

The human eye is one of the most significant organs in the human body. According to the literature, there are 3.7 crores blind people and 12.4 crores with low-vision people present worldwide [1]. The low vision and blindness are heavily dependent on retinal images. The major reasons of worldwide blindness are glaucoma, corneal scarring, age-related macular degeneration and diabetic retinopathy. Diagnosis of these diseases is possible by examining the retinal blood vessel structure [2]. The ophthalmologist examines the patient's retina with the help of high-resolution fundus

P. Yadav (✉) · N. P. Singh
Department of CSE, MMM University of Technology, Gorakhpur 273010, UP, India
e-mail: mail2pratima27@gmail.com

N. P. Singh
e-mail: npscs@mmmut.ac.in

© Springer Nature Singapore Pte Ltd. 2019
A. Khare et al. (eds.), *Recent Trends in Communication, Computing, and Electronics*, Lecture Notes in Electrical Engineering 524,
https://doi.org/10.1007/978-981-13-2685-1_37

387

camera and then inspects the retinal blood vessels disorder to diagnose the retinal vascular diseases. The average cost corresponding with eye-care providers and the growing figures of retinal patients are the key motivations for the acceptance of computer-aided diagnosis of retinal vascular diseases [3]. Therefore, feature extraction of retinal image for computer-aided diagnosis of retinal vascular diseases is a prominent task.

2 Literature Survey

Various authors had given various methodologies to automate the diagnosis of various retinal diseases. The authors Bock et al. [4], Nayak et al. [5], Zhang et al. [6] and Mookiah et al. [7] focused on the retinal disease glaucoma. The authors Acharya et al. [8], Osareh et al. [9], Xiaohui and Chutatape [10], Ravishankar et al. [11], Garcia et al. [12], Sopharak et al. [13], Jayanthi et al. [14], Verma et al. [15], Narasimhan et al. [16], Harangi et al. [17], Mookiah et al. [18], Gandhi et al. [19], Chowdhury et al. [20], Akram et al. [21], Welikala et al. [22] focused on diabetic retinopathy. Some authors focused on other methodologies which are used for retinal diseases prediction, such as Ricci et al. [23], Singh and Srivastava [24–26], and Hoover et al. [27], who worked on retinal vessel segmentation and registration, Soares et al. [28] who worked on automated segmentation of the vasculature, Li et al. [29] who focused on automatic and robust abstraction of features in retinal images and Goldbaum et al. [30] who propagated research paper on automated diagnosis of various retinal diseases.

The rest of the paper is organized as follows: Sect. 2 explains the proposed method in detail. Section 3 contains the experimental results, and discussion is given in Sect. 3. Finally, the conclusion of the proposed work is given in Sect. 4.

3 The Proposed Method

The proposed approach contains three main steps, namely the preprocessing, feature extraction, and classification step, as mentioned in Fig. 1. In a preprocessing step, convert the color retina image into grayscale image by using PCA-based color to grayscale conversion method because it preserves both color discriminability and the texture by using simple linear computations in subspaces with low computational complexity [31], and then, the quality of retinal image is enhanced by applying contrast-limited adaptive histogram equalization (CLAHE). The preprocessing step is beneficial to perform further image processing steps such as segmentation, feature extraction, restoration, registration. In the next step, extract the color as well as grayscale features of the retinal image. Feature extraction is a process of deriving new features for describing and further processing of an image, whereas the features are significant properties of an image which are helpful in feature extraction such as

Fig. 1 Proposed approach

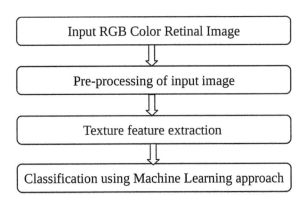

color, shape, edge, texture. This paper focused on a texture feature of an image such as gray-level co-occurrence matrix (GLCM) features, Tamura's features, wavelet features, Laws' texture energy (LTE)-based features, and HOG features, which are described as follows.

Gray-Level Co-occurrence Matrix (GLCM): It is one of the most considerable characteristics of an image texture. This could be obtained by dimensionality of gray levels in its neighborhood. The texture feature contains the properties of each pixel values as well as its neighborhood. The gray-level co-occurrence matrix seems to be a well-known statistical technique for feature extraction. In GLCM, matrix indicates how the pair of pixels (i, j) values varies with respect to spatial relationship of an image. The spatial relationship is defined by angle (·) and the displacement (D), between i and j. For an image, the width and height of the GLCM matrix are equal to the total number of gray levels of image intensity. After creating a gray-level co-occurrence matrix, a total of 22 features are extracted from it, such as energy, entropy, dissimilarity, contrast, inverse difference, correlation, homogeneity, autocorrelation, maximum probability.

Wavelet Feature: With a successive step of averaging and differencing computations, a wavelet can decompose a signal or an image. Decomposition and compression are the areas where wavelets are generally applied. This is the reason that decomposed images are reconstructed by reversing decomposition methods. Wavelets compute average depth features as well as numerous distinctive contrast levels dispensed throughout the photograph. Depending upon the number of levels of average calculated, wavelets can be calculated on various levels of resolution accordingly. They are sensitive to the spatial distribution of gray stage pixels, but can also be able to distinguish and keep information at numerous scales or resolutions. This multi-resolution property allows for the analysis of gray-level pixels irrespective of the dimensions of the community. These qualities lead to the concept that wavelets may want to guide researchers to better texture classification of human organs.

Tamura Feature: Tamura's features are processed on the premise of three key surface features: contrast, coarseness and directionality. Contrast is a measure of intensity of a pixel and its neighbor over the image. It is determined by variation in the color and intensity of the object and background of the object in same FOV. Coarseness is the measure of granularity of an image; accordingly, coarseness can be spoken to utilizing normal size of areas that have a similar power. Directionality is the measure of bearings of the gray values inside the picture.

Laws' Texture Energy (LTE): Laws' texture energy-based features are surface depiction features, developed by Laws [32], where a set of two-dimensional masks derived from five simple one-dimensional filters such as average gray level, edges, spots, waves and ripples. They are L5 = [1 4 6 4 1], E5 = [–1 –2 0 2 1], S5 = [–1 0 2 0 –1], W5 = [–1 2 0 –2 1] and R5 = [1 –4 6 –4 1], which denote average gray level and used for level detection, first difference and used for edge detection, second difference and used for spot detection, wave and ripple detection, respectively. Mutually multiplying these vectors, considering the first term as a column vector and the second term as row vector, then found a 5 × 5 matrix, which is known as Laws' masks. By convoluting the Laws' mask with texture image and calculating energy statistics, a feature vector is derived that can be used for texture description.

Histogram of Oriented Gradients (HOG): The HOG feature is used to characterize the local object and shape of an image. Generally, the HOG features are calculated by dividing the whole image window into small cells. Each cell gathers a local histogram of gradient directions over all the pixels. After that concatenate the cell histogram entries and generate a block which contains the block features. Finally, contrast normalization is applied to achieve better invariance to illumination and shadow, etc. The normalized block features are concatenated into a single HOG feature vector, which will refer for classification of the disease prediction.

After extracting all 108 features, various machine learning-based classifiers such as support vector machine (SVM), k-nearest neighbors (k-NN), decision tree, linear and quadratic discriminant analysis are used for the classification of normal and abnormal retinal images. There is a short discussion about the various used machine learning-based classifiers, which are as follows:

Decision Tree: It is learning of decision trees from class-labeled training tuples. Decision tree is the tree structure, where each internal node represents a test attribute, branch denotes class label. The topmost is called root node.

K-nearest Neighbors (k-NN): k-NN algorithm is easiest and nonparametric learning algorithm. It performs classification based on similarity measures. A case is classified with the help of its neighbor's vote; its k-nearest neighbors is identified by calculating distance such as Euclidean, Manhattan or Minkowski distance function. These distances can be calculated by using the following equations:

$$Euclidean = \sum_{i=1}^{k} (x_i - y_i)^2 \tag{1}$$

$$Manhattan = \sum_{i=1}^{k} |x_i - y_i| \tag{2}$$

$$Minkowski = \left(\sum_{i=1}^{k} \left(|x_i - y_i|^q \right) \right)^{1/q} \qquad (3)$$

Support vector Machine (SVM): SVM follows the concepts of decision planes. Decision planes define decision boundaries. A decision plane is one which groups set of objects with same memberships. SVM performs classification that divides the cases of class labels. Classification is done by building hyper-planes in multi-dimensional space.

Linear Discriminant: One of the popular strategies for the computation of features and reduction in dimension is LDA. It has wide applications in areas such as pattern recognition, face recognition, biometric and image retrieval. LDA technique bunches images of similar classes and isolates images of the distinctive classes. To recognize an input test image, the predicted test picture is contrasted with each predicted training image and test picture is distinguished as the nearest training image.

Quadratic Discriminant: It is quite similar to LDA. There is no premise that the covariance of individual classes is exactly the same. More computations as well as information is required to evaluate parameters required for QDA. Generalization of Bayesian discrimination is QDA.

4   Experimental Results and Discussion

Various retinal image datasets are available online freely for experimental analysis such as DRIVE, STARE, MESSIDOR, HEI-MED, DIARETDB0, ORIGA-light. Based on literature survey, it is observed that an authors use these datasets according to their experimental need. In the proposed approach, two prevailing datasets HRF, and STARE are used. The STARE (STructured Analysis of the Retina) dataset contains approximately 400 retinal images captured by a Topcon TRV-50 fundus camera at 35° FOV, and each image contains 8 bits per color channel having 700 × 605 pixels, out of which 5% are normal and the rest are abnormal retinal images [25]. The high-resolution fundus (HRF) dataset is online freely available database which consists of 45 images out of which 15 images are of healthy patients, rest of the images are of diabetic retinopathy patients and glaucomatous patients. A panel of experts working in the area of retinal image analysis and ophthalmology generated gold-standard data. In order to help in analysis of algorithms which helps in spotting the macula and optic disk or differentiate between arteries and veins, gold-standard data can be further added to the existing images.

The proposed approach has been implemented on 110 different retinal images taken from STARE and HRF datasets. Out of 110 images, 55 is normal (healthy) retinal image and 55 are abnormal (diseases-infected) retinal image. For experimental analysis, total of 108 features are extracted from each image, out of 110 images. These 108 features are collection of 22 GLCM features from F-1 to F-22, 15 LTE features from F-23 to F-37, 32 wavelet features from F-38 to F-69, 03 Tamura features

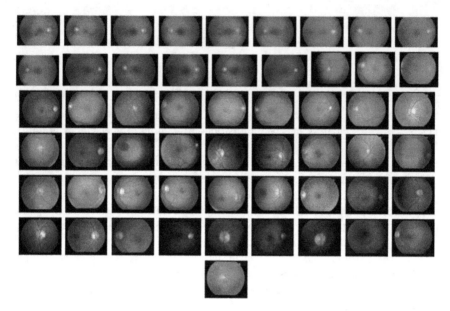

Fig. 2 Normal (healthy) retinal images taken from STARE and HRF datasets for feature extraction and classification

from F-70 to F-72 and 36 HOG features numbered from F-73 to F-108. Similarly, evaluation of all 108 features for each of the normal and abnormal images is shown in Fig. 2 and Fig. 3, respectively. MATLAB R2017a is used for feature extraction and classification. After extracting the features, apply the various machine learning-based approaches together with K-fold and hold-out cross-validation technique. The performance of the proposed approach is measured by evaluating accuracy (ACC) by using Eq^n–4 as given below.

$$Accuracy = \frac{TP + TN}{TP + TN + FP + FN} \tag{4}$$

where TP, TN, FP and FN represent true-positive, true-negative, false-positive and false-negative values of confusion matrix, respectively.

The performance of classifiers is evaluated by applying two different scenarios. In the first scenario, evaluate the performance by applying PCA-based feature selection technique to select the suitable features and then apply k-fold such as 5-fold, 15-fold and 25-fold and hold-out having 10%, 20% and 30% validation techniques on selected features. The results are mentioned in Table 1. In the second scenario, evaluate the performance of classifiers by selecting all extracted features and then apply k-fold by using three different values of k such as 5-fold, 15-fold and 25-fold and hold-out having 10%, 20% and 30% validation techniques. The results are mentioned in Table 2.

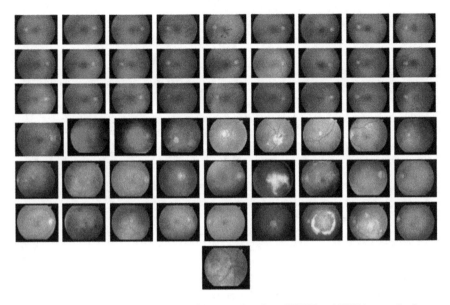

Fig. 3 Abnormal (disease-infected) retinal images taken from STARE and HRF datasets for feature extraction and classification

Table 1 Performance of classifiers by applying PCA-based feature selection technique and k-fold validation technique

	Accuracy					
	5-fold	15-fold	25-fold	10% Hold-out	20% Hold-out	30% Hold-out
Linear discriminant	53.6	53.6	53.6	45.5	50.0	42.4
Quadratic discriminant	53.6	53.6	53.6	45.5	50.0	42.4
SVM	47.3	40.9	50.9	54.5	45.5	51.5
k-NN	50.9	48.1	47.3	54.5	45.5	42.4
Decision tree	51.8	49.1	48.2	36.4	27.3	39.4

After analyzing the results mentioned in Tables 1 and 2, it is observed that the overall performance of SVM classifier is better with respect to the other classifiers as mentioned above. The performance of the proposed approach is 77.3% when we use the SVM classifier with 20% hold-out cross-validation.

Table 2 Performance of classifiers by selecting all extracted features and hold-out validation technique

	Accuracy					
	5-fold	15-fold	25-fold	10% Hold-out	20% Hold-out	30% Hold-out
Linear discriminant	46.2	50.0	53.6	54.5	50.0	45.5
Quadratic discriminant	57.3	58.2	60.9	72.7	63.6	48.5
SVM	69.1	68.2	67.3	72.7	77.3	51.5
k-NN	70.0	60.0	64.5	63.6	68.2	54.5
Decision tree	68.2	61.8	63.6	63.6	72.7	66.7

5 Conclusion

The eye is an important sensory organ of the human visual system. It provides the ability to view the world around us. Therefore, protecting the eye from a serious disease related to vision loss or blindness is a prominent task. The early-stage detection of these diseases could help in curing them completely and prevent from complete blindness. Due to these reasons, this paper focused on developing a method to classify the normal and abnormal retinal images by using retinal image feature-based machine learning classification approach. Based on the experimental result, the performance of the proposed approach was found 77.3%. In the future, the performance of classifier may be improved by applying machine learning approach by using some other features like local binary pattern (LBP), local ternary patterns (LTP) and local tetra pattern (LTrP).

References

1. World Health Organization et al. (2016). Global initiative for the elimination of avoidable blindness: action plan 2006–2011. 2007. F.: *Journal 2*(5), 99–110.
2. Foster, A., & Resnikoff, S. (2005). The impact of vision 2020 on global blindness. *Eye, 19*(10), 1133–1135.
3. Abràmoff, M. D., Garvin, M. K., & Sonka, M. (2010). Retinal imaging and image analysis. *IEEE Reviews in Biomedical Engineering, 3*, 169–208.
4. Bock, R., Meier, J., Michelson, G., Nyul, L. G., & Hornegger, J. (2007). Classifying glaucoma with image-based features from fundus photographs. In *Joint Pattern Recognition Symposium* (pp. 355–364). Springer.
5. Nayak, J., Acharya, R., Bhat, P. S., Shetty, N., & Lim, T. C. (2009). Automated diagnosis of glaucoma using digital fundus images. *Journal of Medical Systems, 33*(5), 337.
6. Zhang, Z., Yin, F. S., Liu, J., Wong, W. K., Tan, N. M., Lee, B. H., et al. (2010). Origa-light: an online retinal fundus image database for glaucoma analysis and research. In *2010 Annual International Conference of the IEEE on Engineering in Medicine and Biology Society (EMBC)* (pp. 3065–3068). IEEE.

7. Mookiah, M. R. K., Acharya, U. R., Lim, C. M., Petznick, A., & Suri, J. S. (2012). Data mining technique for automated diagnosis of glaucoma using higher order spectra and wavelet energy features. *Knowledge-Based Systems, 33,* 73–82.
8. Acharya, R., Chua, C. K., Ng, E. Y. K., Yu, W., & Chee, C. (2008). Application of higher order spectra for the identification of diabetes retinopathy stages. *Journal of Medical Systems, 32*(6), 481–488.
9. Osareh, A., Shadgar, B., & Markham, R. (2009). A computational intelligence-based approach for detection of exudates in diabetic retinopathy images. *IEEE Transactions on Information Technology in Biomedicine, 13*(4), 535–545.
10. Xiaohui, Z., & Chutatape, A. (2004). Detection and classification of bright lesions in color fundus images. In *2004 International Conference on Image Processing, 2004. ICIP'04* (Vol. 1, pp. 139–142). IEEE.
11. Ravishankar, S., Jain, A., & Mittal, A. (2009). Automated feature extraction for early detection of diabetic retinopathy in fundus images. In *2009. CVPR 2009. IEEE Conference on Computer Vision and Pattern Recognition* (pp. 210–217). IEEE.
12. Garcia, M., Sanchez, C. I., Lopez, M. I., Abasolo, D., & Hornero, R. (2009). Neural network based detection of hard exudates in retinal images. *Computer Methods and Programs in Biomedicine, 93*(1), 9–19.
13. Sopharak, A., Dailey, M. N., Uyyanonvara, B., Barman, S., Williamson, T., Nwe, K. T., et al. (2010). Machine learning approach to automatic exudate detection in retinal images from diabetic patients. *Journal of Modern Optics, 57*(2), 124–135.
14. Jayanthi, D., Devi, N., & SwarnaParvathi, S. (2010). Automatic diagnosis of retinal diseases from color retinal images. arXiv:1002.2408.
15. Verma, K., Deep, P., & Ramakrishnan, A. G. (2011). Detection and classification of diabetic retinopathy using retinal images. In *2011 Annual IEEE on India Conference (INDICON)* (pp. 1–6). IEEE.
16. Narasimhan, K., Neha, V. C., & Vijayarekha, K. (2012). An efficient automated system for detection of diabetic retinopathy from fundus images using support vector machine and Bayesian classifiers. In *2012 International Conference on Computing, Electronics and Electrical Technologies (ICCEET)* (pp. 964–969). IEEE.
17. Harangi, B., Antal, B., & Hajdu, A. (2012). Automatic exudate detection with improved naive-Bayes classifier. In *2012 25th International Symposium on Computer-Based Medical Systems (CBMS)* (pp. 1–4). IEEE.
18. Mookiah, M. R. K., Acharya, U. R., Martis, R. J., Chua, C. K., Lim, C. M., Ng, E. Y. K., et al. (2013). Evolutionary algorithm based classifier parameter tuning for automatic diabetic retinopathy grading: a hybrid feature extraction approach. *Knowledge-Based Systems, 39,* 9–22.
19. Gandhi M., & Dhanasekaran, R. (2013). Diagnosis of diabetic retinopathy using morphological process and svm classifier. In *2013 International Conference on Communications and Signal Processing (ICCSP)* (pp. 873–877). IEEE.
20. Roychowdhury, S., Koozekanani, D. D., & Parhi, K. K. (2014). Dream: diabetic retinopathy analysis using machine learning. *IEEE Journal of Biomedical and Health Informatics, 18*(5), 1717–1728.
21. Akram, M. U., Tariq, A., Khan, S. A., & Javed, M. Y. (2014). Automated detection of exudates and macula for grading of diabetic macular edema. *Computer Methods and Programs in Biomedicine, 114*(2), 141–152.
22. Welikala, R. A., Dehmeshki, J., Hoppe, A., Tah, V., Mann, S., Williamson, T. H., et al. (2014). Automated detection of proliferative diabetic retinopathy using a modified line operator and dual classification. *Computer Methods and Programs in Biomedicine, 114*(3), 247–261.
23. Ricci, E., & Perfetti, R. (2007). Retinal blood vessel segmentation using line operators and support vector classification. *IEEE Transactions on Medical Imaging, 26*(10), 1357–1365.
24. Singh, N. P., & Srivastava, R. (2016). Retinal blood vessels segmentation by using gumbel probability distribution function based matched filter. *Computer Methods and Programs in Biomedicine, 129,* 40–50.

25. Singh, N. P., & Srivastava, R. (2017). Weibull probability distribution function-based matched filter approach for retinal blood vessels segmentation. In *Proceedings of International Conference on Computational Intelligence 2015, Advances in Computational Intelligence* (pp. 427–437). Springer.
26. Singh, N. P., & Srivastava, R. (2016). Segmentation of retinal blood vessels by using a matched filter based on second derivative of Gaussian. *International Journal of Biomedical Engineering and Technology, 21*(3), 229–246.
27. Hoover, A. D., Kouznetsova, V., & Goldbaum, M. (2000). Locating blood vessels in retinal images by piecewise threshold probing of a matched filter response. *IEEE Transactions on Medical Imaging, 19*(3), 203–210.
28. Soares, J. V., Leandro, J. J., Cesar, R. M., Jelinek, H. F., & Cree, M. J. (2006). Retinal vessel segmentation using the 2-D gabor wavelet and supervised classification. *IEEE Transactions on Medical Imaging, 25*(9), 1214–1222.
29. Li, H., & Chutatape, O. (2004). Automated feature extraction in color retinal images by a model based approach. *IEEE Transactions on Biomedical Engineering, 51*(2), 246–254.
30. Goldbaum, M., Moezzi, S., Taylor, A., Chatterjee, S., Boyd, J., Hunter, E., et al. (1996). Automated diagnosis and image understanding with object extraction, object classification, and inferencing in retinal images. In *1996 Proceedings, International Conference on Image Processing* (Vol. 3, pp. 695–698). IEEE.
31. Seo, J. W., & Kim, S. D. (2013). Novel PCA-based color-to-gray image conversion. In *2013 20th IEEE International Conference on Image Processing (ICIP)* (pp. 2279–2283). IEEE.
32. Laws, K. I. (1980). Rapid texture identification. In *Image Processing for Missile Guidance, International Society for Optics and Photonics* (Vol. 238, pp. 376–382).

Part VII
Natural Language Processing and Information Retreival

A New Framework for Collecting Implicit User Feedback for Movie and Video Recommender System

Himanshu Sahu, Neha Sharma and Utkarsh Gupta

Abstract In today's digital world due to unlimited content, product and services available online, finding an item that satisfies user requirement and taste by simply web searching is near impossible. Recommender systems are information filtering tool which provides personalized results. Movie and video recommender system is also gaining popularity due to the growth in online streaming video content Web sites and its subscriber. Accuracy and efficiency are two major aspects of a recommendation engine because it is directly related to user experience. To achieve higher accuracy, user feedback is required which can be collected either explicitly or implicitly. Explicit feedback is not always available and not always unbiased, so implicit feedback seems to be a better option for user preference collection. In this paper, a new framework is proposed which collects the implicit user feedback (along with explicit) for a movie and video recommender system. Implicit feedbacks can be converted to explicit feedback using the proposed UARCA which can be used to improve the accuracy of recommendation engine.

Keywords Recommender systems · Feedback learning · Implicit feedback
Explicit feedback · Movie and video recommender system (MVRS)
User action to rating conversion algorithm (UARCA)

1 Introduction

Due to unprecedented growth in internet and communication technology (ICT), the data available online have increased many fold. In the ICT terms, this problem is also known as information overload [1]. It has become a tedious task for a naïve user to find relevant information from the web. For her, simply searching will not

H. Sahu (✉) · N. Sharma · U. Gupta
University of Petroleum and Energy Studies, Dehradun, India
e-mail: himanshu.1689@gmail.com; hsahu@ddn.upes.ac.in

N. Sharma
e-mail: nehasharma@ddn.upes.ac.in

© Springer Nature Singapore Pte Ltd. 2019
A. Khare et al. (eds.), *Recent Trends in Communication, Computing, and Electronics*, Lecture Notes in Electrical Engineering 524,
https://doi.org/10.1007/978-981-13-2685-1_38

399

provide the content, product or item that suits her priority and taste. To resolve the information overload problem, recommender system (RS) [2, 1] came into picture.

The amount of information available on the web increases every day, and this becomes an optimization problem for recommender systems [3]. RS is developed as information filtering system based on different classification techniques [4]. It provides relevant information to the users by using the user preferences [5]. RS has several applications for digital content, online services and e-com retail.

To obtain accurate estimation of users taste and interest is one of the challenges [6, 7]. To provide effective and smooth recommendations to users [8], filtering technique [1] and user feedback are the two main concerns. Various filtering techniques are available such as content-based and collaborative filtering. Feedback plays an important role in attaining high accuracy level for a recommendation engine.

Movie and video recommender system (MVRS) [9] has become important after the success of streaming/on-demand video platform such as YouTube, Amazon Prime Video and Netflix. These sites carrying plethora of content, so for a user, it is not possible to watch everything and decide relevant content. So MVRS helps the user to find the relevant choice. The accuracy of this result will depend on feedback provided by users who have already watched but mostly users skip to rate the content until they highly liked or disliked the content. So, in the present work a new implicit feedback recording platform is proposed that can record the user action on the movie player interface and convert them into user feedback. The remainder of this paper is organized as follows; in Sect. 2, background of RS and feedback technique is provided; in Sect. 3, literature review is provided where the previous works that are related to present work is described; Sect. 4 showcases the proposed framework along with UARCA to convert user actions into rating; Sect. 5 explains the conclusion and possible future work.

2 Background

RS functions as a filter for extracting useful information out of a large chuck of dynamically generated information, based on user preference data or observed behavior for certain products or services [10]. So, RS provides filtered information, product or service as recommendations to the target user. RS is beneficial for the customer as well as the company or the service provider. They reduce cost involved (i.e., searching time) in finding suitable product for a user. Therefore, reliable, accurate and efficient RS are required to cater the need of required information, product, within time bound and suitable for the target users.

2.1 Filtering Techniques

There are different techniques [2, 10, 1] available for recommendation system such as content-based filtering, collaborative filtering and hybrid filtering. All these filtering techniques are described below.

1. *Content-based filtering.* Content-based filtering works on the approach of finding the similar item that a user has used in past. A user is recommended a similar item where the similarity among items is obtained by the properties of an item.
2. *Collaborative filtering.* This filtering technique works on the similarity of users. It finds the set of similar users to a target user and recommends some items to the target user which is being used by other user of the same set.
3. *Hybrid filtering.* It used the combination of both content-based filtering and collaborative filtering.

2.2 User Feedback

RS is developed as information filtering system based on user preferences. So, before filtering information, we must collect the information that will be used as a filtering criterion. This information is also called the user preference or user feedback. RS main task is mining user preference and recommends item to user. Various techniques of feedback collection [7] are described below:

- *Explicit Feedback.* Explicit feedback is a technique in which a user is directly asked (using system interface), i.e., surveyed about the product which she has already used or currently using. The information can be recorded as ratings such as like/dislike or rate on a scale of five/ten. So, availability of explicit feedback is user dependent and the accuracy of a RS depends on the quantity of rating provided by the user. Unavailability of explicit feedback affects the accuracy of the recommendation engine.
- *Implicit Feedback.* Implicit feedback is a technique in which the system monitors the user activity and interaction with the system. These feedbacks are recorded as positive/negative reactions. Implicit feedbacks need to be converted to the explicit feedback before feeding it to RS, known as explicitation. Implicit feedback removes burden from the user to provide feedback, and it also argued to be more objective since it is recorded in unbiased way.
- *Hybrid Feedback.* Hybrid feedback is the combination of the explicit as well as implicit feedback. It removes the limitation of both the system. It can be used as a linear/weighted combination or implicit feedback and can be discarded when explicit feedback is available.

One aspect of RS which is lacking is implementation of information feedback system and collection. The reason for this is dependency on explicit feedback because it relies on users and mostly users avoid rating the content. This hinders the possibility of generation of user profile based on her likes/dislikes. This leads to select user feedback in implicit way. Thus, the dependency on the user to collect feedback is removed and it can be when the user is interacting to the systems, the feedbacks are recorded. The mechanism of recording and converting this to explicit feedback can improve recommender systems functionalities [5].

3 Literature Review

Some research [11, 5, 6, 12] shows hindrance and noise involved in feedback collection. Some implementation [13, 3] is there for the implicit feedback collection for e-book recommendation which is also available.

The efficiency of a RS depends on user preferences. Liu et al. [11] described the difficulty in capturing user's preference due to its multidimensionality and dynamic nature. The multidimensionality refers to the different context in which rating is provided, whereas dynamic nature refers to the timing of the rating.

Núñez-Valdéz et al. [13] collected implicit feedback for different aspects separately in the scenario of e-book recommender system and compared these implicit feedbacks with the prerecorded explicit feedback.

According to Misal and Ganjewar [3] for making RS more efficient, its feedback mechanism needs to be improved. User avoids rating due to time consumption and lack of interest so and author described an implicit feedback collection mechanism for e-book recommender. They proposed UICA algorithm for implicit to explicit feedback conversion.

Further Núñez-Valdez et al. [12] explained the very nature of user which changes while rating and item which incurs due to different mindsets while reading a book and rating a book. Rating breaks the reading flow, and users are forced to rate so there is less chances to receive the accurate user taste. Lerato et al. [5] provided a detailed comparative study of different feedback techniques. They have provided merits and demerits of all feedback learning methods. Amatriain et al. [6] focused on the noise in user feedback due to the inconsistency while rating the content which can be treated as a noise. These inconsistencies will lead to poor accuracy of the RS. They have analyzed and characterized the noise in movies rating system and found the consistency when movies with similar ratings are grouped together.

Konstan and Riedl [7] explained the embedding of the algorithm in the user experience dramatically which affects the value to the user of the recommender. The measurement of user experience requires a framework to be developed which should be a combination of user interface and algorithms running on it. The user interface should capable of collecting user experience in a natural way (i.e., known as implicit feedback).

Based on the above literature review, it can be summarized that it is not always easy to receive explicit feedback from user. So, collecting implicit feedback may be helpful to make recommendation engine more accurate. When both explicit and implicit feedbacks are available, hybrid technique can be used. References [3, 11,

13] showed the way of implicit feedback collection in e-book RS based on which the proposed framework is designed for implicit feedback collection in the scenario of MVRS. In the next section, the design of the proposed framework is explained.

4   Proposed Work

As explained in the previous sections, current MVRS uses the explicit feedback for finding the users interest. But the user interest is very low [13] in most of the cases and they do not rate movies. One more drawback of explicit approach [12] is that actual user preference is not recorded since user has stop watching and rating the movie. This leads to cold-start, sparsity and scalability problems. To avoid this problem, a new framework is proposed capable of capturing implicit feedback in the MVRS which is described in this section.

4.1   Proposed Framework

In the proposed work, a framework will be developed that is basically a streaming web player which will provide recommended videos/movies to users. The framework will be capable of recording user interactions with interface. The UARCA will convert the user interaction to implicit feedback as weighted and positive/negative. The interface can also record explicit feedback. Table 1 describes different user actions and their indication as positive/negative. If both implicit and explicit feedbacks are available, a weighted mean will be calculated. In case of the absence of explicit feedback, the implicit feedback will be directly converted to explicit feedback. Figure 1 illustrates the architecture of the proposed framework.

Components of the proposed framework: These are the component of the proposed framework:

1. Streaming Media Player: This is video streaming media player that will provide the recommended list of videos to a user. This media player is also capable of recording user actions such as play, pause and seek.
2. Implicit Feedback Database: This database will store the recorded implicit feedback received by the user action using the media player.
3. Explicit Feedback Database: This database will store the explicit feedback provide by the user using the same web media player interface.
4. Implicit to Explicit Feedback Conversion: The conversion algorithm converts implicit feedback to explicit feedback.
5. Recommendation Engine: It may use any filtering, most favorable the collaborative filtering. It will provide movie recommendations to the user.

Table 1 User interaction with the interface

S. no.	Name	Type	Indicator	Feedback value
1	Explicit rating of a content	Explicit	–	Direct value
2	Recommend to another user	Explicit	Positive	Weighted direct value
3	Add to favorites list	Explicit	Positive	Weighted direct value
4	Remove from favorites list	Explicit	Negative	
5	Bookmark	Explicit	Positive	Weighted value
6	Time spent watching a content	Implicit	Positive	Calculated value
7	Backward control	Implicit	Positive	Calculated value
8	Forward control	Implicit	Negative	Calculated value
9	Forward seek	Implicit	Negative	Calculated value
10	Backward seek	Implicit	Positive	Calculated value
11	No. of views	Implicit	Positive	Calculated value

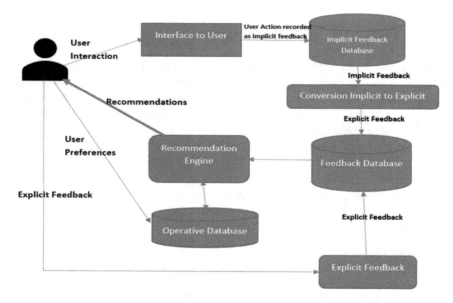

Fig. 1 Proposed methodology: the given figure explains the architecture of the proposed framework. It describes how user is interacting with the interface, i.e., the web media player and each interaction are converted and stored in database. The implicit feedback is converted to the explicit feedback and provided the recommendation engine along with the explicit feedback

4.2 Implicit to Explicit Feedback Conversion

Implicit to explicit feedback conversion requires logging each user's behavior, interactions and watching patterns from the video player and assigning metric for each type of interactions. Tables 2 and 3 show various user actions type whether that action is treated as an explicit or implicit feedback. The indicator field explains whether an action is a positive response indicator. The feedback value column indicates how that explicit or implicit user action can be converted to a direct rating value.

For the conversion of implicit and explicit actions, we will be using user action to rating conversion algorithm (UARCA). User interaction is divided into two categories

1. Explicit action: action that is a direct representation of like/dislike for a content (as given in Table 2).
2. Implicit action: action that is a direct representation of like/dislike for a content (as given in Table 3) and their assigned weight is either directly given or the formula for the calculation is given in the respective tables.

Table 2 Explicit actions and their weights

S. no.	Name	Indicator	Weight
A_1	Explicit rating of a content	Positive	Direct value
A_2	Recommend to another user	Positive	0.1* (#Recommendation of that content)/(total recommendation by that user)
A_3	Add to favorites list	Positive	0.5
A_4	Watch list	Positive	0.1
A_5	Remove from favorites list	Negative	−0.5

Table 3 Implicit actions and their weights

S. no.	Name	Indicator	Weight
A_6	Time spent watching a content	Positive	(Time spend/total content duration)
A_7	Backward control	Positive	MIN (0.1* (#no. of backward control), 0.5)
A_8	Forward control	Negative	MIN (0.1* (#no. of backward control), 0.5)
A_9	Forward seek	Negative	0.5* (seek duration/total content duration)
A_{10}	Backward seek	Positive	0.5* (seek duration/total content duration)
A_{11}	View	Positive	0.5 (view count/total views)

$$FR_{i,j} = E_R + 0.5 * C_{R,j} \quad A1\, if\, is\, available$$
$$= I_R \quad if\, explicit\, rating\, is\, unavailable \qquad (1)$$

where $FR_{i,j}$ is the final estimated rating of movie i by user j, E_R is the explicit rating (using Eq. 2), I_R is the implicit rating (using Eq. 3) and $C_{R,j}$ is correction factor for user j.

$$E_R = A_1 + \sum_{k=2}^{4} WA_k * IA_k \quad if\, E_R < 2.5$$
$$= A_1 - A_5 \quad if\, E_R > 3.0 \qquad (2)$$

where WA_k is the assigned weight and IA_k is assigned indication for an action.

$$I_R = \sum_{k=6}^{12} WA_k * IA_k \qquad (3)$$

$$C_{R,j} = \frac{\sum_{k=0}^{N} E_R - 2 * I_R}{N} \qquad (4)$$

where $C_{R,j}$ is the correction factor for a user calculated as mean of difference of $E_R - 2 * I_R$.

4.3 UARCA

Data: Implicit and Explicit Action Data
      Result: Final Rating$FR_{i,j}$
If (WA_1)
$$FR_{i,j} = E_R + 0.5 * C_{R,j}$$
      Where $FR_{i,j}$ is the final rating of item i user j
              Else
              $$FR_{i,j} = I_R$$
$E_R, FR_{i,j}, C_{R,j}$ are calculated using equation (2), (3)and (4) respectively.
Algorithm 1: UARCA

Algorithm 1 is the user action to rating conversion algorithm that is used to find the final ratings. This algorithm considers both implicit and explicit actions. When the explicit rating is available, the correction factor is added to reduce error by adding mean absolute error between explicit and implicit ratings. Implicit rating is calculated by considering all user actions, and their calculated or direct weight is given in Table 3. In the next section, the conclusion and future scope of the current work is described.

5 Conclusion and Future Scope

In the present paper, a novel approach is proposed to collect implicit user feedback in MVRS and convert them to explicit feedback. The described framework is capable of recording user actions and translating them into user feedback. Since many studies [11, 12, 5] have shown the hindrance and user dependency in collecting explicit feedback from the user, so our present work will help in improving user feedback. The efficiency and accuracy of RS depend on the user preference data available. So, our present work will help in improving the efficiency and accuracy of the MVRS.

In future, we will develop similar implementation for music recommender system, e-book recommender system, e-learning platform, web interface enhancement and many more. We can also use ML techniques to enhance the feedback learning techniques and integrate to our framework. The implicit feedback to explicit feedback weightage needs to be adjusted on different context values so in future we will add context-based feedback learning method in our future work.

References

1. Ricci, F., Rokach, L., & Shapira, B. (2011). Introduction to recommender systems handbook. In *Recommender systems handbook* (pp. 1–35). Springer US.
2. Isinkaye, F. O., Folajimi, Y. O., & Ojokoh, B. A. (2015). Recommendation systems: Principles, methods and evaluation. *Egyptian Informatics Journal, 16*(3), 261–273.
3. Misal, M. V., & Ganjewar, P. D. (2016). Electronic books recommender system based on implicit feedback mechanism and hybrid methods. *International Journal of Advanced Research in Computer Science and Software Engineering, 6*(5).
4. Leskovec, J., Rajaraman, A., & Ullman, J. D. (2014). *Mining of massive datasets.* Cambridge University Press.
5. Lerato, M., Esan, O. A., Ebunoluwa, A.-D., Ngwira, S. M., & Zuva, T. A survey of recommender system feedback techniques, comparison and evaluation metrics, South Africa.
6. Amatriain, X., Pujol, J. M., & Oliver, N. (2009). I like it... i like it not: Evaluating user ratings noise in recommender systems. In *International Conference on User Modeling, Adaptation, and Personalization* (pp. 247–258). Springer, Berlin, Heidelberg.
7. Konstan, J. A., & Riedl, J. (2012). Recommender systems: From algorithms to user experience. *User Modeling and User-Adapted Interaction, 22*(1–2), 101–123.
8. Rana, C., & Jain, S. K. (2015). A study of the dynamic features of recommender systems. *Artificial Intelligence Review, 43*(1), 141–153.

9. Singh M., Sahu H., Sharma N. (2019). A personalized context-aware recommender system based on user-item preferences. In: V. Balas, N. Sharma, A. Chakrabarti (Eds.) Data management, analytics and innovation. *Advances in intelligent systems and computing*, vol. 839. Singapore: Springer.
10. Adomavicius, G., & Tuzhilin, A. (2005). Toward the next generation of recommender systems: A survey of the state-of-the-art and possible extensions. *IEEE Transactions on Knowledge and Data Engineering, 17*(6), 734–749.
11. Liu, X., Wang, G., Jiang, W., & Long, Y. (2016). DHMRF: A dynamic hybrid movie recommender framework. In *Asia-Pacific Services Computing Conference* (pp. 491–503). Springer, Cham.
12. Núñez-Valdez, E. R., Lovelle, J. M. C., Hernández, G. I., Fuente, A. J., Labra-Gayo, J. E. Creating recommendations on electronic books: A collaborative learning implicit approach.
13. Núñez-Valdéz, E. R., Lovelle, J. M. C., Martínez, O. S., García-Díaz, V., De Pablos, P. O., & Marín, C. E. M. (2012). Implicit feedback techniques on recommender systems applied to electronic books. *Computers in Human Behavior, 28*(4), 1186–1193.

EDM Framework for Knowledge Discovery in Educational Domain

Roopam Sadh and **Rajeev Kumar**

Abstract Large volume of data are generated in educational institutions, which are of heterogeneous and unstructured nature. However, there is a dearth of effective data mining tools and techniques which can handle these voluminous academic data and support exploration of essential knowledge. Educational data mining (EDM) is an emerging research area dedicated toward development of tools and techniques for exploring data in educational settings. In this paper, we propose a trusted EDM framework that can deliver multiple academic tasks according to the need of various stakeholders. In order to deliver such purposes, our framework utilizes data mining tools and techniques over unified data collected from institution's databases and various knowledge sources. As an example of the concept, we utilize data provided by National Institutional Ranking Framework (NIRF) for showing how same data can be mined to fulfill different needs of various stakeholders through our proposed framework.

Keywords EDM · Knowledge discovery · Institutional ranking
Academic quality · Academic stakeholders

1 Introduction

Since education is a complex system having strong nonlinear cause–effect relations, hence academic activities and processes in higher educational institutions are interdependent and influence each other [1]. Each of these activities generates large amount of data which are heterogeneous and unstructured [2]. If this whole academic data can be unified and mined properly, then the requirements of several internal and exter-

R. Sadh (✉) · R. Kumar (✉)
School of Computer and Systems Sciences, Jawaharlal Nehru University, New Delhi 110067, India
e-mail: roopam.sadh@gmail.com

R. Kumar
e-mail: rajeevkumar.cse@gmail.com

© Springer Nature Singapore Pte Ltd. 2019
A. Khare et al. (eds.), *Recent Trends in Communication, Computing, and Electronics*, Lecture Notes in Electrical Engineering 524,
https://doi.org/10.1007/978-981-13-2685-1_39

nal academic objectives can be attained easily and effectively, i.e., requirements of different stakeholders can be fulfilled at single place.

In the current academic scenario, rather than adopting unified real-time approach each individual activity is treated separately. This unnecessarily creates huge duplication of information which makes decision-making process complex and cumbersome [3]. Furthermore, due to the lack of dedicated data mining system, data are supplied and analyzed manually each time the need arises, which triggers further challenges of consistency, verification and completeness of data. These issues adversely affect the process of knowledge discovery and impede overall decision-making process [4].

In view of these aforementioned problems, we introduce an educational data mining (EDM) framework that attempts to handle all such issues and provide one unified solution. Emerging research field of EDM targets knowledge discovery, decision making and recommendation inside educational domain [5]. It is concerned with discovering unique data patterns and inferences regarding domain knowledge content, assessments, educational functionalities, etc. [6]. Proposed EDM framework in this paper collects data directly from online repositories of institutions and other publically accessible knowledge sources. First, it unifies data coming from these sources through rigorous cleaning and transformation processes and then applies appropriate data mining tools over preprocessed data to generate new insights according to the needs of multiple stakeholders. Our proposed framework encompasses statistics, data mining and other analytical tools for exploring essential knowledge.

As an example of the concept, we utilize the data provided by NIRF for its institutional ranking of 2017. We depict in our example that institutional ranking data, if collected and mined according to the specification of our proposed framework, can serve the objectives of ranking more effectively. Moreover, our framework enables same data to be used in satisfying the requirements of multiple stakeholders.

The next section describes our proposed framework. Utility of our framework is discussed in Sect. 3 (Experimental Analysis). A short discussion over the findings of analysis is given in Sect. 4. Conclusion of the article is given thereafter.

2   Proposed Framework

Proposed EDM framework in this paper consists mainly of four components (i) data stores, (ii) knowledge sources, (iii) data mining subsystem and (iv) user interface. The architecture of the proposed framework is depicted in Fig. 1.

2.1   Data Stores

Data store component consists of online data repositories established by each participating institution. Required data are fetched from repositories to data mining subsystem on demand. Institutions are required to manage their own active data

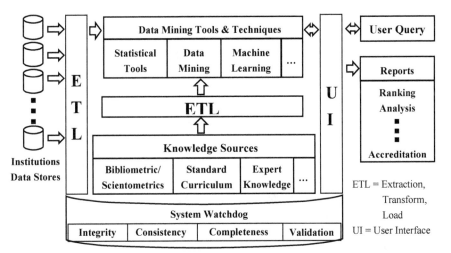

Fig. 1 Proposed EDM framework

management systems to handle data generated from each of the individual academic activities, i.e., repositories are updated each time an academic activity takes place. Structure and format of generated data are previously defined.

2.2 Knowledge Sources

Knowledge sources consist of various generic information sources and previously defined knowledge bases or expert knowledge. Various quantitative parameters required for academic decision making are calculated with the help of these sources. Bibliometric of publications, Scientometrics of Journals and standard Curriculums are some of the required information in this respect. Some well-known publicly available sources of information are Google Scholar, DBLP, SCOPUS, Thomson Reuters JCR.

2.3 Data Mining Subsystem

Data mining subsystem is the core of framework. Subsystem works in different stages such as data preprocessing and data analysis. These stages are defined as follows:

Preprocessing. Preprocessing of data in EDM system is important in two respects. First, it is required for data unification and maintaining system properties such as integrity, consistency and completeness [7]. Second, since traditional data mining approaches cannot be applied directly, each academic task has specific objective and function [8]. Hence, preprocessing is responsible for data transformation according to the task in hand. Preprocessing is carried out in the following rigorous steps.

Extract and Clean. Data of different standard formats such as relational tables and flat files are extracted to a defined space inside the subsystem for necessary integration. After extraction process is over, noise and other irrelevant attributes (not required in relation to user query) are removed.

Transform. Data can be of different structures and formats than that needed to specific data mining tool; hence, conversion of formats and structures is necessary into required specification. Therefore, in this step conversion process is carried out, e.g., formats of bibliometric information can be of different kinds which must be converted to specified format.

Load. After the completion of above required steps, information stored in system-defined intermediate space called local copy is fed to the subsequent data mining system for analysis purpose.

Data Analysis. Provided data from preprocessing stage are thoroughly analyzed for knowledge discovery by various analytical tools such as statistics, data mining algorithms, machine learning and others. Some basic fields of academic requirements are student modeling, student's behavior and performance modeling, various assessments, deployment of curriculum, domain knowledge, support and feedback [9]. Primary task in this phase is to select appropriate analytical tools and techniques according to the objective of analysis. For example, statistical, fuzzy MADM (multi-attribute decision making) and graph theoretic approaches can be used for institutional ranking purpose [10], regression modeling, rough set theory, clustering can be used for student modeling [11], and for prediction and performance modeling artificial intelligence, neural networks and classification can be used [12].

2.4   User Interface

User interface component is responsible for handling end-user query and delivery of the results in reports to end-users. In addition, user interface is equipped with access control mechanism which allows users to access the information according to their role such as administrator, student, faculty and public.

3   Experimental Analysis

Before presenting one of the prospective applications of our framework, we first make some distinctions between the requirements of different academic stakeholders regarding the usage of institutional data. For the sake of simplicity, we take three main categories of academic stakeholders such as administrators, student and faculty [13].

Administrators are responsible for making appropriate policies according to the need of institution and for managing the finances of the institution. In simplest possible case, they may be interested in knowing the fund utilization of the institution.

Students generally use institutional data as a guide for their future course of study and for knowing the possibility of their employment. Students may be interested in the information of particular discipline of the institution for admission purpose.

Institutional data facilitate faculty members to know the strengths and weaknesses of institutional academic performance and research focus regarding the field of study. Faculty may be interested in discipline-wise impact of institutional research.

We will consider three simple queries for the purpose of clarifying our concept and for denoting the requirements of different stakeholders. These queries are classification of institutions based on institutional data, finding the best performing discipline in selected institutes and finding the average impact of institutional research field-wise.

3.1 Data Source

We have taken data provided by National Institutional Ranking Framework (NIRF) of India for showing the utility of our framework. NIRF ranking 2017 has used institutional data of the period 2015–2016 for the assessment of various Indian institutions. We also test in this way, if these data are sufficient to be utilized for satisfying the demands of various stakeholders other than institutional ranking.

NIRF ranking is conducted on the basis of different categories of institutions [14]. These categories are "university" and "college" domain-wise and "engineering," "management," "pharmacy" and "architecture" discipline-wise. One more category "overall" is defined for common ranking of the institutions.

We have considered six diverse institutions from "overall" category as a sample for the sake of understanding. These are Indian Institute of Science Bangalore (IISc), Indian Institute of Technology Kharagpur (IIT-KGP), Jawaharlal Nehru University Delhi (JNU), Jawaharlal Nehru Center for Advanced Scientific Research Bangalore (JNCASR), University of Delhi (DU) and Punjab University Chandigarh (PU).

We have chosen NIRF data such as number of students enrolled, proportion of students in different levels of programs and number of publications.

3.2 Knowledge Source

Information needed for calculating the crucial attributes depends directly on the task at hand or user query. Our framework also uses expert knowledge in the production and modification of knowledge base. System contains a local copy of crucial external information that is used frequently for academic purposes. Examples of some quality factors are impact factor and H-index of individual authors.

NIRF itself has used public databases such as Indian Citation Index (ICI) and SCOPUS for finding the number of publications and their corresponding citations of participating institutions.

3.3 Data Mining Phase

Aforementioned queries demand descriptive analysis of data. Since data provided by NIRF are quantitative in nature and collected at institution level, hence information related to more subtle units of the institutions such as discipline, fields of study, department and individual entity cannot be known. Thus, there is no way to use external knowledge sources for finding additional quality information (impact factor, H-index, etc.) regarding the deep levels of hierarchy. Due to this, we are remained only with NIRF data over which descriptive analysis has very little scope and this limits the suitability of these data for more rigorous knowledge discovery process.

In this phase of the framework, we attempt to explore the solution of aforementioned queries regarding the sample institutions. First query demands the classification of institutions based on data. We seek the evidence inside data that have some implications over distinctive features of institutions in subsequent subsections.

Student Enrolled. One can see huge variation in Fig. 2 regarding the number of students enrolled in sample institutions in the year 2015–2016 in different levels of programs. JNCASR has a total of 313 students in which 95% are Ph.D. students and 5% are PG students. IISC has 61% of research scholars, 25% of PG students and merely 14% of UG students. JNU contains 63% of research students, 26% of PG students and 11% of remaining UG students. With a huge contrast, DU has 21739 total students enrolled in which only 14% are pursuing Ph.D. and 81% are pursuing PG courses. PU has totally different scenario as it contains total of 18,360 students in which only 5% are Ph.D. students and approximately 57% are pursuing UG courses.

Publications. According to Fig. 3, JNCASR has the highest publication per faculty ratio that is approximately 29 research publications per faculty. IISc follows JNCASR with almost 17 research publications per faculty. In case of JNU, publication per faculty is comparatively low approximately three papers per faculty. IIT-KGP has quite decent publication per faculty rate which almost equals to 9 regarding its share of 21% Ph.D. students. Low publication rates 4.5 and 2.5 approx. in DU and PU, respectively, are due to very low proportion of Ph.D. students.

These two evidences variation in number of students and number of publications per faculty suggest that JNCASR, IISC and JNU are research-oriented institutions. DU is mostly related with PG courses, and PU is related mostly to UG courses. Almost balanced trends of IITK assert that it cannot be classified in any of above three categories (research centric, PG centric and UG centric). Hence, it alone stands in fourth category. By this way, sample institutions can be classified in four categories.

For solving second query that resembles the demand of a student "finding the best performing disciplines in sample institutions," there is a need of data availability at

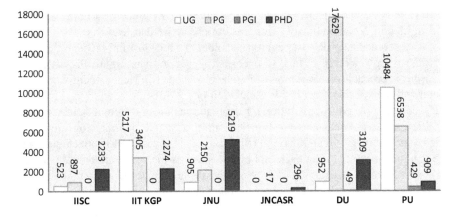

Fig. 2 Number of students enrolled in different levels of programs

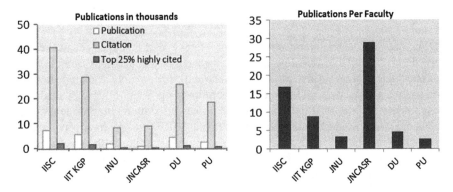

Fig. 3 Number of publications in SCOPUS in thousands and publications per faculty ratio

discipline and field of study levels. Since NIRF collects data at the highest level that is institutional level, hence this query cannot be solved due to unavailability of data.

Third query "calculating the average impact of research in sample institutions field-wise" requires analysis of additional factors, i.e., impact factor and h-index of each individual journal and each author. Finding these factors is possible only when institutional data are available at the lowest level of hierarchy (individual entity).

4   Discussion

NIRF requires most of the institutional data to be fed manually by participating institution. This makes it difficult to verify and validate the information. NIRF also has faced problems regarding data inconsistency and validity while producing ranking in

2016 [15]. For avoiding such issues, our proposed EDM framework uses complete and verified information directly from institutions active data systems.

It is observed that NIRF uses aggregated input of whole institution; hence, it is not possible to apply microlevel analysis over such aggregated data. Due to this, distortion in results is probable and variation cannot be observed [16]. Hence, identification of better and worst performers inside the institutions is difficult. Moreover, this limits the scope of knowledge exploration for the requirements of different stakeholders.

Since academic decisions are typically multi-objective optimization problems, one of such is institutional ranking [17]. Hence, it is necessary to collect and mine the data properly regarding each and every aspect of the academic domain, which requires dedicated data mining tools. We in this paper have developed one such EDM framework that can solve all above-mentioned problems.

5 Conclusion

EDM is the most useful tool for solving the problems of data management and knowledge discovery in educational domain. We have developed an EDM framework in this paper that utilizes verified institutional data and other publicly available information sources for discovering knowledge according to the need of different academic stakeholders. We describe the utility of our framework with respect to the requirements of various academic stakeholders by taking NIRF data. We have found that if the institutional data of NIRF are collected and mined according to the specifications of our proposed framework, then the objectives of ranking can be achieved more efficiently. Furthermore, the same data can be utilized for various other purposes.

References

1. Ketipi, A. K., Koulouriotis, D. E., Karakasis, E. G., Papakostas, G. A., & Tourassis, V. D. (2012). A flexible nonlinear approach to represent cause–effect relationships in FCMs. *Applied Soft Computing, 12*(12), 3757–3770.
2. Moscoso-Zea, O., Sampedro A., & Luján-Mora, S. (2016). Datawarehouse design for educational data mining. In *15th International Conference on Information Technology Based Higher Education and Training (ITHET)* (pp. 1–6). Istanbul.
3. Hussain, M., Al-Mourad, M. B., & Mathew, S. S. (2016). Collect, scope, and verify big data: A framework for institution accreditation. In: *Proceedings of 30th International Conference on Advanced Information Networking and Applications Workshops (WAINA)* (pp. 187–192). Crans-Montana.
4. Fernández, D. B., & Luján-Mora, S. (2017). Comparison of applications for educational data mining in engineering education. In *Proceedings of IEEE World Engineering Education Conference (EDUNINE)* (pp. 81–85). Santos.
5. Peña-Ayala, A. (2014). Educational data mining: A survey and a data mining-based analysis of recent works. *Expert Systems with Applications, 41*(4), 1432–1462.

6. Romero, C., & Ventura, S. (2010). Educational data mining: A review of the state of the art. *IEEE Transactions on Systems, Man and Cybernetics Part C (Applications and Reviews), 40*(6), 601–618.
7. García, S., Luengo, J., & Herrera, F. (2016). Tutorial on practical tips of the most influential data preprocessing algorithms in data mining. *Knowledge-Based Systems, 98,* 1–29.
8. Dutt, A., Ismail, M. A., & Herawan, T. (2017). A systematic review on educational data mining. *IEEE Access, 5,* 15991–16005.
9. Romero, C., & Ventura, S. (2007). Educational data mining: A survey from 1995 to 2005. *Expert Systems with Applications, 33*(1), 135–146.
10. Gambhir, V., Wadhwa, N. C., Grover S., & Goyal, S. (2012). Applying fuzzy MADM approach for the selection of technical institution. In *Proceedings of IEEE International Conference on Industrial Engineering and Engineering Management* (pp. 1405–1408). Hong Kong.
11. Charitopoulos, A., Rangoussi, M., & Koulouriotis, D. (2016). E-Learning platform access and usage statistics through data mining: An experimental study in moodle. In *9th International Conference of Education, Re-search and Innovation (ICERI'16)* (pp. 2958–2967). Seville.
12. Buniyamin, N., Mat, U. B., & Arshad, P. M. (2015). Educational data mining for prediction and classification of engineering student achievement. In *Proceedings of 7th International Conference on Engineering Education (ICEED)* (pp. 49–53). Kanazawa.
13. Jongbloed, B., Enders, J., & Salerno, C. (2008). Higher education and its communities: Inter-connections, interdependencies and a research agenda. *Higher Education, 56*(3), 303–324.
14. NIRF homepage. Retrieved January 15, 2018 from https://www.nirfindia.org/About.
15. Kumar, A., & Tiwari, S. K. (2016). India rankings 2016: Ranking model for Indian higher educational institutions. In *Proceedings of International Conference on ICT in Business Industry and Government* (pp. 1–6). Indore.
16. Abramo, G., D'Angelo, C. A., & Di Costa, F. (2008). Assessment of sectoral aggregation distortion in research productivity measurements. *Research Evaluation, 17*(2), 111–121.
17. Vardy, M. Y. (2016). Academic rankings considered harmful. *Communications of the ACM, 59*(9), 5.

GA-Based Machine Translation System for Sanskrit to Hindi Language

Muskaan Singh, Ravinder Kumar and Inderveer Chana

Abstract Machine translation is the noticeable field of the computational etymology. Computational phonetics has a place with the branch of science which bargains the dialect perspectives with the help of software engineering innovation. In this field, all handling of regular dialect is finished by the machine (PC). Calculation is done by considering all features of the language and in addition vital principal of sentence like its structure semontics and morphology. Machine ought to see all these conceivable parts of the dialect, yet past work does not deal with alternate prerequisites amid machine interpretation. Current online and work area machine interpretation frameworks disregard numerous parts of the dialects amid interpretation. Because of this issue, numerous ambiguities have emerged. Because of these ambiguities, current machine interpreter is not ready to deliver right interpretation. In this proposed work, genetic algorithm-based machine translation system is proposed for the translation of Sanskrit into Hindi language which is more efficient than the existing translation systems.

Keywords Machine translation · GA · Hindi · Sanskrit · NLP

1 Introduction

Machine translation system (MTS) helps the humans to interact with people from other cultures without any language barriers so that they can share their ideas, information and know each other very easily. This is a common application of Natural Language Processing (NLP), and it helps to translate one language into other. As we know, India is very rich for their cultures and all the states have their different languages. So, to provide translation of different languages without any human assistance is the main aim of the MTS. Sanskrit is the primary sacred language of India

M. Singh (✉)
Research Labs, CSED, Thapar University, Patiala, Punjab, India
e-mail: muskaan_singh@thapar.edu

R. Kumar · I. Chana
CSED, Thapar University, Patiala, Punjab, India

© Springer Nature Singapore Pte Ltd. 2019
A. Khare et al. (eds.), *Recent Trends in Communication, Computing,*
and Electronics, Lecture Notes in Electrical Engineering 524,
https://doi.org/10.1007/978-981-13-2685-1_40

419

[1] and official language in one of the states, i.e., Uttarakhand, India, whereas Hindi is also one of the primary languages of India and it is used everywhere in India. So, there is a need for the translation system which can translate these two important languages.

The two important uses of the MTS are assimilation and dissemination. Where is assimilation means one can understand some text which means understand its general meaning but might not grammatically right. Secondly, dissemination means text should be grammatically correct so that it will be treated as publishable content. As different languages have different styles and different structures so the current translation system has to face so many problems. Some of the problems are like:

a. **Meaning of the Word**: Different words have different meanings but sometimes same words have a different meaning when they are translated from one language into other, and it is very difficult to select the right word for correct meaning which is very important part of any translation system.

b. **Order of the Word**: This is also a very important factor while translating one language into other because some of the languages follow the order like Subject (S), Verb (V), and Object (O) and others might have some other order but Sanskrit language can be written using SVO, SOV, and VOS order.

c. **Idioms**: As idioms are the gathering of words set up by utilization as having significance not deducible from those of the individual words. So while translating idiom expressions, it will not convey the original meaning.

2   Related Work

A number of works were performed as MTS. Some of the MTSs are given below.

In 1968, SYSTRAN is established by Dr. Dwindle Toma which is one of the most seasoned machine interpretation organizations. SYSTRAN has done broad work for the US Department of Defense and the European Commission. SYSTRAN gives the innovation to Yahoo! Babel Fish among others. It was utilized by Google's dialect apparatuses until 2007. It is rule-based MTS and deals with 35 different languages. ETSTS [2–4] is a rule-based and example-based approach of MT. Utilizing discourse synthesizer as a module, it changes over target sentence to discourse yield. The outlines of the framework are modularized to text input, grammar and spell check, token generator, translator, parser generator module, RBMT/EBMT engine, its bilingual database, text yield, and a waveform generator.

Google Translate is a free multilingual machine interpretation benefit created by Google, to decipher content from one dialect into another. It is based on statistical approach. It offers a site interface, portable applications for Android and iOS, and an API that enables designers to assemble program augmentations and programming applications. Google Translate bolsters more than 100 dialects at different levels. Khaled Shaalan [5] had given the rule-based approach of machine interpretation for English to Arabic Natural Language Handling and the govern based apparatuses for

Arabic common dialect. It has given the morphological analyzers and generator and syntactic analyzer and generators.

Sandeep Warhade [6] had given a plan of a phrase-based decoder for English-to-Sanskrit interpretation. It portrays the phrase-based measurable machine translation decoder for English as source what's more, Sanskrit as target dialect. They will likely enhance the interpretation quality by improving the interpretation table and by preprocessing the source dialect content research. They examine the significant plan objective for the decoder and its execution with respect to other SMT decoders. ESSS [7] utilize rule-based interpretation system. Subsequent to changing English discourse to content, the sentences shaped are first disintegrated into words. The resultant words are coordinated in database, it likewise separates parts of speech (POS) data utilizing database to characterize each word for a thing, verb, and descriptor and so on, at that point applying sentence structure standards, and Sanskrit content gets created revising the sentence producing target dialect content. This content is given as a contribution to wave shape generator where it gets changed over to Sanskrit discourse.

Microsoft provides a statistical approach-based machine translation system named as Big Translator [8]. E-Trans [6] is a control-based machine interpretation instrument. It is basically in view of plan of synchronous context-free language (SCFG), a subset of context-free grammar (CFG). Dialect portrayal of linguistic structure is finished utilizing SCFG. The process motor created works in two stages, the top–down approach and bottom-to-top examination.

3 Machine Translation Systems

MTSs are categorized into different categories as shown in Fig. 1. They are direct, rule based, corpus based, and knowledge based.

In direct MT [9], there is no middle of the road portrayal of codes. Utilizing bilingual lexicon, there is word-by-word interpretation with the help of bilingual word reference took after by a few syntactic reworking. This technique for interpretation is as it was possible for one dialect match. It requires little examination of content without parsing. Here investigation technique like morphological investigation, preposition handling, syntactic arrangement and morphological generation can be performed.

In rule-based MTS [9], middle portrayal might be created like a parse tree. It depends on rules for morphology, linguistic structure, lexical choice and exchange, semantic examination, and age in this way known as rule based. Rule based can be of two types

(1) Transfer-based
(2) Interlingua

In transfer based, SL to TL is without moderate portrayal while in Interlingua some moderate code portrayal is made through which SL meant TL by means of bury dialect codes.

Fig. 1 Methods of machine
translation systems

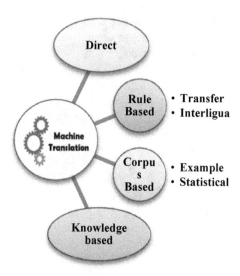

In corpus-based MTS [9, 10], it requires sentence-adjusted parallel content for every dialect combine. It cannot be utilized for dialect sets for which such corpora do not exist. It can be additionally ordered into statistical MT and example-based MT.

In knowledge-based MTS frameworks [11], semantic-based way to deal with dialect investigation is presented by artificial knowledge specialists. It requires vast information base that incorporates both ontological and lexical learning. The fundamental AI approaches incorporate semantic parsing, lexical disintegration into semantic systems and settling ambiguities.

The basic process of the MTS is as given in Fig. 2. The first and foremost step to provide input text to the translation system is according to the user's requirement and also on the basis of the translation system.

The next step of the translation system is tokenization, which is the process of crumbling of given text into smaller units called tokens. Then, grammatical information of these token is generated by using morphological analysis. After that, parse tree will be generated by using parser. It generates grammatical information with respect to context. This will further lead to generator module which takes semantic data from the morphological examination module and does mapping taken after by hunting down the right type of the words from the vocabulary by considering root words to produce a yield of the source dialect. Then, target text will be generated after reformatting.

Fig. 2 MTS

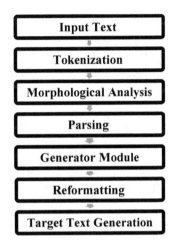

Table 1 Input and tokenization

Input Sentence	Tokens	Category
अहम्अस्ति	1) अहम् 2) अस्ति	Small
अहम्खादामिअत्रभोजनं	1) अहम् 2) खादामि 3) अत्र 4)भोजनं	Large
व्याघ्र:हरिणमखादति	1) व्याघ्र: 2) हरिणम 3) खादति	Large
वृक्षेभ्य: फलं पतति	1) वृक्षेभ्य:2) फलं 3) पतति	Large

4 Proposed Ga-Based Approach for Translator

This proposed system is used to translate the Sanskrit language into Hindi. It is an efficient translator because it uses genetic algorithm -based generator which enhances the mapping process. In this proposed work, we have two main phases: (i) initial phase and (ii) generator phase.

Phase 1: Initial Phase

This phase is responsible for different analysis processes which include acquisition of input, tokenization, morphological analysis, and parsing. These steps play an important role in MTS.

- **Acquisition of Input**: In this step, firstly sentence is taken as an input for translation process. We have divided the sentences into three categories for this proposed work that are (a) small, (b) large, and (c) extra large. So, input sentence may fall into any one of these categories.
- **Tokenization**: This process will divide the input sentence into smaller tokens so that it will be analyzing that sentence falling under which category. For example, some of the sentences and their token values with their category are given in Table 1.

Table 2 Example of morphological analysis

Tokens	Noun (Ablative Case)	Noun(Active Case)	Verb
1) वृक्षेभ्यः 2) फलं 3) पतति	वृक्षेभ्यः	फलं	पतति
Fruit falls from the tree.	From the Tree	Fruit	Fall

- **Morphological Analysis**: This process helps the translation system to collect grammatical information about the input sentence from the tokens. For example, वृक्षेभ्यः फलं पतति). Input sentence have three tokens (1) वृक्षेभ्यः(2) फलं(3) पततिand morphological process provides grammatical information about these tokens which tells that the words belongs to which category means its noun, pronoun or verb as given in Table 2. For example: Vrkssebhyah वृक्षेभ्यःis a noun but it is ablative case of noun, where ablative indicates 'from, on account of, etc.'

- **Parsing**: It is the process of analyzing given input and confirming its grammatical correctness. The parser first scans the words and recognizes them, and then it recognizes the syntactic units. The main three operations performed by the parser are (i) checking and verification of the syntax based on specified syntax rules and (ii) reporting about the errors. In this work, bottom-up parser technique is used which first finds the rightmost derivation in reverse order and then for every token decides when production is found. This parsing approach can handle the largest class of grammars that can be parsed.

Phase 2: Generator Phase

The second phase of the MTS is generator phase. This phase performs the reverse of the initial phase which includes mapping, morphological analysis, and output process. In this work, we proposed a genetic algorithm-based mapper which efficiently maps the tokens with their respective target tokens.

- **Genetic algorithm-based mapper**: Mapping is the process which basically checks the grammatical compatibility of the source and target language. In this work, we generate a GA-based mapper which follows the following steps:

  - Randomly generate a population of 'k' possible solutions where 'k' is the number of inputs.
  - Figure out the degree of acceptance of each solution on the basis of fitness value which is calculated on the basis of grand mean.
  - Select two solutions and perform crossover process. In this, we have used single-point crossover, which generates new solutions.
  - Then, alter these solutions by mutation process.
  - Treat it as the current best solution and repeat until generating new solutions.
  - Repeat until fitness value reaches its maximum value or the number of solutions has reached maximum.

- **Morphological Generator**: The morphological generator takes the input sentence and first looks for its root word from the mapper. This GA-based mapper analyzes the root word with its respective target language and provides the exact match for

Table 3 Example of target tokens

Tokens	Respective Target Word
वृक्षेभ्यः	पेड़
फलं	फल
पतति	गिरता

Table 4 Source and target sentence

Input Sentence	Category	Target Sentence
अहम्अस्ति	Small	मैंहूँ
अहम्खादामिअत्रभोजनं	Large	मैंयहाँभोजनखारहाहूँ
व्याघ्रःहरिणमखादति	Large	हिरणशेरखातीहै
वृक्षेभ्यः फलं पतति	Large	फलपेड़सेगिरताहै

the same in target language. For Example: In (वृक्षेभ्यः फलं पतति)this sentence root word is फलं.The respective target word selected by Mapper is as given in Table 3.

- **Output Process**: This is the last step of this phase which mainly gathers information from the leaves and finally generates the output.

5 Simulation Results

In this work, the genetic algorithm-based machine translation system (GA-MTS) is proposed. The input language for this work is Sanskrit and target is Hindi. Both languages are the part of Indian culture. Sanskrit is the primary sacred language where Hindi is national language of India. This system supports grammar system for both the languages and generates very efficient results. To evaluate the effectiveness of this proposed system, 300 samples of different types of sentences were taken and various analyses were done using proposed GA-MTS. Some of the results are given in Table 4.

The sentences are divided into three different categories according to their size that are (i) small, (ii) large and (iii) very large as given in Table 5, and accuracy of this proposed MTS is analyzed on these different categories. We had taken 200 sentences from small category and 50 each from large and very large category. Accuracy is calculated on the basis of conversions and their matching with actual meaning of the sentences. It means converted samples are matched with the original Hindi samples, and their recognition provides us the accuracy of this proposed convertor. The performance results are as shown in Fig. 3.

Table 5 Different categories of sentences

Category	Parameter	Number of sentences tested	Accuracy (%)
Small	$\geq 1 \,\&<3$	200	98
Large	$\geq 3 \,\&<6$	50	95
Very large	$\geq 6 \,\&\leq 11$	50	90

Fig. 3 Performance of GA-MTS

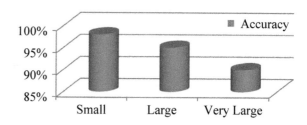

6   Conclusion

This proposed work generates GA-based machine translation system. This GA-MTS has two phases, and both phases perform different tasks. We have added genetic algorithm in generator phase where it helps to map the particular input to its target language and GA performs very efficiently in this proposed MTS. For testing purposes, three different types of sentences were taken. This proposed GA-MTS is tested using 300 different samples and achieves 94% accuracy on an average which is very good in the field of translation for Sanskrit to Hindi language. In future, this proposed system will be tested using a large number of samples of the complex category means samples with sentences of large and very large category. So that more and more efficient machine translation system will be generated for these two prominent languages of India.

References

1. Rathod, S. G. (2014). Machine translation of natural language using different approaches: ETSTS (English to Sanskrit Translator and synthesizer). *International Journal of Computer Applications, 102*(15), 26–31.
2. Zhao, Y., & He, X. (2009). Using N-gram based features for machine translation system combination. In *Proceedings of Human Language Technologies: The 2009 Annual Conference of the North American Chapter of the Association for Computational Linguistics, Companion Volume: Short Papers on—NAACL 09* (pp. 205–208).
3. Zogheib, A. (2009). Genetic algorithm-based multi-word automatic language translation. *Recent Advances in Intelligent Information Systems*, 751–760.
4. Rathod, S. G., & Sondur, S. (2012). English to Sanskrit Translator and synthesizer (ETSTS). *International Journal of Emerging Technology and Advanced Engineering, 2*(12), 379–383.

5. Mane, D. T., Devale, P. R., & Suryawanshi, S. D. (2010). A design towards English To Sanskrit machine translation and sythesizer system using rule base approach. *International Journal of Multidisciplinary Research And Advances In Engineering (IJMRAE), 2*(2), 405–414.

6. Bahadur, P., Jain, A. K., & Chauhan, D. S. (2012). EtranS-A complete framework for English To Sanskrit machine translation. *International Journal of Advanced Computer Science and Applications, 2*(1), 52–59.

7. Tahir, G. R., Asghar, S., & Masood, N. (2010). Knowledge based machine translation. In *International Conference on Information and Emerging Technologies (ICIET)* (pp. 1–5), November 2010.

8. Raulji, J. K., & Saini, J. R. (2016). Sanskrit machine translation systems: A comparative analysis. *International Journal of Computer Applications, 136*(1), 1–4.

9. Mishra, V., & Mishra, R. B. (2008). Study of example based English to Sanskrit machine translation. *Polibits, 37,* 43–54.

10. Patil, S. P., & Kulkarni, P. P. (2014). Online handwritten Sanskrit character recognition using support vector classification. *Internal Journal of Engineering Research and Applications, 4*(5), 82–91.

11. Shahnawaz (2015). Conversion between Hindi and Urdu. In *International Conference on Computing, Communication & Automation* (pp. 309–313).

Part VIII
Advanced Computing and Intelligent Applications

Deep Leaning-Based Approach for Mental Workload Discrimination from Multi-channel fNIRS

Thi Kieu Khanh Ho, Jeonghwan Gwak, Chang Min Park, Ashish Khare and Jong-In Song

Abstract As a non-invasive optical neuroimaging technique, functional near infrared spectroscopy (fNIRS) is currently used to assess brain dynamics during the performance of complex works and everyday tasks. However, the deep learning approaches to distinguish stress levels based on the changes of hemoglobin concentrations have not yet been extensively investigated. In this paper, we evaluated the efficiencies of advanced methods differentiating the rest and task periods during stroop task experiments. First, we explored that the apparent changes of oxy-hemoglobin (HbO) and deoxy-hemoglobin (HbR) concentrations associated with two mental stages did exist across each participant. Then, a novel discrimination framework was studied. Deep learning approaches, including convolutional neural network (CNN), deep belief networks (DBN), have enabled better classification accuracies of $84.26 \pm 9.10\%$ and $65.43 \pm 1.59\%$ as our preliminary study.

T. K. K. Ho · J. Gwak (✉)
Biomedical Research Institute, Seoul National University Hospital, Seoul 03080, Korea
e-mail: james.han.gwak@gmail.com

J. Gwak · C. M. Park
Department of Radiology, Seoul National University Hospital, Seoul 03080, Korea

C. M. Park
Department of Radiology, Seoul National University College
of Medicine, 101, Daehak-ro, Jongno-gu, Seoul 03080, Korea

C. M. Park
Institute of Radiation Medicine, Seoul National University
Medical Research Center, Seoul 03080, Korea

C. M. Park
Seoul National University Cancer Research Institute, Seoul 03080, Korea

A. Khare
Department of Electronics and Communication, University
of Allahabad, Allahabad, Uttar Pradesh, India

T. K. K. Ho · J.-I. Song
School of Electrical Engineering and Computer Science,
Gwangju Institute of Science and Technology, Gwangju 61005, Korea

© Springer Nature Singapore Pte Ltd. 2019
A. Khare et al. (eds.), *Recent Trends in Communication, Computing,
and Electronics*, Lecture Notes in Electrical Engineering 524,
https://doi.org/10.1007/978-981-13-2685-1_41

Keywords Stroop task experiments · Functional near infrared spectroscopy
Convolutional neural networks · Deep belief networks

1 Introduction

Known as a promising non-invasive optical technique to study neurocognitive functions [1, 2], functional near infrared spectroscopy (fNIRS) offers various advantages such as high spatial resolution, continuous monitoring ability, portability and limited physical or behavioral restrictions from subjects [3, 4]. In particular, fNIRS presents information of the temporal hemodynamic oxygenation by the two main parameters: Oxy-hemoglobin (HbO) and deoxy-hemoglobin (HbR). The stimulus responses of these values reflect the changes of neural activations due to the increase in the energy demand which leads to the increase in regional cerebral blood flow in activated areas. With these characteristics, fNIRS constitutes its contribution to working memory and sustained attention studies [5, 6].

We used the prefrontal cortex (PFC) region which has been demonstrated to be the most relevant area for memory and attention-related functions [7] with the aim to classify human mental workload stages (MWS) into discrete levels, ranging from rest to task stages. While there are various estimations of MWS [1] in brain computer interfaces (BCIs), we focused on exploring and proving the correlations between MWS and the changes of HbO, HbR and their differences (HbT). However, unlike speech signals and image domains, fNIRS signals are symmetrically temporal and non-stationary, which makes the tasks, such as stress levels differentiation, difficult for analysis. In particular, collected fNIRS signals tend to have low signal-to-noise ratio (SNR) due to its mixing with much noises such as global interference, instrumental noise, and motion artifacts. In this work, we suggest a framework to evaluate discriminating performances between two mental stage levels through fNIRS measures over PFC from two different deep learning approaches.

Although the machine learning techniques have been successfully applied to signal domains, such as support vector machines [8] and adaptive boosting [9], extraction and selection of features in appropriate ways are still the drawbacks. The cost of these traditional methods increases quadratically with respect to the considered parameters requiring a large amount of labeled training data for supervised learning. Meanwhile, theoretical and biological arguments have strongly recommended building models composed of multilayers of nonlinear processing with few labeled input in correspondence with human brain activity. Therefore, models with deep architectures such as deep belief networks (DBN) and convolutional neural networks (CNN) have unsurprisingly come to the top of our priorities in terms of solving optimization problems for automatic feature extraction and classification tasks.

DBN [10] is constructed by connected simpler probabilistic graphical models, namely restricted Boltzmann machines (RBMs), and interpreted by stochastic binary units. Indeed, each RBM consists of two layers, one bottom visible and one top hidden layer to be fully and symmetrically connected with other RBM via weights.

By stacking layered RBMs to form DBNs, which means the hidden layer of a lower RBM becomes the visible layer for the next RBM. Through this process, higher RBMs can learn more abstract and complex features as activated representations of neurons in brain cortexes. Meanwhile, CNN variants [11, 12] have recently emerged as a robust supervised technique, successfully applied for EEG sleep stage scoring [13] in combination of convolution layers and pooling layers. Units in convolutional layers are organized in feature maps where each unit is linked to local patches in the features maps of the previous layer via the set of the weights, called a filter bank. After being detected local conjunctions of features, pooling layers merge semantically similar features into one, which reduces dimension of representation and creates the features to be more invariant to small shifts and distortions.

The remaining parts of this work are organized as follows. In Sect. 2, we briefly present our experimental setting and fNIRS recording. Then, the models and algorithms of two classifiers are mentioned in Sect. 3. Section 4 systematically describes the MWS classification results by comparing the efficiencies of two classifier performances. The conclusions and future research agendas are summarized in Sect. 5.

2 fNIRS Experiments

2.1 Participants

In this study, 16 approved volunteers including eight males and eight females between the ages of eighteen and thirty are involved. Participants with diagnosed neurological, respiratory, cardiovascular and vision abnormalities are all excluded beforehand.

2.2 Experimental Setup and Data Preprocessing

FNIRS device, Shimadzu FOIRE-3000 using near infrared light (with wavelengths of 780, 805, and 830 nm), was used to record hemodynamic activities with three types of hemoglobin including HbO, HbR and HbT with 7 channels over the PFC region. As depicted in Fig. 1, a 2-by-3 array of T-transmitters (sources) and R-receivers (detectors) with 3 cm separation was attached to each subject's forehead. To stimulate the stress levels of subjects, we used STE experiments [14] which are divided into congruent and incongruent tasks. There were four colors appearing with different conflicted words, e.g., 'RED' was written by the blue color. In the experiments, we checked that the responses of all subjects could be slower and have less accuracy if the words and colors were conflicted during the high incongruent period and vice versa.

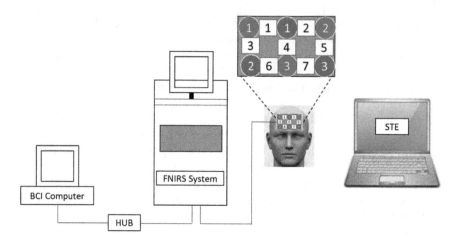

Fig. 1 Experimental setup for FNIRS recording

The raw data contained much noise will significantly affect to the final result. Therefore, we sampled the data with the sampling rate of 18 Hz, filtered the noises and removed artifacts by band-pass filters between 0.3 and 3 Hz. Then, the data were averaged and normalized to baseline. After preprocessing, we segmented dataset into the same length (1×1000 window) without overlapping.

3   Classification Methodology: Deep Models

3.1   Deep Belief Network

DBN allows each RBM receives different representations of data, then the activity of its hidden units driven by the visible units can be training data for learning higher RBM by nonlinear transformation. DBN was characterized as a probabilistic model to encode the input and the features relationships. In this sense, the DBN was made of two main parts in which an encoding part generates the model and the decoding part reconstructs its original data. After training DBN in a layer-by-layer manner, the back propagation algorithm was applied to fine-tune and optimize the weights. The stacking RBMs and parameters of our DBN are illustrated in Fig. 2 and Table 1, respectively.

Fig. 2 DBN model

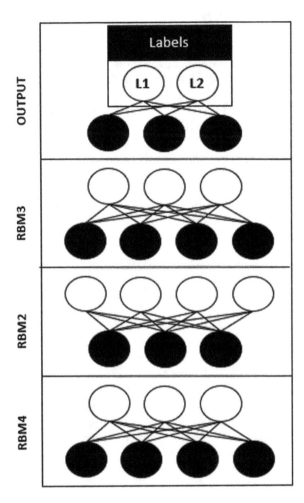

Table 1 DBN model parameters

Parameter	Value
Number of layers	5
Units per layer	1000-1500-2000-3000-2
Number of generative RBMs	3
Learning rate	0.1
Number of epochs	400

3.2 Convolutional Neural Network

CNNs have been proposed as a powerful method to learn interpretable image features by scaling up the network to millions of parameters and massive labeled inputs. Compared to all successful applications of CNNs in image recognition, classification [12, 15] and object detection [16, 17], end-to-end training of CNN for fNIRS signal domains is still challenging. In this work, CNN model was applied to decompose and feed forward 1D fNIRS signal which is known as a precomputed spectrogram or time-frequency, the input then became 2D stacks of frequency-specific activity. As shown in Fig. 3, the CNN model consisted of 9 layers, including two pairs of convolutional and pooling layers (C1-P1 and C2-P2) with a stacked layer S1 between P1 and C2, two fully connected layers (F1 and F2) and a soft-max layer.

4 Experimental Study Results

4.1 Hemodynamic Activity

In Fig. 4, we summarized the changes of hemodynamic responses during rest and task stimulation of the 16 subjects on average values of HbO, HbR and HbT concentrations with seven channels. From the result, we could see that the difference in the reactions of these concentrations did exist. This means that the hemodynamic activation significantly varied and relatively reflected to neural responses among subjects. As our observation, HbO values significantly increased during the task periods and decreased during the rest periods while HbR values were totally contrary. This evidence exactly obeyed to the Beer-Lambert law [18].

4.2 Classification Results

In this study, we alternatively perform two classifiers to discriminate two classes from fNIRS datasets of 16 subjects according to three hemoglobin types for seven channels. We divided datasets into 70% for the training set and 30% for the testing set. PCA was applied to reduce input dimensionality. The results are summarized in Table 2. It should be noted that DBN and CNN obtained average accuracy of $84.26 \pm 9.10\%$ and that of $65.42 \pm 1.58\%$, respectively for 16 Subjects $\times$ 3 Hemoglobin Types $\times$ 7 Channels (Fig. 5).

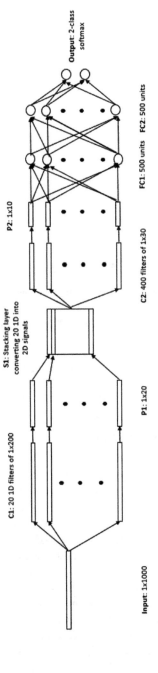

Fig. 3 CNN model and parameters

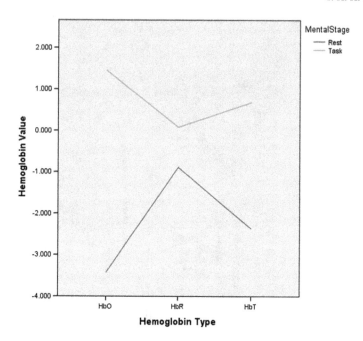

Fig. 4 Average changes of 3 hemoglobin types during stimulus period among 16 subjects

Table 2 Performance comparison of DBN and CNN according to each hemoglobin type (mean ± standard deviation %)

Methods	HbO	HbR	HbT
DBN	85.62 ± 8.64	81.63 ± 9.85	85.53 ± 8.22
CNN	66.41 ± 1.19	63.19 ± 2.14	66.67 ± 1.53

5   Conclusion and Future Work

Despite recent outstanding applications of BCIs, the classification study on hemo-dynamic activities in MWS is still very challenging. One of the challenges is how the labeled inputs can be truly matched to each workload level. In this work, as a preliminary study, we presented two deep learning approaches, DBN and CNN, for MWS classification from multi-channel fNIRS signals recorded over PFC during STEs. For classification results, the DBN model, compared to the proposed CNN, could learn deep features with better accuracy and lower standard deviations. It seems that if CNN could go deeper compared to the proposal, CNN might achieve better results. Therefore, as our future work, we aim to propose a robust deep CNN model with more layers and optimal parameters that can present more complex nonlin-ear structures progressively and capture high-order correlations of activation units from fNIRS inputs. The training with a larger category of MWS datasets is also an indispensable topic.

Fig. 5 Classification results of DBN and CNN

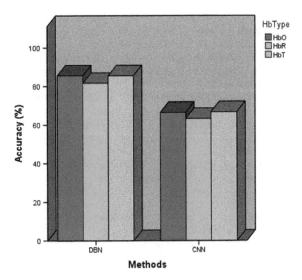

Acknowledgements This work was supported by the Basic Science Research Program through the NRF funded by the Ministry of Education (NRF-2017R1D1A1B03036423) and the Brain Research Program through the National Research Foundation of Korea (NRF) funded by the Ministry of Science, ICT & Future Planning (NRF-2016M3C7A1905477, NRF-2014M3C7A1046050). All procedures performed in studies involving human participants were in accordance with the ethical standards of the institutional and/or national research committee and with the 1964 Helsinki declaration and its later amendments or comparable ethical standards. This article does not contain any studies with animals performed by any of the authors.

References

1. Hoshi, Y., & Tamura, M. (1993). Detection of dynamic changes in cerebral oxygenation coupled to neuronal function during mental work in man. *Neuroscience Letters, 150,* 5–8.
2. Nguyen, H. T., Van Nguyen, H., Truong, K. Q. D., & Van Vo, T. (2013). Analysis of oxy-Hb signals to determine relationship between jaw imbalance and arm strength using fNIRS. *American Journal of Biomedical Engineering, 3,* 107–118.
3. Ferrari, M., & Quaresima, V. (2012). A brief review on the history of human functional near-infrared spectroscopy (fNIRS) development and fields of application. *Neuroimage, 63,* 921–935.
4. Hai, N. T., Cuong, N. Q., Khoa, T. Q. D., & Toi, V. V. (2013). Temporal hemodynamic classification of two hands tapping using functional near infrared spectroscopy. *Frontiers in Human Neuroscience, 7,* Article 516, 1–12.
5. Molteni, E., Baselli, G., Bianchi, A. M., Caffini, M., Contini, D., Spinelli, L. et al. (2009). Frontal brain activation during a working memory task: A time-domain fNIRS study. *Photonic Therapeutics and Diagnostics V,* 71613 N. https://doi.org/10.1117/12.808972.
6. Sassaroli, A., Zheng, F., Coutts, M., Hirshfield, L. H., Girouard, A., Solovey, E. T., et al. (2009). Application of near-infrared spectroscopy for discrimination of mental workloads. In *Proceedings of SPIE 7174, Optical Tomography and Spectroscopy of Tissue VIII* (pp. 71741H).

7. Ramnani, N., & Owen, A. M. (2004). Anterior prefrontal cortex: Insights into function from anatomy and neuroimaging. *Nature Reviews Neuroscience, 5,* 184–194.
8. Suthaharan, S. (2016). Support vector machine. In *Machine learning models and algorithms for big data classification* (pp. 207–235). Springer.
9. Chu, F., & Zaniolo, C. (2004). Fast and light boosting for adaptive mining of data streams. In *Pacific-Asia Conference on Knowledge Discovery and Data Mining* (pp. 282–292). Springer.
10. Hinton, G., Osindero, S., & Teh, Y. W. (2006). A fast learning algorithm for deep belief nets. *Neural Computation, 18,* 1527–1554.
11. LeCun, Y., Kavukcuoglu, K., & Farabet, C. (2010). Convolutional networks and applications in vision. In *Proceedings of the IEEE International Symposium on Circuits and Systems* (pp 253–226).
12. Krizhevsky, A., Sutskever, I., & Hinton, G. (2012). ImageNet classification with deep convolutional neural networks. In *Proceedings of NIPS* (pp. 1097–1105).
13. Tsinalis, O., Matthews, P. M., Guo, Y., & Zafeiriou, S. (2016). Automatic sleep stage scoring with single-channel EEG using convolutional neural networks. arXiv:1610.01683.
14. Schroeter, M. L., et al. (2002). Near-infrared spectroscopy can detect brain activity during a color–word matching Stroop task in an event-related design. *Human Brain Mapping, 17*(1), 61–71.
15. Simonyan, K., & Zisserman, A. (2014). Very deep convolutional networks for large-scale image recognition. *CoRR.* arXiv:1409.1556.
16. Sermanet, P., Eigen, D., Zhang, X., Mathieu, M., Fergus, R., & LeCun, Y. (2014). Overfeat: Integrated recognition, localization and detection using convolutional networks. In *ICLR.*
17. Girshick, R., Donahue, J., Darrell, T., & Malik, J. (2014). Rich feature hierarchies for accurate object detection and semantic segmentation. In *Proceedings of the IEEE Conference on Computer Vision and Pattern Recognition* (pp. 580–587).
18. Baker, W. B., Parthasarathy, A. B., Busch, D. R., Mesquita, R. C., Greenberg, J. H., & Yodh, A. G. (2014). Modified Beer-Lambert law for blood flow. *Biomedical Optics Express, 5,* 4053–4075.

Design and Simulation of Capacitive Pressure Sensor for Blood Pressure Sensing Application

Rishabh Bhooshan Mishra, S. Santosh Kumar and Ravindra Mukhiya

Abstract This paper presents the mathematical modeling-based design and simulation of normal mode MEMS capacitive pressure sensor for blood pressure sensing application. The normal blood pressure of human being is 120/80 mmHg. But this range varies in case of any stress, hypertension and some other health issues. Analytical simulation is implemented using MATLAB®. Basically, normal mode capacitive pressure sensors have a fixed plate and a moveable diaphragm which deflects on application of pressure with the condition that it must not touch the fixed plate. Deflection depends on material as well as thickness, shape and size of diaphragm which can be of circular, elliptical, square or rectangular shape. In this paper, circular shape is chosen due to higher sensitivity compared to other diaphragm shapes. Deflection, base capacitance, change in capacitance after applying pressure and sensitivity are reported for systolic and diastolic blood pressure monitoring application, and study involves determining the optimized design for the sensor. Diaphragm deflection shows linear variation with applied pressure, which follows Hook's law. The variation in capacitance is logarithmic function of applied pressure, which is utilized for analytical simulation.

Keywords Mathematical modeling · Capacitive pressure sensor
Blood pressure measurement

1 Introduction

Microelectro mechanical system (MEMS) is an interesting and popular research area which widely combines the physical, chemical, biological processes on a chip [1]. The single integrated system or device can be analyzed by a specific partial differential equation based on domain of processes [2]. In all the MEMS devices,

R. B. Mishra · S. Santosh Kumar (✉) · R. Mukhiya
Smart Sensors Area, CSIR - Central Electronics Engineering
Research Institute, Pilani 333031, Rajasthan, India
e-mail: santoshkumar.ceeri@gmail.com

© Springer Nature Singapore Pte Ltd. 2019
A. Khare et al. (eds.), *Recent Trends in Communication, Computing,
and Electronics*, Lecture Notes in Electrical Engineering 524,
https://doi.org/10.1007/978-981-13-2685-1_42

except microfluidics area, simple micromechanical and moveable structures like diaphragms and beams are used. These simple microstructures help in measuring the various process variables like pressure, humidity, acceleration, flow, temperature, pH. Micromachining and manufacturing science is the combination of tools and techniques which are adapted from VLSI technology and microelectronic fabrication [3]. In designing the pressure sensors for absolute, gauge or differential pressure measurement such as piezoresistive, capacitive, usually diaphragms are used. In piezoresistive pressure sensors, four piezoresistors are implanted on the diaphragm, and when pressure is applied, diaphragm deflects. Due to change in resistance of piezoresistors, the Wheatstone bridge becomes unbalanced and an output voltage is obtained across the bridge. The applied pressure can be obtained according to the output voltage. The capacitive sensing of pressure offers various advantages over piezoresistive sensing such as higher sensitivity, low power consumption, long-term stability and temperature insensitivity [4, 5].

Human body is a combination of several instrumentation systems like cardiovascular system, respiratory system, nervous system, vascular system. A lot of research in Bio-MEMS for man-instrumentation system is in trend like blood pressure measurement, healthcare application, medical diagnostics, liver on a chip and brain on a chip. Research in cardiovascular psychology and hypertension for various small mammals and human has been carried out using capacitive pressure sensors with new micromachining processes. In these sensors, the capacitive pressure sensors have been designed in an array which gives improved output signals and increases stability. A switched capacitor CMOS circuitry is used to measure the change in capacitance in the sensor array [6]. Therefore, using capacitive pressure sensors, the blood pressure measurement can be carried out. The normal mode capacitive pressure sensor has two parallel plates which are separated by a medium, one of the plates is diaphragm, and the other is a fixed plate. After application of pressure, the diaphragm deflects. This leads to a change in separation between the plates, causing increase in capacitance. Since the capacitance depends upon separation of overlapping area between plates, the change in capacitance varies with the applied pressure on the diaphragm [5, 6]. The sensitivity of sensor is always influenced by the thickness, shape and material of diaphragm as well as the type of diaphragm like slotted, bossed or corrugated. The capacitive pressure sensors are widely accepted in design, manufacturing, consumer electronics, medical and automation industries, etc., for various applications like tire-pressure monitoring, intraocuscholarlar pressure measurement, fingerprint application, blood pressure sensing and barometric pressure measurement [4–8].

The blood pressure is measured in form of systolic and diastolic pressure. For blood pressure measurement, systolic and diastolic pressure must be known. The normal blood pressure is 120/80, where 120 is systolic and 80 is diastolic pressure. The systolic blood pressure varies normally from 120 to 160 mmHg, and diastolic blood pressure varies normally from 80 to 100 mmHg [9].

2 Mathematical Modeling

2.1 Mathematical Modeling of Deflection in Circular-Shaped Clamped Diaphragm

The diaphragm deflection follows Hook's law, i.e., the diaphragm deflection varies linearly when pressure is applied on the diaphragm [10].

If a uniform pressure (P) is applied in normal direction of thin circular flat plate clamped at edges, which is made of elastic, isotropic and homogeneous material, the plate equation can be represented by the following third-order partial differential equation [11]:

$$\frac{1}{r}\frac{\partial}{\partial r}\left[\frac{1}{r}\frac{\partial}{\partial r}\left(r\frac{\partial w}{\partial r}\right)\right] = \frac{P}{2D} \tag{1}$$

Or,

$$\frac{1}{r}\left[\frac{\partial^3 w}{\partial r^3} + \frac{1}{r}\frac{\partial^2 w}{\partial r^2} - \frac{1}{r^2}\frac{\partial w}{\partial r}\right] = \frac{P}{2D} \tag{2}$$

where $D = Et^3/(12 - v^2)$ is flexural rigidity of the diaphragm, deflection at distance r from the center of the diaphragm is w, thickness of diaphragm is t, Poisson's ratio is v, and Young's modulus of diaphragm material is E. To solve plate equation, the following three boundary conditions are used:

$$w(r = a) = 0 \tag{3}$$

$$\frac{\partial w(r = 0)}{\partial r} = 0 \tag{4}$$

$$\frac{\partial w(r = a)}{\partial r} = 0 \tag{5}$$

Here, a is the radius of the diaphragm. Using these three boundary conditions (Eqs. (3), (4) and (5)) the deflection in plate can be given by (Fig. 1):

$$w(r) = w_{max}\left[1 - \left(\frac{r}{a}\right)^2\right]^2 \tag{6}$$

where

$$w_{max} = \frac{Pa^4}{64D} \tag{7}$$

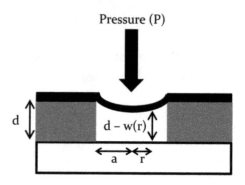

2.2 Mathematical Modeling of Normal Mode Circular-Shaped Capacitive Pressure Sensor

In this paper, we have proposed an equation of capacitance in a circular diaphragm pressure sensor, when pressure is applied. Based on this equation, the MATLAB® is used for predicting the output of a capacitive sensor. If overlapping area between plates, separation gap between plates and permittivity of medium are A, d and ε, respectively, than base capacitance can be given by [3, 6]:

$$C_b = \frac{\varepsilon A}{d} \tag{8}$$

And overlapping area for circular plate can be given by:

$$A = \pi a^2 \tag{9}$$

If pressure P is applied on diaphragm, the changed capacitance can be given by [2, 5]:

$$C_w = \int_0^{2\pi} \int_0^a \frac{\varepsilon r \, dr \, d\theta}{d - w(r)} \tag{10}$$

Substituting $w(r)$ from Eq. (6) in Eq. (10), we get

$$C_w = \frac{2\pi\varepsilon}{d} \int_0^a \frac{r \, dr}{1 - \frac{w_{max}}{d}\left[1 - \left(\frac{r}{a}\right)^2\right]^2} \tag{11}$$

After performing integration, we get,

$$C_w = \frac{\varepsilon A}{2\sqrt{d}\, w_{max}} \ln \left| \frac{\sqrt{d} + \sqrt{w_{max}}}{\sqrt{d} - \sqrt{w_{max}}} \right| \tag{12}$$

2.3 Sensitivity of Clamped Circular-Shaped Capacitive Pressure Sensor

We can define the sensitivity of the sensor as the change in capacitance of the sensor in the given pressure range divided by the change in pressure. Sensitivity of sensor can be given by:

$$S = \frac{C_{max} - C_{min}}{P_{max} - P_{min}} \tag{13}$$

3 Simulation Results and Discussion

3.1 Deflection in Diaphragms

Three different diaphragm thicknesses (4, 5 and 6 μm) and two different separation gaps (1 and 2 μm) are chosen to obtain six different designs. The sensors are designed for a pressure range from 0 to 280 mmHg above atmospheric pressure (760 mmHg), in order to measure the blood pressure. This pressure range equals 1.0–1.4 bar. In each design, the maximum deflection of diaphragm at maximum pressure (1.4 bar) is kept less than one-fourth of the separation gap to ensure good linearity. Based on this consideration, the diaphragm radius is determined. The deflection in 4-μm, 5-μm and 6-μm-thick diaphragms is presented in Figs. 2, 3 and 4, respectively, for two different diaphragm radiuses. These radiuses are obtained for separation gap of 1 and 2 μm for each diaphragm thickness. Diaphragm of particular thickness and smaller radius has less than 0.25-μm deflection at 1.4 bar pressure. On the other hand, for the same thickness of diaphragm, the larger radius has deflection less than 0.5 μm at 1.4 bar pressure. The diaphragm deflection increases as radius of diaphragm increases. In all the simulations, silicon is used as the diaphragm material and the following properties are used: E = 169.8 GPa and v = 0.066.

Fig. 2 Deflection in
4-μm-thick diaphragm

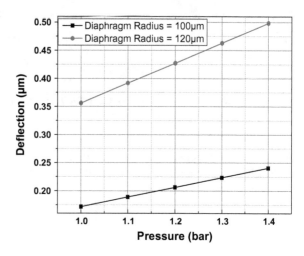

Fig. 3 Deflection in
5-μm-thick diaphragm

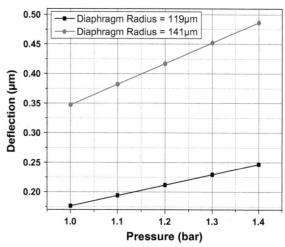

3.2 Base Capacitance and Capacitance Variation with Pressure

The base capacitance (or capacitance at 0 bar pressure) for different diaphragm thicknesses and separation is presented in Table 1. Base capacitance does not depend on diaphragm thickness, but it depends on separation gap and radius of diaphragm.

In the present work, pressure range varies between 1.0 and 1.4 bar. The graphs between pressure and capacitance for the six different designs are shown in Figs. 5, 6 and 7.

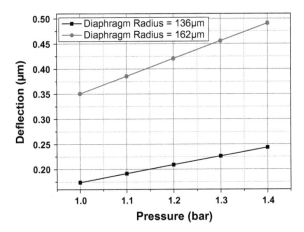

Fig. 4 Deflection in 6-μm-thick diaphragm

Table 1 Base capacitance for different diaphragm thicknesses and separation gaps

Diaphragm thickness (μm)	Separation gap (μm)	Diaphragm radius (μm)	Base capacitance (pF)
4	1	100	0.2782
5	1	119	0.3939
6	1	136	0.5145
4	2	120	0.2003
5	2	141	0.2765
6	2	162	0.365

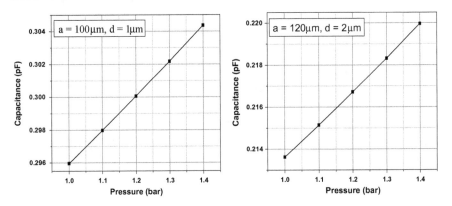

Fig. 5 Capacitance versus pressure for 4-μm-thick diaphragm

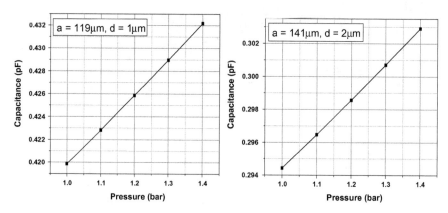

Fig. 6 Capacitance versus pressure for 5-μm-thick diaphragm

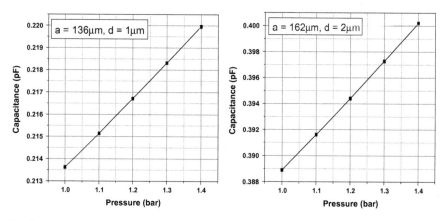

Fig. 7 Capacitance versus pressure for 6-μm-thick diaphragm

Table 2 Sensitivity of different designs

Diaphragm thickness (μm)	Diaphragm radius (μm)	Separation gap (μm)	Sensitivity (fF/bar)
4	100	1	21.0
4	120	2	15.85
5	119	1	30.75
5	141	2	21.2
6	136	1	39.52
6	162	2	28.28

3.3 Sensitivity

The sensitivity of the six different designs is shown in Table 2.

4   Conclusions

In this paper, capacitive pressure sensor is utilized for blood pressure monitoring. In the designs, deflection in diaphragm is less than one-fourth of separation between plates. Using mathematical modeling, the solution of capacitance after applying pressure is obtained which is the logarithmic function. The advantage of capacitive sensor is high pressure sensitivity and insensitivity to temperature change. However, loss in signal due to parasitic capacitance and low change in capacitance are major disadvantages. The circuitry used for measurement must address these issues.

From Table 2, we can conclude that the design having a diaphragm thickness of 6 μm, separation gap of 1 μm and diaphragm radius of 136 μm has the maximum sensitivity among all the six designs. It has a sensitivity of 39.52 fF/bar in the required pressure range. Therefore, this design is best suited for the measurement of blood pressure. The designed sensor can be fabricated with appropriate fabrication process sequence.

Acknowledgements Authors would like to acknowledge the generous support of Director, CSIR—Central Electronics Engineering Research Institute (CEERI), Pilani. And they would like to thank all the scientific and technical staff of Process Technologies Group-SSA for their support and co-operation.

References

1. Kovacs, G. T. A. (1998). *Micromachined transducer sourcebook*. McGraw-Hill.
2. Senturia, S. D. (2001). *Microsystem design*. Springer.
3. Balavelad, K., & Sheeparamatti, B. G. (2015). A critical review of MEMS capacitive pressure sensor. *Sensors and Transducers, 187,* 120–128.
4. Kota, S., Ananthasuresh, G. K., Crary, S. B., & Wise, K. D. (1994). Design and fabrication of microelctromechanical systems. *Journal of Mechanical Design, 116,* 1081–1088.
5. Kumar, S. S., & Pant, B. D. (2012). Design of piezoresistive MEMS absolute pressure sensor. *International Society for Optics and Photonics*, 85491G-85491G-10.
6. Eaton, W. P., & Smith, J. H. (1997). Micromachined pressure sensors: Review and recent developments. *Smart Materials and Structures*, 6(5).
7. Ziaie, B., & Najafi, K. (2001). An implantable microsystem for tonometric blood pressure measurement, *3,* 285–292.
8. Rey, P., Charvet, P., & Delaye, M. T. (1997). High Density capacitive Sensor Array for Fingerprint Sensor Application. In *IEEE, International Conference on Solid State Sensors and Actuators*, Chicago.
9. William, J. S., Brown S. M., & Conlin, P. R. (2009). Blood—pressure measurement. *The New England Journal of Medicine*.
10. Petersen, K. E. (1982). Silicon as a mechanical material. *Proceedings of the IEEE, 70,* 420–457.
11. Timoshenko, S., & Woinowsky, S. K. (1959). *Theory of plates and shells* (2nd ed). New York: McGraw-Hill.

Prospective of Automation for Checkbook Method in Cultivating Allium cepa

Nivedita Kar⑩, Ankita Kar and C. K. Dwivedi

Abstract This paper shows a small experiment for analyzing the yield of Allium cepa with the help of checkbook method. This manual method is used for maintaining the schedule of irrigation in a kitchen garden which is a closed area nearby the living place. The paper also focuses on the careful steps needed in performing the agricultural practices for cultivating any vegetable crop. The result of experiment suggests the need for automation in irrigation as various limitations are observed in following the conventional cultivation practices. Hence for improving the yield of the crop, a scientific automated irrigation system is proposed by harnessing solar energy in the urban and rural sector.

Keywords Allium cepa · Arduino uno · Checkbook method · DAP
Kitchen garden

1 Introduction

Modern living has brought about little changes in the food preferences of our people. During recent years, the interest in vegetable production has increased rapidly. Watching and working with vegetable plants can add a new dimension of enjoyment to life and bring an awareness of nature in the small space having few plants or in large plots. Gardening with vegetables is a healthy habit and provides delicious and highly nutritious fresh food. An area close to our proximity to the living place, e.g., space in the back door, where fresh vegetables, herbs, and fruits can be grown all

N. Kar (✉) · A. Kar · C. K. Dwivedi
Department of Electronics and Communication, (J K Institute of Applied Physics),
University of Allahabad, Allahabad 211002, India
e-mail: nivedita0113@gmail.com

A. Kar
e-mail: ankitakar16@gmail.com

C. K. Dwivedi
e-mail: ckdwivedi@gmail.com

© Springer Nature Singapore Pte Ltd. 2019
A. Khare et al. (eds.), *Recent Trends in Communication, Computing, and Electronics*, Lecture Notes in Electrical Engineering 524,
https://doi.org/10.1007/978-981-13-2685-1_43

the year for houschold use, is known as a kitchen garden. This garden adds special meaning to nature as one can continuously nurture and harvest a small seed into a colorful productive plant with own hands in a reachable area, often by a walk.

But with India's isotropic climate, the cultivators of vegetables in urban–rural sectors are not able to use agricultural resources efficiently due to lack of rains, land and water reservoirs as in [1, 2]. They make continuous efforts for the extraction of water from low water levels of earth, thereby converting fertile lands into the zones of un-irrigated areas as in [1]. At the present era, manually controlled irrigation technique is used at the regular intervals in these gardens. But with these conventional methods, the entire soil surface becomes saturated when over watered, promoting infections by leaf mold fungi and affecting the production. Even if people are using water pump, most of the time, these are kept standstill due to non-availability of grid power in the remote areas where the potential of sunlight availability is tremendous throughout the year as in [3]. If the solar power is harnessed with a proper scientific irrigation method, one can save water, energy, laboring time, and also the production cost if using commercially as in [3]. Through this paper, an attempt has been made to make the readers aware of the steps regarding agricultural practices in cultivating Allium cepa (Onion) and also on the limitations of these conventional methods with the help of an experiment, thus proving a need for a new scientific automation technique for irrigation in the kitchen garden.

2   Steps for Agricultural Practices in Kitchen Garden

Monocotyledoneae Onion (*pyaz*) belongs to family Amaryllidaceae having genus Allium and species cepa as in [4]. It originates from northwest India and spreads around the Mediterranean Sea as in [5, 6]. Most popular varieties of Onion are Agrifound Dark Red and Agrifound Light Red. Onions seeds are sown in a nursery, and then, seedlings are transplanted in both Kharif and Rabi season. Best time for sowing the crop seeds is from middle of October to the end of November. Seedlings are transplanted when they are 8–10 weeks old which means in late December and early January and are extended up to a maximum period of early February. For vegetative propagation, small size bulbs or bottom set of Onion are grown into large ones which are again broken into smaller ones. As Onion crop is a shallow rooted, any root pruning during cultivation affects in reducing the bulb growth. This crop matures when tops just above the bulbs start dropping, while leaves are still green as in [7, 8]. The frequency of irrigation is not definite as it depends on soil and climatic conditions. Irrigation is stopped when the leaf tops mature and starts falling. Onion crop is sensitive to the high acidity of soil, and it is recommended to test the soil sample by optical pH sensors as in [4]. The optimum pH value ranges between 5.8 and 6.5 as in [6]. Because vegetables have 80–95% water with a small amount of flavoring and some vitamins, their yield and quality suffer rapidly when subjected to improper irrigation practices. Thus, for achieving maximum output, some careful steps are needed in agricultural practices for the urban and rural sector. The first step is site selection, i.e., a nearby area is chosen for practicing cultivation. But due to unavailability of the land in the urban sector, container gardening or earthen pots can be an alternate solution for site selection. When planning a vegetable garden, an area is chosen in such a way that there is plenty of morning sunlight and some afternoon shade. The growth of the vegetable crops which bear fruits is the best when exposed to six to eight hours of full sunlight. Leafy vegetables and root type vegetables require partial shade. Most vegetables can easily be grown directly from seeds sown in the garden. Few examples are pumpkin, peas, beans, spinach, tomatoes, and onions.

After site selection, the second step in practicing agriculture is the soil preparation and mulching. Soil preparation is a process to maintain and improve soil conditions by mixing it with organic matter and fertilizers before planting. This preparation and cultivation of the soil are done when it is dry or slightly moist (never when wet). Along with soil preparation, a process known as mulching is done. Mulching is covering the soil with a protective material to improve the nutrient and water-holding capacity, drainage and aeration of the soil. With mulch, very little cultivation effort is needed. Well-rotted manure, compost, barks or wood chips, straw, autumn leaves, shredded newspaper, and grass cuttings are commonly used organic mulches which decompose over time, thereby enriching the soil's organic content. Inorganic mulches such as plastic sheeting, gravel, pebbles, and stones are materials that do not break down but help the soil retain moisture and often creates a buildup of heat in the soil. After mulching is done, for soil preparation trenches are dug in the area chosen for a vegetable garden. Each bed should have narrow pathways in between. The healthy topsoil is kept in one heap, and the poor subsoil in another heap. This trench is now watered. The organic mulch is spread into the trench and is allowed to sink down for a few days. The topsoil is placed back into the trench on the top of the organic mulch. The subsoil is used to make the paths around the beds.

Fertilization is the next step as the kitchen garden needs to be self-reliant for fertility. For checking the selected site's soil, samples should be taken and got tested by the authorized agency as in [4]. This testing helps to gather knowledge about soil deficiencies. The fertilizers should be added to plants occasionally according to test results. For this process, a nutrient solution should be prepared having minor elements such as iron, zinc, boron, and manganese by carefully following the labeled directions and then pouring it or spreading it all over the soil mix. This mixture is highly concentrated and must be diluted before it can be used to fertilize the plants. In urban agriculture, planning the garden design is the next factor. This includes crop selection, crop rotation, and edge planting. Vegetables should be grown densely in the rotation, meaning that the same plant should not be planted wide apart in the same bed season after season. The correct way to do crop rotation is to first grow vegetables which take very little soil nutrients, e.g., beetroot, carrots, onions (low feeders). When these vegetables are harvested, plant vegetables such as cabbage, tomato, sunflower, pumpkins, spinach in the same bed in the next season which take a lot of nutrients (nitrogen) out of the soil. Once those heavy feeders are harvested, plant the givers, e.g., soya beans which make the soil richer by putting nutrients back into it. Maintenance of kitchen garden can be done also by practicing edge planting along with crop rotation in the edges around the garden and its beds. These plants help support the garden by providing mulch, protection from weeds, windbreaks, repelling pests, and producing other useful resources such as the smell of marigold and its leaves around the edge help to repel many types of pest insect. They also produce a substance from their roots which repels damaging soil nematodes.

The most important aspect of urban agriculture is water management, i.e., irrigation. Watering is needed to keep the soil moist in the root zone of the plant throughout the growing season. Excessive fluctuations of soil moisture adversely affect plant growth and quality. The frequency of watering depends on many factors. Vegetables like cabbage, onion, lettuce, corns having shallow roots need a number of irrigations often than deep-rooted vegetables like asparagus, tomato, and watermelon. Coarse-textured soils (sandy loams) need to be irrigated more often than fine-textured (clay or silt loams). Vegetable crops differ in the method of irrigation which can be used economically in their production. The installation of an efficient irrigation system will improve crop performance by ensuring adequate water and nutrients as and when needed. Second last step in cultivation practice is checking for crop diseases. Vegetables grown in any garden are susceptible to same insects and diseases that are common to any agricultural field. One should check the plants periodically for diseases, foliage, and fruit-feeding insects. The distortions in leaf shape and the presence of holes on them or dead dried, powdery or rusty areas are commonly caused due to the presence of insects and fungus, respectively. The use of EPA-approved fungicides and insecticides in a timely manner can reduce the plant disease or harmful insects. The last step is the protection of the selected site, i.e., kitchen garden area. It should not be possible for livestock to enter the area. A permanent fence could be made or thorny plants could be cut and used to make a fence, but the best method is to plant a living fence to protect the garden.

3   Experiment

The checkbook procedure was used to analyze and to study the yield with the help of a small experiment following all the conventional agricultural steps. This method is a soil moisture accounting procedure which checks the crop-water usage values with time as in [6]. For performing this experiment, two similar earth containers, as an alternative to the site selection, having dimensions shown in Table 1, are considered. Instruments used for measurement were standard meter scale and digital weighing balance. The values of these earth pots mentioned in Table 1 are average of a series of three readings taken for calculation during the experiment. The net weight of soil in Pot-1 was 4.75 kg and Pot-2 was 4.55 kg. Inorganic mulch was used for covering the holes in the pots in such a manner that proper water drainage and aeration was maintained.

At first, the soil was prepared for transplanting the seedlings into the container (pot). For this, 8 kg of sandy loam soil and 2 kg of cowdung manure were taken along with 10 gm of DAP and 20 gm of bone ash fertilizers and these were mixed well. Before mixing the cowdung manure, it was ensured manually that the manure

Table 1 Average details of earthen pots

Earth pot no.	Sample pot details				
	Diameter (cm)	Cross-sectional area (sq. cm)	Height (cm)	Tare weight (Kg)	Gross weight (Kg)
Pot-1	24.80	482.81	18.50	1.50	6.25
Pot-2	24.50	471.20	19.80	1.20	5.75

was fresh and free from termites. Diammonium phosphate (DAP) fertilizer, having chemical formula as $(NH_4)_2HPO_4$, is an excellent source of phosphate (P) and nitrogen (N) needed for plant nutrition and the other fertilizer, bone ash $[Ca_5(OH)(PO_4)_3]$ is needed for plant strength, which is a white powdery material produced by the calcinations of bones. This rich soil was divided into two pots, and their gross weight was taken. The filled soil was kept approximately 2.5–3 cm below the height from the top surface of the container to prevent the overflow of water along with nutrients from topsoil while irrigating. For irrigating, a plastic bottle of 250 mL was taken. A line was marked in the middle section of this 250-mL bottle for showing the 125 mL capacity. The initial quantity of water considered for POT-1 and for POT-2 was equal to 250 mL and 125 mL, respectively. By using checkbook method, the initial schedule for watering in both pots at an interval of 10 days considered during the growing season of Allium cepa (from February to May) is specified in Table 2. For the first 50 days, water was supplied in the ratio of 2:1 in POT-1 and POT-2, respectively, as shown in Table 2.

The idle curve for crop-water usage throughout the growing seasons is shown in Fig. 1. This figure demonstrates the consumption of water by a crop increases throughout its initial growing stages, i.e., from germination/emergence stage to vegetative growth stage and up to the reproduction stage. But after this third stage, water requirement gradually decreases in the maturity stage. In the second step, 18 seedlings of Allium cepa ranging from 12.7 to 15.2 cm in size were transplanted in two pots while maintaining an equal gap between the plants. After this process, the initial size of each sample was measured from the surface of the soil as shown in Tables 3 and 4 for POT-1 and POT-2, respectively.

In the third step, plant growth was analyzed for selected 10 healthy samples out of 18 transplanted plants as shown in Figs. 2 and 3. The growth analysis of the samples S1, S2, S4, S5 and S6 for POT-1 and of the samples S3, S7, S8, S9 and S10 for POT-2 shown in Tables 3 and 4, respectively, was done by measuring only the length of the green portion for each sample above the surface of soil. The time period considered for this analysis was, from February to mid of March, divided into four intervals of 10 days, i.e., total 40 days. The actual water consumption for POT-1 and POT-2 at an interval of 10 days from February to early May is shown in Fig. 4.

Out of 5 samples of POT-1, S2 was having a continuous increase in the length of each leaf till 30 days and for samples S4 and S5, the growth continued till 40 days. In this POT-1, only one leaf of S5 and S6 was having dryness in the leaf tip occurring

Table 2 Water schedule during growing season

Month name	Quantity of water (mL)		No. of times in every 10 days	Total amount in Pot-1 for every 10 days (L)	Total amount in Pot-2 for every 10 days (L)
	POT-1	POT-2			
February	250	125	5[a]	1.250	0.625
	250	125	10[b]	2.500	1.250
	250	125	15[c]	3.750	1.875
March	250	125	20[d]	5.000	2.500
	250	125	20	5.000	2.375
	325	250	25[e]	8.125	6.250
April	325	250	30[f]	9.750	7.500
	325	250	25	8.125	6.250
	250	125	20	5.000	2.500
May	250	125	15	3.750	1.875
	250	125	10	2.500	1.250
	250	125	5	1.250	0.625
Total amount consumed in 120 days				56.000	35.000

[a]Once in every alternate day
[b]Once every day
[c]Once in every odd and twice in every even day
[d]Twice every day
[e]Twice in every odd and thrice in every even day
[f]Thrice every day

Fig. 1 Idle curve for crop-water usage during growing seasons

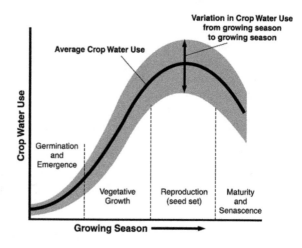

Table 3 Analysis of Pot-1 samples (February to mid-March)

Sample no.	Growth analysis in POT-1						
	Initial no. of dry leaves	Initial size (cm)[a]	Leaf no.[b]	Growth in time interval of 10 days (cm)[c]			
				10	20[d]	30[e]	40[f]
S1	2	11.4	1	28.5	37.5	**40.3**	38.1
			2	21.0	23.6	27.4	**31.9**
			3	10.5	13.4	**15.2**	11.9
S2	1	12.2	1	34.3	40.9	**40.1**	37.9
			2	27.5	30.4	**34.8**	33.6
			3	11.0	14.3	**17.9**	15.8
S4	2	12.7	1	34.5	41.0	**38.8**	36.5
			2	16.0	20.1	25.7	**29.5**
			3	15.1	17.3	**14.7**	12.0
S5	1	9.4	1	26.0	30.4	35.2	**39.7**
			2	23.5	26.8	30.7	**32.2**
			3	7.0	**9.5**	8.9	7.1
S6	1	9.9	1	32.5	36.1	**40.6**	38.0
			2	18.0	23.9	**28.6**	28.4
			3	8.5	**10.4**	8.9	8.0

[a] Sample measured above soil
[b] Selection of the best three leaves
[c] Excluding dry part of leaf tip
[d] Maximum growth in 20 days (bold)
[e] Maximum growth in 30 days (bold)
[f] Maximum growth in 40 days (bold)

after 20 days. But for POT-2, one sample S10 out of 5 samples was showing a decrease in growth after 10 days only. Remaining four samples were having growth in 2 or 3 leaves till 20 days, i.e., dryness was occurring after 20 days. Only one leaf in 3 samples S3, S8 and S9 was showing growth up to 30 days.

4 Results and Discussion

The experiment showed that no sample in POT-2 was having leaf growth up to 40 days. Even in one its sample, dryness was occurring after 10 days. By considering the leaves of all the samples in Tables 3 and 4, it was observed that the maximum leaf growth in POT-1 was 40.6 cm in 30 days and for POT-2 sample was only 36.9 cm in 20 days. This abnormal growth was due to not maintaining the irrigation ratio of 2:1 between the pots as shown in Fig. 4. Because this human error in accounting

Table 4 Analysis of Pot-2 samples (February to mid of March)

Sample no.	Growth analysis in POT-2						
	Initial no. of dry leaves	Initial size (cm)[a]	Leaf no.[b]	Growth in time interval of 10 days (cm)[c]			
				10[d]	20[e]	30[f]	40[g]
S3	1	11.7	1	25.5	**36.9**	29.5	24.8
			2	18.1	20.4	**21.5**	19.2
			3	12.5	**10.3**	8.8	6.4
S7	1	9.4	1	19.0	**28.5**	24.9	20.3
			2	15.5	**14.1**	10.3	7.5
			3	8.6	**9.7**	7.4	5.1
S8	2	9.9	1	24.0	**32.7**	28.7	25.3
			2	14.0	**13.1**	9.9	6.4
			3	8.0	9.5	**11.7**	8.4
S9	2	10.7	1	22.8	**31.0**	26.3	24.4
			2	15.9	**19.9**	17.2	16.8
			3	10.7	12.7	**14.9**	14.0
S10	1	11.2	1	**25.1**	21.6	18.2	13.9
			2	22.4	**25.6**	21.3	16.2
			3	11.8	**14.5**	12.9	9.7

[a]Sample measured above soil
[b]Selection of the best three leaves
[c]Excluding dry part of leaf tip
[d]Maximum growth in 10 days (bold)
[e]Maximum growth in 20 days (bold)
[f]Maximum growth in 30 days (bold)
[g]Maximum growth in 40 days (bold)

water scheduling, crop-water usage throughout the growing seasons was disturbed as compared with Fig. 1. Also from the comparative study of two pots, it was found that the growth response and yield for POT-1 were better than POT-2. For finding the yield, individual sample weight was not counted instead potwise collective measurement was done. It was observed that total yield of five samples in POT-1 was 67.5 g with total water consumption of 46.77 L in 100 days and the yield of all samples in POT-2 was 50 g with total water consumption of 30.12 L for the same period. But even after proper manual scheduling, the growth in both pots was random which means Onion requires continuous dripping of water supply in a controlled manner for its proper growth.

Fig. 2 Samples labeled in
POT-1

Fig. 3 Samples labeled in
POT-2

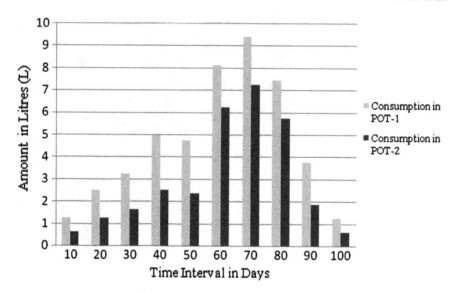

Fig. 4 Water consumption analysis for POT-1 and POT-2 at an interval of 10 days in Liters (February to early May)

5 Suggestions for Future Work

Maintaining the regularity in water supply and in its accounting by the manual check-book method is difficult for a human being due to their physical limitations for retaining the time interval properly. To overcome this, automation may be used for the present crisis of water. In the automated irrigation technique, three LCDs with common data lines are interfaced to an Arduino Uno-based development board having six analog input pins and 14 digital input/output (I/O) pins. These LCDs are used to display the values obtained from optical pH sensor, moisture sensor and temperature sensor which shows the field parameters and weather conditions, respectively, as in [9–11]. The sensed values of all these analog sensors are compared with predefined values. If the sensed value of soil moisture sensor is lower than the earlier set programmed values of the microcontroller, then the water pump is activated. Similarly, if the sensed data of temperature is more than preset value, then again the pump connected to the water reservoir will run. An SPV technology as in [12] is used to run the pump. The solar panel will be disconnected from the circuit if its voltage goes below specified battery voltage, and the battery will be the only energy source to run the pump as in [3, 10]. But if grid power is available, then the SPV will be disconnected from the circuit and only AC power from main supply will be drawn to run the pump as in [3]. This proposed system will be user-friendly and will be more efficient in maintaining a crop-water relationship. Hence it can be proved better for crop growth and yield by harnessing the solar power.

6 Conclusions

Checkbook method, an irrigation scheduling procedure, is used for obtaining maximum crop production. But utilizing this method for conventional agricultural practices does not maintain plant-water balance as the following limitations were observed. The first limitation was that this method was done manually and was more prone to human errors which affect the growth and yield of the plant. Secondly, for measuring the wetness of the soil, feel method was used in this conventional cultivation system. Hence only the topsoil was considered for dryness or wetness which gave the wrong perception about the condition of soil several times. Thirdly, as this manual method is more subjective to irregular water supply, water deficiency caused slow growth rate and lower weighted fruit. Fourth limitation was the persistence of weeds and pest problems in the soil due to overspreading of water in nearby regions of the targeted area which gave a favorable environment for their growth. This also led to the wastage of water during irrigating. Also, the environmental obligations were not met by the used resources as nonrenewable energy resources, i.e., fuel consumption for generators was required which made this old method more costly for the farmers. Lastly, there was a requirement of testing the soil parameters, e.g., electrical conductivity, pH from outside agency before the transplanting process as the condition of the soil helps in improving the plant growth but this extra effort was not done by many people leading to affect the output. Hence for achieving a better result, a user-oriented scientific automated irrigation system is proposed which will be a modern electronic technique for harnessing solar energy in the agricultural sector for improving the yield.

References

1. Mahendra, S., & Bharathy, M. L. (2013). Microcontroller based automation of drip irrigation system. *AE International Journal of Science & Technology, 2*(1).
2. Gokul, M. P., & Naresh, A. K. (2016). Automatic drip irrigation system using solar power. *International Journal of Recent Trends in Engineering & Research, 02*(04), 1–3.
3. Kumar, S., Sethuraman, C., & Srinivas, K. (2017). Solar powered automatic drip irrigation system (SPADIS) using wireless sensor network technology. *International Research Journal of Engineering and Technology, 04*(07), 722–731.
4. Kar, A., & Shukla, N. K. (2017). Use of optical sensors for conserving soil constituents. *International Journal of Emerging Technology and Advanced Engineering, 7*(12), 28–30.
5. Choudhury, B., & Choudhury, D. K. (1996). *Vegetables* (1st ed.). New Delhi, India: National Book Trust.
6. Nath, P., Velayudhan, S., & Singh, D. P. (1994). ICAR low-priced books series: Vegetables for the tropical region. In A. M. Wadhwani (Ed.). New Delhi, India: Publication and Information Division, ICAR.
7. Arakeri, H. R., & Patil, S. S. (1956). Effect of bulb size, spacing and time of planting on the yield of onion seed. *Indian Journal of Agronomy, 1*(2), 75–80.
8. Singh, D. P., & Singh, R. P. (1974). Studies on the effect of time of sowing and age of seedlings on growth and yield of onion. *Indian Journal of Horticulture, 31*(2), 69–73.

9. Janardhanan, A., & Bolar, A. (2015). Automatic drip irrigation using wireless sensor network. *IOSR—Journal of Electronics and Communication Engineering (IOSR-JECE)*, 109–112. e-ISSN: 2278-2834, p-ISSN: 2278-8735.
10. Luciana, M. L., Ramya, B., & Srimathi, A. (2013). Automatic drip irrigation unit using PIC controller. *International Journal of Latest Trends in Engineering and Technology, 2*(3), 108–114.
11. Fang Meier, D. D., Garrote, D. J., Mansion, F., Human, S. H. (1990). Automated irrigation systems using plant and soil sensors. In *Visions of the Future* (pp. 533-537). ASAE Publication 04-90, St. Joseph, Michigan.
12. Kar, N., & Dwivedi, C. K. (2017). Harnessing solar photovoltaic for safe, clean and sustainable development. *International Journal of Emerging Technology and Advanced Engineering, 7*(12), 46–49.

Machine Learning Algorithms for Anemia Disease Prediction

Manish Jaiswal, Anima Srivastava and Tanveer J. Siddiqui

Abstract The remarkable advances in health industry have led to a significant production of data in everyday life. These data require processing to extract useful information, which can be useful for analysis, prediction, recommendations, and decision making. Data mining and machine learning techniques are used to transform the available data into valuable information. In medical science, disease prediction at the right time is the central problem for professionals for prevention and effective treatment plan. Sometimes, in the absence of accuracy this may lead to death. In this study, we investigate supervised machine learning algorithms—Naive Bayes, random forest, and decision tree algorithm—for prediction of anemia using CBC (complete blood count) data collected from pathology centers. The results show that Naive Bayes technique outperforms in terms of accuracy as compared to C4.5 and random forest.

Keywords Anemia · Classification algorithms · Decision making
Complete blood count (CBC)

1 Introduction

The modern health care system generates huge volume of data every day. There is a need to mine and analyze these data to extract useful information and to reveal hidden pattern. Data mining is the process of discovering new patterns from data collected from varying sources. A number of machine learning algorithms have been used successfully in making prediction in various domains such as healthcare, weather

M. Jaiswal (✉) · A. Srivastava · T. J. Siddiqui
Department of Electronics and Communication, University of Allahabad, Allahabad, India
e-mail: manish.jk50@gmail.com

A. Srivastava
e-mail: animasparklestar@gmail.com

T. J. Siddiqui
e-mail: siddiqui.tanveer@gmail.com

© Springer Nature Singapore Pte Ltd. 2019
A. Khare et al. (eds.), *Recent Trends in Communication, Computing,
and Electronics*, Lecture Notes in Electrical Engineering 524,
https://doi.org/10.1007/978-981-13-2685-1_44

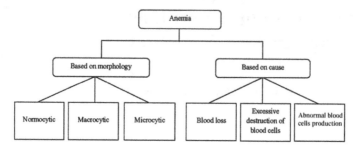

Fig. 1 Classification of anemia

forecasting, stock price prediction, product recommendation. An important aspect of medical science research is the prediction of various diseases and factors that cause them. In medical domain, healthcare data are being used to predict epidemics, to detect disease, to improve quality of life and avoid early deaths [1]. In this work, we investigate three different classification algorithms for its prediction.

Anemia is defined as the decrease in amount of red blood cells (RBCs) or hemoglobin in the blood [2] that has significant adverse health consequences, as well as adverse impacts on economic and social development. Although the most reliable indicator of anemia is blood hemoglobin concentration, there are a number of factors that can cause anemia such as iron deficiency, chronic infections such as HIV, malaria, and tuberculosis, vitamin deficiencies, e.g., vitamins B12 and A, cancer, and acquired disorders that affect red blood cell production and hemoglobin synthesis.

Anemia causes fatigue and low productivity [3–5] and, when it occurs in pregnancy, may be associated with increased risk of maternal and perinatal mortality [6, 7]. According to World Health Organization (WHO), maternal and neonatal mortality were responsible for 3.0 million deaths in 2013 in developing countries.

Anemia disease prediction plays a most important role in order to detect other associated diseases. Anemia disease is classified on the basis of morphology or on the basis of its underlying cause (Fig. 1).

Based on the morphology, anemia is divided into three types, which are normocytic, microcytic and macrocytic. Based on cause, anemia is classified into three types, namely blood loss, inadequate production of normal blood and excessive destruction of blood cells.

In this paper, we attempt to investigate the performance of Naive Bayes, random forest and decision tree algorithm for anemia disease prediction on dataset collected from local pathology centers. The need of this investigation arises from the fact that the underlying cause of the disease varies from one region to another. Although random forest classifier has been earlier investigated for predicting heart and chronic kidney disease, to the best of our knowledge it has not been investigated for anemia disease prediction. This adds novelty to the work.

The rest of the paper is organized as follows:

Section 2 introduces and briefly reviews existing related work. In Sect. 3, we discuss various types of anemia diagnosis tests. Section 4 presents proposed methodology. Section 5 presents experimental details and discussion. Finally, we conclude in Sect. 6.

2 Related Works

In the last decade, numerous data mining and machine learning techniques have been used for anemia disease. Most noted ones are the following:

In [8], SMO support vector machine and C4.5 decision tree algorithm have been used for the prediction of anemia and the performance comparison of the two algorithms is done.

In [9], WEKA is used to get a suitable classifier for developing a mobile app, which can predict and diagnose hematological data comments. The authors compared neural network classification algorithms with J48 and Naive Bayes classifier. The results show that J48 classifier exhibits maximum accuracy.

Dogan and Turkoglu [10] developed a decision support system for detecting iron deficiency anemia using the decision tree algorithm. The algorithm uses three hematology parameters, serum iron, serum iron-binding capacity and ferritin. The evaluation is done on data of 96 patients, and the results were successfully matched with physician's decision.

Abdullah and Al-Asmari [11] experimented with WEKA algorithms: Naive Bayes, multilayer perception, J48 and SMO in an attempt to predict anemia types using CBC reports. The evaluation was done on real data constructed from CBC reports of 41 anemic persons. Similar to [9], J48 decision tree algorithm along with SMO was the best performer with an accuracy of 93.75%.

Unlike the work in [9, 11], we have chosen a different set of classifier and local data in our work.

3 Diagnostic Tests Classification

There are four main tests that are ordered to diagnose anemia disorder which are complete blood count (CBC), ferritin, PCR (polymerase chain reaction) and hemoglobin electrophoresis.

- CBC test is the most frequently blood test to measure overall health and determine a wide range of diseases [8] including anemia, infection and leukemia. A complete blood count test measures almost 15 parameters including: hemoglobin (Hb), red blood cells (RBC), hematocrit (HCT), mean corpuscular hemoglobin (MCH), mean corpuscular volume (MCV), and so on [8].

- A ferritin test measures the amount of iron store in the body. High levels of ferritin indicate an iron storage disorder, such as hemochromatosis. Low levels of ferritin indicate iron deficiency, which causes anemia.
- PCR test is a molecular test, which is used to diagnose genetic disorder.
- A hemoglobin electrophoresis test is a blood test used to measure and identify the different types of hemoglobin in the bloodstream.

4   Methodology

We have used three classifiers, namely random forest, Naive Bayes and decision tree C4.5 algorithm. Figure 2 depicts the flowchart of the proposed method.

4.1   Random Forest Algorithm

Random forest (RF) algorithm derives from decision tree classifier. It is a combination of tree predictors, which aggregates the results of all the trees in the collection and uses majority voting in prediction.

Fig. 2 Flowchart of proposed model

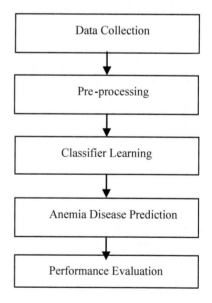

4.2 Decision Tree Algorithm

A decision tree is a tree in which each branch node represents a choice between a number of alternatives, and each leaf node represents a decision. It has been extensively used in various fields [12, 13]. C4.5 (J48 in WEKA) is a decision tree developed by Ross Quinlan.

4.3 Naive Bayes Algorithm

Naive Bayes algorithm is based on Bayes rule of conditional probability. It uses all the attributes contained in the data and analyzes them individually as they are equally important and independent of each other. It requires very less amount of training data.

5 Experimental Results and Discussion

5.1 Dataset

We collect data from different pathology centers and laboratory test centers in nearby area. The collected dataset consists of 200 test samples. These are CBC test data. The dataset contains 18 attributes out of which we have selected only those that are required for anemia disease detection. These are age, gender, MCV, HCT, HGB, MCHC and RDW.

5.2 Experimental Setup

The proposed method uses CBC test values. First, the data are pre-processed to extract the seven attributes as mentioned in Sect. 5.1. Then, we apply the random forest, decision tree and NB classifier on it. The performance evaluation is done in terms of accuracy and mean absolute error (MAE). The mean absolute error (MAE) measures how close the predictions are to the eventual outcomes. Table 1 shows the results of the three classifiers. Tenfold cross-validation has been used to obtain accuracy.

The comparative performance of each classification algorithm based on accuracy and MAE is shown in Figs. 3 and 4, respectively. The Naive Bayes classifier exhibits the best performance on our dataset, which is unlike [9] and [11]. It is not surprising because the dataset being used in these works is different and the cause of disease in different countries might be different. We achieve a maximum accuracy of 96.09%

Table 1 Comparision of algorithms

	Random forest	Naive Bayes	C4.5
Mean absolute error	0.0332	0.0333	0.0347
Accuracy	95.3241	96.0909	95.4602

Fig. 3 MAE using each algorithm

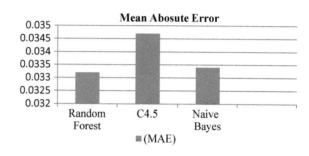

Fig. 4 Comparison of accuracy using each algorithm

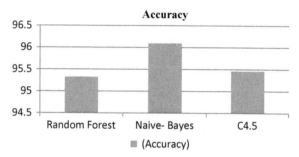

with NB classifier which is better than the best performing classifiers—SMO and J48 with an accuracy of 93.75%—reported in [11].

6 Conclusion and Future Work

In this paper, we have compared the performance of three different classifiers in the prediction of anemia disease. The experimental result on a sample dataset suggests that Naive Bayes classification algorithm provides the best performance in terms of accuracy as compared to C4.5 and random forest. Automatic prediction can reduce manual effort involved in diagnosis. In the future, automated tools can be developed which can help the prediction results to suggest further diagnosis. Such automated tools can prove valuable in timely detection of more serious diseases. Furthermore, such disease prediction system can be extended to recommend a treatment plan.

References

1. Arun, V., et al. (2015). Privacy of health information in telemedicine on private cloud. *International Journal of Family Medicine & Medical Science Research*.
2. Provenzano, R., Lerma, E. V., & Szczech, L. (2018). *Management of anemia*. Springer.
3. Ezzati, M., Lopez, Ad, Rodgers, A., & Murray, C. J. L. (2004). *Comparative quantification of health risks: Global and regional burden of disease attributable to selected major risk factors*. Geneva: World Health Organization.
4. Balarajan, Y., et al. (2011). *Anaemia in low-income and middle-income countries*.
5. Haas, J. D., Brownlie, T. (2001). Iron deficiency and reduced work capacity: A critical review of the research to determine a causal relationship. *The Journal of Nutrition*.
6. Kozuki, N., Lee, A. C., & Katz, J. (2012). Child health epidemiology reference group. Moderate to severe, but not mild, maternal anemia is associated with increased risk of small-for-gestational-age outcomes. *The Journal of Nutrition*.
7. Steer, P. J. (2000). Maternal hemoglobin concentration and birth weight. *The American Journal of Clinical Nutrition*.
8. Shilpa, S. A., Nagori, M., & Kshirsaga, V. (2011). Classification of anemia using data mining techniques. In *Swarm, evolutionary, and memetic computing* (pp. 113–121). Springer.
9. Amin, N., & Habib, A. (2015). Comparison of different classification techniques using WEKA for hematological data. *American Journal of Engineering Research, 4*(3), 55–61.
10. Dogan, S., & Turkoglu, I. (2008). Iron deficiency anemia detection from hematology parameters by using decision tree. *International Journal of Science and Technology*, 85–92.
11. Abdullah, M., & Al-Asmari, S. (2016). Anemia types prediction based on data mining classification algorithms. In *Communication, management and information technology*. London: Taylor & Francis Group.
12. Jerez-Aragonés, J. M., et al. (2003). A combined neural network and decision trees model for prognosis of breast cancer relapse. *Artificial Intelligence in Medicine*, 45–63.
13. Podgorelec, V., et al. (2002). Decision trees: An overview and their use in medicine. *Journal of Medical Systems*, 445–463.

Learning Pattern Analysis: A Case Study of Moodle Learning Management System

Rahul Chandra Kushwaha⊙, Achintya Singhal and S. K. Swain

Abstract This paper presents the learning pattern analysis of online learning management system Moodle. The experimental work was carried out in Banaras Hindu University, India on the students of three years post graduate course on Computer Application. The comparative study has done on learning patterns of the students on tradition pedagogy with online pedagogy through the learning management system. The Moodle data analytics tool was used for the purpose of reporting students' data. The results of students learning performance were compared and analyzed using t-test, content analysis and various other mining and analytics tools.

Keywords Learning analytics · Educational data mining
Learning pattern analysis · Moodle · Learning management system

1 Introduction

The learning management system is a popular platform for information sharing and social interaction. There are many learning management system like BlackBoard, Moodle, WebCT etc. Now a day, Moodle is the most popular open source Learning Management System (LMS). The LMS has powerful database to store its data. Every click of the user date is stored in Moodle database [1, 2]. The data mining techniques can be used to analyze on these data to find out useful information. The learning

R. C. Kushwaha (✉)
DST-Centre for Interdisciplinary Mathematical Sciences, Institute
of Science, Banaras Hindu University, Varanasi, India
e-mail: rahulncert@gmail.com

A. Singhal
Department of Computer Science, Institute of Science, Banaras Hindu University, Varanasi, India
e-mail: achintya.singhal@gmail.com

S. K. Swain
Faculty of Education, Banaras Hindu University, Varanasi, India
e-mail: skswain59@gmail.com

© Springer Nature Singapore Pte Ltd. 2019   471
A. Khare et al. (eds.), *Recent Trends in Communication, Computing,
and Electronics*, Lecture Notes in Electrical Engineering 524,
https://doi.org/10.1007/978-981-13-2685-1_45

analytics is future of education [3–5]. The students using virtual learning like LMS, MOOCs and social media are generating large amount of data, which can be used to analyze student's activities and their learning performance [1–11].

Due to the digitization of education in academic institutions, huge amounts of students' related data are generating [6]. These data can be converted into meaningful information, which can be helpful for the teachers, academician and administrator for decision making. The main purpose of data mining methods is to extract meaningful knowledge from data [2]. The application of data mining methods to educational data is referred to as Educational Data Mining (EDM) [5]. Baker [5, 12] has proposed five primary categories for Educational Data Mining [13]. These are as followed. (1) Prediction (2) Clustering (3) Relationship mining (4) Discovery with models (5) Distillation of data for human judgment.

In Prediction, the goal is to predict the class or label of data object.

In Clustering, the goal is to find data points that naturally group together, splitting the group data into a set of clusters.

In relationship mining, the goal is to discover the relationship between variables. In discovery with a model, a model of a phenomenon is developed via prediction, clustering, or in some cases knowledge engineering. This model is then used as a component in another analysis, such as prediction or relationship mining.

Distillation of data for human judgement is another EDM technique, which means the human inferences about data.

Learning analytics is the measurement, collection, analysis and reporting of data about learners and their contexts, for understanding and optimizing learning and the environments in which it occurs [3]. Haythornthwaite et al. [4] discussed that learning analytics is a multidisciplinary area which includes the system development to support, display and analyze online learning; Course design analysis and resulting outcomes; Development of learning communities; Following, Supporting, and Evaluating progress and practice of learning communities; Design and development of social learning systems; Analysis of formal and informal learning settings and design features in terms of networked learning practices.

Academic analytics is the term used to describe the intersection of technology, information management culture and application of information to manage academic enterprise. It was derived from business intelligence [3]. It is a new tool to respond accountabilities in higher education and to develop actionable intelligence to improve student success and learning environment [11, 14, 15].

These are the three peek research interest due to the emergence of online learning and Big Data [13].

This research works finds out the following research questions:

Research questions:

RQ1: What is the impact of Learning Management System on students at Post Graduate level in Computer Applications?
RQ2: What is the learning pattern of students in online learning?
RQ3: How does identify the students learning pattern by learning management system?

Objectives:

(1) To analyze the learning pattern of the students' performance by Moodle Learning Management system.
(2) To examine the online learning pattern of the university students.
(3) To study the impact of the students learning performance by Moodle Learning Management System.

The research study shows the some findings of the research work done on educational data mining and learning analytics. Baker (2010) has started the work on EDM, and since then its growing rapidly [12, 16].

2 Data and Experimental Setup

The research work has implemented on Moodle Learning Management System (LMS) at Banaras Hindu University, Varanasi, India. The Moodle based LMS has established and various courses related to the post-graduation in Computer Applications program of the university has run through this system. MCA (Master of Computer Application) is three year (Six Semester) full time course running at the university campus. The students and teachers of MCA course has enrolled to the LMS and teaching–learning through Moodle has performed. Various course activities were planned by the teachers through Moodle platform for students' learning engagement. The course learning contents were developed and uploaded to the course. The video lectures are provided to the students and the online assignments are given to the students for online submission within the given time period. Various other online learning activities like forum discussions, chat, workshop etc. has performed through the LMS. The online course examinations are conducted through Moodle platform and their performance are evaluated and provided to the students' online learning platform. The research work is based on pre-test post-test study. The data were collected from various subjects which are listed bellowed in Table 1. In Pre-test the students were instructed by the traditional pedagogy without using online learning platform and data on various subjects were recorded. In Post-test the students were instructed by the online learning platform Moodle and data were recorded.

It is revealed from Table 1, the list of different courses conducted at MCA program. The variable is the short name of the courses. The short name is used in the data sets. These are the various courses running in master's programme in Computer Application. Both the offline as well as online pedagogy has been planned for the above courses and students' performance has been recorded. MCA is three years (six semesters) programmes, these courses are running in various semesters during the three years programme.

Table 1 Description of data set variables

Variables	Descriptions
DSA	Data Structure and Algorithm
DBMS	Database Management System
DCN	Data Communication and Networks
SE I	Software Engineering I
CG	Computer Graphics
CD	Compiler Design
SE II	Software Engineering II
OS	Operating System
CN	Computer Networks
IR	Information Retrieval

3 Applying Data Mining and Learning Analytics Techniques on Moodle Data

Moodle is very rich in storing usage data into its database. Every user's clicks are stored in its database. These data can be obtained from Moodle in many formats (like xls, csv etc.) which can be used for various data analytics tools. Moodle has also data analytics and visualization tool for generating reports and graphical representation. Some of these are discussed here.

3.1 Students' Performance

Moodle has very good data analytics and visualization tools for student's participation in course and their activities. The student participation in various courses and their grades are stored in Moodle database which can be access by the Moodle.

It is revealed from Table 2, the students' score in various courses when online learning is not yet implemented. The result shows the Mean, Median, Mode and Standard Deviation of the sample data on traditional methods without using online learning pedagogy.

Table 2 Semester-wise grade report of students without using Moodle LMS

Index	OS: Grade/20.00	CN: Grade/20.00	IR: Grade/20.00
Students	15	18	18
Mean	6.5333	6.4444	7.1111
Median	6	6	7
Mode	6	6	10
S.D.	2.0656	2.0065	3.0848

Table 3 Subject wise statistics of MCA SEM II students with Moodle online learning Performance

Index	DSA: Grade/20.00	DBMS: Grade/20.00	DCN: Grade/20.00	SE: Grade/20.00	Total: Grade/80
Students	15	15	15	15	15
Mean	8.1333	8.4	7.4	7.6667	31.6
Median	8	9	6	7	32
Mode	5	4	5	8	43
S.D.	2.7740	3.5415	3.5816	3.5989	9.2874

Table 4 Subject wise t-Test comparison of students' performance MCA SEM II

DSA:DBMS	DSA:DCN	DSA:SE
0.364180294*	0.272993358*	0.350980574*
DBMS:DSA	**DBMS:DCN**	**DBMS:SE**
0.364180294*	0.164894998*	0.274981824*
DCN:DSA	**DCN:DBMS**	**DCN:SE**
0.272993358*	0.164894998*	0.378089949*
SE:DSA	**SE:DBMS**	**SE:DCN**
0.350980574*	0.274981824*	0.378089949*

* $p < 0.05$

Table 5 Grades statistics in various subjects of MCA SEM IV students

Index	CG: Grade/20.00	CD: Grade/20.00	SE: Grade/20.00	Total: Grade/60
Students	18	18	18	18
Mean	13.8333	10.9444	7.6111	32.3889
Median	14	10	8	29.5
Mode	15	9	8	27
S.D.	3.0726	3.9478	2.3298	7.5626

It is revealed from Table 3, the statistics of scores of the students after Moodle LMS is implemented. In Table 3, the statistics of four subjects of Semester II students are given. The table shows the grade score of the students of MCA Semester II in various subjects.

On comparing the statistics of Tables 2 and 3, it is found that the meaningful increment in scores of the students has been achieved. Thus The Moodle LMS is useful for the teaching–learning at the post graduate level. It may be also compare the statistics of Tables 2, 3, 4 and 5, and again the meaningful increment of the scores after the implementation of Moodle LMS has achieved. This is result of our Research Question 1.

It is revealed Table 4, the t-test [17] among the grade score various courses. On comparing the t-test score with t-Test table, it was observed that the t-test score is significant. Thus, it was observed that the online learning with Moodle LMS provides positive impact on these courses. This is the result of the Research Question 2. Similarly the following table, extends the results.

Table 6 T-test among course scores of MCA Semester IV

CG:CD	CG:SE	CD:CG
0.000518026*	0.000260613*	0.000518026*
CD:SE	**SE:CG**	**SE:CD**
0.000559254*	0.000260613*	0.000559254*

* p<0.05

Similarly it is revealed from the Table 5, the statistics of the grade scores of MCA semester IV on various courses e.g. Computer Graphics, Compiler Design and Software Engineering. On comparing the statistics of Tables 2 and 5, it is found that the meaningful increment in scores of the students has been achieved.

It is revealed from the Table 6, the t-test [17] among the grade score of various courses in MCA semester IV. On comparing the t-test score with t-Test table, it was observed that the t-test score is significant. Thus, it was observed that the online learning with Moodle LMS provides positive impact on these courses.

3.2 Learning Analytics

It is revealed from the Fig. 1, the graph of the students' learning patterns in a specific course 'Database Management System'. The vertical axis is represents the students' percentage score in the subject as well as the horizontal axis represents the scores of the participatory students. The red line represents the percentage scores of the particular student while blue line represents the time taken by the students (in minutes) to performed the quiz. On observations, it was found that the high achiever students take average time while slow responder students takes maximum and low achievers. It can be observed that fast responders are high achievers. This is the result of Research Question 3.

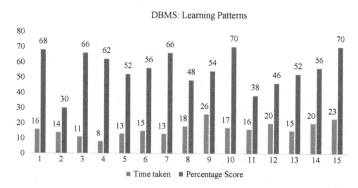

Fig. 1 Learning patterns in a specific course DBMS: Time taken versus percentage score

Table 7 Moodle statistics on quiz analysis

Number of complete graded first attempts	15
Total number of complete graded first attempts	15
Average grade of first attempts	55.60%
Average grade of all attempts	55.60%
Average grade of last attempts	55.60%
Average graded of highest graded attempts	55.60%
Median grade (for all attempts)	56.00%
Standard deviation (for all attempts)	11.86%
Score distribution skewness (for all attempts)	−0.6549%
Score distribution kurtosis (for all attempts)	−0.0364%
Coefficient of internal consistency (for all attempts)	72.42%
Error ratio (for all attempts)	52.52%
Standard error (for all attempts)	6.23%

3.3 Content Analysis

The Teacher can create course in Moodle and upload the course contents like Videos, Books, PDF and PPT etc. They can also plan various activities like Quiz, Assignment and Forum for learning engagement to the students. The Moodle has powerful analytics tool to report on these activities.

Table 7 shows, the statistics of Quiz analysis reported by the Moodle analytics tool for the specific course on 'Database Management System'. In this course quiz, only single attempt is allowed therefore the first, all and average grades are the same value. The other statistics can be observed by the table.

4 Conclusion and Future Works

The research study has investigated all three research questions and results will be helpful for the teachers, academicians and the administrators. The first research question concerns the impact of online pedagogy in higher education in university campuses. It is revealed from the study that the Moodle based online Learning Management System might be implemented for virtual learning environment in higher education system. The research question 2 concerns the online learning patterns. It is revealed from the study that there are three group of student: the high achiever students, the low achiever students and the average students. The study might help to analyze the learning patterns in the online environments. The third research question concerns about LMS analytic tools to analyze students' learning patterns. It is revealed from the study that the slow responder students are normally low achiever while the fast responder students are higher achievers. The Moodle analytics tools provide the various learning pattern about online learning. The results might help to understand learning patterns. The results satisfy all three research questions. The major educational data mining techniques has applied on Moodle data and achieved the significant results.

The important future works is to applying data mining and analytics for the students' cognitive study. The machine learning techniques can be implemented to learn and improve the learning patterns of the students. The machine learning techniques like clustering, classification, association rule mining techniques might be implemented for the cognitive study.

Acknowledgements This research study has done on MCA students of Banaras Hindu University, Rajiv Gandhi South Campus, at Barkachha, Mirzapur, India. The authors express their acknowledgement to the Course Coordinator, MCA programme (Dr. Achintya Singhal); Head of the Department of Computer Science (Prof. S. Kartikeyan (current) and Prof. S. K. Basu (former)); Professor Incharge, Rajiv Gandhi South Campus (Prof. Saket Kushwaha (current) and Prof. R. P. Shukla (former)) for providing Computer Laboratory Infrastructure and other facilities for conducting research study. The author extends their acknowledgement to the participants MCA Students for their active participation and engagement for the study. The authors also thanks to the teachers and computer laboratory staffs for their help and supports.

References

1. https://moodle.com/2015/02/05/7-ways-to-get-started-with-analytics-reports-in-moodle/.
2. Romero, C., Ventura, S., & Garcia, E. (2008). Data mining in course management systems: Moodle case study and tutorial. *Computers & Education, 51,* 368–384.
3. Siemens, G. (2013). Learning analytics: The emergence of a discipline. *American Behavioral Scientist, 57*(10), 1380–1400.
4. Haythornthwaite, C., Laat, D. M., & Dawson, S. (2013). Introduction to the special issue on learning analytics. *American Behavioral Scientist, 57*(10), 1371–1379.
5. Baker, R. S., & Inventado, P. S. (2014). *Educational data mining and learning analytics, learning analytics* (pp. 61–75).
6. Park, Y., & Hyun, J. (2017). Using log variables in a learning management system to evaluate learning activity using the lens of activity theory. *Assessment & Evaluation in Higher Education, 42*(4), 531–547. https://doi.org/10.1080/02602938.2016.1158236.
7. Park, Y., Yue, J. H., & Jo, I. H. (2016). Clustering blended learning courses by online behavior data: A case study in a Korean higher education institute. *Internet and Higher Education, 29,* 1–11.
8. You, J. W. (2016). Identifying significant indicators using LMS data to predict course achievement in online learning. *Internet and Higher Education, 29,* 23–30.
9. Romero, C., Lopez, M. I., Luna, J. M., & Ventura, S. (2013). Predicting students' final performance from participation in on-line discussion forums. *Computers & Education, 68,* 458–472.
10. Alexandron, G., et al. (2017). Copying@Scale: Using harvesting accounts for collecting correct answers in a MOOC. *Computers & Education, 108,* 96–114.
11. Park, Y., & Jo, H. I. (2017). Using log variables in a learning management system to evaluate learning activity using the lens of activity theory. *Assessment & Evaluation in Higher Education, 42*(4), 531–547.
12. Baker, E. (2010). *International Encyclopaedia of Education* (3rd ed.). Oxford, UK: Elsevier.

13. Asif, R., Merceron, A., Ali, S., & Haider, N. (2017). Analyzing undergraduate students' performance using educational data mining. *Computers & Education, 113,* 177–194.
14. Rodriguez, T. E. (2012). The acceptance of Moodle technology by business administration students. *Computers & Education, 58,* 1085–1093.
15. Teresa, M., & Fernandez, A. S. (2009). The role of new technologies in the learning process: Moodle as a teaching tool in Physics. *Computers & Education, 52,* 35–44.
16. Cerezo, R., Santill, M. S., Ruiz, M. P. P., & Núnez, J. C. (2016). Students' LMS interaction patterns and their relationship with achievement: A case study in higher education. *Computers & Education, 96,* 42–54.
17. T Value Table. http://www.ttable.org/.

Analysis of Electroencephalogram for the Recognition of Epileptogenic Area Using Ensemble Empirical Mode Decomposition

Gurwinder Singh, Birmohan Singh and Manpreet Kaur

Abstract Recognizing the epileptogenic area of a brain is done by analyzing the electroencephalogram signal. This area is responsible for the occurrence of seizure activity in a brain. In this paper, a methodology has been presented for the analysis of electroencephalogram to recognize epileptogenic area of brain. Ensemble empirical mode decomposition (EEMD) has been used for the estimation of intrinsic mode functions (IMFs), and six parameters consisting of statistical and frequency-based feature have been extracted from first ten IMFs. The ReliefF algorithm has been used to select the relevant features for the training of artificial neural network (ANN) for recognition of epileptogenic area. The methodology has been evaluated based on accuracy, specificity and sensitivity. The comparison has also been made with other methods of epileptogenic area detection where it has been observed that the proposed method outshines other.

Keywords Epileptogenic · Ensemble empirical mode decomposition
Intrinsic mode function · ReliefF · Artificial neural network

1 Introduction

Epilepsy is a serious neurological disease which is identified by a repeated occurrence of seizures. These seizure activities occur due to firing of multiple neurons in discontinuous fashion, at same time [1]. About 50 million people are currently

G. Singh · B. Singh
Department of CSE, SLIET, Longowal, India
e-mail: kareergurwinder@hotmail.com

B. Singh
e-mail: birmohans@gmail.com

M. Kaur (✉)
Department of EIE, SLIET, Longowal, India
e-mail: aneja_mpk@yahoo.com

© Springer Nature Singapore Pte Ltd. 2019
A. Khare et al. (eds.), *Recent Trends in Communication, Computing, and Electronics*, Lecture Notes in Electrical Engineering 524,
https://doi.org/10.1007/978-981-13-2685-1_46

481

suffering from epilepsy, and every year 2.4 million people are diagnosed positive with epilepsy [2]. Electroencephalogram (EEG) is widely used for the diagnosis of an epilepsy as it covers the detail information of dendrites of the neurons of the brain [3]. EEG is recorded by placing electrodes on the scalp or inside the scalp [4]. The functional magnetic resonance imaging (fMRI) is also used to diagnose epilepsy which measures the brain activity on the basis of blood flow [5], but it is much expensive and has inferior resolution than EEG [6].

When seizure concentrates on small portion of a brain, that epilepsy is known as focal epilepsy or partial epilepsy. The surgery of the responsible portion (epileptogenic area) of brain is the only option left when a patient does not respond to medication [7]. The identification of the epileptogenic area plays an important step prior to surgery because wrong identification may affect the important part of brain [7].

Several techniques have been proposed by the researchers to detect the focal area by using continuous wavelet transformation (CWT) and discrete wavelet transformation (DWT), where different entropy measure, statistical and amplitude-based parameters have been used to train the classifiers [8–12]. The effectiveness of these techniques is based on choice of wavelet functions, decomposition level and quality factor in DWT and CWT. The empirical mode decomposition (EMD) is nonparametric and data-driven method for sub-band estimation. The entropy measures, central tendency measures, frequency- and amplitude-based features of intrinsic mode functions (IMFs) from EMD have been used for discriminating non-focal and focal class EEG signals [13–16]. The application of EMD on EEG signal results in mixing of the different oscillations into one mode and dissemination of oscillations belonging to the same mode to different modes which is known as mode mixing problem. Recently, ensemble EMD (EEMD) and complete ensemble EMD (CEEMD) have also been employed on EEG signals for the classification of seizure activity and focal area EEG signals where spectral features and entropy measures have been used as feature set to discriminate normal and abnormal EEG signals [3, 17].

In this paper, EEG signal has been classified into normal and focal class. The EEMD has been used for estimation of sub-bands (Sect. 3) and various features have been extracted from sub-bands (Sect. 4). The ReliefF algorithm (Sect. 5) is used the selection of relevant features which have been used to train artificial neural network (Sect. 6) for classification. Figure 1 shows the flowchart of the proposed methodology.

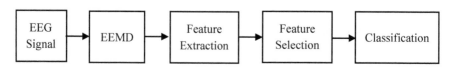

Fig. 1 Flowchart of the proposed methodology

2 Dataset

The publicly available EEG dataset has been used in this work [18]. It consists of EEG signals belonging two classes non-focal and focal where focal EEG data are from of five patients who were suffering from pharmacoresistant temporal lobe epilepsy. The dataset consists of total 3750 pairs of focal and non-focal EEG signals which are denoted as x, y. Each signal is of duration 20 s and has been sampled at 512 Hz. For focal class EEG signals, x has been collected from those channels from which seizure has been detected and y has been collected from one of the adjacent channels of the x. The signals have been filtered by employing the band-pass filter of range 0.5–150 Hz. The more detail of the dataset is available in [19]. In this work, 50 EEG signals from non-focal and focal class have been used. In this paper, difference between x and y represented by $x - y$ has been used to design the methodology for the detection of epileptogenic area of a brain from an EEG signal. The EEG epoch originated from epileptogenic area is known as focal EEG sample, whereas EEG epoch originated from normal area is known as non-focal EEG signal. The usage of $x - y$ has been motivated from the work presented in [20]. One sample epoch of x, y and $x - y$ belonging to non-focal class is shown in Fig. 2.

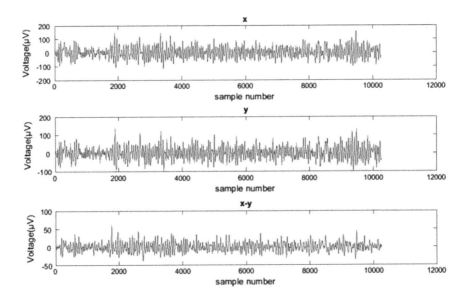

Fig. 2 EEG epochs belonging to non-focal class (x signal, y signal and $x - y$ signal)

3 Ensemble Empirical Mode Decomposition (EEMD)

A time series ($x(t)$) decomposes into intrinsic mode functions (IMFs) by employing EEMD, where time series after decomposition is shown by Eq. 1, $r_n(t)$ denotes the residual after N number of IMFs have been extracted [21].

$$x(t) = \sum_{i=1}^{N} IMF_i(t) + r_n(t) \tag{1}$$

Each empirically decomposed IMF satisfies two conditions: (a) the mean value of both minima and maxima defined by the envelope must be zero at any point of time and (b) the number of maxima and number of zero crossing must be the same or differ by one at the most, over the full length of IMF. In EEMD, each IMF consists of additional white noise of finite amplitude which is different in each trial. This added white noise forces to assemble the part of signal of comparable scale and exhaust all possible solution in sifting process. The addition of white noise to IMF eliminates the mode mixing problem of EMD. In this work, first ten IMFs estimated by EEMD have been used for the feature extraction from $x - y$. The first ten IMFs of normal class EEG epoch are shown in Fig. 3.

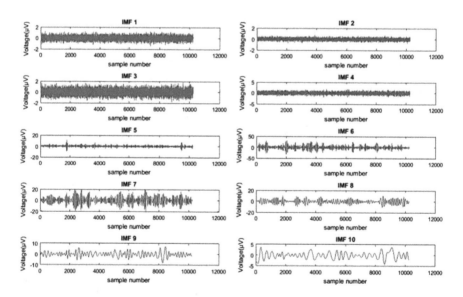

Fig. 3 Ten IMFs estimated using EEMD of normal class EEG epoch

4 Feature Extraction

The parameter extraction is one of the most important steps for the detection seizure activity where relevant features are extracted from the signal. These features are fed to classifier to discriminate samples belonging to different class. In this paper, combination of statistical parameters and zero crossing parameter has been used to determine the EEG signal belonging to epileptogenic area of a brain. These statistical parameters of sub-bands resulted of the DWT have been for the same problem [10]. Six statistical parameters, namely maximum, minimum, standard deviation, mean, skewness and kurtosis, have been extracted from first ten IMF signals. Frequency has been one of the widest used features for the detection abnormal EEG signals [16, 22]. The number of times a signal crosses zero axis line is known as zero crossing (ZC) [23, 24]. In this paper, ZC has been used for the detection of EEG signals belonging to focal class. The six statistical parameters and ZC has been extracted from first ten IMFs which is further input to ReliefF algorithm for the selection of feature set of concise size and relevant nature.

5 Feature Selection

ReliefF is a feature ranking algorithm which is preferred for correlated and noisy features space [25]. Features are ranked by their ability to differentiate the instances by how much they are nearer to each other in two or more classes. An instance (R) is selected from a class in a random fashion. The k number of instances are selected from same class which are known as nearest hits (H), and k number of instances are selected from other classes which are known as nearest misses $(M(C))$. The weight (W) of a feature (A) of randomly selected sample is updated according to the nature of corresponding feature in differentiating the samples of nearest misses and nearest hits (Eq. 2). The $diff(A, R, H)$ calculates the difference between the value of A for instance R and H [25]. The m represents the number of time whole process is repeated. After that, features are ranked in descending order of their respective weights as features with higher weights represent better discrimination ability. In this paper, ranked features have been used to train the classifier one by one. Initially, first ranked feature has been fed to classifier and then next ranked feature along with the previous features have been used to train classifier. The procedure has been used to find the best as well as smallest size of feature set in this paper.

$$W[A] = W[A] - \frac{diff(A, R, H)}{m} + \sum_{C \neq class(R)} \frac{[P(c) \times diff(A, R, M(C))]}{m} \quad (2)$$

6   Classification

The artificial neural network (ANN) has been trained after selecting the relevant features because ANN has been proved as best classier in solving the numerous signal processing problems [26, 27]. It is self-adaptive and robust technique which has been the choice in many researchers for information processing [12, 16, 28]. The rapid testing mechanism of ANN after its weight updation during training is main advantage of its usage. It consists of multiple layers where each layer comprises of neurons. The number of neurons in the input and output layer is equal to the number of features which have been extracted from the sample and number of classes in which sample need to be classified. The hidden layer neurons produce output by multiplying the input with respective weights of input layer neurons. Similarly, the output of output layer is calculated and error is propagated back to update the weights of neurons. The error is calculated by differencing the sum of squared of desired value and output value.

7   Results and Discussion

The publicly available dataset has been used in this paper to design the detection of epileptogenic area of a brain. The brief information about the dataset has been provided in Sect. 2. Each EEG signal has been decomposed into set of IMFs by using EEMD, and features have been extracted from first ten IMFs. Estimated IMFs of the one EEG epoch are shown in Fig. 2. Total seven features comprising of six well-known statistical parameters and ZC have been extracted from each of the ten IMFs. After parameter extraction phase, feature matrix consists of 70 features of 100 EEG epochs. The extracted features have been ranked by using the ReliefF algorithm which ranked the features according to their relevance as explained in Sect. 5. In this paper, the value of k is fixed to 10 as suggested in [25]. ANN with 10 hidden neurons and tan sigmoid activation function has been trained with obtained features to classify the EEG epoch into normal and focal class where focal class depicts the EEG originated from epileptogenic area. Initially, the first ranked feature has been applied to the ANN and the accuracy has been noted down. The next ranked features (one by one) along with all previous ranked features have been applied to the ANN for the classification of normal and focal EEG signals. While training of ANN, 60% dataset has been used for training, 5% has been used for validation, and 35% has been used for testing of the methodology. By experimentation, it has been observed that first five ranked features, which consist of ZC of IMF1, IMF3 and IMF4 and mean of IMF3 and IMF4 have shown maximum accuracy 94.30%, sensitivity 90% and specificity 100%. The validation of the features has also been evaluated using the Kruskal–Wallis test, and p values are shown in Table 1 along with their respective range in normal class and focal class EEG signal. The IMF1 denotes the first estimated IMF, Avg. denotes the mean value, and Std. denotes the

Table 1 p values of first five ranked features using Kruskal–Wallis test and their respective range

S. No.	Feature (IMF)	P value	Avg. ± Std. (normal)	Avg. ± Std. (focal)
1	ZC (IMF1)	2.014e−07	5350±723.51	5989±356.03
2	ZC (IMF4)	3.763e−10	612.38±89.86	728±83.26
3	ZC (IMF3)	2.529e−10	1286±194.06	1494±137.21
4	Mean (IMF4)	0.003	10.53±5.92	7.62±5.14
5	Mean (IMF3)	0.0084	6.61±3.83	5.03±3.79

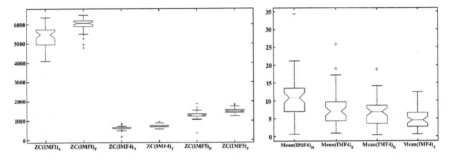

Fig. 4 Box plot of selected features for the recognition of epileptogenic area

Table 2 Comparison of accuracy of the proposed methodologies with other methodologies for the detection of focal class EEG signals

S. No.	Researchers (Year)	Size of dataset	Accuracy (%)
1	Sharma et al. [13] (2014)	50 normal and 50 focal	85.00
2	Sharma et al. [11] (2015)	50 normal and 50 focal	84.00
3	Sharma et al. [14] (2015)	50 normal and 50 focal	87.00
4	Proposed method (2018)	50 normal and 50 focal	94.30

standard deviation of the extracted features. The box plots of the selected features are shown in Fig. 4, where ZC(IMF1)$_N$ denotes ZC feature of IMF1 from normal class, ZC(IMF1)$_F$ denotes ZC feature of IMF1 from focal class and so on. The performance has been compared with the performances of other methodologies which have used the same dataset which clearly showed that the proposed methodology outshines other as shown in Table 2.

8 Conclusion

The epileptogenic are has been determined by classifying the focal class EEG signals which may assist the neurologist for the surgery of the epileptic patients. Ensemble empirical mode decomposition has been employed on each EEG signal for the estimation of intrinsic mode functions. The six statistical parameters and one frequency-based parameters, namely zero crossing, has been extracted from intrinsic mode functions. The features have been used to train artificial neural network after selecting the relevant features using ReliefF algorithm. The zero crossing parameters obtained from first four intrinsic mode functions with artificial neural network have shown good results. In the future, the performance of the technique will be measured on long-term EEG signals.

References

1. Dastidar, S. G., Adeli, H., & Dadmehr, N. (2007). Mixed band wavelet chaos neural network methodology for epilepsy and epileptic seizure detection. *IEEE Transactions on Biomedical Engineering, 54*, 1545–1551.
2. Epilepsy. http://www.who.int/mediacentre/factsheets/fs999/en/.
3. Hassan, A. R., & Subasi, A. (2016). Automatic identification of epileptic seizures from EEG signals using linear programming boosting. *Computer Methods and Programs in Biomedicine, 136*, 65–77.
4. Greenfield, L. J., Geyer, J. D., & Carney, P. R. (2012). Reading EEGs: A practical approach. Lippincott Williams & Wilkins.
5. Menon, V., & Crottaz-Herbette, S. (2005). Combined {EEG} and f{MRI} studies of human brain function. *International Review of Neurobiology, 66*, 291–321.
6. Zanzotto, F. M., & Croce, D. (2010). Comparing EEG/ERP-like and fMRI-like techniques for reading machine thoughts. In *International Conference on Brain Informatics* (pp. 133–144).
7. Das, A. B., & Bhuiyan, M. I. H. (2016). Discrimination and classification of focal and non-focal EEG signals using entropy-based features in the EMD-DWT domain. *Biomedical Signal Processing and Control, 29*, 11–21.
8. Sharma, R., Kumar, M., Pachori, R. B., & Acharya, U. R. (2017). Decision support system for focal EEG signals using tunable-Q wavelet transform. *Journal of Computer Science, 20*, 52–60.
9. Bhattacharyya, A., Pachori, R. B., & Acharya, U. R. (2017). Tunable-Q wavelet transform based multivariate sub-band fuzzy entropy with application to focal EEG signal analysis. *Entropy, 19*.
10. Chen, D., Wan, S., & Bao, F. S. (2017). Epileptic focus localization using discrete wavelet transform based on interictal intracranial EEG. *IEEE Transactions on Neural Systems and Rehabilitation Engineering, 25*, 413–425.
11. Sharma, R., Pachori, R. B., & Rajendra Acharya, U. (2015). An integrated index for the identification of focal electroencephalogram signals using discrete wavelet transform and entropy measures. *Entropy, 17*, 5218–5240.
12. Singh, G., Kaur, M., & Singh, D. (2016). Detection of epileptic seizure using wavelet transformation and spike based features. In *2015 2nd International Conference on Recent Advances in Engineering and Computational Sciences, RAECS 2015* (pp. 1–4).
13. Sharma, R., Pachori, R. B., & Gautam, S. (2014). Empirical mode decomposition based classification of focal and non-focal EEG signals. In *2014 International Conference on Medical Biometrics* (pp. 135–140).
14. Sharma, R., Pachori, R. B., & Acharya, U. R. (2015). Application of entropy measures on intrinsic mode functions for the automated identification of focal electroencephalogram signals. *Entropy, 17*, 669–691.
15. Rai, K., Bajaj, V., & Kumar, A. (2015). Features extraction for classification of focal and non-focal EEG signals. *Lecture Notes in Electrical Engineering, 339*, 599–605.

16. Kaur, M., & Singh, G. (2017). Classification of seizure prone EEG signal using amplitude and frequency based parameters of intrinsic mode functions. *Journal of Medical and Biological Engineering, 37,* 540–553.
17. Das, A. B., & Bhuiyan, M. I. H. (2016). Discrimination of focal and non-focal EEG signals using entropy-based features in EEMD and CEEMDAN domains. In *9th International Conference on Electrical and Computer Engineering* (pp. 435–438).
18. Andrzejak, R. G., Schindler, K., & Rummel, C. (2012). Nonrandomness, nonlinear dependence, and nonstationarity of electroencephalographic recordings from epilepsy patients. *Physical Review E, 86,* 046206.
19. The Bern-Barcelona EEG database. http://ntsa.upf.edu/downloads/andrzejak-rg-schindler-k-rummel-c-2012-nonrandomness-nonlinear-dependence-and.
20. Yadav, R., Shah, A. K., Loeb, J. A., Swamy, M. N. S., & Agarwal, R. (2012). Morphology-based automatic seizure detector for intercerebral EEG recordings. *IEEE Transactions on Biomedical Engineering, 59,* 1871–1881.
21. Wu, Z., & Huang, N. E. (2005). Ensemble empirical mode decomposition: A noise-assisted data analysis method. *Advances in Adaptive Data Analysis, 1,* 1–41.
22. Bajaj, V., & Pachori, R. B. (2012). Classification of seizure and nonseizure EEG signals using empirical mode decomposition. *IEEE Transactions on Information Technology in Biomedicine, 16,* 1135–1142.
23. van Putten, M. J., Kind, T., Visser, F., & Lagerburg, V. (2005). Detecting temporal lobe seizures from scalp EEG recordings: A comparison of various features. *Clinical Neurophysiology, 116,* 2480–2489.
24. Borbély, A. A., & Neuhaus, H. U. (1979). Sleep-deprivation: Effects on sleep and EEG in the rat. *Journal of Comparative Physiology, 133,* 71–87.
25. Kononenko, I. (1994). Estimating attributes: Analysis and extensions of RELIEF. In *European Conference on Machine Learning* (pp. 171–182). Berlin, Heidelberg: Springer.
26. Weng, W., & Khorasani, K. (1996). An adaptive structure neural networks with application to EEG automatic seizure detection. *Neural Networks, 9,* 1223–1240.
27. Hazarika, N., Chen, J. Z., Tsoi, A. C., & Sergejew, A. (1997). Classification of EEG signals using the wavelet transform. *Signal Processing, 59,* 61–72.
28. Kumar, Y., Dewal, M. L., & Anand, R. S. (2014). Epileptic seizures detection in EEG using DWT-based ApEn and artificial neural network. *Signal, Image and Video Processing, 8,* 1323–1334.

Forensic Investigation Framework for Complex Cyber Attack on Cyber Physical System by Using Goals/Sub-goals of an Attack and Epidemics of Malware in a System

Shivani Mishra

Abstract A cyber attack on critical infrastructure differs from attack on general information and communication systems. Recent trends of cyber attacks on critical infrastructure are found to be complex cyber attacks (CCA) because they are multistage, multi-phase and multi-pace. Detection of these complex cyber attacks is yet a challenging problem because they are intractable to describe and analyze. In this paper, complex cyber attacks are analyzed and as a response to detection of an attack, a forensic investigation framework for CCA is proposed. This paper focuses on forensic investigation framework for CCA in cyber physical system, which is large and geographically distributed. A model for forensics investigation process is proposed which is based on goals and sub-goals of an attack. This helps to reconstruct the event and collect data for evidence. Since complex cyber attacks are constructed with a variety of malwares and some of them show the property of self-propagation, an epidemic analysis in forensic investigation process determines the spread of infection in large infrastructures. Addition of epidemic behavior of malware in forensic investigation process is helpful to understand the dynamics of infection in a large, heterogeneous infrastructure.

Keywords Critical infrastructure · SCADA · Stuxnet · Malware
Complex cyber attack

1 Introduction

The advancement in information and communication technology (ICT) has introduced numerous vulnerabilities that are exploited by cyber attackers. Depending upon the method of attacks, various categories of cyber attacks are discovered and

S. Mishra (✉)
Motilal Nehru National Institute of Technology Allahabad, Allahabad, India
e-mail: shivanialld@gmail.com

© Springer Nature Singapore Pte Ltd. 2019
A. Khare et al. (eds.), *Recent Trends in Communication, Computing, and Electronics*, Lecture Notes in Electrical Engineering 524,
https://doi.org/10.1007/978-981-13-2685-1_47

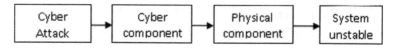

Fig. 1 Direction of attack in a cyber physical system

reported in the literature [1–4]. Cyber attacks on critical infrastructures (CI) are found to be most complicated and devastated attacks because disrupting its function would create significant socioeconomic crisis [4, 5].

Critical infrastructures are integrated cyber and physical systems that deliver necessary services to the society. Cyber physical systems (CPS) are incorporated in electricity generation and distribution, oil and gas pipelines, nuclear power plants, aviation, transportation, telecommunication systems, etc. A typical example of cyber physical system includes an electric grid system which contains transformers, power lines, controllers and communicating computational units such as embedded computers. Cyber systems aim to monitor and control the behavior of physical processes. As the interaction of cyber and physical units is increased, the susceptibility to exploit physical resources is also increased. Recent trend of attacks on cyber physical systems indicates that attackers exploit cyber entities to target damage in physical units [5–9]. Figure 1 shows direction of attack in a cyber physical system.

Figure 1 shows, through cyber attacks on a system, attackers target physical resources of a system which makes the systems unstable and vulnerable for cyber attacks. Attacks of this kind are termed as cyber-to-physical attacks.

An attack on critical infrastructure differs from general information technology (IT) systems. However, the techniques applied to attack are similar to traditional IT systems [3]. Some common categories of cyber attacks are as follows [10, 11]: Web defacement, denial of service attacks, protocol attacks and malware attacks.

Malwares are malicious set of instructions, intended to disrupt the normal functioning of computer system and networks. For some malicious intents, this software is designed to perform delete, block, modify or copy data [12, 13] Various applications of malwares include stealing information, spreading misinformation, enabling remote login and replicating malware payload for self-propagation. Common categories of malwares are described on the basis of its functioning; they are as follows [14]:

(a) Virus, Worms, Trojans Horse, Spywares, Rootkits, Key loggers, Trapdoors, Adware, backdoors, Bots are the famous known forms of Trojan horses.
(b) Blended Attacks: Recent trend of malwares shows that programs combine the properties of more than one category of malware.

A significant increase in cyber-to-physical attacks is seen in critical infrastructures [5, 11, 13, 15]. The attackers introduce a malicious program in a cyber physical network, which spreads in a network and finds a physical resource to damage. Stuxnet [16, 17] was the first discovered malware that targets cyber-to-physical resources. The technical analysis reveals that Stuxnet malware is most complex and sophisticated among another malwares [18, 19].

In [20], this malware attack is defined as complex cyber attack (CCA), which is multistage, multi-phase and multi-pace. Duqu [21, 22], Flame and Red October are other similar complex cyber attacks [23], which follow Stuxnet attack strategy and similar technical details [18, 24, 25].

The objective of the research work in this paper is to perform forensic analysis in a cyber physical system for complex cyber attacks. A forensic investigation in a system is applied in response to an attack, to figure out, how it happened? This process requires

(a) Determination of forensic investigation framework for cyber physical system.
(b) Since cyber physical systems are large infrastructures, epidemic analysis of malware determines the dynamics of infection in a system.

Based on these objectives, the following contributions are made:

(1) A complex cyber attack is found multistage, multi-phase and multi-pace. Complexity of malware attack is discussed in terms of interactions of malicious components [20].
(2) A multi-tree view of complex cyber attack is presented and analyzed. Attack tree security modeling approach is used to depict and analyze the attack [20].
(3) Analysis of multi-tree view of complex cyber attack represents discovery of sequence of coordinated attacks [20].
(4) A forensic investigation framework is proposed for complex cyber attacks.
(5) Epidemic analysis of malware infection in closed system is used to determine the dynamics of infection in a system.

This paper aims to provide a model for digital forensic investigation process for complex cyber attack. These attacks are found to be self-propagating; therefore, an epidemic analysis of complex cyber attack is introduced in forensic investigation process to understand the spread of infection in large infrastructure.

1.1 Digital Forensic Investigation Process

Digital forensic investigation is the process of identifying, collecting, examining, preserving and presenting data to establish evidences in courts of law [26–28]. Among the four phases of forensic process, collection and examination phases are most important for supporting investigations, establishing evidences and suggesting preventive measures in a system, in an intrusion-affected system [29, 30]. Figure 2 presents four phases of digital investigation process.

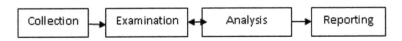

Fig. 2 Major steps in a forensic process

In this work, the focus is on identification, collection and examination of data from cyber physical systems (CPS). A CPS requires continuous monitoring and control of various operations for which nonstop availability of corresponding component is must. Collecting the required data at regular intervals from field devices situated at remote locations is another challenging issue in forensic investigation. Further, advance malware, constituting complex cyber attacks, is hard to detect in heterogeneous infrastructure. This research work determines the following objectives for forensic investigation:

- To model forensic investigation process, for complex cyber attacks, and data extraction process for reconstruction of attacking events.
- To analyze epidemic nature of malware to determine event reconstruction with reference to time.
- To indentify and analyze suspicious and infected objects in all the components of a CPS.

Based on these objectives, the following contributions are made in this research work

- A goal and sub-goal of an attack-based forensic investigation framework are proposed to understand how an attack was carried out.
- A susceptible–infected–recovered model of epidemics is used to understand the dynamics of infection in closed sets of CPS.
- Statistics of susceptible, infected and recovered objects of a system are identified.
- An exhaustive list of data to trace the complete attack scenario is proposed.

In this work, digital forensic procedure is assumed to be applied on CPS after the detection of a complex cyber attack in a system. Forensic analysis of CPS is not similar to cyber forensic applied on cyber entities because devices used in this system are continuously operational and cannot stop for the forensic investigation [30]. Malware intrusions in case of complex cyber attack pose bulk of abnormal activities that may be dependent on each other; therefore, forensic investigations for malware attack have also different approach. In the following subsection, we will discuss forensic challenges in CPS and forensic challenges for malware intrusion investigation.

1.2   Digital Forensic Investigation Process in CPS

Due to unique infrastructure architecture, continuous availability and large-scale networks, the following requirements are chosen for the conduct of forensic analysis in this chapter:

- Real-time forensic investigation procedure is required in view of the continuous operation of the system.
- A widespread investigation procedure covering all the cyber and physical components is needed for the accurate result analysis.

- Field devices contain volatile data; therefore, a forensic analysis should not highly dependent on such volatile data. Various devices have less amount of memory; therefore, the data fetched from them need special consideration in drawing conclusions for forensic analysis.
- **Cyber Physical Systems are Closed Systems**: In this research work, CPSs are considered as closed systems. Closed systems contain a set of objects which are repositories of information and ensure that information flow is generated from the specified set and consumed in the same set.

2 Proposed Work

Two approaches are identified for malware investigation in a system: a process-driven approach [26] and goal-oriented approach [27]. A goal-oriented approach is chosen, to determine the complete intrusion scenarios in a system.

In this section, we describe a forensic investigation framework of complex cyber attack on cyber physical system using goals, sub-goals and epidemics of malware. A forensic model gives the framework to understand the system, determine the intrusion scenarios, event reconstruction and analysis of an attack. Here, we focused only on critical elements for forensic analysis and considered the constraints of cyber physical system for forensic analysis. The model is based upon the recent works presented in [20].

A digital forensic investigation process model for malware attack on CPS is defined as follows:

$$\text{DFI} = (G_0, SG, S, f, Q, E, Trace\ set, Collect),$$

where

G_0 Target of an attacker, final goal
SG   Set of sub-goals to achieve final goal
S Set of system containing various objects forming closed set
f Relation which locates objects representing goal and sub-goals in a system defined as follows $f : (G_0 \cup SG) \rightarrow S$
Q Set of objects for investigation, forming closed system
E Epidemic analysis of malware infection spread

Further explanation of above parameters is as follows:

1. **G_0: Goal/sub goal of an Attack**: The Goal of an attack is the target fixed for exploitation of the system by an attacker. Sub goals are used to successfully reach to the target location and exploit the targeted end device through a mechanism applied by the attacker to exploit the end target device. A set of goals for a complex cyber attack is determined by using Attack tree method and following goals are discovered.

$$Goals = \left\{ \begin{array}{c} Sabotage\ the\ facility,\ Install\ the\ worm,\ spread\ the\ worm, \\ load\ the\ worm, \\ search\ siemens\ SCADA\ software, \\ Reprogram\ the\ PLC \end{array} \right\}$$

For each goal we have a set of sub goals, here, we are showing set of sub goals only for first goal. I.e. Sabotage the facility.

2. **SG**:

$$Sub\ Goals = \left\{ \begin{array}{c} Gain\ SCADA\ center\ system\ access, \\ disrupt\ communictaion\ system, \\ disrupt\ field\ data\ interface \end{array} \right\}$$

Goals and sub goals of an attack are discovered to understand the attack and how it was carried out in a system. With each attack following informations are revealed for forensic investigation process.

3. **Set of objects for investigation**: for each identified goal and sub goal of an attacker, collect all objects from a closed system.
4. **Epidemics**: An epidemic of malware is studied to understand the spread dynamics of cyber worm. Our studies results the closed set having large number of infections, rate of infection and rate of recovery. These parameters are helpful to relate the sequence of events in a system.
5. **Trace set**: Trace sets represent the path of attacks. An event reconstruction is done at this juncture. A function is used to map goals and sub goals of an attack to the set of method of attack. This determines pre and post condition of an attack and reconstructs the events in a system.
6. **Collect**: Attack signatures are collected from different locations of a system. Preserve, Analysis and Representation phase of digital forensics are similar to normal forensic procedure, as described in [26].

In the next subsection, we discovered that once an entity of a system is compromised or infected by an attack, how this infected entity infects other objects of a system.

2.1 Epidemics of Malware in Closed Systems

In this section, epidemic behavior of malware that constitutes complex cyber attack is analyzed. Modeling and analyzing malware spreading gives results for forensic analysis and is helpful to determine the location of countermeasure in a system.

The epidemic models are inspired from biology, where spread of infectious disease is modeled and analyzed. Application of epidemic theory is seen in modeling growth of Internet worms at the early stage in [31]. A susceptible–infected (SI) model is used

Fig. 3 Whole population in closed boundaries, divided into three compartments

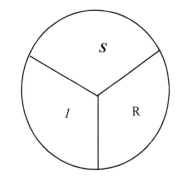

Fig. 4 Direction of flow of disease in a population with α contact rate and β recovery rate

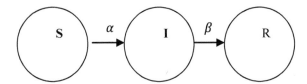

to demonstrate the growth of infection. A susceptible–infected–recovered model is used to model mobile virus propagation [32]. In the basic form SIR model, a population is divided into three compartments, namely susceptible (S), infected (I) and recovered (R), where

Susceptible (S): Individuals are susceptible and not being infected but able to catch disease. Such individuals then move to infected compartment [32].
Infected (I): Infected individuals can spread the disease to susceptible individuals; if they are recovered, they enter recovered compartment.
Recovered (R): Individuals in recovered compartment are immune for life or proper functioning.

The entire population is divided into three categories; Fig. 3 shows the population in a closed boundary, and Fig. 4 shows the direction of flow of disease. The disease spreads in the population with contact rate as α and β as recovery rate.

The SIR model has three basic equations, and they are as follows:

$$\frac{dS(t)}{dt} = -\alpha \cdot S(t) \cdot I(t) \tag{1}$$

$$\frac{dI(t)}{dt} = \alpha \cdot S(t) \cdot I(t) - \beta \cdot I(t) \tag{2}$$

$$\frac{dR(t)}{dt} = \beta \cdot I(t). \tag{3}$$

In addition, with an assumption that total number of population N is constant, i.e., population is closed set, we have

$$N = S(t) + I(t) + R(t). \tag{4}$$

Table 1 Epidemic attributes for forensic analysis

Notation	Description
$I(t)$	No. of infected objects at time t
$S(t)$	No. of susceptible objects at time t
$R(t)$	No. of recovered hosts at time t
N	Total number of objects under consideration
α	Infection transmission rate at time t
β	Rate of recovery

Here, in this work, we have applied susceptible–infected–recovered (SIR) model on the closed set of information flows. Table 1 shows epidemic attributes that are used in SIR model for forensic analysis.

Here, SIR model is used to understand the spread statistics of an infection in the system. Let us consider a closed community of objects in the system. A closed set is chosen by locating compromised objects by using $f : (G_0 \cup SG) \rightarrow S$. Each closed set in a system contains a number of susceptible objects and a number of infected objects by analyzing insecure information flows in a system. Initial values of S and I at time t can be determined for a closed set. A spread of infection in a closed set is analyzed according to SIR model. At each time $t = 1, 2, 3, 4, \ldots$ *units (days)*, the following steps are applied in accordance with the SIR model.

- Spread of infection at $t = 0$, i.e., from a selected closed set, all objects of a set and its associated information flow, is secure, except one object which is vulnerable for some threat. A zero-day attack, which exploits new unidentified vulnerability, is considered at this stage.
- For the spread of infection at time $t = 1, 2, 3, \ldots$ unit/day, it is found that numbers of susceptible objects are decreasing and numbers of infected objects are increasing.
- Rate of recovery in SIR model is directly proportional to the number of infected objects in a set, while from infection to recovered states some preventive measures are applied and it will take some unit of time for recovery.

The following case study demonstrates the application of SIR model on CPS.

2.2 Case Study

For a cyber physical system, we have N total number of objects in system. For a population size k, which denotes total number of objects in a set, all the compartments and their statistics are presented in Table 1. Here, $\frac{dS(t)}{dt}$, $\frac{dI(t)}{dt}$ and $\frac{dR(t)}{dt}$ gives the rates at which the number of population in compartments is changing.

According to SIR model, the following data are analyzed, for k = 100; initially, there is no infection, and then the number of susceptible objects initially is 100. After

Table 2 Size of compartments in each day

T (days)	S	I
0	100	0
1	90	10
2	80	20
3	70	30
4	60	20
…	….	…

one day, the infection starts and ten objects were moved to infected compartment. The following data are listed for day-wise increase in infection and decrease in susceptible compartments (Table 2).

If rate of change is fixed for the number of objects in S compartment, then value of S after t days is specified as follows: $S = 100 - 10 \cdot t$.

Similarly, we can find the rate of recovery, from infection to recovered compartment change in the recovered objects

$$\frac{dR(t)}{dt} = \frac{I}{2} objects\, per\, day,$$

where numeric value $\frac{1}{2}$ denotes objects in infected compartment were recovered after 2 days.

If $\beta = \frac{1}{2}$, then

$$\frac{dR(t)}{dt} = \beta I(t),$$

gives recovery rate of objects from infected to recovered compartment.

Now, the rate of transmission α is used to find the number of contacts by infected objects to susceptible objects per day, i.e., net growth rate of infected compartment by contacts by calculating

$$\frac{I_{t+1}}{I_t} = \frac{20}{10} = 2\, contacts\, per\, day.$$

The forensic investigation with these parameters is helpful to analyze the rate at which the number of objects in a closed set is moving from susceptible to infected compartment and then to recovered compartment over the course of time. A graph-based representation of disease flow during epidemic is as follows (Figs. 5, 6 and 7):

The following quantitative questions can be investigated for forensic analysis

1. When does the infection have maximum value?
2. What is the rate of infection?
3. How many susceptible converted into infected in some unit of time?

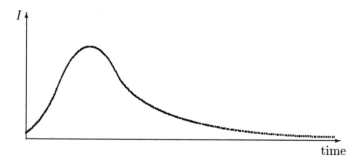

Fig. 5 During the epidemics, variation in I compartment

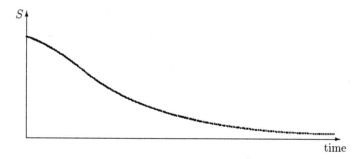

Fig. 6 During the epidemics, the number of susceptible objects becomes infected

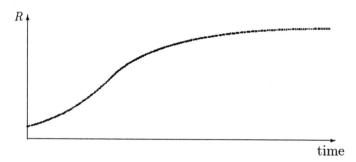

Fig. 7 During the epidemics, the number of infected objects is recovered

Thus, from a single closed set to entire lattice of closed set can be analyzed with the help of SIR model, to answer the above-stated questions.

Trace set: In this, phase data are collected according to goals and sub-goals of an attack. A function to reconstruct (event) is defined to map goals and sub-goals of an attack to a method of security violation. We define

reconstruct : *Set of* (Goals/SubGoals) → *set of security violation method*.

A set of security violation is identified for each set of goals in attack tree analysis of complex cyber attack.

$$
Set\ of\ attack = \left\{ \begin{array}{c} Man\ in\ the\ Middle\ Attack(MITM),\\ Denial\ of\ service\ against\ Network,\\ Privilege\ escalation,\ Unauthorized\ access,\\ Flooding\ attack\ for\ PLC's,\\ Root\ kit\ Attack,\ Buffer\ overflow\ Attack,\\ Export\ hooking\ to\ gather\ PLC\ information,\\ Resource\ Exhaustion \end{array} \right\}.
$$

With varying severity of an attack, all these methods of security violation are analyzed to reconstruct the attacking method. This method helps to determine the exploitation of subjects, objects and secure information flow in a system. Anomaly records are created and correlate with a reason of insecure information flow to the suspected intrusion activity. An audit record is created which stores subjects, objects for an event, exception condition and time stamps that are recorded.

Collect: In this phase, corresponding to each insecure flow an intrusion scenario is analyzed exhaustively. For insecure information flows, the following operations are explored (Table 3).

Similarly, all the resources are analyzed for a number of insecure information flows in a system (Table 4).

Table 3 Information flow operations and generated actions

Operations	Action generated
Read	Read
Write	Create, modify append
Exec	Invoke, exec, system Call, function call
Communication	Send, receive
Block	Stop the process, spoof

Table 4 Information flow operations and generated actions

Resources	Information stored
Hardware system	CPU, network inter face, switches, workstation, terminal unit device. SCADA central units, PLC device
Software system	Application software, database software, network software
Account information	Uid, gid, mode, USB enable mode
Operating system	OS version available patches, vulnerability list
Field device software system	Display software, computing software
Files	Directories in a computing system, link, path, timestamp

Thus, detailed and exhaustive lists of resource are created to correlate the sequence of attack, and thus event is reconstructed, and intrusions are determined and analyzed. Based on the analysis of an attack, a preventive measure and its exact location are chosen for entire infrastructure.

2.3 Comparison with Other Existing Forensic Investigation Process Model

For a typical SCADA, forensic framework is proposed in [30] and cyber investigation phases were proposed as depicted in Table 5.

Identification and preparation phases aimed to understand the entire system. This step is problematic for large and distributed networked SCADA system. An appropriate strategy is needed to understand the system and gather information from the system such as software, hardware, vendors and contractors.

Comparing with our proposed approach of forensic investigation, we have provided a system model of secure information flow. Any closed set in a system that represents a site contains several objects. Thus, a set is chosen for a forensic investigation process.

Data sources that contain forensic evidence are discovered after analyzing the attack. The detection of attack at this phase is necessary to discover potential source of evidence.

In our proposed model, an attack representation is used to locate potential sources of evidence. Mapping of goals and sub-goals to system exactly locates the principal source of evidence and is governed by $f : (G_0 \cup SG) \rightarrow S$.

The analysis phase aims to reconstruct the event and concludes the results for forensic investigation.

Comparing with our proposed approach, here a statistics of malware infection spread is provided. This helps to understand how far malware infection spreads through the connected objects in a system. Event reconstruction and logical conclusion can be drawn with reference to time.

Other phases are common for all digital forensic investigation process; therefore, we are not emphasizing on these phases. Thus, a systematic way of forensic

Table 5 Phases of forensic investigation for SCADA		
	Phase 1	Identification and preparation
	Phase 2	Identifying data sources
	Phase 3	Preservation, prioritization and collection
	Phase 4	Examination
	Phase 5	Analysis
	Phase 6	Reporting and presentation
	Phase 7	Reviewing results

investigation framework is proposed and compared with the existing SCADA forensic architecture. In this comparison, we have found for a large infrastructure our approach of forensic framework is more efficient and focused on forensic evidence collection and analysis.

3   Conclusion

Modeling forensic investigation process for a CPS is a major objective of this paper. An informational flow model for CPS is used here; this facilitates to investigate insecure information flows in a system. Epidemic model is applied in informational closed system for CPS and to determine the rate of contact between infectious and susceptible objects of a system. This helps to understand propagation of malware in entire system. A set of goals and sub-goals of an attack help to reconstruct the attack scenario and collect the data from specified locations.

References

1. Sandip, P., & Zaveri, J. (2010). A risk-assessment model for cyber attacks on information systems. *Journal of Computers, 5*(3), 352–359.
2. Zhu, B., Joseph, A., & Sastry, S. (2011). A taxonomy of cyber attacks on SCADA systems. In *4th International Conference on Cyber, Physical and Social Computing Internet of Things (iThings/CPSCom)*. IEEE.
3. Kang, D. J., et al. (2009). Analysis on cyber threats to SCADA systems. In *Transmission & Distribution Conference & Exposition: Asia and Pacific, 2009*. IEEE.
4. Virvilis, N., Gritzalis, D., & Apostolopoulos, T. (2013). Trusted computing vs. advanced persistent threats: Can a defender win this game?. In *10th International Conference on Ubiquitous Intelligence and Computing, Autonomic and Trusted Computing (UIC/ATC)*. IEEE.
5. Virvilis, N., & Gritzalis, D. (2013). The big four-what we did wrong in advanced persistent threat detection?. In *Eighth International Conference on Availability, Reliability and Security (ARES)*. IEEE.
6. Sheng, S., Yingkun, W., Yuyi, L., Yong, L., & Yu, J. (2011). Cyber attack impact on power system blackout. In *IET Conference on Reliability of Transmission and Distribution Networks (RTDN 2011)*, 22–24 November 2011 (pp. 1–5).
7. Pasqualetti, F., Dorfler, F., & Bullo, F. (2013). Attack detection and identification in cyber-physical systems. *IEEE Transactions on Automatic Control, 58*(11), 2715–2729
8. Ten, C-W., Manimaran, G., & Liu, C-C. (2010). Cyber security for critical infrastructures: Attack and defense modeling. *IEEE Transactions on Systems, Man and Cybernetics, Part A: Systems and Humans, 40*(4), 853–865.
9. Cardenas, A. A., et al. (2011). Attacks against process control systems: Risk assessment, detection, and response. In *Proceedings of the 6th ACM Symposium on Information, Computer and Communications Security*. ACM.
10. Govindarasu, M., Hann, A., & Sauer, P. (2012). Cyber physical systems security for smart grid. In *The future grid to enable sustainable energy systems*. PSERC Publication.
11. Wong, T. P. (2011). *Active cyber defense: Enhancing national cyber defense*. Naval postgraduate school, Monterey, December 2011.
12. What is a malware virus?. http://www.kaspersky.com/threats/what-is-malware.

13. Vondra, T. (2013). Master Thesis. Czech Technical University in Prague, Department of Cybernetics. May 2013.
14. Malwares. https://www.csiac.org/sites/default/files/malware.pdf.
15. Zhioua, S. (2013). The middle east under malware attack dissecting cyber weapons. In *International Conference on "Distributed Computing Systems" Workshops (ICDCSW)* (pp. 11–16). IEEE.
16. Chen, T. M. (2010). Stuxnet, the real start of cyber warfare? [Editor's Note]. *IEEE Network, 24*(6), 2–3.
17. Chen, & Abu-Nimeh, S. (2011). Lessons from stuxnet. *IEEE Computer, 44*(4), 91–93.
18. Falliere, N., Murchu, L., & Chien, E. (2011). W32.Stuxnet dossier. Symantec, February 2011.
19. Matrosov, A., Rodionov, E., Harley, D., & Malcho, D. (2011). Stuxnet under the microscope. ESET, January 2011.
20. Mishra, S., Kant, K., & Yadav, R. S. (2012). Multi tree view of complex attack–stuxnet. In *Advances in computing and information technology* (pp. 171–188). Berlin, Heidelberg: Springer.
21. Symantec. (2011). W32.Duqu—The precursor to the next stuxnet. Symantec.
22. Bencsath, B., Pek, G., Buttyan, L., & Felegyhazi, M. (2012). Duqu: Analysis, detection, and lessons learned. In *Proceedings of the 2nd ACM European Workshop on System Security*.
23. Bencsath, B., Pek G., Buttyan L., & Felegyhazi M. (2012). The cousins of stuxnet: Duqu, flame, and gauss. In *Proceedings of the Future Internet*.
24. Falliere, N., Murchu, L., & Chien, E. (2011). W32 Stuxnet dossier, Version 1.4, technical report. Symantec Corporation, February 2011.
25. Kaspersky Lab. Gauss. (2012). Abnormal distribution; Technical report. Kaspersky Lab: Moscow, Russia.
26. Almarri, S., & Sant, P. (2014). Optimised malware detection in digital forensics. *International Journal of Network Security & Its Applications, 6*(1).
27. Hellany, A., Achi, H., & Nagrial, M. (2008). An overview of digital security forensics approach and modelling. In *International Conference on Computer Engineering & Systems, 2008. ICCES 2008*. IEEE.
28. Altschaffel, R., Kiltz, S., & Dittmann, J. (2009). From the computer incident taxonomy to a computer forensic examination taxonomy. In *Fifth International Conference on IT Security Incident Management and IT Forensics, 2009. IMF'09*. IEEE.
29. Denning, D. E. (1987). An intrusion-detection model. *IEEE Transactions on Software Engineering, (2)*, 222–232.
30. Fabro, M., & Cornelius, E. (2008). *Recommended practice: Creating cyber forensics plans for control systems*. Department of homeland security.
31. Zou, C. C., Gong, W., Towsley, D., & Gao, L. (2005). The monitoring and early detection of internet worms. *IEEE/ACM Transactions on Networking, 13*(5), 961–974.
32. Daley, D. J., & Gani, J. (1999). *Epidemic modelling: An introduction*. Cambridge University.

An Improved BPSO Algorithm
for Feature Selection

Lalit Kumar and Kusum Kumari Bharti

Abstract In machine learning and data mining tasks, feature selection has been used to select the relevant subset of features. Traditionally, high-dimensional datasets have so many redundant and irrelevant features, which degrade the performance of clustering. Therefore, feature selection is necessary to improve the clustering performance. In this paper, we select the optimal subset of features and perform cluster analysis simultaneously using modified-BPSO (Binary Particle Swarm Optimization) and K-means. Optimality of clusters is measured by various cluster validation indices. By comparing the overall performance of the modified-BPSO with the BPSO and BMFOA (Binary Moth Flame Optimization Algorithm) on six real datasets drawn from the UC Irvine Machine Learning Repository, the results show that the performance of the proposed method is better than other methods involved in the paper.

Keywords BPSO · BMFO · Cluster validation index · Data clustering
Feature selection · Swarm intelligence

1 Introduction

Clustering is a type of unsupervised classification method for statistical data analysis, which is used when we have unlabeled datasets and classifies the unlabeled datasets on the basis of some distance measure such as the Euclidean.

Feature selection (FS) is one of the most important steps in data mining that reduces the dimensionality of huge scientific datasets. Traditionally, the high-dimensional datasets consist of noisy, irrelevant and redundant features which lead to degrading the performance of clustering measures. Therefore, feature selection is

L. Kumar (✉) · K. K. Bharti
Design and Manufacturing, PDPM-Indian Institute of Information Technology,
Dumna Airport Road, P.O. Khamaria, Jabalpur, MP, India
e-mail: 1611007@iiitdmj.ac.in

K. K. Bharti
e-mail: kusum@iiitdmj.ac.in

© Springer Nature Singapore Pte Ltd. 2019
A. Khare et al. (eds.), *Recent Trends in Communication, Computing,
and Electronics*, Lecture Notes in Electrical Engineering 524,
https://doi.org/10.1007/978-981-13-2685-1_48

required to select a subset of proper features from the set of original features. We already know that major types of feature selection methods are filter and wrapper. In filter method, a subset of features is selected based on the statistical tests for their correlation (without any clustering algorithm) and remove the irrelevant features from the subset [1, 2], whereas the wrapper method uses the pre-decided clustering algorithm for selecting the subset of features. Though wrapper method is computationally expensive, it achieves better result compared with the filter method [3].

Nature-inspired algorithms (NIA) are the most famous techniques for feature selection process, and these techniques are observed from nature. In recent years, plenty of attention has been received for meta-heuristic algorithms like genetic algorithm (GA) [4], simulated annealing (SA) [5], particle swarm optimization (PSO) [5], ant colony optimization (ACO) [5] and differential evolution (DE) [5, 6] due to their good performance in solving the feature selection problem. PSO and ACO have greater accuracy compared with other meta-heuristic algorithms for selecting the subset of features [7, 8]. PSO is an easy-to-implement algorithm, has less adjustable parameters and is also computationally less expensive than analogous algorithms. Therefore, the PSO has been used as a major technique in engineering problems and for selecting a subset of relevant features [9, 10].

The main objective of this paper is to propose an FS method in wrapper approach that uses the modified-BPSO and the Silhouette index (fitness function) as an evaluator. The rest of the paper is planned as follows: Sect. 2 describes the BPSO and BMFO algorithms. In Sect. 3, the details of the proposed approach are discussed. In Sect. 4, the experimental results are presented while the conclusions and future direction are outlined in Sect. 5.

2 Algorithms Background

2.1 BPSO

Particle swarm optimization (PSO) was proposed by Kennedy and Eberhart in 1995 [9, 10]. PSO is a population-based algorithm, inspired by the grouping and schooling patterns of fish and birds. In PSO, each particle has its fitness which is evaluated by a fitness function. The fitness value of a particular particle is called as a personal best (pbest) solution achieved so far. The particle who has the best solution among all pbest, called the global best particle (gbest). The journey of particle is organized by its own position (pbest and gbest) in the swarm. Updation of velocity and position of particle can be described as in Eqs. (1) and (2), respectively.

$$v_i^d(t+1) = v_i^d(t) + c_1r_1 * (pbest_i^d(t) - x_i^d(t)) + c_2r_2 * (gbest^d(t) - x_i^d(t)) \qquad (1)$$

$$x_i^d(t+1) = x_i^d(t) + v_i^d(t+1) \qquad (2)$$

where d is the dimension of particle, t is the iteration, r_1 and r_2 are random numbers in the interval (0, 1), and c_1 and c_2 are positive acceleration constants. PSO was proposed for real search spaces. So, by extending the PSO, Kennedy and Eberhart introduced BPSO [11]. For this, a sigmoidal function [12] is used to convert the real search space position into the new position in binary search space.

$$v_i^d(t+1) = w * v_i^d(t) + c_1 r_1 * (pbest_i^d(t) - x_i^d(t)) + c_2 r_2 * (gbest^d(t) - x_i^d(t)) \quad (3)$$

$$x_i^d = \begin{cases} 1 \text{ if } \quad r \text{ and } < \frac{1}{1+e^{-v_i^d}} \\ 0 \text{ otherwise} \end{cases} \quad (4)$$

where w is inertia weight, $rand$ is a random number selected in an interval (0, 1) and x_i^d is the position of particle i at dimension d.

2.2 BMFO

Moth Flame Optimization (MFO), proposed by Seyedali Mirjalili in 2015 [13], is a new optimization algorithm which is inspired by the moths special navigation method in nature called transverse orientation. Moths are fancy insects, which attract toward the moonlight or artificial lights made by humans. According to MFO algorithm, main modules of MFO algorithm moths (actual search agents) and flames (best position of moths). The position of distinct moth is updated with respect to a flame as:

$$M_i = S(M_i, F_j) = D_i * e^{bt} * cos(2\pi t) + F_j \quad (5)$$

where M_i indicates the ith moth, F_j indicates the jth flame, S is the logarithmic spiral function, D_i indicates the distance between M_i and F_j, b is a constant which define the shape of the logarithmic spiral and t is a random number in $[-1, 1]$.

For binary search space, the position of moths is limited to binary variable. Therefore, sigmoidal function is used to transform the position of each moth which is updated with respect to a flame (refer Eq. (5)) into the new position in binary search space (refer Eq. (6)) for BMFOA [14].

$$M_i = \begin{cases} 1 \text{ if } \quad r \text{ and } < \frac{1}{1+e^{-(X(I+1))}} \\ 0 \text{ otherwise} \end{cases} \quad (6)$$

where $X(I+1)$ is the real-valued location update of the moth for $(I+1)$th iteration and $rand$ is a random number selected in interval (0, 1).

3 Proposed Method

3.1 Initialization Method of Swarm Population

In this paper, we symbolize a particle of size *nvar* as a binary vector, where *nvar* is original set of features (except id number and label) in the dataset. In that symbolization, 1 indicates the existence of the feature and 0 indicates the nonexistence of that feature. In this work, we take n population size of swarm but generate $2n$ swarm randomly within search boundaries. After that, we perform K-means on generated swarm and evaluate them by fitness function. However, we have to take n population so we take the best n evaluated swarm from the $2n$ swarm with the help of fitness function.

3.2 Evaluation Criteria/Fitness Function

Here, we discuss the fitness function (Silhouette index) by which we measure the quality of the selected subset of feature. The Silhouette index (SI) is introduced by Kaufman and Rousseeuw [15] which is a measure of how similar an object is to its own cluster (cohesion) compared with other clusters (separation). Various clustering measures are also used in this paper such as Dunn index (DI) and Davies–Bouldin Index (DBI). DI introduced by Dunn in 1974 [16] assesses the goodness of a clustering by measuring the maximal diameter of clusters and relating it to the minimal distance between clusters. The DI is the division of minimum inter-cluster distance and the maximum cluster size. DBI is introduced by Davies and Bouldin in 1979 [17] by which quality of clustering is measured.

3.3 Proposed Algorithm

In respect to applying modified-BPSO, first, we select the best swarm of size n with the help of fitness function as discussed in Sect. 3.1. After that, we apply K-means algorithm [18, 19] on selected features of dataset for each particle in the swarm. Here, particle's quality is evaluated by fitness function (refer Sect. 3.2). Further, on the basis of quality of particles, pbest and gbest are assigned and updated regularly. If particular particle continues to generate same results for some iterations, then we re-initialize that particle randomly or by taking the complement of that particle. This work analyzes the proposed modified-BPSO with linearly increasing inertia weight ($\uparrow w$) [20] which is calculated in each iteration.

Inputs: *n* number of particles (swarm size);
T number of iteration;
Variables Number of clusters, count, max count;
Output: gbest and subset of relevant features;
Proposed Algorithm:
1)   Initialize population of swarm of size *2n* randomly in search space;
2)   *For* distinct particle
3)   Calculate particle fitness by fitness function (Silhouette Index);
4)   Setpbest;
5)   *Endfor*
6)   Setgbest (best solution among pbest in the swarm);
7)   *While* maximum iteration is not touched
8)   *For*distinctparticle
9)   Update its velocity and position by using equations (3) to (4);
10) Calculate fitness of new updated particle;
11) *If* (fitness of prev. particle >= fitness of new particle)
12) Increase count value;
13) *Else* set count value to zero;
14) *If* (count >= max_count)
15) Re-initialize or perform complement of that particle and set count value to zero;
16)   *End if*
17)   *End if*
18) Update pbest of particle;
19)   *End for*
20) Update gbest accordingly;
21)   *End while loop*

4   Experimental Results and Discussion

4.1   *Datasets and Parameters Setup*

Table 1 shows the six used datasets drawn from the UCI data repository [21]. The parameters for the three different optimization algorithms namely BPSO, BMFO and modified-BPSO are defined in Table 2.

4.2   *Experiment Results*

Here, we compare the performance of the modified-BPSO with BPSO and BMFO. Tables 3, 4 and 5 show the length of selected relevant features and evaluation of clustering done by Silhouette index, Dunn index and Davies–Bouldin index, respectively. The higher value of SI and DI indicates a more relevant subset of features while a low value of DBI indicates more relevant subset of features, eventually leading to the increased accuracy of the clustering task.

Table 1 Datasets description

Dataset	Number of features	Number of samples
Ionosphere	34	351
Breast Cancer Wisconsin (BCW)	10	699
Connectionist Bench (Sonar, Mines vs. Rocks)	60	208
Iris	4	150
Statlog (Vehicle Silhouettes)	18	946
Parkinson	23	197

Table 2 Parameter settings

Algorithm	Parameter	Value
BPSO	Population size (swarm size)	50
	c_1 and c_2	1.5
	$[V_{min}, V_{max}]$	$[-6, 6]$
	T (number of iterations)	150
BMFO	No. of search agents (moths)	50
	b defining the shape of the logarithmic spiral in MFO	1.0
	T (number of iterations)	150
Modified-BPSO	Population size (swarm size)	50
	c_1 and c_2	1.5
	w_{max}	0.9
	w_{min}	0.4
	T (number of iterations)	150
	max_count	10

Table 3 SI of selected features

Algorithm	Dataset	Length of selected features	Average of gbest	Best gbest	Worst gbest	Standard deviation
BPSO	Ionosphere	1	0.5396	1.0000	0.2692	0.2366
	BCW	2	0.7912	0.8789	0.6896	0.0523
	CB	1	0.4837	0.9597	0.2742	0.2383
	Iris	2	0.8145	0.8916	0.6450	0.0872
	Vehicle	1	0.8282	0.9881	0.6214	0.1040
	Parkinson	2	0.8274	0.9459	0.5665	0.1186

(continued)

Table 3 (continued)

Algorithm	Dataset	Length of selected features	Average of gbest	Best gbest	Worst gbest	Standard deviation
BMFO	Ionosphere	20	0.5118	0.5743	0.4885	0.0257
	BCW	6	0.8641	0.8729	0.8399	0.161
	CB	43	0.4590	0.5090	0.4363	0.0234
	Iris	1	0.8916	0.8916	0.8916	$2.3406\,e^{-16}$
	Vehicle	8	0.8661	0.9485	0.8539	0.0296
	Parkinson	14	0.9459	0.9459	0.9459	$1.2879\,e^{-06}$
Modified-BPSO	Ionosphere	2	0.6909	0.8262	0.6059	0.0800
	BCW	2	0.8789	0.8789	0.8789	$1.1703\,e^{-16}$
	CB	12	0.7755	0.8041	0.7245	0.0224
	Iris	1	0.8916	0.8916	0.8916	$2.3406\,e^{-16}$
	Vehicle	2	0.9881	0.9881	0.9881	0
	Parkinson	2	0.9479	0.9511	0.9459	0.0027

Table 4 DI of selected features

Algorithm	Dataset	Length of selected features	Average of gbest	Best gbest	Worst gbest	Standard deviation
BPSO	Ionosphere	17	0.0701	0.0971	0.0285	0.0162
	BCW	7	0.1040	0.1770	0.0602	0.0376
	CB	33	0.1563	0.2746	0.0572	0.0689
	Iris	2	0.1529	0.3192	0.0392	0.1080
	Vehicle	4	0.1566	1.1551	0.0421	0.2175
	Parkinson	20	0.1994	0.3409	0.0082	0.1472
BMFO	Ionosphere	20	0.0950	0.0985	0.0902	0.0026
	BCW	6	0.1745	0.1770	0.1620	0.0053
	CB	42	0.2674	0.2763	0.2512	0.0092
	Iris	3	0.3192	0.3192	0.3192	$5.8514\,e^{-17}$
	Vehicle	9	0.6304	1.7418	0.2051	0.4827
	Parkinson	19	0.3409	0.3409	0.3409	$1.0790\,e^{-06}$
Modified-BPSO	Ionosphere	18	0.5582	0.5856	0.5340	0.0161
	BCW	7	0.1770	0.1770	0.1770	$2.9257\,e^{-17}$
	CB	26	0.3330	0.3415	0.3106	0.0031
	Iris	1	0.3192	0.3912	0.3912	$5.8514\,e^{-17}$
	Vehicle	1	3.2821	3.3803	2.3981	0.3106
	Parkinson	20	0.3409	0.3409	0.3409	$1.2411\,e^{-14}$

Table 5 DBI of selected features

Algorithm	Dataset	Length of selected features	Average of gbest	Best gbest	Worst gbest	Standard deviation
BPSO	Ionosphere	1	1.3257	0.2960	1.8773	0.4843
	BCW	2	0.6359	0.4046	0.8289	0.1357
	CB	2	1.4125	0.2575	1.9634	0.5335
	Iris	2	0.4090	0.2654	0.7931	0.1572
	Vehicle	2	0.4720	0.1045	1.0968	0.2311
	Parkinson	2	0.4986	0.3513	0.9947	0.1865
BMFO	Ionosphere	22	0.2977	0.2521	0.3237	0.0199
	BCW	6	0.4654	0.4046	0.5374	0.0486
	CB	36	1.3787	1.3131	1.4308	0.0361
	Iris	1	0.2654	0.2654	0.2654	$5.8514\,e^{-17}$
	Vehicle	9	0.4021	0.3970	0.4094	0.0044
	Parkinson	14	0.3513	0.3513	0.3513	$2.8570\,e^{-06}$
Modified-BPSO	Ionosphere	2	0.6037	0.3967	0.7263	0.0984
	BCW	2	0.4046	0.4046	0.4046	0
	CB	13	0.6892	0.6035	0.7984	0.0669
	Iris	1	0.2654	0.2654	0.2654	$5.8514\,e^{-17}$
	Vehicle	1	0.1135	0.1045	0.1946	0.0285
	Parkinson	1	0.3200	0.3197	0.3212	$5.5535\,e^{-04}$

5 Conclusion and Future Directions

This paper proposed a new wrapper feature selection using the modified-BPSO algorithm. The main goal of the proposed FS approach is to identify the subset of relevant features that could obtain better results than using all features of the dataset and we also show that if initialization is better than results are better. A set of well-known FS datasets from UCI data repository are used to evaluate the proposed approach, and the results are compared with the BPSO and BMFO algorithms. The results show superior performance for the modified-BPSO approach compared to other approaches. As future work, we plan to update the parameters to improve feature selection in order to explore the NIA.

References

1. Guyon, I., & Elisseeff, A. (2003). An introduction to variable and feature selection. *The Journal of Machine Learning Research, 3,* 1157–1182.
2. Kabir, M., Shahjahan, M., & Murase, K. (2012). A new hybrid ant colony optimization algorithm for feature selection. *Expert Systems with Applications, 39*(3), 3747–3763.
3. Prakash, J., & Singh, P. (2015). Particle swarm optimization with k-means for simultaneous feature selection and data clustering. In *IEEE 2015 Second International Conference on Soft Computing and Machine Intelligence (ISCMI)*, Nov 23 (pp. 74–78).
4. Gwak, J., Jeon, M., & Pedrycz, W. (2016). Bolstering efficient SSGAs based on an ensemble of probabilistic variable-wise crossover strategies. *Soft Computing, 20*(6), 2149–2176.
5. Yang, S.: Nature-inspired optimization algorithms. Elsevier (2014).
6. Ali, M., Awad, N., Suganthan, P., & Reynolds, R. (2017). An adaptive multipopulation differential evolution with dynamic population reduction. *IEEE Transactions on Cybernetics, 47*(9), 2768–2779.
7. Moradi, P., & Gholampour, M. (2016). A hybrid particle swarm optimization for featuresubset selection by integrating a novel local search strategy. *Applied Soft Computing, 43,* 117–130.
8. Nanda, S., & Panda, G. (2014). A survey on nature inspired metaheuristic algorithms for partitional clustering. *Swarm and Evolutionary Computation, 16,* 1–8.
9. Kennedy, J. (2011). Particle swarm optimization. In *Encyclopedia of machine learning* (pp. 760–766). US: Springer.
10. Marini, F., & Walczak, B. (2015). Particle swarm optimization (PSO). A tutorial. *Chemometrics and Intelligent Laboratory Systems, 149,* 153–165.
11. Kennedy, J., & Eberhart, R. (1997). A discrete binary version of the particle swarm algorithm. In *1997 IEEE International Conference on Systems, Man, and Cybernetics. Computational Cybernetics and Simulation* (Vol. 5). IEEE.
12. Liu, J., Mei, Y., & Li, X. (2016). An analysis of the inertia weight parameter for binary particle swarm optimization. *IEEE Transactions on Evolutionary Computation, 20*(5), 666–681.
13. Mirjalili, S. (2015). Moth-flame optimization algorithm: A novel nature-inspired heuristic paradigm. *Knowledge-Based Systems, 89,* 228–249.
14. Reddy, S., Panwar, L., Panigrahi, B., & Kumar, R. (2017). Solution to unit commitment in power system operation planning using modified moth flame optimization algorithm (MMFOA): A flame selection based computational technique. *Journal of Computational Science,* 1–18.
15. Kaufman, L., & Rousseeuw, P. (2009). Finding groups in data: an introduction to cluster analysis. *Probability and statistics.* Wiley.
16. Bezdek, J., & Pal, N. (1995). Cluster validation with generalized Dunn's indices. In *Proceedings of the Second New Zealand International Two-Stream Conference on Artificial Neural Networks and Expert Systems*, Nov 20 (pp. 190–193). IEEE.
17. Davies, D., & Bouldin, D. (1979). A cluster separation measure. In *IEEE Transactions On Pattern Analysis And Machine Intelligence*, Apr 2 (pp. 224–227).
18. Jain, A., & Dubes, R. (1988). *Algorithms for clustering data* (Vol. 6). Prentice Hall Englewood Cliffs.
19. Prakash, J., & Singh, P. (2012). An effective hybrid method based on de, ga, and k-means for data clustering. In *Proceedings of the Second International Conference on Soft Computing for Problem Solving (SocProS 2012)*, Dec 28–30 (pp. 1561–1572). Springer.
20. Jain, I., Jain, V., & Jain, R. (2018). Correlation feature selection based improved-binary Particle Swarm Optimization for gene selection and cancer classification. *Applied Soft Computing, 62,* 203–215.
21. Frank, A. *UCI Machine Learning Repository.* http://archive.ics.uci.edu/ml.

Part IX
Electronic Theories and Applications

A Comparative Analysis of Asymmetrical and Symmetrical Double Metal Double Gate SOI MOSFETs at the Zero-Temperature-Coefficient Bias Point

Amrish Kumar, Abhinav Gupta and Sanjeev Rai

Abstract The silicon-on-insulator (SOI) technology provides the higher current driving capability, low power consumptions, reduced SCEs and extensive scaling of the channel length. But SOI-based MOSFETs are weak in thermal stability like self-heating. In this paper, a comparative analysis of asymmetrical double metal double gate (ADMDG) and symmetrical double metal double gate (SDMDG) at the zero-temperature-coefficient (ZTC) bias point is proposed. ZTC is the bias point where the device constraints become free of variation in temperature. ADMDG and SDMDG devices are simulated by 2-D Atlas simulator. 2D-Atlas simulation revealed the figure of merit (FOMs) such as transconductance (g_m), output conductance (g_d), intrinsic gain (Av), on-current (I_{on}), off-current (I_{off}), on–off current ratio (I_{on}/I_{off}) and cutoff frequency (f_T). The simulation results give the presence of inflection point of the devices. The variation of ZTC point for transconductance (ZTC_{gm}) and drain current (ZTC_{IDS}) for ADMDG and SDMDG MOSFETs is compared.

Keywords ADMDG · SDMDG · Gate engineering · SCEs · Analog · RF FOMs

A. Kumar · S. Rai
Department of Electronics and Communication Engineering,
Motilal Nehru National Institute of Technology Allahabad, Allahabad 211004, India
e-mail: amrish1288@gmail.com

S. Rai
e-mail: srai@mnnit.ac.in

A. Gupta (✉)
Electronics Engineering Department, Rajkiya Engineering College, Sonbhadra 231206, India
e-mail: abhinavkit87@gmail.com

© Springer Nature Singapore Pte Ltd. 2019
A. Khare et al. (eds.), *Recent Trends in Communication, Computing, and Electronics*, Lecture Notes in Electrical Engineering 524,
https://doi.org/10.1007/978-981-13-2685-1_49

517

1 Introduction

Recently, industries are interested in designing such type of transistors or integrated circuits which can be efficiently operated in various range of temperature. The designers of integrated circuits (ICs) designed such type which can be used in both low-temperature and high-temperature application [1, 2]. These days trend of ICs designed is moved from conventional CMOS to SOI-based MOSFETs. There are most of the points behind it, for example, junction capacitance and various short-channel effects (SCEs) are minimized in SOI MOSFETs. At high temperature, the problem of leakage current and latch up affect the performances of conventional CMOS and the problem of latch up removed in SOI MOSFETs and leakage current also minimized [3]. All the digital and analog circuits are always biased about a point where the I–V characteristics indicate no or little deviation with respect to temperature. This point is called ZTC point. The ZTC point for SOI MOSFETs of single gate and double gate for 100–400 K temperature range has been identified in [4]. Both investigational and logical result for ZTC point over a varied range of temperature for partially depleted (PD) SOI MOSFET and fully depleted (FD) SOI n-MOSFET are presented in [5, 6]. The value of drain current (I_D) is different in earlier and later definite bias point with deviation in temperature. The value of current depends on temperature because of mobility and high field effect produced by bias voltage [7]. This is the recent approach to find out ZTC point of ADMDG and SDMDG by using 2-D Atlas simulator [8]. Here we compared various performances of ADMDG and SDMDG likewise SCEs electrostatic, analog and RF performance parameters are observed by temperature variation and final conclusion is specified in Sect. 4.

2 Device Structure and Simulation

The different structure of ADMDG and SDMDG is presented here through their 2-D numerical simulation and schematic diagram in Fig. 1. The structure adopted different physical dimension that is given here with following specification. The thickness of silicon oxide is 1.1 nm with channel length of 40 nm technology trend [9] and source and drain length taken here is 20 nm which are vertically placed in this structure, the thickness of silicon body is 10 nm with width of 0.5 μm. The metal gate work function is regulated in the middle of 4.6 and 4.8 eV with fixed threshold voltage of 0.4 V at room temperature [10]. Metal gate technology is used here due to their unique property of eliminating degradation in transistor performance. One of the best material used here for metal gate technology is molybdenum because its work function is adjusted between desired range of 4.5–5 eV [11]. Drain and source (D and S) are highly doped (n-type $= 1 \times 10^{20}$ cm^{-3}) due to minimized effect of mobility degradation by coulombs scattering [12]. Drain bias is fixed at numerical value VDD $= 1.0$ and VDS $= 0.5$ V, gate-to-source voltage selected as a variable value from Vgs $= 0-1.0$ V. The transfer characteristics provide the value of threshold voltage

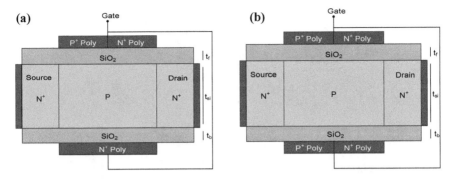

Fig. 1 Schematic of simulated device, **a** asymmetric double metal double gate (ADMDG), **b** symmetrical double metal double gate (SDMDG) SOI MOSFET

(Vth). The simulation is performed with 2-D numerical simulation using FLDMOB, CONMOB and Lombardi models (CVT), Shockley–Read–Hall (SRH) model along with Auger recombination model for varying operating temperature (100–400 K).

3 Simulation Result and Discussion

The analog/RF performance of the ADMDG and SDMDG MOSFET structures is studied with varying operating temperature (100–400 K). Analysis of electrostatic performance parameters such as the on-state drive current (I_{on}), off-state leakage current (I_{off}) and (I_{on}/I_{off}) ratio is used in nanoscale device design. The evaluation of important analog circuit/RF performance is by parameters like transconductance (g_m), output conductance (g_d), intrinsic gain (A_v) and cutoff frequency (f_T).

As we know that the variation in temperature changes various physical parameters like carrier mobility reduces with the increases in temperature as

$$\mu(T) = \mu(T_0)(\frac{T}{T_0})^{-n} \tag{1}$$

where n varies from 1.6 to 2.4 [13]. From Eq. 1, threshold voltage (Vth) is varied with temperature. The drain current I_D affected by both parameters mobility and threshold voltage Vth as [12]

$$I_D(T) \propto \mu(T)[V_{GS} - Vth(T)] \tag{2}$$

Both the terms mobility and threshold voltage have opposite behavior with respect to temperature about a constant gate-to-source bias voltage. The effect of both terms is canceled at constant value of bias voltage, which is called as ZTC bias point.

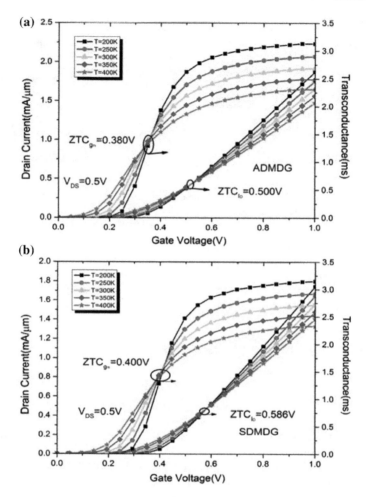

Fig. 2 Variation of both the drain current (I_D) and transconductance (g_m) w.r.t. the gate voltage at the various temperature of **a** ADMDG, **b** SDMDG

At V_{GS} > Vth, the behavior of gm with temperature is just opposite to that of V_{GS} < Vth because of mobility degradation. The Vth reduces with rise in temperature gm, while the degradation of mobility decreases g_m. These two occurrences produce the desire effects each to increase the ZTC bias point for g_m. From Fig. 2, we can evaluate that the value of gm at the ZTC point (0.38 V) is lesser than the drain current (I_D) at the ZTC bias point (0.50 V). ZTC I_D and $ZTCg_d$ bias points are two main constraints in analog and digital circuit design for extensive variety of temperature as shown in Fig. 3.

The intrinsic gain can be calculated ($A_V = g_m/g_d$) and expressed in early voltage V_{EA} as shown in Eq. 3.

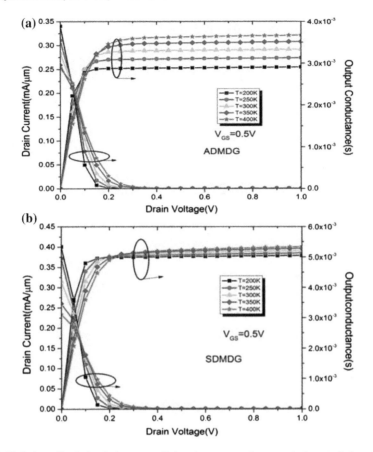

Fig. 3 Variation of both the drain current (I_D) and output conductance (g_d) w.r.t. drain voltage at various temperatures of **a** ADMDG, **b** SDMDG

$$A_V \cong \frac{g_m}{g_d} \cong \frac{g_m}{I_D} V_{EA} \tag{3}$$

From Eq. (3), high value of intrinsic gain of device is also liable for lower value of gd or higher value of V_{EA}. From Fig. 4, it has been observed that intrinsic gain of SDMDG device is higher (about 42 dB) as compared to ADMDG device (about 40 dB) at same conditions.

The deviation of Ion, Ioff and Ion/Ioff ratio with change in temperature is shown in Fig. 5. When we rise the temperature, the charge carriers are getting excited and more improve the carrier transport efficiency. It means the on-current and leakage current (Ioff) with temperature are increased as shown in Fig. 5. The Ion/Ioff ratio is an effective constraint for justifying the switching ability. That should be high for a good switching application. The Ioff current is considerably less in SDMDG structure as compared to its equivalent ADMDG structure as shown in Fig. 5. Simi-

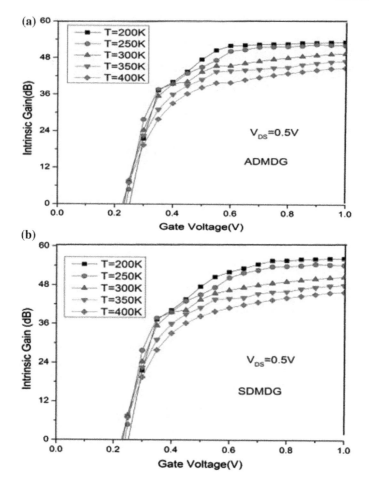

Fig. 4 Intrinsic gain (A_v) versus gate voltage at various temperatures of **a** ADMDG, **b** SDMDG

larly, the SDMDG configurations exhibit higher on-current is due to improved carrier transport efficiency and low leakage current as shown in Fig. 5, it means that the value of Ion/Ioff ratio is considerable higher (i.e., between 10^9 and 10^{10}) for SDMDG configurations as compared to that the value (i.e., between 10^8 and 10^9) for the ADMDG configuration. From Fig. 5, it observed that Ion/Ioff drops with rise in temperature. This is due to more leakage current at higher temperatures. The Ioff current indicates a less value at 200 K and after growing with rise in temperature above 200 K, this is because of low subthreshold slope (SS) and high Vth values at low temperatures.

Figure 6a illustrates the intrinsic capacitances (C_{gs} and C_{gd}) versus gate-to-source voltage for both subthreshold (weak inversion) and superthreshold (strong inversion) regions. The C_{gs} and C_{gd} capacitances are calculated for all devices at a single AC frequency (i.e., 1 MHz) and a gate voltage varies from 0 to 1 V with a period of

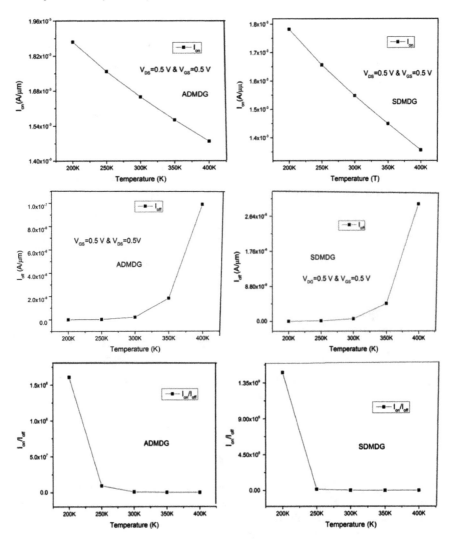

Fig. 5 Variation of on-state current (Ion), off-state current (Ioff) and on–off current ratio (Ion/Ioff) w.r.t. temperature of ADMDG and SDMDG SOI MOSFETs

0.025 V applied. The ZTC C_{gs} points are recognized through a lesser value as going from ADMDG (0.40 V) to SDMDG (0.45 V). In case of subthreshold region (i.e., under ZTC point), at higher value of T then the value of C_{gs} is high and it slowly drops as lowering the value of T. But in superthreshold region (beyond ZTC point) condition, the opposite situations have been observed.

Cutoff frequency (f_T) versus gate voltage (V_G) is shown in Fig. 6b which is evaluated from Eq. 5. f_t is largely affected by g_m and has slight variation of C_{gg}.

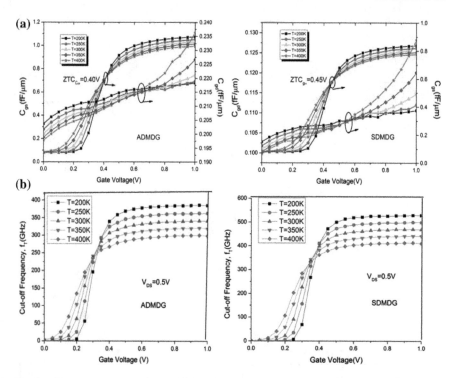

Fig. 6 a Gate-to-source capacitance (C_{gs}) and gate-to-drain capacitance (C_{gd}) and **b** f_T versus V_{GS} with different value of operating temperature

$$f_T = \frac{g_m}{2\pi \left(C_{gs} + C_{gd} \right)} \qquad (5)$$

where g_m, C_{gs} and C_{gd} are the transconductance and the C_{gg} is the sum of gate-to-source capacitance(C_{gs}), and gate-to-drain capacitance C_{gd}. At low temperature, there is enhancement in cutoff frequency.

The important parameters for SCEs, Ion/Ioff ratio and analog/RF performance are shown in Table 1. As of the observation, SDMDG has edge over ADMDG for all the performances for wide range of temperature.

Table 1 The important parameters for SCEs, Ion/Ioff ratio, A_V (dB) and $f_T(GH_Z)$ of SDMDG and ADMDG

Temperature (K)	I_{on}/I_{off}		A_V (dB)		$f_T(GH_Z)$	
	ADMDG	SDMDG	ADMDG	SDMDG	ADMDG	SDMDG
400	1.0×10^4	1.8×10^4	40.6	42.3	301.5	400.6
350	1.0×10^4	1.2×10^5	45.8	46.5	312.4	440.9
300	1.5×10^5	1.6×10^6	47.0	52.2	333.2	455.3
250	1.0×10^7	1.0×10^8	51.9	53.7	350.7	500.5
200	1.3×10^8	1.4×10^9	54.0	58.0	380.6	550.0

4 Conclusion

This paper presents a comparative analysis of the ADMDG and SDMDG SOI MOSFETs on the basis of the ZTC bias points using the 2-D Atlas simulation. The simulation outcomes give the comprehensive information about the ZTC bias point for electrostatics, analog and RF parameters with variation of the temperature. Also the comparison is performed between several parameters like Ion/Ioff, A_V and f_T for ADMDG and SDMDG over a varied range of temperatures. It is observed that at a room temperature, the SDMDG structure exhibits better as compared to ADMDG structure and this is due to 10 times higher in I_{on}/I_{off} ratio, the intrinsic gain A_V is increased 11.07% and also f_T 37% is increased. Hence, SDMDG MOSFET leads to a major enhancement in DC, analog and RF performances and can be used for various choice of temperature applications.

References

1. Scheinert, S., Paasch, G., & Schipanski, D. (1995). Analytical model and temperature dependence of the thin film SOI FET. *Solid-state Electronics, 38*(5), 949–959.
2. Kumari, V. et al. (2012). Temperature dependent drain current model for Gate Stack Insulated Shallow Extension Silicon On Nothing (ISESON) MOSFET for wide operating temperature range. *Microelectronics Reliability, 52.6*, 974–983.
3. Sahu, P. K., Mohapatra, S. K., & Pradhan, K. P. (2015). Zero temperature-coefficient bias point over wide range of temperatures for single-and double-gate UTB-SOI n-MOSFETs with trapped charges. *Materials Science in Semiconductor Processing, 31*, 175–183.
4. Mohapatra, S. K., Pradhan, K. P., & Sahu, P. K. (2015). Temperature dependence inflection point in Ultra-Thin Si directly on Insulator (SDOI) MOSFETs: An influence to key performance metrics. *Superlattices and Microstructures, 78*, 134–143.
5. Pradhan, K. P., & Sahu, P. K. (2016). Temperature dependency of double material gate oxide (DMGO) symmetric dual-k spacer (SDS) wavy FinFET. *Superlattices and Microstructures, 89*, 355–361.

6. Pradhan, K. P., et al. (2015). Reliability analysis of charge plasma based double material gate oxide (DMGO) SiGe-on-insulator (SGOI) MOSFET. *Superlattices and Microstructures, 85,* 149–155.
7. Roy, N. C., Gupta, A., & Rai, S. (2015). Analytical surface potential modeling and simulation of junction-less double gate (JLDG) MOSFET for ultra-low-power analog/RF circuits. *Microelectronics Journal, 46.10,* 916–922.
8. ATLAS Device Simulator Software. (2015). *Silvaco.* CA, USA: Santa Clara.
9. International Technology Roadmap for Semiconductors. http://public.itrs.net.
10. Kumar, A. (2016). Analog and RF performance of a multigate FinFET at nano scale. *Superlattices and Microstructures, 100,* 1073–1080.
11. Rai, S.. (2017). Reliability analysis of Junction-less Double Gate (JLDG) MOSFET for analog/RF circuits for high linearity applications. *Microelectronics Journal, 64,* 60–68.
12. Kumar, A., Gupta, A., & Rai, S. (2017). Charge plasma based graded channel with dual material double gate JLT for enhance analog/RF performance. In *4th International Conference on Power, Control & Embedded Systems (ICPCES)* (pp. 1–6–2017/11). IEEE.
13. Tan, T. H., & Goel, A. K. (2003). Zero-temperature-coefficient biasing point of a fully-depleted SOI MOSFET. *Microwave and Optical Technology Letters, 37*(5), 366–370.

Level-Wise Scheduling Algorithm for Linearly Extensible Multiprocessor Systems

Abdus Samad and Savita Gautam

Abstract The valuable treating of parallelism on an interconnection network entails optimizing inconsistent performance indices, such as the reduction of communication and scheduling overheads and also uniform distribution of load among the nodes. In this kind of a system a number of nodes process the numerous jobs concurrently. A novel dynamic scheduling scheme that supports task unbiased structure approach is proposed for a particular class of multiprocessor networks known as linearly extensible multiprocessor networks. The significance of proposed scheduling scheme is remedying the communication overhead, delay in task execution and efficient processor utilization, which ultimately improves the total execution time. The proposed algorithm is implemented on a set of processors known as nodes which are linked through certain interconnection network. In particular, the performance is evaluated for linear type of multiprocessor architectures. In addition, a comparison is also made by implementing standard scheduling algorithm on same architectures with same number of nodes. The metrics used for comparison are Load Imbalance Factor (LIF), which represents the deviation of load among processors after achieving load balancing and execution time. The comparative simulation study shows that the proposed scheme gives better performance in terms of task scheduling and execution time when implemented on various linearly extensible multiprocessor networks.

Keywords Linearly extensible network · Load imbalance factor
Level scheduling algorithm · Interconnection network · Execution time

A. Samad (✉) · S. Gautam
University Women's Polytechnic, F/O Engineering & Technology,
Aligarh Muslim University, Aligarh 202002, India
e-mail: abdussamadamu@gmail.com

S. Gautam
e-mail: savvin2003@yahoo.co.in

A. Khare et al. (eds.), *Recent Trends in Communication, Computing,
and Electronics*, Lecture Notes in Electrical Engineering 524,
https://doi.org/10.1007/978-981-13-2685-1_50

1 Introduction

In the past decade, interconnection networks have been studied and often used to execute parallel applications. Many different topologies have been designed which provide high-speed services in supercomputers, clusters of machines and all those systems where parallel applications are required. In order to evaluate the performance of an interconnection network, it is not sufficient that they possess attractive topological properties but also they must be performed on equal footing when working with parallel applications. Scheduling of load on such multiprocessor networks is considered as one of the performance components in parallel systems. Task scheduling problem has been tackled using different approaches on different interconnection networks. The main motive is to ensure that no processor should remain overloaded or underloaded after scheduling the tasks on the network of processors. For instance, the scheduling algorithms are designed for hypercube, tree and mesh networks [1–3]. Algorithms are also designed for a new class of multiprocessor architectures known as Linearly Extensible architectures [4]. These networks have the desirable properties of hypercube architecture while eliminating the drawbacks of cube-based topologies. Such systems may consists of a variety of parallel systems such as simple Binary Tree (BTree), Ternary Tree (TTree), Linearly Extensible Tree (LET) [5], Linearly Extensible Triangle (LEΔ) [6], Linearly Extensible Cube (LEC) [7] and Linear Cross Cube (LCQ) [8]. For effective utilization of nodes in any network, efficient scheduling algorithm plays a key role in the computation of parallel applications on multiprocessor networks. The common objective of research in this area is to reduce error or more specifically known as load balancing error. An efficient algorithm tries to make even distribution of load with minimum complexity and execution time. Although number of algorithms has been designed to produce efficient schedulers, there seems to still exist room for further improvement in this direction. The present work is an effort to investigate an appropriate scheduling algorithm for a particular class of networks known as Linearly Extensible Networks (LEN).

The rest of the paper is organized as follows. Section 1 is introduction. Section 2 describes the past work in brief. In Sect. 3 the existing algorithm and proposed algorithm for solving the problem are described. The results obtained by implementing the proposed algorithm are presented and discussed in Sect. 4. The last section is written for conclusion.

2 Past Research on Scheduling Algorithms

The literature is full with such research works where main objective is to maximize the execution speed and resourceful utilization of nodes in the network. These algorithms may be classified on the basis of many parameters, namely static or dynamic, local/global, sender initiated or receiver initiated [9]. The performance of static algorithms will be much better for predefined load or fixed load, whereas these are not suited for unpredictable type of load. The dynamic algorithms on the other hand take decision on fly and allocate/re-allocate the load in real time without priory information of load.

The scheduling problem has been studied for a variety of application models. These applications can be categorized based on the architecture involved. In general, architectures of these processor networks are either tree-based or cube-based. The scheduling approaches for tree type network are based on stored and forward technique where decision is taken based on local information. Scheduling algorithms that utilize local information has good scalability; however, it does not cover the entire network. Problem of task scheduling on tree-based networks in general is NP-hard [10–12]. Dimension Exchange Method [DEM] algorithm is used for cube-based networks as it balances a small domain of network first and consecutively combines the larger domain to ultimately cover the entire network [13, 14]. However, DEM results high communication cost when number of processors are high. It is because nodes are connected in the same direction at different level of architecture. Therefore, more steps are required to cover the entire network particularly when large numbers of nodes are available in the network. Similarly, a modified version of DEM is reported where lesser number of steps is required to facilitate exchanging the tasks between to cover the entire network [15]. This algorithm is efficient particularly for cube-based architectures where number of processors increase rapidly at higher levels. System performance can also be improved by non-contiguous processor allocation, especially in multi-computing system [16]. However, the author suggests great scope in the designing of efficient scheduling approaches as well as selection of interconnection topologies.

3   Proposed Algorithm

For deterministic execution, dynamic allocation techniques are best suited and can be applied to real-world applications. Static techniques on the other hand always have a prior knowledge of all the tasks to be executed. The objective is to derive optimal task assignment so that task execution time and communication cost could be minimized. This paper considers the task assignment with the following scenario.

- The system consists of a set of homogeneous processors or nodes
- The sequence of communication with the nodes is to be selected based on communication links
- The load may be migrated between nodes in different chunk sizes.

There is different criterion in different algorithms that decide the sequence of communication. One such criterion is based on minimum distance property which allows communication between those nodes which are having direct communication links. Initially, we attempt to execute the performance of linearly extensible networks by applying minimum distance algorithm (MDS). This method is described in detail in [3], and pseudo-code for the same is given below:

```
MDS (Output: Balanced Network)
Generate Load at all processors
/* Consider P is total number of Processors*/
For HeadProcessor obtain Connected processors
Calculate total load (sum), IdealLoad for all procesors
Migrate Load such that Load(HeadProccessor)==Ideal
If Load(HeadProcessor)> IdealLoad)
 Send excess load to other connected processor
else if(Load(HeadProcessor)< IdealLoad)
 Receive excess load from other Connected processor
Calculate Load Imbalance Factor(LIF)
/*Connected processor collection*/
For all processors check connectivity
if (connectivtymatrix[HeadProcessor][processor]==1)
processor is connected to HeadProcessor
```

Mostly scheduling algorithms are designed to equally spread the tasks among available processors and to minimize the total execution time. Therefore, the two performance metrics, namely Load Imbalance Factor (LIF) and execution time, are taken into consideration. In this paper, the dynamic task scheduling algorithm is examined for linearly extensible multiprocessor systems. The purpose of our study is not only to propose an algorithm for these types of networks but also to make comparative study of the quality of solution.

The proposed algorithm is developed to produce efficient task scheduling with lesser or equal execution time as required when MDS is implemented. The algorithm decides different sets of processors at each level in which migration of tasks can take place. As the level increases, the number of processors also increases. Therefore, when task volume is high, the algorithm produces larger number of processors to communicate which improve the performance of entire network. However, going to higher level may increase the execution time. Thus, to achieve optimum performance the level decision must be taken intelligently. One method to decide the depth of level may be based on the volume of tasks. The complete algorithm is described below:

```
Proposed Algorithm:
I Processor Load Balancing
Generate Load at all processors
  /* Consider P is total number of Processors*/
For HeadNode perform First Level processor collection
Calculate total load(sum) on all processors
Compute IdealLoad
IdealLoad= sum/No. Of Processors
Migrate Load such that  Load(HeadProccessor)==IdealLoad
If Load(HeadProcessor)> IdealLoad)
Send excess load to First Level processor
 else if(Load(HeadProcessor)< IdealLoad)
Receive excess load from First Level processor
Repeat steps of Level one For secondLevel processor
Calculate LIF after computing max load on a processor
LIF=(max-IdealLoad)/IdealLoad
  II First Level processor collection
For all processors check connectivity
if (connectivityMatrix[HeadProcessor][processor]==1)
processor is connected to HeadProcessor
  III Second Level processor collection
 For all processors at level one find out second level
processors which are connected to these processors.
 IV Find Execution Time
Obtain value of StartTime(T1) when Load starts Migrate
Obtain value of finishedTime(T2) at Migrate finished
ExecutionTime=T2-T1
```

4   Simulation Results

In this section, performance of the proposed dynamic scheduling algorithm and MDS algorithms is evaluated and compared. The performance of these algorithms when applied on linearly extensible networks is shown in diagrams in which the horizontal axis represents the number of tasks and the vertical axis is the average LIF (%) and execution time. For simulation purpose a total of four networks, namely Linearly Extensible Tree (LET), Linearly Extensible Triangle (LEΔ), Linearly Extensible Cube (LEC) and Linear Cross Cube (LCQ), are taken into consideration. Each network used in simulation consists of 6 processors (nodes). The communication between nodes depends upon the available link entered through adjacency matrix. Initially, the tasks are generated randomly at level 0. Subsequently, tasks are doubled at the next level and so on. Both the scheduling algorithms are tested for the same set of task structure with the same number of nodes in each network. The simulation results are presented in two data sets of task volumes. The first set consists of tasks in thousands, whereas the second data set has high volume of tasks that may go up to ten

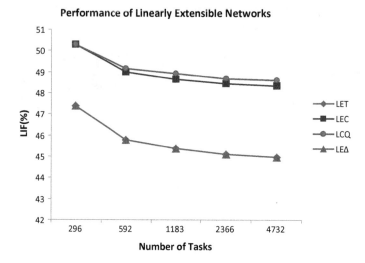

Fig. 1 Performance of LEN for moderate load with MDS

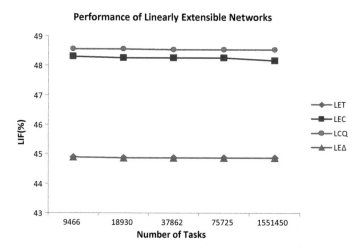

Fig. 2 Performance of LEN for heavy load with MDS

thousands or more. These data sets are framed to monitor the performance difference under medium and heavily loaded systems particularly in an unpredictable manner. The simulation results of these algorithms for various interconnection networks are obtained and presented in Figs. 1 and 2.

The results obtained for medium load show lesser variations in LIF on different interconnection networks. However, the initial value of LIF in LEC and LCQ is greater than that obtained for LET and LEΔ networks. On the other hand when the

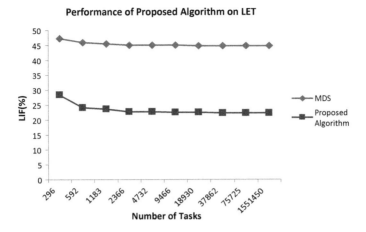

Fig. 3 Performance of proposed algorithm on LET

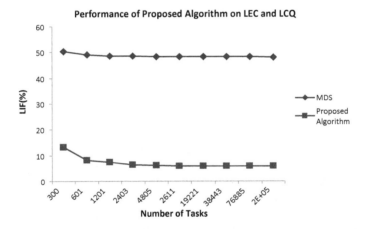

Fig. 4 Proposed algorithm on LEC and LCQ

network is heavily loaded, the LIF in LEC and LCQ is further reduced and became constant in all networks at higher level of task volumes. This trend is shown in Fig. 2.

In order to create a comprehensive study of experimental results obtained by implementing proposed scheduling algorithm, the curves are drawn between LIF and number of tasks for each multiprocessor network depicted in Figs. 3 and 4.

The value of LIF is decreasing in general with the same pattern for all the networks we considered and producing similar behavior. For instance, the value of LIF in LET network with the proposed algorithm as shown in Fig. 3 is reduced from 48 to 28%.

In case of LEC and LCQ networks similar performance is observed and the value of LIF is reduced from 50 to 12% and shown in Fig. 4. Similar results are obtained

Table 1 Execution time for different networks

Algorithms	Execution time (ms)			
Networks	LEC	LCQ	LET	LEΔ
MDS	126	184	206	206
Proposed algorithm	455	454	475	532

for LEΔ network with the proposed algorithm. The LIF in case of LEΔ is improved significantly from 48 to 20%.

Another parameter to monitor the performance of proposed scheduling algorithm is execution time. The results are also evaluated for execution time and demonstrated in Table 1. Since the algorithm forming a set of processors for performing migration the tasks at a particular level, it is obvious that it will consume little bit greater time as those in MDS algorithm. However, the execution time could be reduced if migration of tasks is carried out in terms of packets. For better comparison the results shown in Table 1 are obtained without packet formation.

5   Conclusion

In this paper, a dynamic method for scheduling of tasks on a new class of interconnection network known as linearly extensible multiprocessor networks is proposed. The algorithm works level-wise and schedules the tasks on a particular set of processors at each level. The proposed algorithm is implemented with the same set of task volumes and tested for medium to high load. The performance of the proposed algorithm on these networks is evaluated based on two parameters namely LIF and execution time. The simulation results obtained show that the proposed algorithm is performing better particularly on linearly extensible multiprocessor networks when the system is heavily loaded.

The aim is to minimize the load imbalance with minimum communication cost. The proposed algorithm works level-wise which results effective utilization of nodes especially under heavy load. The algorithm is best suited particularly for a network with lesser number of nodes and node degree.

References

1. Birmpilis, S., & Aslanidis, T. (2017). A critical improvement on open shop scheduling algorithm for routing in interconnection networks. *International Journal of Computer Networks & Communications (IJCNC), 9*(1), 1–19.
2. Barbosa, G., & Moreira, B. (2011). Dynamic scheduling of batch of parallel task jobs on heterogeneous clusters. *Journal of Parallel Computing, 37,* 428–438.

3. Prasad, N., Mukkherjee, P., Chattopadhyay, S., Chakrabarti, I. (2018). Design and evaluation of ZMesh topology for on-chip interconnection networks. *Journal of Parallel and Distributed Computing*, 17–36.
4. Samad, A., Rafiq, M. Q., & Farooq, O. (2012). Two round scheduling (TRS) scheme for linearly extensible multiprocessor systems. *International Journal of Computer Applications, 38*(10), 34–40.
5. Khan, Z. A., Siddiqui, J., & Samad, A. (2013). Performance analysis of massively parallel architectures. *BVICAM's International Journal of Information Technology (IJIT), 5*(1), 563–568.
6. Manullah, (2013). A Δ-based linearly extensible multiprocessor network. *International Journal of Computer Science and Information Technology, 4*(5), 700–707.
7. Khan, Z. A., Siddiqui, J., & Samad, A. (2016). Properties and performance of cube-based mutiprocessor architectures. *International Journal of Applied Evolutionary Computation (IJCNIS), 7*(1), 67–82.
8. Khan, Z. A., Siddiqui, J., Samad, A. (2015). Linear Crossed Cube (LCQ): A new interconnection network topology for massively parallel architectures. *International Journal of Computer Network and Information Science (IJCNIS), 7*(3), 18–25.
9. Seth, A., & Singh, V. (2016). Types of scheduling in parallel computing. *International Research Journal of Engineering and Technology (IRJET), 3*(5), 521–526.
10. Mahafzah, B. A., & Jaradat, B. A. (2010). The hybrid dynamic parallel scheduling algorithm for load balancing on chained-cubic tree. *The Journal of Supercomputing, 52*, 224–252.
11. Ding, Z., Hoare, R. R., Jones, K. A. (2006). Level-wise scheduling algorithm for fat tree interconnection networks. In *Proceedings of the 2006 ACM/IEEE SC\06 Conference (SC'06)* (pp. 9–17).
12. Alebrahim, S., & Ahmad, I. (2017). Task scheduling for heterogeneous computing systems. *The Journal of Supercomputing, 73*(6), 2313–2338.
13. Abdelkader, D. M., & Omara, F. (2012). Dynamic task scheduling algorithm with load balancing. *Egyptian Informatics Journal, 13*, 135–145.
14. Nayak, K., Padhy, S. K., & Panigrahi, S. P. (2012). A novel algorithm for dynamic task scheduling. *Future Generation Computer System, 28*, 709–717.
15. Samad, A., Khan, Z. A., & Siddiqui, J. (2016). Optimal dynamic scheduling algorithm for cube based multiprocessor interconnection networks. *International Journal of Control Theory and Applications, 9*(40), 485–490.
16. Mohammad, S. B., Ababneh, I. (2108). Improving system perfromance in non-contiguous processor allocation for mesh interconnection networks. *Journal of Simulation Modeling and Practice*. 80, 19–31.

Performance Evaluation
of Multi-operands Floating-Point Adder

**Arvind Kumar, Sunil Kumar, Prateek Raj Gautam, Akshay Verma
and Tarique Rashid**

Abstract In this paper, an architecture is presented for a fused floating-point three operand adder unit. This adder executes two additions within a single unit. The purpose of this execution is to lessen total delay, die area, and power consumption in contrast with traditional addition method. Various optimization techniques including exponent comparison, alignment of significands, leading zero detection, addition, and rounding are used to diminish total delay, die area, and power consumption. In addition to this, the comparison is described of different blocks in term for die area, total delay, and power consumption. The proposed scheme is designed and implemented on Xilinx ISE Design 14.7 and synthesized on Synopsis.

Keywords Floating-point adder · Significand bits · Exponent bits · Total delay
and Xilinx

1 Introduction

The use of floating-point arithmetic, which is according to IEEE-754 standard [1], is to make general-purpose application specific processor. Floating-point number contains three components: exponent bits, the sign bit, and significand bits that are

A. Kumar (✉) · S. Kumar · P. Raj Gautam · A. Verma · T. Rashid
Motilal Nehru National Institute of Technology Allahabad, Allahabad, India
e-mail: arvindk@mnnit.ac.in

S. Kumar
e-mail: rel1516@mnnit.ac.in

P. Raj Gautam
e-mail: prateekrajgautam@gmail.com

A. Verma
e-mail: rel1602@mnnit.ac.in

T. Rashid
e-mail: rel1404@mnnit.ac.in

© Springer Nature Singapore Pte Ltd. 2019
A. Khare et al. (eds.), *Recent Trends in Communication, Computing,
and Electronics*, Lecture Notes in Electrical Engineering 524,
https://doi.org/10.1007/978-981-13-2685-1_51

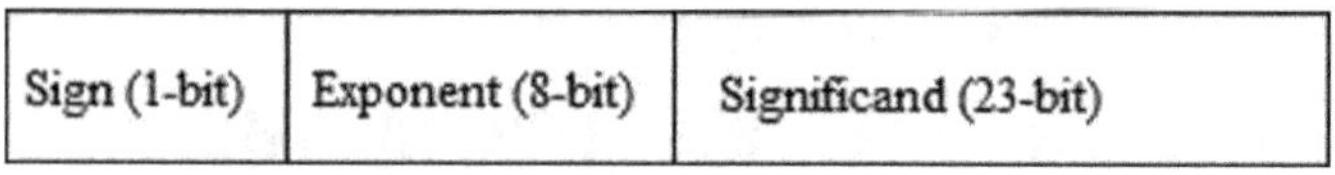

Sign (1-bit)	Exponent (8-bit)	Significand (23-bit)

Fig. 1 Representation of single precision floating-point number [1]

Sign (1-bit)	Exponent (11-bits)	Significand (52-bits)

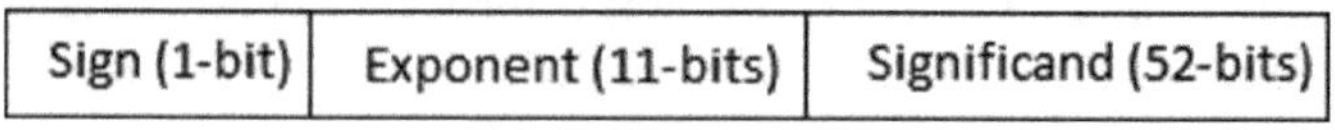

Fig. 2 Representation of double precision floating- point number [1]

Fig. 3 Discrete versus fused floating-point adder [2]

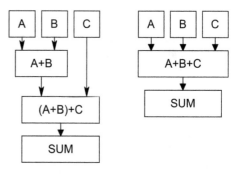

shown in Fig. 1. There are two standard floating-point representations [1]: single precision and double precision representation. In single precision representation, there are one sign bit, eight exponent bit, and twenty-three significand bits. However, in double precision representation, there are one sign bit, eleven exponent bit, and fifty-two significand bits that are shown in Figs. 1 and 2.

An addition is more significance in arithmetic, and it is widely used operation in various applications. Discrete floating-point adder uses two operands at a time which is well optimized. In order to use multiple operands for the addition, we have to use the multiple traditional adder ones after other because it can use only two operands at a time. Discrete floating-point adders degrade accuracy owing to the multiple rounding one after in each addition. Due to this die area, total delay and power consumption become larger. To improve quality, the fused floating-point adder is used. It executes two additions in a single unit so that only single rounding is required which reduces die area and power consumption. The comparison between discrete and fused floating-point adder [2, 3] is shown in Fig. 3.

The proposed adder performed addition of three floating-point operands and executed additions as

$$S = A \pm B \pm C$$

There are many fused floating-point units that are presented: fused multiply-add (FMA), fused add subtract (FAS), fused dot product (FDP), and a fused three-term adder (FTA) [4, 5].

2 Methodology

The algorithm of three terms is given in Fig. 4 is represented as [6–8]

1. Unpacking each of the three floating-point numbers A, B, and C to obtained sign bit (1 bit), exponent (8 bits), and significand (23 bit +1 bit hidden).
2. In order to find the maximum exponent from the three exponents and calculate the exponent difference.
3. Arrange the significands right shift according to their respective exponent difference.
4. Sign logic determines the sign of A, B, and C according to op-codes op1 and op2.
5. Invert the significands according to their respective sign obtained from the sign logic.
6. Significand addition is performed by using 3:2 CSA (carry-save adder).
7. Leading zero detector is to compute leading zero of the output of CSA, and accordingly, significand is shifted by the same amount and exponent is also adjusted.
8. Rounding operation is performed to round off the resultant significand.
9. If output of CSA produces carry, then right shift the significand by 1, and accordingly, exponent will increment by 1.
10. Pack the resultant sign bit, exponent bits, and significand bits to produce the resultant floating-point number.

3 Proposed Design and Implementation

3.1 Exponent Comparison and Alignment of Significand

For floating-point addition [9, 10], that is essential to compute maximum exponent from the three exponents. Exponent difference is performed by subtracting the respective exponents from the maximum exponent. Significands are aligned by right shifting the significand by the amount of the respective exponent difference. All the arrangements of six subtractions of exponent differences ($\exp_a - \exp_b$, $\exp_b - \exp_a$, $\exp_b - \exp_c$, $\exp_c - \exp_b$, $\exp_a - \exp_c$, and $\exp_c - \exp_a$) are performed to calculate.

The differences of each pair, an absolute value is adopted based on the exponent comprising results that enables skipping the complementation after the subtractions.

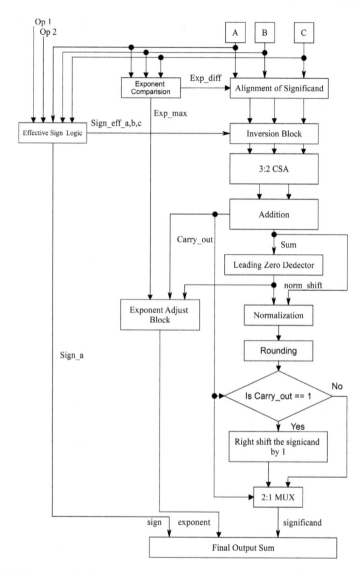

Fig. 4 Multi-operands floating point adder

An exponent comparison and significand arrangement logic is shown in Fig. 5.

The control logic estimates the largest exponent and arranged the significands based on the exponent comprising results as shown in Table 1. In order to guarantee the significand precision, the aligned significands become $2f + 6$ bits wide, including two overflow bits, round bits, guard bits and sticky bits. Where f is the significand bits can be seen in Fig. 6.

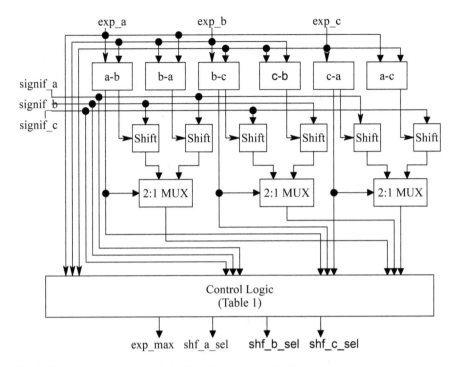

Fig. 5 Exponent comparison and significand arrangement logic

Table 1 Exponent comparison control logic [2]

$A \geq B$	$B \geq C$	$C \geq A$	exp_max	shf_a_sel	shf_b_sel	shf_c_sel
0	0	0	NA	NA	NA	NA
0	0	1	exp_c	shf_ca	shf_bc	signif_c
0	1	0	exp_b	shf_ab	signif_b	shf_bc
0	1	1	exp_b	shf_ab	signif_b	shf_bc
1	0	0	exp_a	signif_a	shf_ab	shf_ca
1	0	1	exp_c	shf_ca	shf_bc	signif_c
1	1	0	exp_a	signif_a	shf_ab	shf_ca
1	1	1	any	signif_a	signif_b	signif_c

3.2 Effective Sign Logic

Sign logic determines the three effective sign bits ($sign\_eff\_a$, $sign\_eff\_b$ and $sign\_eff\_c$) on the basis of the three sign bits and two op-codes as

$$sign\_eff\_a = sign\_a$$

$$sign\_eff\_b = sign\_a \oplus (sign\_b \oplus op1)$$

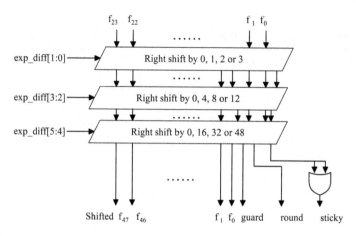

Fig. 6 Significand shifter is shown for single precision [2]

$$sign\_eff\_c = sign\_a \oplus (sign\_c \oplus op2)$$

where $\oplus$ is the sign of exclusive-OR operation.

3.3 Inversion Block

Inversion block complements the significand on the basis of their respective effective sign.

Up to two significands are complimented with the help of three operand subtraction (e.g., $A - B - C = A + B' + 1 + C' + 1 = A + B' + C' + 2$). Increments are avoided after inverters and 2 bits are extended to the LSB of the significands as shown in Table 2.

Table 2 2-bit extended LSBs for complementation [2]

s_eff_a	s_eff_b	s_eff_c	$a_{-1}a_{-2}$	b_1b_{-2}	$c_{-1}c_{-2}$	sum0
0	0	0	00	00	00	0
0	0	1	10	00	10	1
0	1	0	00	10	10	1
0	1	1	10	11	11	2
1	0	0	10	10	00	1
1	0	1	11	10	11	2
1	1	0	11	11	10	2
1	1	1	00	00	00	0

3.4 Carry-Save Adder (CSA)

Each significand is passed to the 3:2 reduction tree. Carry save-adder (CSA) is used to perform the reduction that reduces the three significands with respect to two and then performed the addition. The advantage of using CSA is that it does not propagate carry. It saves the carry which minimizes the total delay in performing addition operation as compared to carry propagate adder.

3.5 Leading Zero Detector and Normalization

This block determines a position of the leading zero from the MSB of the output of the CSA. Significand becomes normalized significand based on the amount of left shift obtained from the leading zero detectors. An exponent is also adjusted by the amount obtained from leading zero detector block. Significand addition with normalization is the highest bottleneck of fused floating-point adder. To diminish the overhead, normalization is used.

3.6 Exponent Adjust Block

The largest exponent (exp max) determined by the exponent compare logic is adjusted by subtracting the shift amount from LZA and adding the carry out of the significand addition as shown in Fig. 7

Fig. 7 Exponent adjust block

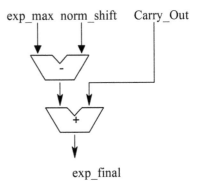

Table 3 Table for rounding operation

G	R	S	Operation performed
0	0	X	No changes in the LSB
0	1	X	No changes in the LSB
1	1	X	Add 1 to LSB Bit
1	0	0	Round to nearest even
1	0	1	Add 1 to the LSB

Significand bits(23 bits)	G	R	S

Fig. 8 Figure shows the position of significand and guard, round, and sticky bit

3.7 Rounding

In order to truncate the significand, we have to perform the rounding operation with the help floating-point multiplier; significand is round off based on guard bit (G), round bit (R), and sticky bit (S) as shown in Fig. 7 and Table 3. Rounding is determine to rounded floating the value of carry, guard, LSB, round, and sticky bits (Fig. 8).

Here, least significand bit (LSB) bit is just left of the guard bit as shown above.

4 Result Analysis

In this section, modules of proposed architecture are designed in Xilinx 14.7 and synthesis on synopsis tool. Their corresponding results are shown, respectively.

The result of the addition of three floating-point numbers is shown in Fig. 9.

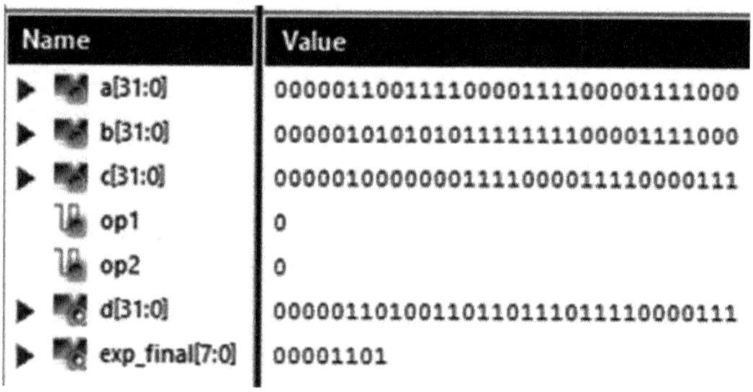

Name	Value
▶ a[31:0]	00000110011110000111100001111000
▶ b[31:0]	00000101010101111111100001111000
▶ c[31:0]	00000100000001111000011110000111
op1	0
op2	0
▶ d[31:0]	00000110100110110111011110000111
▶ exp_final[7:0]	00001101

Fig. 9 Simulation result

Table 4 Comparison between a proposed paper with implementation of fused floating-point three-term adder unit [3]

Modules	Implementation of fused floating-point three-term adder unit		Performance evaluation of multi-operands floating-point adder	
	Number of Slices LUT used	Delay (ns)	Number of Slices LUT used	Delay (ns)
Exponent comparison and alignment of significand	519	5.965	1035	4.086
Carry save adder (CSA)			2	0.893
Effective sign logic	2	3.696	2	0.889
Inversion Logic			318	2.942
Leading zero detector and normalization	28	8.524	1052	12.999
Rounding	4	3.809	48	2.788
Control logic	3	3.809	165	0.751
Overall output			2316	18.993

Comparison between execution of fused floating-point three-term adder unit [3] and performance evaluation of multi-operands floating-point adder on the basis numbers of slices LUT used and delay are shown in Table 3. The fundamental difference between proposed and conventional design is alignment of significand bits and rounding. The proposed design executes the lesser significand bits addition compared to conventional designs. Further, the proposed design executes the significand bits addition and rounding at the same time so that the delay is diminished significantly.

The synthesis result obtained from synopsis tool is shown in Table 4.

5 Conclusion

In this paper, we have introduced an improved architecture for three-term adder with a fused floating point which is used to diminish die area, total delay, and power consumption in i with the discrete floating point adder. Further, this paper also compares the different performance of proposed architecture for Implementation of three-term adder unit with fused floating point in terms of delay and number of slices LUT used. In addition, die area and power consumption of different optimized blocks are provided by synthesis result. The optimization blocks are exponent comparison, alignment of significand, CSA, effective sign logic, inversion logic, leading zero detector, normalization, rounding, and control logic. In future, we will design architecture in order to obtain high-speed adder (Table 5).

Table 5 Synthesis result analysis on synopsis

Modules	Power (mW)	Area (μm^2)
Exponent comparison and alignment of significand	1.6009	14411.488
Carry-save adder (CSA)	0.6439	2472.736
Effective sign logic	0.00589	31.36
Inversion logic	0.398	3395.50472
Leading zero detector and normalization	1.0558	5357.856
Rounding	0.006759	700.1120
Control logic	0.5263	3206.5601
Overall output	6.6984	26683.440262

References

1. Zuras, D., Cowlishaw, M., Aiken, A., Applegate, M., Bailey, D., Bass, S., et al. (2008). IEEE standard for floating-point arithmetic. *IEEE Standards, 754–2008*, 1–70.
2. Sohn, J., & Swartzlander, E. E. (2014). A fused floating-point three-term adder. *IEEE Transactions on Circuits and Systems I: Regular Papers, 61*(10), 2842–2850.
3. Popalghat, M., & Palsodkar, P. (2016). Implementation of fused floating point three term adder unit. In *2016 International Conference on Communication and Signal Processing (ICCSP)* (pp. 1343–1346). IEEE.
4. Drusya, P., & Jacob, V. (2016). Area efficient fused floating point three term adder. In International Conference on Electrical, Electronics, and Optimization Techniques (ICEEOT) (pp. 1621–1625). IEEE.
5. Sohn, J., & Swartzlander, E. E. (2012). Improved architectures for a fused floating-point add-subtract unit. *IEEE Transactions on Circuits and Systems I: Regular Papers, 59*(10), 2285–2291.
6. Tenca, A.F. (2009). Multi-operand floating-point addition. In *2009 19th IEEE Symposium on Computer Arithmetic, ARITH 2009.* (pp. 161–168). IEEE
7. Seidel, P. M., & Even, G. (2004). Delay-optimized implementation of IEEE floating-point addition. *IEEE Transactions on Computers, 53*(2), 97–113.
8. Tao, Y., Deyuan, G., Xiaoya, F., & Xianglong, R. (2012). Three-operand floating-point adder. In *2012 IEEE 12th International Conference on Computer and Information Technology (CIT)* (pp. 192–196). IEEE.
9. Underwood, K. (2004). Fpgas vs. cpus: trends in peak floating-point performance. In *Proceedings of the 2004 ACM/SIGDA 12th international symposium on Field programmable gate arrays* (pp. 171–180). ACM.
10. Monniaux, D. (2008). The pitfalls of verifying floating-point computations. *ACM Transactions on Programming Languages and Systems (TOPLAS), 30*(3), 12.

An Explicit Cell-Based Nesting Robust Architecture and Analysis of Full Adder

Bandan Kumar Bhoi, Tusarjyoti Das, Neeraj Kumar Misra and Rashmishree Rout

Abstract Moving towards micrometre scale to nanometre scale device shrinks down emerging nanometre technology such as quantum-dot cellular automata as a nesting success. The introduced architecture is robust where the explicit design of full adder and full subtraction uses for Ex-OR design. A new architecture of Ex-OR based on one majority gate is proposed, which its most optimized architecture and its placement of cells from the novel design. The analysis based on simulation showed that the introduced Ex-OR and full adder makes only 11 and 46 cells count, respectively. In proposed Ex-OR design, first output is received with no any latency which can be a suitable design for implementation of the high-speed full adder design. In addition, power estimation results are obtained after simulation of proposed designs in QCAPro tool. Therefore, the novel designs improve the energy dissipation parameters such as mean leakage energy dissipation, mean switching energy dissipation and total energy dissipation 75, 11.28 and 82.19% in comparison with the most robust design in existing.

Keywords Nanometre scale · Full adder · Quantum-dot cellular automata Complexity · Majority gate

1 Introduction

Quantum-dot cellular automata (QCA) is an emerging technology, which has the features to overcome the limitation of CMOS technology in scaling [1]. This promising technology is very simple as its basic building block in a QCA cell, where interaction between cells is purely Columbia rather than transportation of charge. Hence, there

B. K. Bhoi (✉) · T. Das · R. Rout
Department of Electronics & Telecommunication, Veer Surendra Sai
University of Technology, Burla 768018, India
e-mail: bkbhoi_etc@vssut.ac.in

N. K. Misra
Bharat Institute of Engineering and Technology, Hyderabad, India

© Springer Nature Singapore Pte Ltd. 2019
A. Khare et al. (eds.), *Recent Trends in Communication, Computing, and Electronics*, Lecture Notes in Electrical Engineering 524,
https://doi.org/10.1007/978-981-13-2685-1_52

547

is absent of leakage current. This technology also provides excellent optimization for complex circuits. Any Boolean function can be implemented by a different arrangement of these cells in QCA Technology [2–4]. Hence, QCA is becoming one of the most promising areas of research in the field of the era of nanoelectronics.

In this work efficient 1-bit, full adder is introduced and analyses the performance parameters. The full adder uses two Ex-OR gates [5, 6], which takes three inputs, namely A, B and Ci. These two Ex-OR gates are connected in series that gives the final sum output ($S = A \oplus B \oplus Ci$), and another three number inputs majority gate is used to produce the carry out ($Co = AB + BC + AC$). The Ex-OR gate used here is the most efficient one as it requires only 11 number of QCA cells and occupies a very small area of 0.02 μm^2. This Ex-OR gate comprises of only one two-input majority gate, and an inverter is used to design the proposed circuit that occupies an area of 0.06 μm^2. The newly introduced design is a coplanar type circuit. The logical operation of the proposed work is verified using QCADesigner Simulator tool [6]. In this design, the carry out is produced without having any delay and the sum output is produced after one clock delay. Hence, the proposed 1-bit full adder is very efficient in the area, cell count and clock delay.

The remaining part of this paper is arranged in the following manner. Section 2 describes the basic idea and operating principle of QCA. Section 3 shows the QCA implementation and simulation result of the proposed full adder circuit. The comparison result of the proposed and other existing full adder circuits is tabulated in Sect. 4. Finally, the paper is concluded in Sect. 5.

2 Basics of QCA

In a QCA circuit design, the basic component is a QCA cell. The QCA cell is a square box having four quantum dots located at its four corners. Each cell has two electrons placed in two quantum dots those are diagonally opposite to each other because of coulombs interaction. The cells have two basic binary states, i.e. logic '0' and logic '1' which is determined by the location of electrons. Unlike CMOS, transportation of charge between cells does not occur in QCA. It is purely Columbia in case of QCA. State of a cell at a given time is determined by the state of its nearby cells during a previous clock cycle.

In QCA, there are four clock zones, namely clock-zone 0, clock-zone 1, clock-zone 2, clock-zone 3. Clock-zone 0 is referred as SWITCH state and is indicated by green colour. Similarly, clock-zone 1 is referred as HOLD and indicated by magenta colour, clock-zone 2 is referred as RELEASE and indicated by blue colour, and clock-zone 3 is referred as RELAX and indicated by white colour. Two types of crossovers are allowed in QCA design, namely coplanar crossover and multilayer crossover. Coplanar crossover involves only a single layer and uses two types of cells, i.e. regular cell and rotated cell, whereas the multilayer crossover involves more layers of cells.

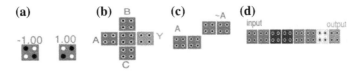

Fig. 1 QCA preliminary **a** QCA cell. **b** Majority gate. **c** Inverter **d** QCA wire

2.1 Basic QCA Elements

By doing various physical arrangements of several numbers of QCA cells, we can get different QCA devices such as (a) QCA wire (b) QCA inverter (c) QCA majority gates.

Binary logic '0' and logic '1' are represented by QCA cells which are represented in Fig. 1a.

QCA Majority Gate: A three-input majority gate can be implemented by arranging the cells in the way shown in Fig. 1b. Majority gate gives the majority of inputs at its output. The output function of the majority gate is M(A, B, C) = AB + BC + AC where A, B, C are the three inputs. If we set any of its inputs to logic '0' or logic '1' then the majority gate will function as AND gate and OR gate, respectively. Then by connecting a QCA inverter at its output, we can achieve NAND and NOR gate also. Therefore, by using this majority gate we can design any arithmetic circuit.

QCA inverter: To design a QCA inverter, cells should be arranged in the fashion shown in Fig. 1c. According to the working principle of a NOT gate or an inverter, it flips the provided input and gives its complement as output.

QCA wire: It is the series arrangement of QCA cells (Fig. 1d). In a QCA wire, the cells are clocked in a regular manner.

3 Proposed Work

In this section delay, efficient full adder circuit has been detailed. This full adder is implemented using an efficient Ex-OR gate. The Ex-OR gate used here comprises only 11 number of cells. It also requires an area of 0.02 μm^2. The Ex-OR gate is made up of only one majority gate and an inverter. The proposed design of logic design, cell layout, and simulation result are presented in Fig. 2a, b, c, respectively.

Here in the design of the 1-bit full adder circuit, we have used two Ex-OR gates. These two Ex-OR gates are cascaded, and the output of this cascade connection serves as the sum of the full adder. Another majority gate is used to implement the carry out of the full adder circuit. So the proposed design comprises a total of three majority gates and two inverters. The full adder circuit requires only 46 number of cells and an area of 0.06 μm^2. The logical diagram, QCA structure and simulation result of the full adder are shown in Fig. 3a, b, c, respectively.

Fig. 2 The proposed
Ex-OR. **a** logical truth table
b QCA implementation. **c**
Simulation result

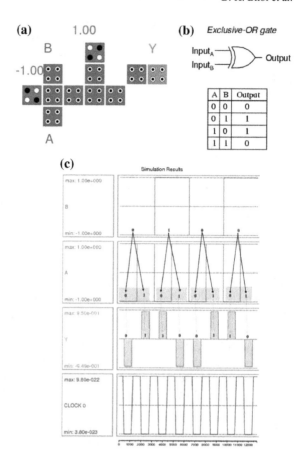

A	B	Output
0	0	0
0	1	1
1	0	1
1	1	0

This simulation result implemented using QCADesigner shows the functionality result of the proposed 1-bit full adder. Here, from the output waveform shown in Fig. 3c, it can be observed that the carry out (Co) of the full adder is produced without any clock delay, whereas the sum output (S) is having a single clock delay.

4 Comparison

The proposed 1-bit full adder is compared with various existing full adders on the basis of parameters such as area, cell count, delay, number of majority gates and inverters used. The comparison result is shown in Table 2. In few cases, present full adder designs have lesser area [7], but in overall, the proposed one has better performance than others, which is clearly indicated by the tabulation. So we can conclude that the proposed design has higher efficiency than the existing circuits. For energy dissipation, related parameters are estimated by QCAPro tool by the QCA design.

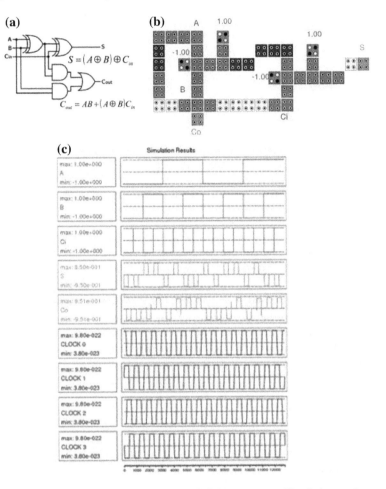

Fig. 3 The proposed full adder. **a** Block diagram. **b** QCA structure. **c** Simulation result

Table 1 presents the energy-related parameters such as mean leakage energy dissipation, mean switching energy dissipation and total energy dissipation. Total energy dissipation is the sum of mean leakage and mean switching. The extracted energy dissipation-related parameters are estimated by distinct energy levels such as 0.5, 1, 1.5 Ek and the 2 K selected for operating temperature. Further, the energy dissipation thermal map of Ex-OR and full adder designs at a 2 K temperature and 0.5 Ek is presented in Fig. 4a, b, respectively. In thermal map, darker the cell is presented the more power dissipation in the design. Table 1 presents the energy dissipation comparisons parameters results comparison to existing designs. These tables can evidence the efficient energy dissipation parameters of our introduced designs. As per comparisons table evidence that the design achieves the lower energy dissipation at different tunnelling energy level in comparisons to the existing designs. The

(a)   (b)

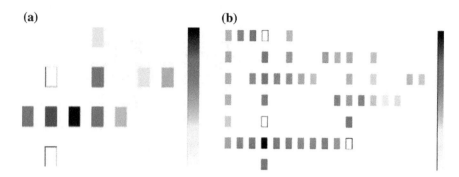

Fig. 4 Thermal layout map of proposed designs with 0.5 Ek at 2 K temperature. **a** Ex-OR. **b** Full adder

better energy dissipation results are achieved by nonrotating cell and no-crossover in the proposed design. This is the cause to achieve lower energy dissipation. In a real comparison of energy dissipation parameters result in existing and presented design in Table 1, the proposed design achieves on mean leakage energy dissipation, mean switching energy dissipation and total energy dissipation 75, 11.28 and 82.19% less energy as compared to existing at 0.5 Ek, respectively, whereas mean leakage energy dissipation, mean switching energy dissipation and total energy dissipation 77.51, 85.16 and 81.34% less energy as compared to existing at 1 Ek, respectively, similarly mean leakage energy dissipation, mean switching energy dissipation and total energy dissipation 78.72, 85.71 and 81% less energy as compared to existing at 1 Ek, respectively.

5 Conclusion

In this article, the new architecture of Ex-OR was presented with the intention of being robust and no latency-based design. In this way, robust and no latency-based Ex-OR design was proposed to be used as an efficient full adder design. Further, the explicit cell approach was introduced in QCA for the physical realization of the robust full adder design. The comparative analysis of the new full adder with the prior designs showed that the introduced are optimized in different parameters intention to the complexity, area and latency. For future work in this way, the new architecture of Ex-OR and full adder modules can be extended for synthesizing robust multiplier, and divider unit.

Table 1 Power estimation comparison results

New Design	Mean Leakage energy dissipation (eV) 0.5, 1, 1.5 Ek			Mean switching energy dissipation (eV) 0.5, 1, 1.5 Ek			Mean Energy consumption (eV) 0.5, 1, 1.5 Ek		
Ex-OR	0.00408	0.01049	0.01733	0.00748	0.00625	0.00523	0.01156	0.01674	0.02256
References [8]	0.068	0.209	0.376	0.269	0.236	0.203	0.337	0.445	0.579
FA	0.017	0.047	0.080	0.042	0.035	0.029	0.060	0.083	0.110

Table 2 Comparison of proposed QCA design and existing designs

Full adders	# Maj	# Inverter	Area (in μm$^{2)}$	Cell count	Latency	Layer type
Reference [7]	3	2	0.038	52	1 clk	Coplanar
Reference [9]	3	1	0.127	53	1 clk	Multilayer
Reference [10]	3	2	0.097	102	2 clk	Coplanar
Reference [11]	3	2	0.16	145	–	Coplanar
Reference [12]	2 (one 3i/p, one 5i/p)	2	0.16	145	2 clk	Multilayer
Reference [13]	3	1	0.07	70	1 clk	Multilayer
Proposed	3	2	0.06	46	1 clk	Coplanar

References

1. Lent, C. S., Tougaw, P. D., Porod, W., & Bernstein, G. H. (1993). Quantum cellular automata. *Nanotechnology., 4,* 49–57.
2. Orlov, A. O., Amlani, I., Bernstein, G. H., Lent, C. S., & Snider, G. L. (1997). Realization of a functional cell for quantum-dot cellular automata. *Science, 277,* 928–930.
3. Tougaw, P. D., & Lent, C. S. (1994). Logical devices implemented using quantum cellular automata. *Journal of Applied Physics, 75*(3), 1818–1825.
4. Lent, C. S., & Tougaw, P. D. (1997). A device architecture for computing with quantum dots. *Proceedings of the IEEE, 85*(4), 541–557.
5. Bhoi, B. K., Misra, N. K., & Pradhan M. (2018). Novel robust design for reversible code converters and binary incrementer with quantum-dot cellular automata. In S. Bhalla, V. Bhateja, A. Chandavale, A. Hiwale, & S. Satapathy (Eds.), *Intelligent computing and information and communication. Advances in intelligent systems and computing* (Vol 673). Singapore: Springer.
6. Walus, K., Dysart, T. J., Jullien, G., & Budiman, A. R. (2004). QCADesigner: A rapid design and simulation tool for quantum-dot cellular automata. *IEEE Transactions on Nanotechnology, 3*(1), 26–31.
7. Ramesh, B. & Rani, M. A. (2016). Implementation of parallel adders using area efficient quantum dot cellular automata full adder. In *2016 10th International Conference on Intelligent Systems and Control (ISCO),* (pp. 1–5). IEEE.
8. Taherkhani, E., Moaiyeri, M. H., & Angizi, S. (2017). Design of an ultra-efficient reversible full adder-subtractor in quantum-dot cellular automata. *Optik-International Journal for Light and Electron Optics, 142,* 557–563.
9. Sonare, N., & Meena, S. (2016). A robust design of coplanar full adder and 4-bit Ripple Carry adder using quantum-dot cellular automata. In *IEEE International Conference on Recent Trends in Electronics, Information & Communication Technology (RTEICT),* May 2016 (pp. 1860–1863). IEEE.
10. Hanninen, I., & Takala, J.: Robust adders based on quantum-dot cellular automata. In *2007 IEEE International Conference on Application-specific Systems, Architectures and Processors, ASAP.* (pp. 391–396). IEEE, July 2007.

11. Wang, W., Walus, K., & Jullien, G. A.: Quantum-dot cellular automata adders. In *2003 Third IEEE Conference on Nanotechnology. IEEE-NANO 2003*, August 2003 (Vol. 1, pp. 461–464). IEEE.
12. Bishnoi, B., Giridhar, M., Ghosh, B. & Nagaraju, M. (2012). Ripple carry adder using five input majority gates. In *2012 IEEE International Conference on Electron Devices and Solid State Circuit (EDSSC)*, December 2012, (pp. 1–4). IEEE.
13. Chudasama, A., & Sasamal, T. N. (2016). Implementation of 4×4 vedic multiplier using carry save adder in quantum-dot cellular automata. In *2016 International Conference on Communication and Signal Processing (ICCSP)*, April 2016 (pp. 1260–1264). IEEE.

Low Power SAR ADC Based on Charge Redistribution Using Double Tail Dynamic Comparator

Sugandha Yadav

Abstract In this paper I have designed a low power SARADC based on charge redistribution. In the proposed design, I have designed a successive approximation register ADC using low power double tail dynamic comparator with control transistors (MC1 & MC2). The peculiar advantage of low power double tail dynamic comparator with control transistor (MC1 & MC2) is that the power consumption is reduced by three times as compared with the existing low power double tail dynamic comparator. The proposed comparator having the power consumption 58 μW can be operated at maximum frequency 16 MHz. Best efforts has been made to design an efficient SAR and control unit in the proposed design. In this 4-bit SAR ADC which can be operated at clock frequency of 16.5 MHz and having sampling frequency of 3.3 MHz is designed. The power consumption of the proposed circuit is 113 μW. In the paper quantization noise, SNR, SNDR and ENOBs are also obtained which are better in performances than existing ADCs.

Keywords SAR ADC · Charge redistribution · Quantization noise
Double tailed dynamic comparator · SNR · SNDR and ENOBs

1 Introduction

Analog to digital converters are very useful component in many applications. In most of the applications of ADC, power consumption, chip area and speed (i.e. sampling frequency) have become major constraints in today's era. Some application of analog to digital convertors are useful in the area of speech processing systems, biometric devices and RFID (radio frequency identification devices) etc. Also ADCs are widely used in various sensing circuits and transducers.

S. Yadav (✉)
School of VLSI Design & Embedded System, National Institute of Technology Kurukshetra, Kurukshetra, India
e-mail: sugandhayadav555@gmail.com

© Springer Nature Singapore Pte Ltd. 2019
A. Khare et al. (eds.), *Recent Trends in Communication, Computing, and Electronics*, Lecture Notes in Electrical Engineering 524,
https://doi.org/10.1007/978-981-13-2685-1_53

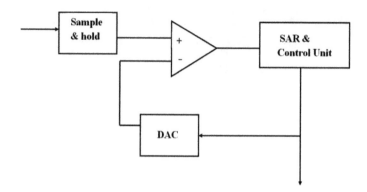

Fig. 1 Block diagram of SAR ADC

There are various types of ADCs viz. SAR ADC, flash ADC and dual slope integrating type ADC. SAR ADC is very important because of its high resolution of 8–18 bits, suitability for data acquisition and its ability to work at higher sampling rates ranging from 50 kHz to 50 MHz. SAR ADC requires conversion time of n $\times$ T_{clk} to do digitization of any analog signal where n corresponds to n bit ring counter used in SAR ADC and T_{clk} is the time period of the clock signal used. SAR ADC follows binary search algorithm for converting any analog waveform into its equivalent discrete digital representation.

Many researchers have worked on SAR ADC so the constraints discussed earlier can be achieved. In [1], James L. McCreary discussed some analog to digital conversion techniques. In [2], Silvia Dondi et al. proposed a 6-bit time-Interleaved analog to digital converter for wide band application Mallik Kandala et al. reduced the power consumption of charge redistribution SAR ADC by splitting MSB capacitor into a binary scale sub array in [3].

In my design, I have proposed a low power SAR ADC based on charge redistribution, a ring counter designed with novel D flip-flops having peculiar advantage of less area requirement to reduce static power consumption. As well as sample and hold circuit is embedded with DAC as a single block to get further reduction in required power. So in this way I primarily focused on reducing power requirement of conventional SAR ADC and further I analyzed SNR, SNDR, SFDR and ENOBs.

2 Proposed Architecture

Basic block diagram of SAR ADC is shown in Fig. 1. There are four basic blocks: sample & hold circuit, DAC, comparator, SAR & control unit. I will discuss about these blocks in following sections.

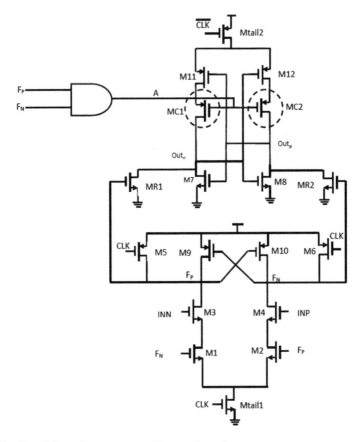

Fig. 2 Double tail dynamic comparator with control transistor

2.1 Comparator

In the proposed comparator, two transistors are added in series with M11 & M12 (Fig. 2). This results in low power consumption and speed increment. The operation of the proposed comparator is given as follows:

During reset phase, when CLK = 0, transistor Mtail1 & Mtail2 are off. So there is no direct path from VDD to GND i.e. no static power is there. In this phase, transistors M5 & M6 are on. Due to this, F_N & F_P nodes are charged to VDD and this makes M9 & M10 off. Hence transistors MR1 & MR2 pull down Out_n & Out_p to zero.

Now during decision making phase, when CLK = 1, transistors Mtail1 & Mtail2 are ON as well as M5 & M6 are OFF. According to (INN & INP) input voltages, F_P & F_N start discharging with different rates. Let us say VINP > VINN. So F_N starts decreasing faster than F_P. Because of this, transistor M9 starts conducting and F_P again charges to VDD. There is a direct path from VDD to GND i.e. from the transistor Mtail2 to transistor MR2 passing through the transistor M12 in which

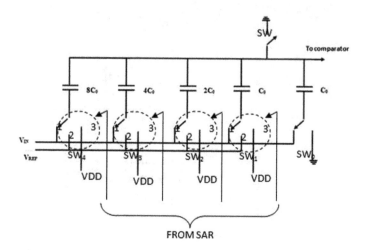

Fig. 3 Sample and hold with DAC

during the decision making phase, maximum current flows. So to prevent from this direct current path, two transistors (i.e. MC1 & MC2) are added in series with M11 & M12. These transistors are controlled by an AND gate which produces the output A on giving the inputs F_P & F_N. This makes the current path disconnects when $A = 1$. Hence the average current flowing through this path reduces which results in low power consumption.

2.2 Sample and Hold Combined with DAC

For sample and hold: In the proposed SAR ADC, I used sample and hold circuit combined with DAC using comparator array. The circuit for this purpose is given in Fig. 3. In the figure, binary weighted capacitors are used i.e. C0, C1 = C0, C2 = 2C0, C3 = 4C0, C4 = 8C0.

In sample mode, switch (SW) is close and all the switches SW0, SW1, SW2, SW3 and SW4 are in position 1. So the voltage vin is sampled by these capacitors.

In hold mode, switch (SW) is open and switches SW4, SW3, SW2, SW1 and SW0 are switched to position 3. Thus nodeA charges to vin.

In DAC mode, switch SW4, SW3, SW2 and SW1arecontrolled by SAR output and switch SW0 is connected to GND in this mode. Now first of all, in first cycle 1000 is stored in SAR, according to this, switch SW4 is connected to VREF and the remaining switches are connected to GND.

According to the output of comparator, Q_3 gets either reset or remains unchanged. In second cycle, Q_2 is forced to 1 and accordingly SW_3 is connected to V_{REF}. Similar to first cycle, Q2 gets also either reset or remains unchanged. This procedure is repeated four times for 4 bit SAR ADC.

Fig. 4 Block diagram of
SAR

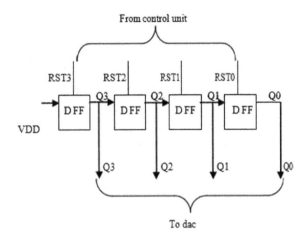

2.3 SAR

SAR is used for storing the output. SAR consists storing element like flip flops. Every
bit uses one flip flop so in our 4-bit SAR ADC 4 D flip-flops are used. So a novel DFF
is proposed for this ADC. The block diagram of SAR is shown in Fig. 4. There are
four D flip-flops connected in cascade for the purpose of generating appropriate bits
in SAR. These flip flops use synchronous clock. During the conversion time these
flip flops are needed to be set or reset according to the input. The signal for making
the flip-flops set or reset is given by control unit and the output of SAR is given to
the DAC.

The low power D flip-flop is placed in the convertor. These D flip-flops contain
controlled inverter. So basically the transmission gates of conventional D flip-flop
are discarded and the inverters are replaced with controlled inverter. This controlled
inverter has a control pin to make the inverter connected or disconnected. The circuit
diagram of D flip-flop is shown in Fig. 5 in which there are six inverters i.e. four controlled inverters and two conventional inverters. By replacing conventional inverter
and transmission gate with controlled inverter, the power consumption is reduced to
a great extent. So the overall power consumption is also reduced.

2.4 Control Unit

The last and most important block of SAR ADC is control unit. The control unit is
used for the successful execution of the conversion. This unit executes the program
in a pre-defined sequence. The main function of control unit is to generate the set
or reset signal according to the output of comparator. The other function is to give

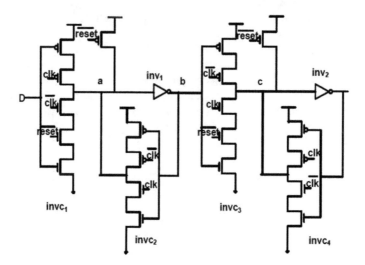

Fig. 5 Low power D-FF

the clock signal to each D flip-flop at proper timing so the ring counter operation is executed properly.

The proposed diagram for control unit is illustrated in Fig. 6. The output of comparator should be given to the D flip-flops (SAR) according to the current state. For this purpose one 1:4 de-multiplexer is used. And a 2-bit counter is used for tracking current state and for generating the outputs at that state. The outputs of counter are connected to the select lines of de-multiplexer and the data input of de-multiplexer is connected to the output of comparator. The output of de-multiplexer is connected to the reset pin of flip-flops. So, according to the output of counter and comparator the appropriate flip-flop is set or reset. So the desired output can be obtained easily.

A NOR gate based circuit is used for the purpose of producing clocks to D flip-flops. In this circuit only one pulse is given to the bit Q3 because after one clock cycle Q3 achieves its final state, there is no need of change the output.

Similarly, bits Q2, Q1 and Q0 receive 2, 3 and 4 pulses respectively.

3   Simulation Results

I designed all the building blocks like comparator, DAC, sample and hold circuit and SAR and control unit individually using gpdk180 library of CADENCE design tool. After that I combined all these building blocks and designed complete 4-bitSAR ADC. Then I simulated the complete circuit using analog design environment (ADE) and obtained the appropriate output. I added a DAC at the output of SAR ADC to

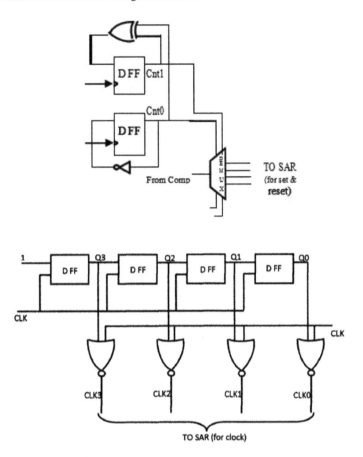

Fig. 6 Proposed control unit

convert the output word into its analog equivalent. The input waveform of SAR ADC and output waveform of DAC is shown in Fig. 7 for ramp input.

To determine various performance parameters the DFT of input and output is obtained as per given in Fig. 8. So the maximum clock frequency is 40 MHz and sampling frequency is 8 MHz for the proposed ADC. The other parameters of the circuit are given in the Table 1.

3.1 Comparison of Proposed Results with Existing ADC's

I have designed different blocks of SAR ADC like D flip-flop and comparator with some previous technique. So a comparative study of SAR ADC design using different techniques is given in Table 2.

Fig. 7 Input and output
waveforms, time versus
voltage (input is green and
output is red)

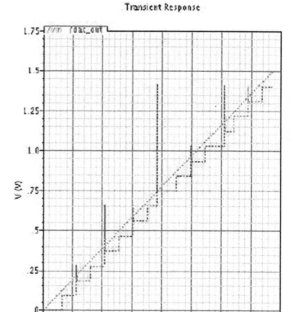

4   Conclusions

The main concern in designing of low power SAR ADC is the design of D-flip-flop
for SAR and comparator. So a low power D flip-flop is used. The average power
consumed by D flip-flop is 50.9 μW and the maximum operating frequency of the
proposed D flip-flop is 3.5 GHz. A low power double tail dynamic comparator with
control transistors is also proposed in which the leakage current is reduced by adding
control transistors and the stored charge is reused by adding two transistors so that
total power of comparator is reduced to 48 μW and the maximum operating frequency is 16 MHz. A control unit comprises of a de-multiplexer and a ring counter
is designed. A sample and hold circuit embedded with the capacitor array based
DAC is designed. Then I combined all these blocks and designed a low power 4-bit
SAR ADC which consumes total power of 113 μW. The maximum clock frequency
of proposed SAR ADC is 16.5 MHz and sampling frequency is 3.3 MHz. Other
performance parameters like quantization noise voltage, signal-to-noise plus distortion ratio (SNDR), effective number of bits (ENOB) are also improved as 2.04 V,
24.15 dB and 3.72 bits respectively.

Fig. 8 DFT of input and output signal when ramp signal is applied to the input

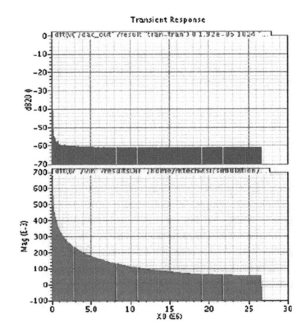

Table 1 Performance parameters of proposed circuit

Parameter	Pro.	[4]	[5]	[6]	[7]
VDD (in Volts)	1.8	1.8	1.2	1.8	1.8
Technology (nm)	180	180	130	180	180
Average power	113.6 μW	30 mW	32 mW	7.8 mW	23.3 mW
Sampling Freq. (MHz)	3.3	40	1250	500	710
Clock Freq. (MHz)	16.5	–	–	–	–
Quantization noise (mV)	2.04	–	–	–	–
SNDR (DB)	24.15	–	–	–	–
ENOB (in bits)	4.21	–	–	–	3.68

5 Future Scope

Low power SAR ADC has a wide future scope in wireless biometric sensors for heart beat measurement, blood pressure measurement, body temperature measurement etc. The other use of low power ADCs is in space communication for transmission of

Table 2 Comparison between different SAR ADCs

Parameters	Type-I	Type-II	Type-III
Power	3.4 mW	236.6 μW	113.6 μW
Quantization noise	1.96 mV	1.89 mV	2.04 mV
SNDR	24.31 dB	24.48 dB	24.15 dB
ENOB	3.75 bits	3.77 bits	3.72 bits

atmospheric parameters like temperature, pressure and humidity etc. The power consumption of SAR ADCs can be further reduced by the use of dynamic logic circuit in place of CMOS circuit in control unit. I can reduce the power by using some more appropriate technique so the charge which is stored in one cycle can be used in next cycle. I can use low power GaAs comparator and bulk tuned comparator so the power consumption and area can be reduced further. For SAR, I can use low power GaAs D flip-flop and low power dual edge triggered D flip-flop so the power consumption can be reduced and operating frequency can be increased. I have to look after to design low power, less area and high speed SAR ADCs for the above discussed future applications.

References

1. McCreary, J. L., & Gray, P. R. (1975). AII-MOS charge redistribution analog-to-digital conversion techniques—Part I. *IEEE Journal of Solid-State Circuits*.
2. Dondi, S., Vecchi, D., Boni, A., & Bigi, M. (2006). *A 6-bit, 1.2 GHz interleaved SAR ADC in 9 Onm CMOS*. IEEE. ISBN 1-4244-0157-7.
3. Kandala, M., Sekar, R., Zhang, C., & Wang, H. (2010). *A low power charge redistribution ADC with reduced capacitor array*. IEEE. ISBN 978-1-4244-6455-5.
4. Banik, S., Gangopadhyay, D., & Bhattacharya, T. K. (2006). A low power 1.8 V 4-bit 400-MHz flash ADC in. 18 μm Digital CMOS. In *Proceedings of IEEE VLSID*.
5. Ginsburg, B. P., & Chandrakasan, A. P. (2005). Dual scalable 500MS/s5b time Intervealed SARADCs for UWB application. In *Proceedings of IEEE CICC*.
6. Caol, Z., Yan, S., & Li, Y. (2008). A32 mW 1.25GS/s6b/step SARADC in 0.13 μm CMOS. In *Proceedings of IEEE ISSCC*.
7. Talekar, S. G., & Ramasamy, S. (2009). A low power 700 MSPS4-bit Time Intervealed SAR ADC in 0.18 μm *CMOSIEEE*.
8. van Elzakker, M., van Tuijl, Ed., Geraedts, P., Schinkel, D., E. A. M. Klumperink, & Bram, N. (2010). A 10-bit charge-redistribution ADC consuming 1.9 μW at 1 MS/s. *IEEE Journal of Solid-State Circuits, 45*(5).
9. Xiaozong, H., Jing, Z., Weiqi, G., Jiangang, S., & Hui, W. *A 16-bit, 250 ksps successive approximation register ADC based on the charge-redistribution technique*. IEEE. ISBN 978-1-4577-1997-4/2011.
10. Zhu1, X., Chen2, Y., Tsukamoto2, S., & Kuroda1, T. (2012). *A 9-bit 100 MS/s tri-level charge redistribution SAR ADC with asymmetric CDAC array*. IEEE. ISBN 978-1-4577-2081-9/2012.
11. Villanueva, I. C., & Lopez-Martin, A. (2013). *An ultra low energy 8-bit charge redistribution ADC for wireless sensors*. IEEE. ISBN 978-1-4673-5221-5.

12. Goll, B., & Zimmermann, H. (2009). A comparator with reduced delay time in 65-nm CMOS for supply voltages down to 0.65‖. *IEEE Transactions on Circuits and Systems II, Express Briefs, 56*(11), 810–814.
13. Freitas, D., & Current, K. (1983). CMOS current comparator circuit. *Electronics Letters, 19*(17), 695–697.

Stabilization and Control of Magnetic Levitation System Using 2-Degree-of-Freedom PID Controller

Brajesh Kumar Singh⑩ and Awadhesh Kumar

Abstract The magnetic levitation (maglev) system has become a very efficient technology in the rapid mass transportation system due to its frictionless motion. It is an open-loop unstable system, so it requires a controller implementation for its stabilization and position-tracking. Since it is inherently a nonlinear system, its controller design is a challenging problem. This paper presents a 2-degree-of-freedom PID controller designed to stably levitate the object in the magnetic field as well as for the position-tracking. All the simulation works are performed under MATLAB environment, and the simulation results have been discussed at the end of this paper.

Keywords Magnetic levitation system · Maglev · PID · Stability
2-Degree-of-Freedom PID controller

1 Introduction

The magnetic levitation system plays a vital role in various industrial processes due to its ability of removing the friction between moving and stationary parts. It has some other abilities that include the working in high temperature range and high rotation speed. These abilities make it very efficient for many mechatronics systems. There are many applications [1, 2] of the maglev system that includes flywheel system, high-speed maglev passenger train, frictionless magnetic bearing, maglev wind turbine, maglev heart pump (cardiopulmonary bypass), electromagnetic aircraft launch system, maglev rocket-launching system.

B. K. Singh (✉) · A. Kumar
Department of Electrical Engineering, Madan Mohan Malaviya
University of Technology, Gorakhpur, UP, India
e-mail: brajesh.k.singh91@gmail.com

A. Kumar
e-mail: akee@mmmut.ac.in

© Springer Nature Singapore Pte Ltd. 2019
A. Khare et al. (eds.), *Recent Trends in Communication, Computing,
and Electronics*, Lecture Notes in Electrical Engineering 524,
https://doi.org/10.1007/978-981-13-2685-1_54

There are many techniques available in the literature to design the controller for stabilization of the magnetic levitation system. The controller is also able to regulate the position of the levitated ball as per the desired specification. Various controller design techniques available in the literature are feedback linearization, state feedback control, conventional proportional integral plus derivative control (PID), fractional-order PID controller, linear quadratic Gaussian (LQG), linear quadratic regulator (LQR), sliding mode controller (SMC), back-stepping control design technique, artificial intelligence-based techniques, optimization techniques, etc. Charara et al. in [3] have proposed a nonlinear controller based on input–output feedback linearization which has been successfully implemented and tested on magnetic levitation system. A fractional-order proportional plus integral and derivative controller with anti windup (FOCAW) scheme in [4] has been successfully implemented and tested on maglev system by taking consideration of actuator saturation. An adaptive nonlinear controller-based back-stepping approach has been successfully designed and tested in [5] for a magnetic levitation system. A suboptimal controller based on model order reduction technique has been proposed in [6, 7]. The advantage of using a suboptimal controller is that it does not require a costly reconstructive filter to measure the missing states. An adaptive robust controller based on nonlinear back-stepping design has been presented in [8]. Huang et al. in [9] have proposed a two-degree-of-freedom (DOF) PID controller for stabilization based on two electromagnets to improve the balance range of the magnetic levitation system. A 2-DOF PID controller has been implemented in [10] which provides superior robustness with suitable choice of the feed forward gain parameters. A 2-DOF controller based on linear quadratic Gaussian (LQG) design has been reported in [11] for a magnetic suspension system. Ahsan et al. in [12] have presented different controllers based on linear and nonlinear design tools to ensure the robustness and stability of the magnetic levitation system. A LQR theory-based tuning of PID controller has been presented in [13]. In this control design, a new method of selecting the weighing matrices based on natural frequency and damping ratio has also been proposed. A sliding mode controller (SMC) has been successfully designed and tested in [14] for a current-controlled magnetic levitation system. A model predictive controller (MPC) has been proposed in [15] which is based on nonlinear SISO model of magnetic levitation system. Artificial intelligence (AI)-based techniques are also available in the literature in which PD fuzzy logic controller, neural network, and feedback error learning (FEL)-based controllers are designed in [16, 17] for the magnetic levitation system. The investigator intends to design and implement a 2-DOF PID controller for the maglev system with an objective of stabilization and position tracking.

2 Mathematical Modeling of Magnetic Levitation System

Block diagram of the magnetic levitation system depicting the different components is shown in Fig. 1.

Fig. 1 Block diagram for magnetic levitation system

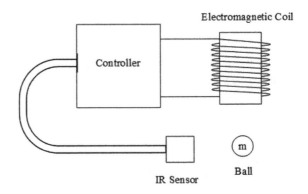

Fig. 2 Modeling of electrical subsystem for the electromagnetic coil

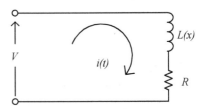

The magnetic levitation model consists of an electromagnet, a levitating ball (ferromagnetic), and an IR photosensor to measure the position of the moving ball. The modeling can be divided into two subsystems, i.e., electrical subsystem (electromagnet) and mechanical subsystem (motion of the ball in the magnetic field). In Fig. 2, the model of the electrical subsystem is shown.

Mathematically, the model can be expressed with the following Eq. (1).

$$V = Ri(t) + \frac{d\varphi}{dt},\qquad(1)$$

where V is the voltage supplied to the electromagnet, $i(t)$ is the current through the coil, R is resistance of the coil, and $\varphi = L(x).i$ is magnetic flux linkage. $L(x)$ is the inductance of the coil which is dependent upon the ball position x. The inductance $L(x)$ can be expressed by (2) as reported in [12]:

$$L(x) = L_1 + \frac{L_0}{1 + (x/a)},\qquad(2)$$

where L_1, L_0 and a are positive constants. The inductance has maximum value when the ball is nearest to the coil and reduced to a constant minimum value when the ball is away from the coil. The energy stored in the electromagnet is given by (3):

$$E = \frac{1}{2}L(x)i^2.\qquad(3)$$

The electromagnetic force to the ball is given as (4):

$$F(x, i) = \frac{\delta E}{\delta x} = \frac{L_0 i^2}{2a(1 + x/a)^2}. \tag{4}$$

Now the modeling of the mechanical subsystem (motion of the ball) can be expressed by the following dynamic Eq. (5):

$$m\ddot{x} = -k\dot{x} + mg + F(x, i), \tag{5}$$

where m is mass of the ferromagnetic ball, g is acceleration due to gravity, k is viscous friction coefficient. For this system, three state variables are defined as follows:

$$x_1 = x \ (position)$$
$$x_2 = \dot{x} \ (velocity)$$
$$x_3 = i \ (current).$$

The control input (supply voltage) is defined as $u = V$.

Now, the detailed nonlinear state-space model for the magnetic levitation system can be expressed as (6):

$$\left. \begin{array}{l} \dot{x}_1 = x_2 \\[2mm] \dot{x}_2 = g - \dfrac{k}{m}x_2 - \dfrac{L_0 a x_3^2}{2m(a + x_1)^2} \\[3mm] \dot{x}_3 = \dfrac{1}{L(x_1)}\left[-Rx_3 + \dfrac{L_0 a x_2 x_3}{(a + x_1)^2} + u \right] \end{array} \right\} \tag{6}$$

The values of the modeling parameters [12] are given in Table 1.

Table 1 Values of different parameters of the maglev system

Parameter	Description	Value
m	Mass of the ball	0.1 kg
R	Resistance of the coil	1 Ω
L_0	Positive constant	0.01 H
L_1	Positive constant	0.02 H
a	Positive constant	0.05 m
g	Acceleration due to gravity	9.81 m/s^2
k	Viscous friction coefficient	0.01 N/m/s

3 Control Design Strategies

3.1 Implementation of 1-DOF PID Controller

The PID control strategy is the most commonly used control scheme in the process industry. The block diagram for 1-DOF PID controller is shown in Fig. 3.

Different control objectives can be achieved by controlling three parameters: proportional gain K_P, differential gain K_D, and integral gain K_I.

The actuating signal for PID control consists of three error signals: proportional error, integral error, and derivative error. So, for PID control the actuating signal can be given by (7):

$$e_a(t) = K_P e(t) + K_I \int e(t)dt + K_D \frac{d}{dt} e(t). \tag{7}$$

A PID controller has been designed to stabilize the magnetic levitation system. To design the PID controller, firstly the system has been linearized around an equilibrium point. The equilibrium position of the ball is considered to be $x = 0.05$ m. The steady-state current I_{ss} and voltage V_{ss} can be obtained as described in (8):

$$\left. \begin{array}{l} I_{ss} = \sqrt{\dfrac{2mg(a + x_1)^2}{L_0 a}} = 6.26A \\[4mm] V_{ss} = RI_{ss} = 6.26V \end{array} \right\} \tag{8}$$

The origin of the system has been shifted to

$$(x_1 - 0.05, \ x_2, \ x_3 - 6.26).$$

The linearized state-space model for the original system can be obtained as in (9):

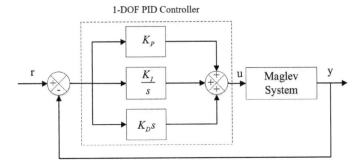

Fig. 3 Block diagram for 1-DOF PID controller

$$A = \begin{bmatrix} 0 & 1 & 0 \\ 195.938 & -0.1 & -3.13 \\ 0 & 12.52 & -40 \end{bmatrix}; \ B = \begin{bmatrix} 0 \\ 0 \\ 40 \end{bmatrix} \Bigg\}$$

$$C = \begin{bmatrix} 1 & 0 & 0 \end{bmatrix}; \ D = [0]$$

(9)

The tuning of the parameters for the PID controller has been done based on model-based tuning algorithm which uses rise time and damping ratio to obtain a better tuning of the parameters of the controller. The PID parameters are tuned for the mathematical model (9) of the maglev system. After carrying out the simulation, the parameters of 1-DOF PID controller were found to be $K_P = -480$, $K_I = 3.5$, and $K_D = 0.05$.

3.2 Implementation of 2-DOF PID Controller

A 2-DOF PID controller provides an extra degree to improve upon the transient response as obtained with 1-DOF PID controller. The regulatory control performance of the closed-loop system along with its robustness has been improved considerably with suitable choice of extra parameters of the controller.

The block diagram for the 2-DOF PID controller has been shown in Fig. 4. The actuating signal for 2-DOF PID controller can be given by following expression (10):

$$e_a(t) = (b.r - y)K_P + K_I \int (r - y)dt + K_D \frac{d}{dt}(c.r - y). \tag{10}$$

The three parameters of the controller K_P, K_I, and K_D have been obtained based on the same tuning algorithm as applied in 1-DOF PID control, by setting the additional parameters $b = 1$ and $c = 1$.

With $b=c=1$, the controllers parameters K_P, K_I, and K_D as obtained are set. Further changing the value of b a little and then applying the same tuning method for

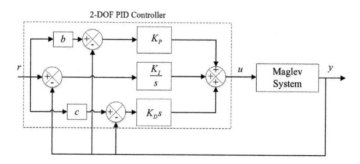

Fig. 4 Block diagram for 2-DOF PID controller

the readjustment of the values of K_P, K_I and K_D. If the obtained response is not satisfactory, we go on changing the value of c also. The iterative process is repeated till the response better than the one obtained with 1-DOF PID control is achieved. After carrying out the exhaustive iterative simulations, finally, we arrived at the following parameter values: $K_P = -289.28$, $K_I = -793.58$, $K_D = -22.86$, $b = 0.45$, $c = 0.35$.

4 Simulation and Results

After carrying out the simulations, the following results have been obtained. The step response for the maglev system with 1-DOF PID controller has been shown in Fig. 5.

The rise time and settling time of the response obtained are quite less; means system response is fast enough, but the overshoot is high. The steady-state response of the system is good enough, but the transient response is not satisfactory. The response specifications obtained are given in Table 2.

Now the reference input of the system has been changed to square wave input. The reference input tracking of the levitated ball for square wave reference input with 1-DOF PID controller has been shown in Fig. 6.

From Fig. 6, it has been shown clearly that, 1-DOF PID controller is able to track the square wave input satisfactorily in steady state, but the transient performance

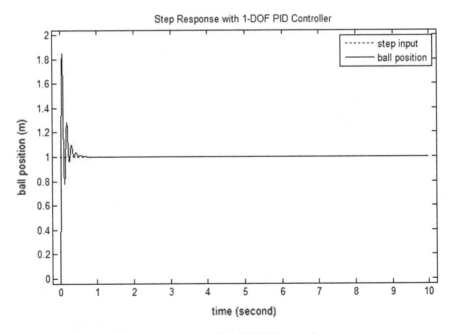

Fig. 5 Step response of the maglev system with 1-DOF PID control

Table 2 Time response specifications for the maglev system with 1-DOF PID controller	Specifications	Value
	Rise time	0.02 s
	Peak-time	0.06 s
	Settling time	0.35 s
	Maximum overshoot	84%

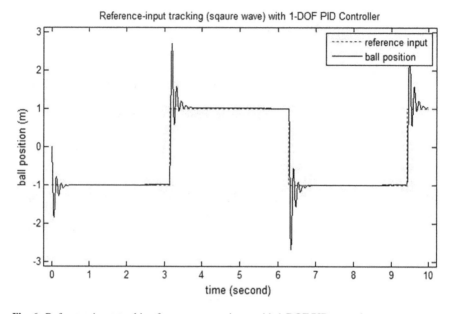

Fig. 6 Reference input tracking for square wave input with 1-DOF PID control

is not up to the mark. To meet out this challenge, a 2-DOF PID controller can be applied which uses two additional free parameters. The step response for the maglev system with 2-DOF PID controller has been shown in Fig. 7.

From Fig. 7 it is clearly visible that with the application of 2-DOF PID controller, the improvement in the transient response has been obtained over the 1-DOF PID controller as shown in Fig. 5. The reference input tracking of the levitated ball for square wave reference input with 2-DOF PID controller has been shown in Fig. 8.

From Figs. 5 to 8, it is quite clear that the PID control is able to stabilize the magnetic levitation system with the step input satisfactorily and shows a reference tracking for a square wave input as well. The PID controller is simple in design and proves to be reasonably a good stabilizing controller.

The 2-DOF PID controller uses two additional parameters for the tuning which gives it superiority over 1-DOF PID controller. The settling time and rise time are increased a little bit for a 2-DOF PID controller, but at the same time, overshoot has been reduced significantly and the steady-state error has been reduced almost to zero.

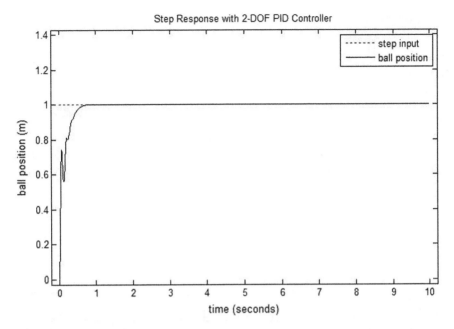

Fig. 7 Step response of the maglev system with 2-DOF PID control

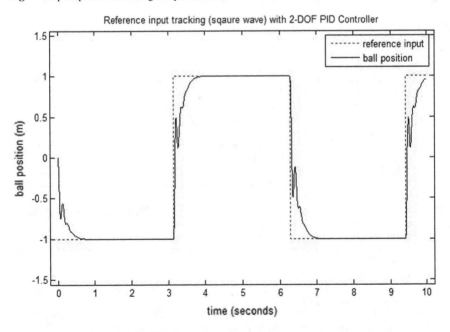

Fig. 8 Reference input tracking for square wave input with 2-DOF PID control

5 Conclusion

In this paper, a 1-DOF PID controller and 2-DOF PID controller have been designed and successfully implemented which are able to stabilize the magnetic levitation system. After carrying out simulations, it was found that the response obtained with 2-DOF controller on the magnetic levitation system is far superior to the results obtained with 1-DOF PID controller in both the transient and steady-state domain. The 2-DOF PID controller uses five free parameters which makes it more efficient in tuning. The proposed controller was found to show excellent position tracking with square wave input; however, a nonlinear control strategy may be implemented which is open for investigation.

References

1. Yaghoubi, H. (2013). The most important maglev applications. *Journal of Engineering, 2013*.
2. Rote, D. M., & Cai, Y. (2002). Review of dynamic stability of repulsive-force maglev suspension systems. *IEEE Transactions on Magnetics, 38*(2), 1383–1390.
3. Charara, A., De Miras, J., & Caron, B. (1996). Nonlinear control of a magnetic levitation system without premagnetization. *IEEE Transactions on Control Systems Technology, 4.5*, 513–523.
4. Pandey, Sandeep, Dwivedi, Prakash, & Junghare, Anjali. (2017). Anti-windup Fractional Order PI^{λ}-PD^{μ} controller design for unstable process: A magnetic levitation study case under actuator saturation. *Arabian Journal for Science and Engineering, 42*(12), 5015–5029.
5. Green, Scott A., & Craig, Kevin C. (1998). Robust, digital, nonlinear control of magnetic-levitation systems. *Journal of Dynamic Systems, Measurement, and Control, 120*(4), 488–495.
6. Pati, A., & Negi, R. (2014). Suboptimal control of magnetic levitation (Maglev) system. In *2014 3rd International Conference on Reliability, Infocom Technologies and Optimization (ICRITO)* (Trends and Future Directions). IEEE.
7. Pati, A., Kumar, A., & Chandra, D. (2014). Suboptimal control using model order reduction. *Chinese Journal of Engineering, 2014*, Article ID 797581, 5 p. Hindawi Publishing Corporation. http://dx.doi.org/10.1155/2014/797581.
8. Yang, Z.-J., & Tateishi, M. (2001). Adaptive robust nonlinear control of a magnetic levitation system. *Automatica, 37*(7), 1125–1131.
9. Huang, H., Du, H., Li, W. (2015). Stability enhancement of magnetic levitation ball system with two controlled electromagnets. In *2015 Australasian Universities Power Engineering Conference (AUPEC)*. IEEE.
10. Arun, G. et al. (2014). Design and implementation of a 2-DOF PID compensation for magnetic levitation systems. *ISA Transactions, 53.4*, 1216–1222.
11. Ibraheem, Ibraheem Kasim. (2017). Design of a two-Degree-of-Freedom controller for a magnetic levitation system based on LQG technique. *Al-Nahrain Journal for Engineering Sciences, 16*(1), 67–77.
12. Ahsan, M., Masood, N., & Wali, F. (2013). Control of a magnetic levitation system using non-linear robust design tools. In *2013 3rd International Conference on Computer, Control & Communication (IC4)*. IEEE.
13. Kumar, E. V., & Jerome, J. (2013). LQR based optimal tuning of PID controller for trajectory tracking of magnetic levitation system. *Procedia Engineering, 64*, 254–264.
14. Fallaha, C., Kanaan, H., & Saad, M. (2005). "Real time implementation of a sliding mode regulator for current-controlled magnetic levitation system. In *Proceedings of the 2005 IEEE International Symposium on, Mediterranean Conference on Control and Automation Intelligent Control*. IEEE.

15. Lukáš, R., Krhovják, A., & Bobál, V. (2017). Predictive control of the magnetic levitation model. In *2017 21st International Conference on Process Control (PC)*. IEEE.
16. Sharkawy, A. B., & Abo-Ismail, A. A. Intelligent Control of Magnetic Levitation System.
17. Aliasghary, M. et al. (2008). Magnetic levitation control based-on neural network and feedback error learning approach. In *PECon 2008. IEEE 2nd International Power and Energy Conference, 2008*. IEEE.

Design of High-Gain CG–CS 3.1–10.6 GHz UWB CMOS Low-Noise Amplifier

Dheeraj Kalra, Manish Kumar and Abhay Chaturvedi

Abstract A high-gain low-power CMOS low-noise amplifier is simulated using TSMC 0.18-μm CMOS technology. The cascade topology is used to get the high-gain and low-noise figure value. The source degeneration technique is used for the wideband matching. The circuit is simulated for 3.1–10.6 GHz in ultawideband. The simulated results show the maximum gain of 21.574 dB at 6.378 GHz and positive gain maintained during the entire frequency range. The highest noise figure value is 4.311 dB at 7.662 GHz, and the lowest value is 2.477 dB at 3.1 GHz. The matching circuit at input and output terminals shows the input return loss of 22.262 dB at 10.38 GHz, while the output return loss of 30.936 dB at 4.1 GHz. The circuit is simultaed at 1.2 V which draws the power consumption of 17.734 mW. The designed circuit shows the optimum value of gain, noise figure, matching and power consumption.

Keywords Low-Noise amplifier · Noise figure · S parameters

1 Introduction

As per the norms of FCC (Federal Communications Commission), the unlicensed frequency band is 3.1–10.6 GHz [1]. The requirement of the high data rate in any communication is the primary focus which can be achieved using UWB band. The limitation of UWB is to maintain the low power so that the signal should not get interfered with the licensed band signal. UWB system provides the transmission of signal for the short range at high data rate [2]. CMOS technology is a broad area of circuit designing due to its low cost, low area and low power consumption. The block diagram of any receiver system consists of an antenna, low-noise amplifier,

D. Kalra · M. Kumar (✉) · A. Chaturvedi
GLA University, Mathura, India
e-mail: manish.kumar@gla.ac.in

D. Kalra
e-mail: dheeraj.kalra@gla.ac.in

© Springer Nature Singapore Pte Ltd. 2019
A. Khare et al. (eds.), *Recent Trends in Communication, Computing, and Electronics*, Lecture Notes in Electrical Engineering 524,
https://doi.org/10.1007/978-981-13-2685-1_55

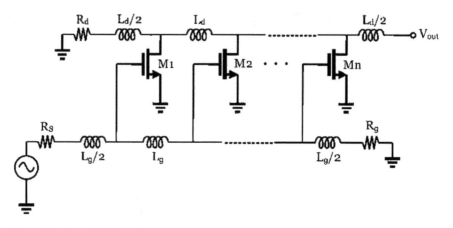

Fig. 1 Distributed LNA topology

mixer and then the IF stage. The RF front-end low-noise amplifier is the key block of any receiver systems as its function is to increase the signal-to-noise ratio of the received signal, i.e., improving the noise figure [3]. The various parameters of LNA are gain, noise figure, input impedance, output impedance matching and power consumption. The design of LNA is typical because there is always trade-off between LNA parameters [4].

So there are various topologies to improve the LNA parameters. In the distributed topology, MOSFETs connected in the cascade configuration result in the multiplication of individual gain, which help in getting the high gain, also provide the wideband matching, and help in getting the higher bandwidth. The schematic of the distributed topology is shown in Fig. 1 [6].

But the disadvantage of this topology is that it requires more chip area; hence, the power consumption is high. The source degeneration topology is used to linearize the circuit. The inductor is connected at the source terminal, which helps in cancelling the high frequency capacitances of MOSFET, i.e., C_{gs} and C_{gd} [5, 6], and makes the input terminal purely resistive for wideband frequency (Fig. 2).

Common-gate topology gives the inherent wideband matching at low power consumption and high value of IIP3 [7]. But a single common-gate topology is not enough to give the sufficient amount of gain. The noise figure cannot be flat by common-gate amplifier (Fig. 3).

The simulated circuit shows that various topologies have been used to improve one or many parameters.

Fig. 2 Source degeneration
LNA topology

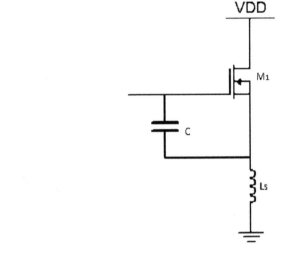

Fig. 3 Common-gate LNA
topology

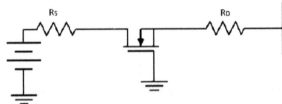

2   Circuit Design

The circuit is simulated in 0.18-μm CMOS TSMC 0.18-μm technology. The
designed circuit is shown in Fig. 4. The π network matching circuit is connected
which comprises L_7, C5 and L_8 components at the input port, while L_{10}, C_6 and L_9 at
the output port. The transistor T_1 connected in common-gate configuration provides
the good wideband matching, but is not able to provide the sufficient amount of gain.
Transistors T_2 and T_5 are connected in the cascade configuration whose output is
cascaded with T_4 and T_3. Transistor T_4 connected in the common drain configuration, i.e., source follower, provides the good impedance matching so as to get the
maximum power at the output port. Capacitors C_3 and C_4 act as the coupling capacitors which help in getting the dc stability of the circuit. Two resonant circuits R_1, L_1,
C_1 and R_3, L_5, C_2 provide the resonance at high and low frequencies. Inductors L_2
and L_6 behave in the source degeneration topology which is used to provide the flat
gain value. While simulating the circuit, various points have been noted regarding
the effect of changing the values of component on LNA parameters. Capacitor C_1
has no effect on the value of gain, but noise figure gets degraded on increasing its
value.

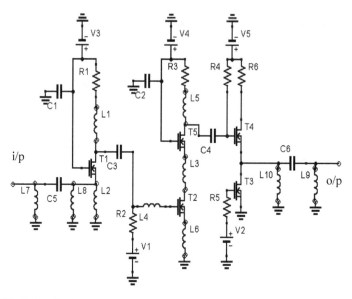

Fig. 4 LNA design circuit

On increasing the R_1 resistance value, the gain becomes more flat, gain increases with the increase in the value of R_1, and capacitor C_5 helps in the input impedance matching. Capacitor C_3 helps in improving the gain value, increasing the value of inductors L_1 and L_2 increases the power gain, capacitor C_2 helps in decreasing the noise figure, and inductors L_3, L_4 and L_5 provide the flat gain. The circuit is simulated at 1.2 V; by increasing the biasing voltage, the gain increases, but at the same time the power consumption goes on increasing. So the circuit is optimized with the values of components.

3 Simulation Results

The various parameters of the design circuit have been calculated. The simulation result for S_{21} is shown in Fig. 5. The maximum value of the gain is 21.574 dB at 6.378 GHz, and the gain is more than 9 dB for the entire range of frequency. The positive value of S_{21} shows the good matching at port 2.

The simulation result for the noise figure is shown in Fig. 6. For frequency range of 3.1–10.6 GHz, the maximum value is 4.311 dB at 7.662 GHz and the minimum value is 2.477 dB at 3.1 GHz. At 7.662 GHz, the resistance is less, so the value of noise figure is high. The good value of noise figure improves signal-to-noise ratio.

Simulation results of S_{11} and S_{22} are shown in Figs. 7 and 8 which show the good impedance matching at input and output ports.

The power consumption of the designed circuit is 18.05 mW at voltage of 1.2 V.

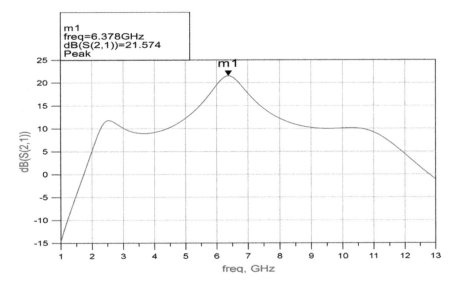

Fig. 5 Simulation of S$_{21}$ versus frequency

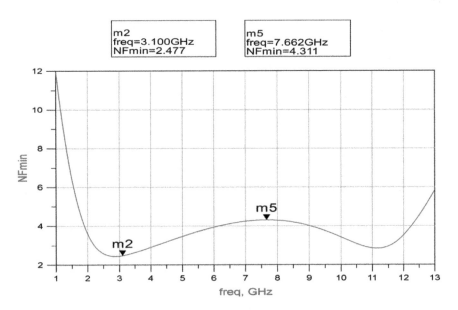

Fig. 6 Simulation of noise figure versus frequency

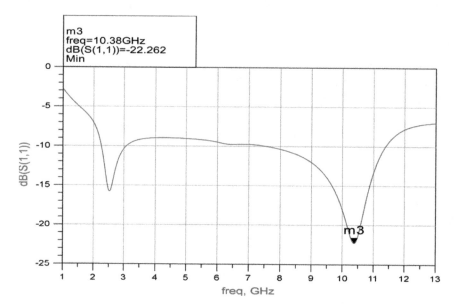

Fig. 7 Simulation of S_{11} versus frequency

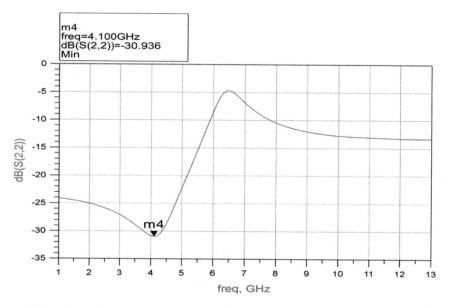

Fig. 8 Simulation of S_{22} versus frequency

4 Conclusion

In the designed LNA for wideband, a common-gate stage followed by two common source stages is connected in the cascade configuration for high gain. A proper LC wideband matching is done at input terminal for low-noise figure. The simulated circuit is optimized for gain, noise figure, input and output return losses. The circuit is simulated for 1–13 GHz. The simulation results show the gain value of 21.574 dB, the noise figure of 2.477 dB, the input return loss of −22.262 dB, the output return loss of −30.936 dB and the power consumption of 18.05 mW. So, the circuit has good values of its parameter for the UWB system.

References

1. Heinrichs, M., Bark, N. & Kronberger, R. (2018). Decreasing size, increasing battery life: A wide-band, low-noise amplifier for 5G multistandard broadband communications. *IEEE Microwave Magazine, 19*(2), 77–82.
2. Neeraja, A. R., & Yellampalli, S. S. (2017). Design of cascaded narrow band low noise amplifier. In *2017 International Conference on Electrical, Electronics, Communication, Computer, and Optimization Techniques (ICEECCOT)*, Mysuru (pp. 1–4).
3. Zulkifli, T. Z. A., Marzuki, A., & Murad, S. A. Z. (2017). UWB CMOS low noise amplifier for mode 1. In *2017 IEEE Asia Pacific Conference on Postgraduate Research in Microelectronics and Electronics (PrimeAsia)*, Kuala Lumpur (pp. 117–120).
4. Saied, A. M., Abutaleb, M. M., Ibrahim, I. I., & Ragai, H. (2017). Ultra-low-power design methodology for UWB low-noise amplifiers. In *2017 29th International Conference on Microelectronics (ICM)* (pp. 1–3).
5. Shaeffer, D. K., & Lee, T. H. (1997). A 1.5-V, 1.5-GHz CMOS low noise amplifier. *Journal of Solid-State Circuits, 32*.
6. Yu, Y., et al. (2017). Analysis and design of inductorless wideband low-noise amplifier with noise cancellation technique. *IEEE Access, 5*, 9389–9397.
7. Kumar, M., Kumar Deolia, V., & Kalra, D. (2016). DESIGN and simulation of UWB LNA using 0.18 μm CMOS technology. In *2016 2nd International Conference on Communication Control and Intelligent Systems (CCIS)*, Mathura (pp. 195–197).

Development of Nano-rough Zn$_{0.92}$ Fe$_{0.08}$ O Thin Film by High Electro-spin Technique via Solid-State Route and Verify as Methane Sensor

Brij Bansh Nath Anchal, Preetam Singh and Ram Pyare

Abstract In this paper, development of nano-rough Zn$_{0.92}$ Fe$_{0.08}$O thin film (NFZO) on glass substrate and application in the detection of methane are reported. The deposition of film used technique is high electro-spin 500–3500 rpm pattern. Doping of 8% iron in zinc oxide synthesized by solid-state route and chemical route is applied in spin coating. Study of thin film characterizions by HR-SEM, EDX and AFM, while sensor response for 100–300 ppm methane at 150 °C. There are iron ions like ferrous and ferric improved to the gas sensing properties. Overall sensitivity rapidly increases with concentrations of methane at 150 °C. The deposition of NFZO by the high rpm spin pattern is low cost and simple operation.

Keywords HR-SEM · EDX · AFM · Thin film · Sensor

1 Introduction

Zinc oxide is large band gap as 3.37 eV semiconductor materials [1, 2]. It gap effected by the transitions metal doping, causing increase in the negative charge carriers. Charge carrier moves from conduction to valance band and increases the conductivity [1, 3–5]. Zinc oxide is the hexagonal wurtzite and thermal stabled crystal structure [1, 2]. Zinc oxide-based thin-film methane sensors are developed [6–10]. Metal oxides as tin oxides, tungsten oxide-based methane sensors also developed. Mechanism of sensing depends on the adsorption of gas molecules and reduction in surface in the target gas [6, 10–13]. The iron doping effect is improved to the gas sensing properties of zinc oxide thin film. High electro-spin technique is applicable for the development of smooth surface [4]. Nano-roughness provide to spe-

B. B. N. Anchal (✉) · P. Singh · R. Pyare
Department of Ceramic Engineering, Indian Institute of Technology (BHU),
Varanasi 221005, India
e-mail: bbnanchal.rs.cer12@iitbhu.ac.in

© Springer Nature Singapore Pte Ltd. 2019
A. Khare et al. (eds.), *Recent Trends in Communication, Computing,
and Electronics*, Lecture Notes in Electrical Engineering 524,
https://doi.org/10.1007/978-981-13-2685-1_56

cific area for adsorption. Its all properties improved by the nano-grain size, nano-roughness, film smoothness and fabrication of nano-wires, micro flowers, etc. [1, 2, 14]. Formation of oxygen species due to the chemi-adsorb of oxygen at different temperatures and reduction processed in the presence of the reducing gases [4, 15]. There are 8% iron doping not changing in crystal structure and substitute in the host lattice [16]. Doping effect correlated to the improved of the sensing properties. Iron-doped zinc oxide is promising material in the sensing of CH_4 [17].

Electro-spin deposition is easily operated and minimum expensive over the physical depositions as sputters, evaporations, etc. Spray and dip coating are easy method, but preferential thickness controlled in the spin coating [1–5]. So spin coating process is valuable in the deposition research. Morphology and compositions are explained by high resolution electron microscope (HR-SEM) and energy dispersive X-ray spectrometry (EDX) respectively. Smoothness and roughness are explained by atomic force microscopy (AFM).

2 Materials and Methods

Zinc oxide (ZnO, 99% pure powder, Sigma-Aldrich purchases) and ferric oxide (Fe_2O_3, 99% pure powder, sigma-Aldrich purchases) with stoichiometric doping weight in ball mill (planetary) at 350 rpm for 1.5 h. Thereafter they were dried in oven at 150 °C over night and after grinding they were kept in furnace at 850 °C for 10 h. Calcined material dissolved in aqua regia solution and stirred for 1 h at 50 °C. Glass substrate was cut in the sizes of 2.5 cm × 2.5 cm and cleaned in ultrasonic bath. Substrate was held on spin vacuum plate and started rpm at 500 rpm. Caste was dropped and spun for 1 min, similarly drop on 1500, 2500, 3500 rpm and spun at the same time; after this process sample was kept on hot plate at 250 °C for one min. The same process was repeated three times and kept in furnace at 500 °C for one half hour. Finally, the metallization of silver contacts with the metal mask. The thickness of the grown film was 1.6 μm measured by stylus profilometer. Nova Nano SEM 450, provided HR-SEM and EDX provided by EVO-18 Research ZEISS used in surface morphology and compostions analysis. NT-MDT-NTEGRA, provided AFM used in smoothness and roughness analysis. Here planetary mill used FRITSCH-PULVERISETTE 5/4 CLASSIC LINE and spin coating by DELTA SCIENTIFIC EQUIPMENT-SPIN COATER MODEL DELTA SPIN-1 and resistance measured by KEITHELY 195A digital multimeter and assembled gas sensing set up in laboratory (Fig. 1).

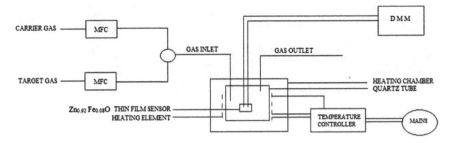

Fig. 1 Schematic diagram of assembled gas sensing set up

3 Results and Discussion

High resolution scanning electron microscope (HR-SEM) image is shown in Fig. 2, cleared to film is uniform and porous. Grains in linear situated and sizes are 58, 82, 131 nm, etc.,. These grain sizes are in nano and micro. Film is highly dense and porous in nano and micro. These conditions increase to gas adsorption and enhance sensitivity of sensor. Oxygen species formation in porous created space charge layer same as on surface [15]. EDX shown in Fig. 3 confirms to the peaks of iron and zinc oxide on the scale of binding energy in the range of 10 keV (Fig. 3). Overlapped iron ions indicated to the effect of doping in zinc oxide. Figures 4 and 5 shown to 2D, 3D AFM of $Zn_{0.92}$ $Fe_{0.08}$ O thin film and confirm to the film is smooth and uniform with the nano-roughness is 41 nm determined from NT-MDT-NTEGRA. The large specific area is provided for adsorption on the surface in the different sides. In AFM image roughness in uniform plane forms, which is suitable to gas adsorption. Ferrous and ferric ions substitute into host lattice of zinc oxide without any change in crystal structure and decrease potential barrier and role as catalytic in the gas sensing properties [3–5]. All parameters confirm to that film are sensitive for gas detection. There is response of sensor for methane at 150 °C; points A, D, G show gas flow is open, and points C, F, I show gas flow is closed, while A to B, D to E, G to H sensor responses for 100, 200 and 300 ppm methane, respectively (Fig. 6). Points B to C, E to F, H to I show sensor is stable in the presence of methane (Fig. 6). Points C to D, F to G, I to J show sensor recover state after methane flow closed (Fig. 6). When methane flow in sensing set up and expose on sensor then resistance of film decreases sharply and stabled after some time at 150 °C, resistance increasing in the condition of gas flow is closed and methane removed from sensing setup then sensor recovers in original resistance and stable. When there is an increase in concentration of methane, sensitivity increases and decreases the response time in this experiment. Average response time is 58 s, recovery time is 30 s, sensitivity is 48.33% for 100–300 ppm concentration of methane at 150 °C, and highest sensitivity is 58.81% for 300 ppm at same operating temperature (Fig. 7). Sensitivity determined by Eq. (1) [6, 7, 9, 12].

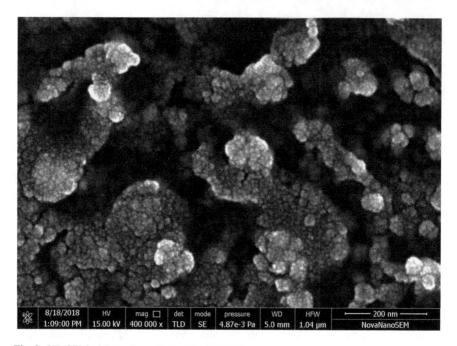

Fig. 2 HR-SEM of $Zn_{0.92}Fe_{0.08}O$ thin film (NFZO)

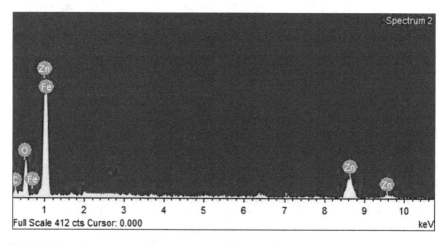

Fig. 3 EDX of $Zn_{0.92}Fe_{0.08}$ O thin film (NFZO)

$$\text{Sensitivity (\%) } S = \left[(R_a - R_g)/R_a \right] \times 100 \tag{1}$$

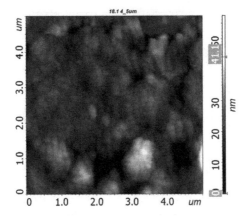

Fig. 4 2D-AFM of $Zn_{0.92}Fe_{0.08}O$ thin film (NFZO)

Fig. 5 3D-AFM of $Zn_{0.92}Fe_{0.08}O$ thin film (NFZO)

where R_a is resistance of thin film in the air and R_g in the presence of gas.

4 Conclusions

In summary, nano-rough $Zn_{0.92}Fe_{0.08}O$ thin film (NFZO) is developed by high electro spin via solid-state route method. NFZO thin film is uniform and nano-rough as shown in AFM. Grain sizes in nano–micro confirmed by HR-SEM and sensitivity increases with concentrations 100–300 ppm of methane at 150 °C. Over all NFZO is applicable in the detection of 100–300 ppm methane at 150 °C, and highest sensitivity is 58.81% for 300 ppm at 150 °C. So nano-rough $Zn_{0.92}Fe_{0.08}O$ thin film (NFZO) is verified as methane sensor. The deposition of $Zn_{0.92}Fe_{0.08}O$ thin film (NFZO) is done by using high rpm electro-spin technique. It is low-cost and simple operated process.

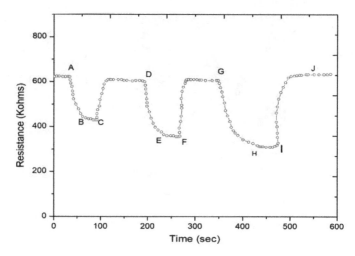

Fig. 6 Response of nano-rough $Zn_{0.92}Fe_{0.08}O$ thin film (NFZO) sensor for 100, 200 and 300 ppm methane at 150 °C

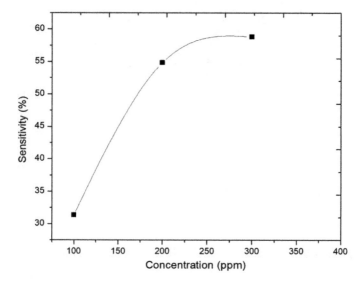

Fig. 7 Sensitivity of NFZO thin film sensor for 100–300 ppm methane at 150 °C

Acknowledgements This work is supported by Department of Ceramic Engineering of Indian Institute of Technology (BHU), Varanasi, India and Central Instrument Facility Centre-IIT (BHU), Varanasi, India.

References

1. Zhu, L., & Zeng, W. (2017). Room-temperature gas sensing of ZnO-based gas sensor: A review. *Sensors and Actuators A: Physical, 267,* 242–261.
2. Kumar, R., Al-Dossary, O., Kumar, G., & Umar, A. (2015). Zinc oxide nanostructures for NO2 gas–sensor applications: A review. *Nano-Micro Letters, 7,* 97–120.
3. Singh, K., Devi, V., Dhar, R., & Mohan, D. (2015). Structural, optical and electronic properties of Fe doped ZnO thin films. *Superlattices and Microstructures, 85,* 433–437.
4. Rambu, A. P., Doroftei, C., Ursu, L., & Iacomi, F. (2013). Structure and gas sensing properties of nanocrystalline Fe-doped ZnO films prepared by spin coating method. *Journal of Materials Science, 48,* 4305–4312.
5. Gao, F., Liu, X. Y., Zheng, L. Y., Li, M. X., Bai, Y. M., & Xie, J. (2013). Microstructure and optical properties of Fe-doped ZnO thin films prepared by DC magnetron sputtering. *Journal of Crystal Growth, 371,* 126–129.
6. Mitra, P., & Mukhopadhyay, A. (2007). ZnO thin film as methane sensor. *Bulletin of the Polish Academy of Sciences-Technical Sciences, 55,* 281–285.
7. Basu, P. K., Jana, S. K., Saha, H., & Basu, S. (2008). Low temperature methane sensing by electrochemically grown and surface modified ZnO thin films. *Sensors & Actuators, B: Chemical, 135,* 81–88.
8. Motaung, D. E., Mhlongo, G. H., Kortidis, I., Nkosi, S. S., Malgas, G. F., Mwakikunga, B. W., et al. (2013). Structural and optical properties of ZnO nanostructures grown by aerosol spray pyrolysis: Candidates for room temperature methane and hydrogen gas sensing. *Applied Surface Science, 279,* 142–149.
9. Bhattacharyya, P., Basu, P. K., Saha, H., & Basu, S. (2007). Fast response methane sensor using nanocrystalline zinc oxide thin films derived by sol-gel method. *Sensors & Actuators, B: Chemical, 124,* 62–67.
10. Bhattacharyya, P., Mishra, G. P., & Sarkar, S. K. (2011). The effect of surface modification and catalytic metal contact on methane sensing performance of nano-ZnO-Si heterojunction sensor. *Microelectronics Reliability, 51,* 2185–2194.
11. Mitra, P., Chatterjee, A. P., & Maiti, H. S. (1998). ZnO thin film sensor. *Materials Letters, 35,* 33–38.
12. Choudhary, M., Mishra, V. N., & Dwivedi, R. (2013). Effect of temperature on Palladium-doped Tin Oxide (SnO_2) thick film gas sensor. *Advanced Science, Engineering and Medicine, 5,* 932–936.
13. Basu, S., & Basu, P. K. (2009). Nanocrystalline metal oxides for methane sensors: Role of noble metals. *Journal of Sensors, 2009.*
14. Cui, J., Shi, L., Xie, T., Wang, D., & Lin, Y. (2016). UV-light illumination room temperature HCHO gas-sensing mechanism of ZnO with different nanostructures. *Sensors & Actuators, B: Chemical, 227,* 220–226.
15. Sadek, A. Z., Choopun, S., Wlodarski, W., Ippolito, S. J., & Kalantar-zadeh, K. (2007). Characterization of ZnO nanobelt-based Gas Sensor for H_2, NO_2, and hydrocarbon sensing. *IEEE Sensors Journal, 7,* 919–924.
16. Baranowska-Korczyc, A., Reszka, A., Sobczak, K., Sikora, B., Dziawa, P., Aleszkiewicz, M., et al. (2012). Magnetic Fe doped ZnO nanofibers obtained by electrospinning. *Journal of Sol-Gel Science and Technology, 61,* 494–500.
17. Vyas, R., et al. (2015). Comparative study of Fe–doped ZnO nanowire bundle and their thin film for NO_2 and CH_4 gas sensing. *Macromolecular Symposia, 357* (1), 99–104.